UNDERSTANDING ELEMENTARY ALGEBRA

UNDERSTANDING ELEMENTARY ALGEBRA

A Text/Workbook

- **Arthur Goodman**
  Queens College of The City University of New York

- **Lewis Hirsch**
  Rutgers University

West Publishing Company

- St. Paul
- New York
- Los Angeles
- San Francisco

Interior design: Nancy Blodget
Cover design: Kathi Townes
Copy editing: Susan Reiland
Artwork: Ben Turner Graphics
Composition: Jonathan Peck Typographers, Ltd.
Production management: Susan Reiland
Cover: Miklos Pogany, PAGE FROM *DEATH IN VENICE* III, 1984.
Collage with mixed media, $16\frac{3}{4} \times 12$ inches. Courtesy of Victoria
Munroe Gallery, New York City.

ISBN 0-314-28493-1

Table of Contents

The Fundamentals

0

The Integers

1

Algebraic Expressions

2

First-Degree Equations and Inequalities

3

Rational Expressions

4

Exponents and Polynomials

5

Factoring

6

More Rational Expressions

7

Graphing and Systems of Linear Equations

8

Radical Expressions

9

Quadratic Equations

10

PREFACE to the Instructor

Purpose *Understanding Elementary Algebra—A Text/Workbook* is an attempt on our part to offer a worktext which reflects our philosophy—that students can *understand* what they are doing in algebra and why.

We offer a view of algebra which takes every opportunity to explain why things are done in a certain way, and to show how supposedly "new" topics are actually just new applications of concepts already learned.

This book assumes only a basic knowledge of arithmetic.

Pedagogy We believe that a student can successfully learn elementary algebra by mastering a few basic concepts and being able to apply them to a wide variety of situations. Thus, each section begins by relating the topic to be discussed to material previously learned. In this way the students can see algebra as a whole rather than as a series of isolated ideas.

Basic concepts, rules, and definitions are motivated and explained via numerical and algebraic examples. Formal proofs have been avoided except for those occasions when they illuminate the discussion.

Concepts are developed in a series of carefully constructed illustrative examples. Through the course of these examples we compare and contrast related ideas, helping the student to understand the sometimes subtle distinctions among various situations. In addition, these examples strengthen a student's understanding of how this "new" idea fits into the overall picture.

Every opportunity has been taken to point out common errors often made by students and to explain the misconception that often leads to a particular error.

Basic rules and/or procedures are highlighted so that students can find important ideas quickly and easily.

Chapter 0 is primarily intended as a review of the arithmetic of fractions, decimals, and percents. However, the first section of Chapter 0 is quite important as it contains much basic notation and terminology. Many instructors will want to cover Section 0.1 even if they choose to skip the remainder of the chapter.

A spiral approach has been used for the presentation of some more difficult topics. That is, a topic is first presented at an elementary level and then returned to at increasing levels of complexity. For example, simple rational expressions are covered in Chapter 4, while more complex rational expressions are dealt with in Chapter 7. Verbal problems are covered in Sections 2.5, 3.3, 4.6, 7.6, and 10.6. Factoring is covered in Section 2.3, and in Chapters 6 and 7.

Features   The various steps in the solutions to examples are explained in detail. Many steps appear with annotations (highlighted in a second color) which involve the student in the solution. These comments explain why a solution is proceeding in a certain way.

- Each illustrative example is accompanied by a **Learning Check** in the margin which enables the student to work out a similar example and in so doing check his or her understanding of the concepts involved. Moreover, while many texts encourage students to read a textbook with pencil and paper in hand, the Learning Checks make reading the textbook a much more active learning process. The answers to all the Learning Checks appear at the end of the section just before the exercises.

- We have also included **Progress Checks** at various strategic points. Each of these is a short set of exercises which reviews material that is essential for the new subject matter immediately following. The answers to these appear on the spot, and if the student does not get at least 75% correct, he or she is advised to review the relevant sections before proceeding.

- One of the main sources of students' difficulties is that they do not know how to study algebra. In this regard we offer a totally unique feature. Each section in Chapters 1–4 concludes with a **Study Skill.** This is a brief paragraph discussing some aspect of studying algebra, doing homework, or preparing for or taking exams. Our students who have used the preliminary version of this book indicated that they found the Study Skills very helpful.

- There are almost 2,800 homework exercises. Not only have the exercise sets been matched odd/even, but they have also been designed so that, in many situations, successive odd-numbered exercises compare and contrast subtle differences in applying the concepts covered in the section. Additionally, variety has been added to the exercise sets so that the student must be alert as to what the problem is asking. For example, the exercise sets in Sections 4.3 and 7.3, which deal primarily with adding rational expressions, also contain some exercises on multiplying and dividing rational expressions. The exercise set in Section 4.4 on solving fractional equations also asks the student to combine rational expressions. The exercise set in Section 10.1, which deals primarily with quadratic equations, contains some linear equations as well.

- Almost every exercise set contains **Questions for Thought,** which offer the student an opportunity to *think* about various algebraic ideas. They may be asked to compare and contrast related ideas, or to examine an incorrect solution and explain why the solution is wrong. The Questions for Thought are intended to be answered in complete sentences and in grammatically correct English. The answers to all the Questions for Thought in a given chapter appear at the end of the chapter following the Chapter Summary.

- Each chapter contains a **Chapter Summary** describing the basic concepts in the chapter. Each point listed in the summary is accompanied by an example illustrating the concept or procedure.

- There are over 750 review exercises. Each chapter contains a set of **Chapter Review Exercises** and a **Chapter Practice Test.** Additionally, there are three **Cumulative Review Exercise Sets** and three **Cumulative**

Practice Tests following Chapters 3, 6, and 9, along with two **Sample Final Exams**, offer the student more opportunities to practice choosing the appropriate procedure in a variety of situations.

- Many sections contain **Calculator Exercises.** This allows those instructors who so desire to integrate the use of the calculator into the text without interrupting the flow of the text for those who do not.

- The answer key contains answers to all the odd-numbered exercises, as well as to *all* the review exercises and practice test problems. The answer to each verbal problem contains a description of what the variable(s) represent and the equation (or system of equations) used to solve it. In addition, the answers to the cumulative review exercises and cumulative practice tests contain a reference to the section in which the relevant material is covered.

- Throughout the text a second color has been used to highlight important ideas, definitions, and procedural outlines.

Supplements   An instructor's manual contains the answers to *all* the exercises, five additional chapter tests for each chapter, two additional cumulative tests for every three chapters, and two final exams. The instructor's guide also contains some additional suggestions on using the Study Skills.

A computer software package offers the student the opportunity to work out a variety of multiple-choice exercises and receive diagnostic computer responses keyed to the student's choice.

A series of videotapes is also available to supplement the textbook. Information on the computer software and videotapes is available from the publisher.

Acknowledgments   The authors would like to sincerely thank the following reviewers for their thoughtful comments and numerous suggestions: Andrew Aheart, Western Virginia State College; Daniel Anderson, University of Iowa; Ron Beeler, East Central College; Susan Bordon, Oregon State; Allen Christian, Eastfield College; Terry Czerwinski, University of Illinois, Chicago Circle; Betsy Darken, University of Tennessee at Chattanooga; Lloyd Davis, College of San Mateo; Frank Demana, Ohio State University; Carol Edwards, St. Louis College at Florissant Valley; Richard Faber, Boston University; S. Gendler, Clarion University of Pennsylvania; Pat Gilbert, Diablo Valley College; Curtis Gooden, Cuyahoga Community College; George Grisham, Illinois Central College; Shirley Hagewood, Austin Peay State University; George Henderson, University of Wisconsin, Eau Claire; Herbert Hooper, Chattanooga State Technical Community College; Corinne Irwin, Austin Community College; Jane Jameson, Northern Michigan University; Charles Maderer, Indiana University of Pennsylvania; Carla Martin, Middle Tennessee State University; John Monroe, University of Akron; Richard Negley, North Georgia College; Lois Norris, Northern Virginia Community College; Mary Ann Pitts, Montgomery Community College; Douglas Robertson, University of Minnesota; Ann Smallen, Mohawk Valley Community College; James Thorpe, Saddleback Community College; John Tobey, North Shore Community College; Wesley Tom, Chaffey College; Lynn Tooley, Bellevue Community College; George Wales, Ferris State College; and Bill White, University of South Carolina, Spartanburg.

The authors would also like to thank Malka Cymbalista, who helped us use TeX to prepare the original manuscript for class testing at Queens College and Rutgers University. Obviously, the production of a textbook is a collaborative effort and we must thank our editor Pat Fitzgerald for his indefatigable support, Susan Reiland for her expert copy editing, and Jacqueline Rothstein for her assistance in checking the solutions. Of course, any errors that remain are the sole responsibility of the authors, and we would greatly appreciate their being called to our attention. Finally, we would like to thank our wives, Sora and Cindy, and our families for their constant encouragement during the two years they each endured an absentee husband and father.

PREFACE to the Student

This text is designed to help you understand algebra. We are convinced that if you understand what you are doing and why, you will be a much better algebra student. (Our students who have used this book in its preliminary form seem to agree with us.) This does not mean that after reading each section you will understand all the concepts clearly. Much of what you learn comes through the course of doing lots and lots of exercises and seeing for yourself exactly what is involved in completing an exercise. However, if you read the textbook carefully and take good notes in class you will find algebra not quite so menacing.

Here are a few suggestions for using this textbook:

- Always read the textbook with a pencil and paper in hand. Reading mathematics is not like reading other subjects. *You* must be involved in the learning process. Work out the examples along with the textbook, and *think* about what you are reading. Make sure you understand what is being done and why.

- You must work homework exercises on a daily basis. While attending class and listening to your instructor are important, do not mistake understanding someone else's work for the ability to do the work yourself. (Think about watching someone else driving a car as opposed to driving yourself.) Make sure *you* know how to do the exercises.

- Read the Study Skills which appear at the end of each section in Chapters 1–4. They discuss the best ways to use the textbook and your notes. They also offer a variety of suggestions on how to study, do homework, prepare for and take tests.

- Do not get discouraged if you have difficulty with some topics. Certain topics may not be absolutely clear the first time you see them. Be persistent. We all need time to absorb new ideas and become familiar with them. What was initially difficult will become less so as you spend more time with a subject. Keep at it and you will see that you are making steady progress.

The Fundamentals

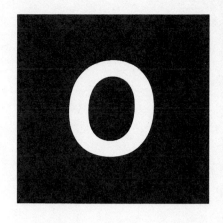

What Is Algebra? Introduction and Basic Notation

0.1

Ask most students, "What is algebra?" and *they* get a glassy-eyed look, but *you* will most likely get no response. It is not a very difficult question, and it has a very straightforward answer. Algebra is a language. It happens to be the language of mathematics.

We are going to learn this language in the same way we would learn any new language. We will begin by learning the alphabet (that is, the symbols) we will be using. For the most part our alphabet consists of letters and symbols we are already familiar with, such as the letters of the English alphabet, Arabic numerals, and the basic symbols of arithmetic. Next, we will learn the "grammar" of our new language, that is, the rules for putting the symbols together and manipulating them. After we learn the structure of the language, we can begin to actually use algebra to solve problems.

In some sense we can say that algebra is the generalization of arithmetic. We are going to let letters represent numbers and state our rules and our conclusions using letters, so that they will be valid for all numbers.

Sets

One concept which is used frequently is the idea of a set. The word *set* is used in mathematics in much the same way it is used in everyday life. A *set* is simply a well-defined collection of objects. The phrase *well-defined* means that there are clearly determined criteria for membership in the set. The criteria can be a list of those objects in the set, called the *elements* or *members* of the set, or it can be a description of those objects in the set.

For example, it is not sufficient to say "the set of all tall people in the class." *Tall* is a subjective criterion. It is possible to say "the set of all people in the class more than 6 feet tall."

One way to represent a set is to list the elements of the set, and enclose the list in "set brackets" which look like { }.

We often designate sets by using capital letters such as A, B, C. For example:

$$A = \{3, 4, 5, 8\}$$
$$B = \{a, e, i, o, u\}$$
$$C = \{red, white, blue\}$$

are three sets.

The symbol we use to indicate that an object is a member of a particular set is $\in$. Thus, $x \in S$ is a symbolic way of writing that x is a member or an *element* of S. We use the symbol $\notin$ to indicate that an object is *not* an element of a set. [In general, when we put a "/" through a mathematical symbol it means *not*. For example, "$\neq$" means *not equal*.] For example, using the sets A, B, and C listed above we have:

$5 \in A$ 5 is an element of A.

$p \notin B$ p is not an element of B.

Sometimes, if there are many, or an infinite number, of objects in a set, we use a variation on the listing method. The set of even numbers greater than

0 and less than or equal to 100 can be written as {2, 4, 6, . . . , 100}. The three dots mean that the set continues according to the same pattern. Similarly, the set O = {1, 3, 5, . . .} means the set of all odd numbers greater than 0. There is no number after the dots because this set is infinite; it has no last element.

Of course, this method of listing a set can be used only when the first few elements clearly show the pattern for *all* the elements in the set.

Certain frequently used sets of numbers are given special names.

The set of numbers we use for counting is called the set of **natural numbers**, and is usually denoted with the letter N:

$$N = \{1, 2, 3, . . .\}$$

If we add the number 0 to this set it is called the set of **whole numbers** and is denoted with the letter W:

$$W = \{0, 1, 2, 3, . . .\}$$

Often when we describe a set we use the word "between," which can be ambiguous. When we say "the numbers between 5 and 10" do we mean to include or exclude 5 and 10? Let's agree that when we say "between" we mean "in between" and we do *not* include the first and last numbers.

EXAMPLE 1 List the following sets:

(a) The set A of whole numbers between 6 and 30

(b) The set B of odd numbers greater than 17

Solution

(a) The whole numbers are the same as the natural numbers, except that the whole numbers include 0. Note that 6 and 30 are not included.

$$A = \{7, 8, 9, . . . , 29\}\ *$$

(b) Since no upper limit to this set is given, the answer is

$$B = \{19, 21, 23, 25, . . .\}$$ Note that 17 is not included.   ∎

Sometimes we cannot *list* the elements of a set, but rather we must describe the set. When this is the case we use what is called *set-builder notation*. *Set-builder notation* consists of the set braces, a **variable** that acts as a place holder, a vertical bar (|) which is read "such that," and a sentence which describes what the variable can be. This part is called the **condition** on the variable. For example:

{ x | x is an even number greater than 0 and less than 10}
   ↑   ↑   ↑
 Variable Such that   *Condition on the variable*

This is read "the set of all x such that x is an even number greater than 0 and less than 10," which is the set {2, 4, 6, 8}.

LEARNING CHECKS

1. List the following sets:

  (a) The set of natural numbers between 4 and 19.

  (b) The set of even numbers greater than or equal to 11.

*Throughout the text we will use color boxes to indicate the final answer to an example.

2. List the following set:

$\{t \mid t$ is a whole number divisible by 5$\}$

EXAMPLE 2   List the following set:

$$\{x \mid x \text{ is a whole number divisible by 3}\}$$

Solution   The number 0 is included because $0 \div 3 = 0$. Thus, our answer is

$$\boxed{\{0, 3, 6, 9, \ldots\}}$$

■

It is possible to place a condition on a set which no elements satisfy, as, for example,

$$F = \{x \mid x \text{ is an odd number divisible by 2}\}$$

Since it is impossible for an odd number to be divisible by 2, the set F has no members. It is called the **empty set** or the **null set** and it is symbolized by $\varnothing$. Thus, we have $F = \varnothing$.

Before we can continue we must introduce some terminology and notation.

Sums, Terms, Products, and Factors

The most frequent error made by students in algebra is that of confusing terms with factors and factors with terms. We can do some things with factors that we cannot do with terms, and vice versa. We will define them now for arithmetic expressions, and point out the differences throughout the book.

Sum is the word we use for addition. In an expression involving a sum, the numbers to be added in the sum are called the **terms**. The symbol used to indicate a sum is the familiar "+" sign.

Product is the word we use for multiplication. In an expression involving a product, the numbers being multiplied are called the **factors**. Saying that "a is a **multiple** of b" is equivalent to saying that "b is a **factor** of a." For example,

20 is a multiple of 5 because 5 is a factor of 20

48 is a multiple of 8 because 8 is a factor of 48

Thus, a factor of n is a number that divides exactly into n, whereas a multiple of n is a number that is exactly divisible by n.

In algebra, we generally use the symbol "·" to indicate multiplication. We do not use the "×" to indicate multiplication because we very often use x in our work as a variable. Frequently we will also indicate a product simply by writing numbers or expressions next to each other with the appropriate "punctuation." For example, if we let x represent a number, then

Sums can be written as:		*Products can be written as:*	
$3 + 4$	(the sum of 3 and 4)	$3 \cdot 4$	(the product of 3 and 4)
$7 + x$	(the sum of 7 and x)	$7 \cdot x$	(the product of 7 and x)
		$7x$	(also the product of 7 and x)

If there is no operation symbol between two variables or between a number and a variable, such as in xy or $7a$, multiplication is understood. However, we cannot write "3 times 4" as 34! If we want to indicate multiplication of *numbers*, there must be some punctuation between the two numbers. We have already said that we can use the "·" and write $3 \cdot 4$. Alternatively, we can write "3 times 4" as (3)(4) or 3(4) or (3)4; in this way, the 3 and 4 are next to each other to indicate multiplication but the parentheses show us that 3 and 4 are two separate numbers.

<hr/>

EXAMPLE 3 List the following sets:

(a) $\{x \mid x$ is a whole number less than 30 and a multiple of 5$\}$

(b) $\{y \mid y$ is a natural number multiple of 7$\}$

Solution

(a) The answer is

$$\{0, 5, 10, 15, 20, 25\}$$

Note that 0 is included because 0 is a multiple of 5, since $0 = 0 \cdot 5$.

(b) The answer is

$$\{7, 14, 21, \ldots\}$$

∎

If set A is contained in set B, then we say that set A is a **subset** of set B. Thus, the set of even numbers greater than 0 is a subset of the natural numbers.

There is another subset of the natural numbers which is very important. It is called the set of *prime numbers*.

> **DEFINITION** A *prime number* is a natural number (excluding 1) which is divisible only by itself and 1. In other words, a prime number is a natural number greater than 1 whose only factors are itself and 1.
>
> A natural number greater than 1 which is not prime is called **composite**.

For example, the numbers 5 and 13 are prime numbers because they are not divisible by any number other than themselves and 1. The number 12 is composite (not prime) because it is divisible by other numbers, such as 3 and 4.

The set of prime numbers is a perfect example of the necessity for set-builder notation. If we simply start listing the set of prime numbers, we would write

$$P = \{2, 3, 5, 7, 11, 13, 17, 19, 23, \ldots\}$$

Since the prime numbers have no pattern, unless someone knows this set is the set of prime numbers, they cannot tell what the next number in the set will be. On the other hand, having the definition of a prime number, we can write

$$P = \{m \mid m \text{ is a prime number}\}$$

and then we can determine whether or not a number is in the set.

Every composite number can be broken down (the word that is usually used is **decomposed**) into its prime factors. For example, we can break down 30 as follows:

$$30 = 2 \cdot 15$$
$$= 2 \cdot 3 \cdot 5$$

3. List the following sets:

(a) $\{n \mid n$ is a natural number less than 40 and a multiple of 8$\}$

(b) $\{s \mid s$ is a whole number multiple of 9$\}$

Basically, we can pick *any* two factors we recognize and start with them. Then we continue breaking the factors down until *all* the factors are prime numbers.

4. Give the prime factorization of 72.

EXAMPLE 4   Decompose the number 48 into its prime factors.

Solution   We have many choices for the first two factors. We will illustrate just two of the possible paths to the answer:

$$48 = 2 \cdot 24 \qquad \text{\textit{Break down} 24.}$$
$$= 2 \cdot 2 \cdot 12 \qquad \text{\textit{Break down} 12.}$$
$$= 2 \cdot 2 \cdot 2 \cdot 6 \qquad \text{\textit{Break down} 6.}$$
$$= \boxed{2 \cdot 2 \cdot 2 \cdot 2 \cdot 3}$$

or

$$48 = 8 \cdot 6 \qquad \text{\textit{Break down} 8 and 6.}$$
$$= 2 \cdot 4 \cdot 2 \cdot 3 \qquad \text{\textit{Break down} 4.}$$
$$= \boxed{2 \cdot 2 \cdot 2 \cdot 2 \cdot 3}$$

No matter which factors you decide to start with, the final answer (since it involves *prime* factors only) will be the same.   ∎

The Number Line

The number line gives us a very useful geometric representation of the various sets of numbers with which we will be working.

Since our basic set of numbers so far is the set of whole numbers, we will, for the time being, associate the set of whole numbers with points on the number line in the following way. First, we draw a horizontal line. We mark off some point on the line and label it 0. Then we mark off another point to the right of 0 and label it 1. The distance between 0 and 1 is called the **unit length**.

We continue to mark off points 1 unit length apart moving toward the right. The arrow at the end of the number indicates that the numbers are getting *larger* as we move in the direction of the arrow—to the right:

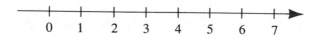

Thus, we have associated each whole number with the point on the number line *that* many units to the right of 0.

If we are asked to "graph" a set on the number line, it means that we want to indicate those points on the number line which are in the set. We usually indicate this by putting a heavy dot at those points which are in the given set.

EXAMPLE 5 Graph the set {3, 4, 6, 9} on the number line.

Solution

5. Graph the set {2, 5, 7, 8} on a number line.

The number line gives us a very simple way of defining the idea of "order" on the number line. For example, 3 is less than 7 because 3 is to the left of 7 on the number line. In general, we define a to be **less than** b if a is to the left of b on the number line. The symbol we use for "less than" is "$<$."

$$\boxed{a < b \text{ means that } a \text{ is to the left of } b}$$

$4 < 9$ is the symbolic statement for   "4 is less than 9"

which *means* that 4 is to the left of 9 on the number line.

Similarly, the symbol "$>$" is used for the expression *greater than*.

$a > b$ is the symbolic statement for   "a is greater than b"

which means that a is to the right of b on the number line.

The symbols "$<$" and "$>$" are called **inequality symbols**.

$$\text{lesser} < \text{greater} \qquad \text{greater} > \text{lesser}$$

Note that the inequality symbol always points toward the *smaller* number.

The accompanying box contains a list of all our equality and inequality symbols, what each means, and an example of a *true* statement using each symbol.

EQUALITY AND INEQUALITY SYMBOLS

$a = b$	a "equals" b	$7 + 3 = 12 - 2$
$a \neq b$	a "is not equal to" b	$4 + 5 \neq 10$
$a < b$	a "is less than" b	$3 < 8$
$a \leq b$	a "is less than *or* equal to" b	$3 \leq 8$
$a > b$	a "is greater than" b	$7 - 2 > 4$
$a \geq b$	a "is greater than *or* equal to b"	$6 \geq 6$

Note that $a \leq b$ means that *either* $a < b$ or $a = b$, and similarly for $a \geq b$.

EXAMPLE 6 List each of the following sets and represent them on the number line.

(a) $\{x \mid x \in N \text{ and } x \leq 5\}$ = {1, 2, 3, 4, 5}

(b) $\{a \mid a \in W \text{ and } a \geq 3 \text{ and } a < 9\}$

{3, 4, 5, 6, 7, 8}

6. List each of the following sets and represent them on a number line.

(a) $\{t \mid t \in W \text{ and } t < 4\}$

(b) $\{u \mid u \in N \text{ and } u \geq 6 \text{ and } u \leq 10\}$

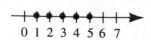

0 1 2 3 4 5 6 7

0 1 2 3 4 5 6 7 8 9

Solution

(a) The letter N stands for the set of natural numbers; therefore, the answer is

$$\boxed{\{1, 2, 3, 4, 5\}} \quad \textit{Note that 5 is included, but 0 is excluded. Why?}$$

(b) The letter W stands for the set of whole numbers. Therefore, the answer is

$$\boxed{\{3, 4, 5, 6, 7, 8\}} \quad \textit{Note that 3 is included but 9 is excluded.} \quad \blacksquare$$

Answers to Learning Checks in Section 0.1

1. (a) $\{5, 6, 7, \ldots, 18\}$ **(b)** $\{12, 14, 16, \ldots\}$ **2.** $\{0, 5, 10, 15, \ldots\}$

3. (a) $\{8, 16, 24, 32\}$ **(b)** $\{0, 9, 18, 27, 36, \ldots\}$ **4.** $2 \cdot 2 \cdot 2 \cdot 3 \cdot 3$

5.

0 1 2 3 4 5 6 7 8

6. (a) $\{0, 1, 2, 3, 4\}$ **(b)** $\{6, 7, 8, 9, 10\}$

0 1 2 3 4 5

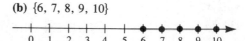

0 1 2 3 4 5 6 7 8 9 10

Exercises 0.1

In Exercises 1–10, *indicate whether the given statement is true or false.*

1. $3 \in \{1, 3, 5, 8\}$

2. $5 \notin \{1, 3, 5, 8\}$

3. $8 \notin \{2, 3, 5, 7, 9\}$

4. $7 \in \{2, 4, 6, 4, 17\}$

5. $a \in \{b, c, d, a\}$

6. $g \notin \{r, g, c, b\}$

7. $17 \in N$

8. $0 \in N$

9. $24 \in W$

10. $0 \in W$

In Exercises 11–30, *list each of the following sets. Unless otherwise specified, assume that all numbers are whole numbers.*

11. $\{x \mid x \text{ is a natural number less than 8}\}$

12. $\{x \mid x \in N \quad \text{and} \quad x < 8\}$

13. $\{y \mid y \text{ is an even number less than 20}\}$

14. $\{m \mid m \text{ is an odd number greater than 7}\}$

15. $\{x \mid x \leq 6\}$

16. $\{n \mid n < 9\}$

17. $\{x \mid x < 6\}$

18. $\{n \mid n \geq 9\}$

19. $\{x \mid x \geq 6\}$

20. $\{n \mid n > 9\}$

21. $\{x \mid x > 6\}$

22. $\{n \mid n \leq 9\}$

23. $\{a \mid a \text{ is greater than 2 and less than 6}\}$

24. $\{a \mid a \text{ is greater than 6 and less than 2}\}$

25. $\{m \mid m \text{ is a multiple of 4}\}$

26. $\{n \mid n \text{ is a multiple of 5}\}$

27. $\{m \mid m \text{ is a multiple of 3 and a multiple of 4}\}$

28. $\{n \mid n \text{ is a multiple of 2 and a multiple of 5}\}$

29. $\{x \mid x \text{ is a multiple of 10 and not divisible by 5}\}$

30. $\{y \mid y \text{ is a multiple of 5 and not divisible by 10}\}$

ANSWERS

1. _____
2. _____
3. _____
4. _____
5. _____
6. _____
7. _____
8. _____
9. _____
10. _____
11. _____
12. _____
13. _____
14. _____
15. _____
16. _____
17. _____
18. _____
19. _____
20. _____
21. _____
22. _____
23. _____
24. _____
25. _____
26. _____
27. _____
28. _____
29. _____
30. _____

ANSWERS

31. _____
32. _____
33. _____
34. _____
35. _____
36. _____
37. _____
38. _____
39. _____
40. _____
41. _____
42. _____
43. _____
44. _____
45. _____
46. _____
47. _____
48. _____
49. _____
50. _____
51. _____
52. _____
53. _____
54. _____
55. _____
56. _____
57. _____
58. _____
59. _____
60. _____
61. _____
62. _____
63. _____
64. _____

In Exercises 31–40, fill in the appropriate ordering symbol: either $<$, $>$, or $=$.

31. 4 ___ 2

32. 8 ___ 20

33. 7 ___ 7

34. 5 ___ 0

35. 19 ___ 24 − 10

36. 18 − 5 ___ 10 + 4

37. 19 + 53 ___ 72

38. 16 ___ 12 + 4

39. 8 − 8 ___ 4 · 0

40. 8 + 0 ___ 4 · 2

In Exercises 41–50, fill in as many of the following ordering symbols as make the statement true. Choose from $<$, $\leq$, $>$, $\geq$, $=$, $\neq$.

41. 5 _____ 8

42. 17 _____ 12

43. 43 _____ 43

44. 5 _____ 2 + 4

45. 4 · 2 _____ 4 + 2

46. 1 · 2 _____ 1 + 2

47. 6 · 0 _____ 6 + 0

48. 25 − 5 _____ 14 + 3

49. 16 _____ 12 − 5

50. 4 − 4 _____ 17 · 0

In Exercises 51–64, decompose the given number into its prime factors. If the number is prime, say so.

51. 14

52. 26

53. 33

54. 35

55. 30

56. 50

57. 37

58. 80

59. 64

60. 41

61. 96

62. 120

63. 87

64. 91

QUESTIONS FOR THOUGHT

1. What is the difference between the set N (the set of natural numbers) and the set W (the set of whole numbers)?

2. What is the difference between a *term* and a *factor*? Give an example to illustrate the difference.

3. What is the difference between a *factor* and a *multiple*? Give an example to illustrate the difference.

Fractions

0.2

In Section 0.1 we mentioned that there are various ways to indicate multiplication. Similarly, there are two popular ways to indicate division. If we wish to indicate "20 divided by 4" we may write

$$20 \div 4 \quad \text{or} \quad \frac{20}{4}$$

The fraction bar is an alternative way of indicating division, and is the notation most frequently used in algebra.

In general, a **fraction** is an expression of the form $\frac{p}{q}$, which means p divided by q. In the expression $\frac{p}{q}$, p is called the **numerator** of the fraction and q is called the **denominator** of the fraction.

We can easily locate a fraction on the number line. For example, if we want to locate $\frac{3}{4}$ we would simply divide the interval from 0 to 1 into 4 equal parts and count 3 of the 4 equal parts as shown here:

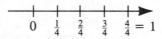

While this book assumes no previous knowledge of algebra, we do assume a basic familiarity with fractions, decimals, and percent. Nevertheless, in the remainder of this chapter we will review the arithmetic of fractions, decimals, and percent for those of you who need to brush up, as these skills will be used consistently throughout your study of algebra.

We know from previous experience with fractions that two fractions may look different but in fact have the same value. This is illustrated in Figure 0.1. The figure illustrates that the fractions $\frac{3}{4}$ and $\frac{6}{8}$ both represent the same amount. Having 3 out of 4 equal parts of a certain whole is the same as having 6 out of 8 equal parts of the same whole.

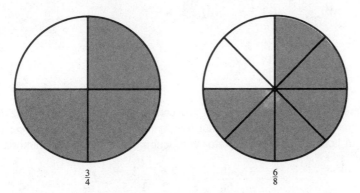

Figure 0.1 *Equality of two fractions*

Instead of drawing pictures to determine if two fractions are equivalent, we can use the Fundamental Principle of Fractions, stated in the box at the top of page 12.

> **FUNDAMENTAL PRINCIPLE OF FRACTIONS**
>
> The value of a fraction is unchanged if the numerator and denominator are multiplied or divided by the same nonzero number.

Algebraically we can write this as follows:

> **FUNDAMENTAL PRINCIPLE OF FRACTIONS**
>
> $$\frac{a}{b} = \frac{a \cdot k}{b \cdot k} \qquad b, k \neq 0$$

Two fractions which have the same value are called *equivalent fractions.* When we multiply both the numerator and denominator by the same number we say that we are *building fractions,* while when we divide both numerator and denominator by the same number we say that we are *reducing fractions.* For example:

$$\frac{3}{4} = \frac{3 \cdot 2}{4 \cdot 2} = \frac{6}{8} \qquad \textit{This is an example of building fractions.}$$

$$\frac{6}{8} = \frac{3 \cdot \cancel{2}}{4 \cdot \cancel{2}} = \frac{3}{4} \qquad \textit{This is an example of reducing fractions.}$$

The slashes indicate that the numerator and denominator were both divided by 2.

DEFINITION   We say that a fraction is reduced to *lowest terms* if the numerator and denominator have no common factor other than 1.

We will always expect our final answers to be reduced to lowest terms.

✔ **LEARNING CHECKS**

1. (a) $\dfrac{6}{22}$

   (b) $\dfrac{32}{80}$

   (c) $\dfrac{15}{90}$

   (d) $\dfrac{20}{27}$

EXAMPLE 1   *Reduce to lowest terms.*

(a) $\dfrac{10}{35}$ (b) $\dfrac{48}{60}$ (c) $\dfrac{14}{42}$ (d) $\dfrac{18}{25}$

Solution

(a) $\dfrac{10}{35} = \dfrac{2 \cdot \cancel{5}}{\cancel{5} \cdot 7} = \boxed{\dfrac{2}{7}}$

(b) There are several ways we can proceed. Since the Fundamental Principle of Fractions tells us that we can reduce only *common factors,* one way to begin is to factor the numerator and denominator into prime factors:

$$\frac{48}{60} = \frac{2 \cdot 2 \cdot 2 \cdot 2 \cdot 3}{2 \cdot 2 \cdot 3 \cdot 5}$$

$$= \frac{\cancel{2} \cdot \cancel{2} \cdot 2 \cdot 2 \cdot \cancel{3}}{\cancel{2} \cdot \cancel{2} \cdot \cancel{3} \cdot 5}$$

$$= \frac{2 \cdot 2}{5}$$

$$= \boxed{\frac{4}{5}}$$

Alternatively, we may see that 12 is the greatest common factor of both the numerator and denominator:

$$\frac{48}{60} = \frac{4 \cdot 12}{5 \cdot 12} = \frac{4 \cdot \cancel{12}}{5 \cdot \cancel{12}} = \boxed{\frac{4}{5}}$$

This is often written in the following shorthand fashion:

$$\frac{48}{60} = \frac{\overset{4}{\cancel{48}}}{\underset{5}{\cancel{60}}} = \boxed{\frac{4}{5}}$$

Note that this shorthand method "hides" the fact the the common factor is 12. Nevertheless, because it is very efficient it is often used.

(c) $\dfrac{14}{42} = \dfrac{2 \cdot 7}{6 \cdot 7} = \dfrac{\cancel{2} \cdot \cancel{7}}{\underset{3}{\cancel{6}} \cdot \cancel{7}} = \boxed{\dfrac{1}{3}}$

Do not forget the understood factor of 1 in the numerator.

(d) $\dfrac{18}{25} = \dfrac{2 \cdot 3 \cdot 3}{5 \cdot 5}$ cannot be reduced since there are no common factors.   ∎

At this point it is important to point out again that we can reduce only common *factors* and *not* common *terms*. Avoid the following type of common error:

$$\frac{20 + 4}{4} \neq \frac{20 + \cancel{4}}{\cancel{4}} \neq 20 \quad \text{because} \quad \frac{20 + 4}{4} = \frac{24}{4} = 6$$

REMEMBER

The Fundamental Principle of Fractions allows us to reduce common *factors*, not common terms.

The process of building fractions is the reverse of reducing. We obtain an equivalent fraction by multiplying both the numerator and the denominator by the same number. For example:

$$\frac{2}{3} = \frac{2 \cdot 5}{3 \cdot 5} = \frac{10}{15}$$

$$\frac{2}{3} = \frac{2 \cdot 11}{3 \cdot 11} = \frac{22}{33}$$

$$\frac{2}{3} = \frac{2 \cdot 2 \cdot 7}{3 \cdot 2 \cdot 7} = \frac{2 \cdot 14}{3 \cdot 14} = \frac{28}{42}$$

Hence, $\dfrac{2}{3}, \dfrac{10}{15}, \dfrac{22}{33},$ *and* $\dfrac{28}{42}$ *all represent the same number.*

2. Express $\frac{4}{9}$ as an equivalent fraction with a denominator of 45.

EXAMPLE 2 **EXAMPLE 2**   Write $\frac{5}{12}$ as an equivalent fraction with a denominator of 72.

Solution

$$\frac{5}{12} = \frac{?}{72} \qquad \text{We notice that } 72 = 12 \cdot 6.$$

$$\frac{5}{12} = \frac{?}{12 \cdot 6} \qquad \begin{array}{l}\textit{Since the denominator was multiplied by 6, the Fundamental}\\ \textit{Principle requires us to multiply the numerator by 6.}\end{array}$$

$$\frac{5}{12} = \frac{5 \cdot 6}{12 \cdot 6}$$

$$\frac{5}{12} = \boxed{\frac{30}{72}} \qquad \blacksquare$$

3. A quality inspector finds 45 defective items out of a total of 900 items. What fractional part of the total is defective?

EXAMPLE 3   An archer shoots 80 arrows at a target and hits the target 72 times. What fractional part of her total shots hit the target?

Solution

The archer hit the target 72 out of 80 times.

$$\frac{72}{80} = \frac{\cancel{8} \cdot 9}{\cancel{8} \cdot 10} = \boxed{\frac{9}{10}}$$

The archer hit the target 9 out of 10 times. $\blacksquare$

✔ *Answers to Learning Checks in Section 0.2*

1. (a) $\frac{3}{11}$ **(b)** $\frac{2}{5}$ **(c)** $\frac{1}{6}$ **(d)** Cannot be reduced   **2.** $\frac{20}{45}$ **3.** $\frac{1}{20}$

Exercises 0.2

In Exercises 1–10, write two fractions equivalent to the given fraction.

1. $\dfrac{6}{9}$ 3 $\dfrac{2}{3}$

2. $\dfrac{12}{15}$

3. $\dfrac{14}{35}$ $\left(\dfrac{2}{5}\right)$

4. $\dfrac{21}{18}$

5. $\dfrac{33}{44}$

6. $\dfrac{39}{65}$

7. 5

8. 9

9. 1

10. 0

In Exercises 11–26, reduce the given fraction to lowest terms.

11. $\dfrac{8}{10}$

12. $\dfrac{15}{18}$

13. $\dfrac{6}{42}$

14. $\dfrac{9}{45}$

15. $\dfrac{18}{36}$

16. $\dfrac{32}{48}$

17. $\dfrac{20}{49}$

18. $\dfrac{25}{36}$

19. $\dfrac{28}{72}$

20. $\dfrac{24}{32}$

1. _____

2. _____

3. _____

4. _____

5. _____

6. _____

7. _____

8. _____

9. _____

10. _____

11. _____

12. _____

13. _____

14. _____

15. _____

16. _____

17. _____

18. _____

19. _____

20. _____

21. $\dfrac{90}{15}$

22. $\dfrac{80}{16}$

23. $\dfrac{220}{80}$

24. $\dfrac{175}{30}$

25. $\dfrac{81}{100}$

26. $\dfrac{30}{77}$

In Exercises 27–34, fill in the question mark to make the fractions equivalent.

27. $\dfrac{5}{6} = \dfrac{?}{24}$

28. $\dfrac{3}{8} = \dfrac{?}{40}$

29. $\dfrac{15}{7} = \dfrac{45}{?}$

30. $\dfrac{28}{9} = \dfrac{56}{?}$

31. $\dfrac{16}{25} = \dfrac{?}{100}$

32. $\dfrac{5}{12} = \dfrac{?}{96}$

33. $\dfrac{1}{18} = \dfrac{?}{108}$

34. $\dfrac{1}{14} = \dfrac{?}{42}$

35. A businesswoman spent \$1,500 at a computer store. She spent \$1,250 on a computer and \$250 on software. What fractional part of the total did she spend on the computer? What fractional part of the total did she spend on the software?

21. _____

22. _____

23. _____

24. _____

25. _____

26. _____

27. _____

28. _____

29. _____

30. _____

31. _____

32. _____

33. _____

34. _____

35. _____

Multiplying and Dividing Fractions

0.3

We know that when we want to compute twice a number we multiply the number by 2. Similarly, if we want to compute $\frac{1}{2}$ of a number we multiply by $\frac{1}{2}$. How then should we compute $\frac{1}{2}$ of $\frac{1}{3}$ (which we would write as $\frac{1}{2} \cdot \frac{1}{3}$)?

We can visualize $\frac{1}{2}$ of $\frac{1}{3}$ as shown in Figure 0.2.

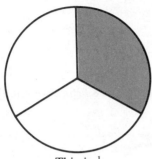 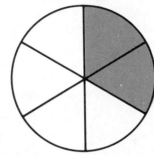

This is $\frac{1}{3}$.

$\frac{1}{3} = \frac{2}{6}$
so we see that $\frac{1}{2}$ of $\frac{1}{3}$ is $\frac{1}{6}$.
In other words, we see that $\frac{1}{2} \cdot \frac{1}{3} = \frac{1}{6}$.

Figure 0.2

Note that we obtain $\frac{1}{6}$ if we multiply $\frac{1 \cdot 1}{2 \cdot 3}$. We can generalize this to the multiplication of any two fractions, as indicated in the box.

RULE FOR MULTIPLYING FRACTIONS

$$\frac{a}{b} \cdot \frac{c}{d} = \frac{a \cdot c}{b \cdot d} \qquad b, d \neq 0$$

In other words, the product of two fractions is the product of the numerators over the product of the denominators.

EXAMPLE 1 *Multiply.* $\frac{3}{7} \cdot \frac{9}{10}$

Solution

$$\frac{3}{7} \cdot \frac{9}{10} = \frac{3 \cdot 9}{7 \cdot 10} = \boxed{\frac{27}{70}} \qquad \blacksquare$$

EXAMPLE 2 *Multiply.* $\frac{5}{8} \cdot \frac{16}{25}$

✔ **LEARNING CHECKS**

1. $\frac{4}{5} \cdot \frac{3}{11}$

2. $\frac{6}{25} \cdot \frac{15}{8}$

Solution

$$\frac{5}{8} \cdot \frac{16}{25} = \frac{5 \cdot 16}{8 \cdot 25} = \frac{80}{200} = \frac{2 \cdot 40}{5 \cdot 40} = \boxed{\frac{2}{5}}$$

In obtaining this answer we followed the rule for multiplying fractions as stated. However, the rule for multiplying fractions says in effect that $\frac{a}{b} \cdot \frac{c}{d}$ is really just one fraction with the factors of the numerator and denominator separated by a multiplication sign, Thus, it is easier to reduce before we multiply, and so we could have proceeded as follows:

$$\frac{5}{8} \cdot \frac{16}{25} = \frac{\cancel{5}}{8} \cdot \frac{\overset{2}{\cancel{16}}}{\underset{5}{\cancel{25}}} = \frac{2}{5}$$

In general, it is usually to our advantage to try to reduce *before* we multiply.

■

3. $\dfrac{12}{18} \cdot \dfrac{30}{48}$

EXAMPLE 3   *Multiply.* $\dfrac{49}{24} \cdot \dfrac{32}{28}$

Solution

$$\frac{49}{24} \cdot \frac{32}{28} = \frac{\overset{7}{\cancel{49}}}{\underset{3}{\cancel{24}}} \cdot \frac{\overset{4}{\cancel{32}}}{\underset{4}{\cancel{28}}} = \boxed{\frac{7}{3}}$$

■

EXAMPLE 4   *Multiply.* $8 \cdot \dfrac{5}{12}$

4. $6 \cdot \dfrac{4}{9}$

Solution   It helps to think of 8 as $\dfrac{8}{1}$.

$$8 \cdot \frac{5}{12} = \frac{8}{1} \cdot \frac{5}{12} = \frac{\overset{2}{\cancel{8}}}{1} \cdot \frac{5}{\underset{3}{\cancel{12}}} = \boxed{\frac{10}{3}}$$

■

Division

In order to understand the rule for dividing fractions, let's keep in mind that division is defined in terms of multiplication. For example, $20 \div 4 = 5$ because $5 \cdot 4 = 20$.

RULE FOR DIVIDING FRACTIONS

$$\frac{a}{b} \div \frac{c}{d} = \frac{a}{b} \cdot \frac{d}{c} \qquad b, c, d \neq 0$$

EXAMPLE 5 *Divide.* $\dfrac{4}{7} \div \dfrac{3}{5}$

5. $\dfrac{8}{27} \div \dfrac{20}{63}$

Solution Following the rule for dividing fractions, we get

$$\frac{4}{7} \div \frac{3}{5} = \frac{4}{7} \cdot \frac{5}{3} = \boxed{\frac{20}{21}}$$

Note that in order to check this division we would compute

$$\frac{20}{21} \cdot \frac{3}{5} \stackrel{?}{=} \frac{4}{7}$$

$$\frac{\overset{4}{\cancel{20}}}{\underset{7}{\cancel{21}}} \cdot \frac{\cancel{3}}{\cancel{5}} \stackrel{?}{=} \frac{4}{7}$$

$$\frac{4}{7} \stackrel{\checkmark}{=} \frac{4}{7}$$

In other words, the rule for dividing fractions is formulated in such a way that the multiplication checks. ∎

EXAMPLE 6 *Divide.* $\dfrac{16}{45} \div \dfrac{12}{25}$

6. $\dfrac{27}{40} \div \dfrac{9}{16}$

Solution

$$\frac{16}{45} \div \frac{12}{25} = \frac{16}{45} \cdot \frac{25}{12}$$

$$= \frac{\overset{4}{\cancel{16}}}{\underset{9}{\cancel{45}}} \cdot \frac{\overset{5}{\cancel{25}}}{\underset{3}{\cancel{12}}}$$

$$= \boxed{\frac{20}{27}}$$

∎

EXAMPLE 7 *Divide.* $\dfrac{8}{9} \div 72$

7. $\dfrac{5}{6} \div 30$

Solution We think of 72 as $\dfrac{72}{1}$.

$$\frac{8}{9} \div 72 = \frac{8}{9} \div \frac{72}{1} \qquad \textit{Follow the rule for division.}$$

$$= \frac{8}{9} \cdot \frac{1}{72}$$

$$= \frac{\overset{1}{\cancel{8}}}{9} \cdot \frac{1}{\underset{9}{\cancel{72}}}$$

$$= \boxed{\frac{1}{81}}$$

∎

There is a very useful idea related to the division of fractions.

DEFINITION   The *reciprocal* of a number x is $\dfrac{1}{x}$.

Thus, we have the following:

The reciprocal of 8 is $\dfrac{1}{8}$.

The reciprocal of $\dfrac{3}{5}$ is $\dfrac{1}{\dfrac{3}{5}} = 1 \div \dfrac{3}{5} = 1 \cdot \dfrac{5}{3} = \dfrac{5}{3}$.

In general, we have that the reciprocal of $\dfrac{a}{b}$ is $\dfrac{b}{a}$.

In light of this definition, we can state the rule for dividing fractions as "to divide by a fraction, multiply by its reciprocal."

✔   *Answers to Learning Checks in Section 0.3*

1. $\dfrac{12}{55}$ **2.** $\dfrac{9}{20}$ **3.** $\dfrac{5}{12}$ **4.** $\dfrac{8}{3}$ **5.** $\dfrac{14}{15}$ **6.** $\dfrac{6}{5}$ **7.** $\dfrac{1}{36}$

Exercises 0.3

Perform the indicated operations and simplify as completely as possible.

1. $\dfrac{2}{3} \cdot \dfrac{5}{7}$

2. $\dfrac{3}{7} \cdot \dfrac{2}{5}$

3. $\dfrac{1}{4} \cdot \dfrac{1}{9}$

4. $\dfrac{1}{3} \cdot \dfrac{1}{8}$

5. $\dfrac{8}{9} \cdot \dfrac{3}{4}$

6. $\dfrac{9}{16} \cdot \dfrac{4}{3}$

7. $\dfrac{5}{6} \cdot \dfrac{4}{15}$

8. $\dfrac{6}{7} \cdot \dfrac{14}{9}$

9. $\dfrac{5}{6} \div \dfrac{4}{15}$

10. $\dfrac{6}{7} \div \dfrac{14}{9}$

11. $\dfrac{15}{18} \cdot \dfrac{24}{25}$

12. $\dfrac{12}{18} \cdot \dfrac{20}{36}$

13. $\dfrac{15}{18} \div \dfrac{24}{25}$

14. $\dfrac{12}{18} \div \dfrac{20}{36}$

15. $\dfrac{3}{5} \cdot \dfrac{5}{3}$

16. $\dfrac{4}{7} \div \dfrac{7}{4}$

17. $\dfrac{3}{5} \div \dfrac{5}{3}$

18. $\dfrac{4}{7} \cdot \dfrac{7}{4}$

19. $18 \cdot \dfrac{3}{2}$

20. $12 \cdot \dfrac{3}{4}$

1. _____

2. _____

3. _____

4. _____

5. _____

6. _____

7. _____

8. _____

9. _____

10. _____

11. _____

12. _____

13. _____

14. _____

15. _____

16. _____

17. _____

18. _____

19. _____

20. _____

21. $18 \div \dfrac{3}{2}$ **22.** $12 \div \dfrac{3}{4}$ **23.** $\dfrac{3}{2} \div 18$ **24.** $\dfrac{3}{4} \div 12$

25. $\dfrac{28}{45} \cdot \dfrac{63}{40}$ **26.** $\dfrac{28}{45} \div \dfrac{63}{40}$ **27.** $\dfrac{3}{5} \cdot \dfrac{12}{7} \cdot \dfrac{10}{9}$ **28.** $\dfrac{4}{15} \cdot \dfrac{6}{13} \cdot \dfrac{26}{36}$

29. $\dfrac{12}{25} \cdot \dfrac{5}{2} \div \dfrac{6}{5}$ **30.** $\dfrac{18}{35} \div \dfrac{9}{14} \cdot \dfrac{5}{4}$

31. What is the reciprocal of $\dfrac{5}{9}$?   **32.** What is the reciprocal of $\dfrac{3}{10}$?

33. $\dfrac{4}{7}$ is the reciprocal of what number?   **34.** 8 is the reciprocal of what number?

21. _____

22. _____

23. _____

24. _____

25. _____

26. _____

27. _____

28. _____

29. _____

30. _____

31. _____

32. _____

33. _____

34. _____

Adding and Subtracting Fractions

0.4

We begin by defining addition and subtraction of two fractions with the same denominator. We can state the rule as follows:

RULE FOR ADDING AND SUBTRACTING FRACTIONS

$$\frac{a}{c} + \frac{b}{c} = \frac{a+b}{c} \qquad \frac{a}{c} - \frac{b}{c} = \frac{a-b}{c} \qquad c \neq 0$$

In other words, this rule says that we can add or subtract fractions with common denominators by adding or subtracting the numerators and putting the result over the common denominator.

EXAMPLE 1   *Add.* $\dfrac{5}{14} + \dfrac{7}{14}$

Solution

$$\frac{5}{14} + \frac{7}{14} = \frac{5+7}{14} = \frac{12}{14} = \frac{\overset{6}{\cancel{12}}}{\underset{7}{\cancel{14}}} = \boxed{\frac{6}{7}}$$

Remember that final answers should be reduced to lowest terms.   ∎

In order to add or subtract two fractions with different denominators we first use the Fundamental Principle of Fractions to change each fraction into an equivalent fraction with the same denominator. In general, we will find it most convenient to use the smallest possible common denominator—called the **least common denominator** or LCD for short.

We will find the following outline useful for finding the LCD.

OUTLINE FOR FINDING THE LCD

Step 1 Factor each denominator as completely as possible.

Step 2 The LCD consists of the product of each *distinct* factor the *maximum* number of times it appears in any one denominator.

Let's illustrate how to use this outline in Example 2.

EXAMPLE 2   *Add.* $\dfrac{7}{12} + \dfrac{1}{30}$

Solution   We need to convert both fractions into equivalent fractions with the same denominator. Using our outline, we proceed as follows:

Step 1 $\left.\begin{array}{l} 12 = 2 \cdot 2 \cdot 3 \\ 30 = 2 \cdot 3 \cdot 5 \end{array}\right\}$ *Notice that the **distinct** factors are 2, 3, 5.*

✔ LEARNING CHECKS

1. $\dfrac{4}{15} + \dfrac{8}{15}$

2. $\dfrac{3}{16} + \dfrac{7}{10}$

Step 2 We choose each distinct factor the *maximum* number of times it appears in any one denominator.

$$\text{In } 12 = 2 \cdot 2 \cdot 3 \begin{cases} 2 \text{ appears as a factor twice.} \\ 3 \text{ appears as a factor once.} \end{cases}$$

$$\text{In } 30 = 2 \cdot 3 \cdot 5 \begin{cases} 2 \text{ appears as a factor once.} \\ 3 \text{ appears as a factor once.} \\ 5 \text{ appears as a factor once.} \end{cases}$$

Following our outline, we need 2 as a factor twice, 3 as a factor once, and 5 as a factor once.

Thus, the LCD is $2 \cdot 2 \cdot 3 \cdot 5 = 60$.

Now we use the Fundamental Principle to build fractions into equivalent fractions with LCD of 60 as denominator. We examine each denominator and determine which factors are missing from the LCD:

$$\frac{7}{12} = \frac{7}{2 \cdot 2 \cdot 3} = \frac{7 \cdot \boxed{5}}{2 \cdot 2 \cdot 3 \cdot \boxed{5}} \qquad \textit{The missing factor was 5.}$$

$$\frac{1}{30} = \frac{1}{2 \cdot 3 \cdot 5} = \frac{1 \cdot \boxed{2}}{\boxed{2} \cdot 2 \cdot 3 \cdot 5} \qquad \textit{The missing factor was 2.}$$

Therefore, we have

$$\frac{7}{12} + \frac{1}{30} = \frac{7 \cdot 5}{60} + \frac{1 \cdot 2}{60}$$

$$= \frac{35}{60} + \frac{2}{60}$$

$$= \boxed{\frac{37}{60}} \qquad\qquad\qquad \blacksquare$$

3. $\dfrac{1}{6} + \dfrac{4}{15}$

EXAMPLE 3   *Add.* $\dfrac{3}{10} + \dfrac{15}{4}$

Solution   You can probably "see" that the LCD is 20. If not, you can use the outline to obtain $2 \cdot 2 \cdot 5 = 20$ as the LCD. Try it!

$$\frac{3}{10} + \frac{15}{4} = \frac{3(2)}{20} + \frac{15(5)}{20}$$

$$= \frac{6 + 75}{20}$$

$$= \boxed{\frac{81}{20}} \qquad\qquad\qquad \blacksquare$$

4. $\dfrac{5}{6} \cdot \dfrac{4}{15}$

EXAMPLE 4   *Multiply.* $\dfrac{3}{10} \cdot \dfrac{15}{4}$

Solution   Be careful! This is a multiplication problem and so no LCD is needed.

$$\frac{3}{10} \cdot \frac{15}{4} = \frac{3}{\underset{2}{\cancel{10}}} \cdot \frac{\overset{3}{\cancel{15}}}{4} = \boxed{\frac{9}{8}} \qquad\qquad \blacksquare$$

Looking at the answer to Example 4, it seems like an appropriate time to comment on the idea of an "improper" fraction versus a mixed number. Some texts refer to a fraction in which the numerator is greater than or equal to the denominator as an *improper fraction*. For our purposes, however, all we require is that the fraction be reduced to lowest terms.

Thus, in the last example, our answer of $\frac{9}{8}$ is acceptable. On some occasions we may prefer to express $\frac{9}{8}$ as a **mixed number** by dividing 8 into 9, giving $1\frac{1}{8}$.

EXAMPLE 5 *Combine.* $\frac{5}{4} - \frac{8}{9} + \frac{13}{18}$

5. $\frac{2}{9} - \frac{4}{15} + \frac{3}{5}$

Solution We begin by using our outline for finding the LCD:

$$\left.\begin{array}{l} 4 = 2 \cdot 2 \\ 9 = 3 \cdot 3 \\ 18 = 2 \cdot 3 \cdot 3 \end{array}\right\} \quad \textit{The distinct factors are 2 and 3.}$$

2 appears as a factor at most twice and 3 appears as a factor at most twice; therefore, the LCD is $2 \cdot 2 \cdot 3 \cdot 3 = 36$.

$$\frac{5}{4} = \frac{5}{2 \cdot 2} = \frac{5\,(3 \cdot 3)}{2 \cdot 2 \cdot 3 \cdot 3} = \frac{45}{36}$$

$$\frac{8}{9} = \frac{8}{3 \cdot 3} = \frac{8\,(2 \cdot 2)}{2 \cdot 2 \cdot 3 \cdot 3} = \frac{32}{36}$$

$$\frac{13}{18} = \frac{13}{2 \cdot 3 \cdot 3} = \frac{13\,(2)}{2 \cdot 2 \cdot 3 \cdot 3} = \frac{26}{36}$$

Therefore, we have

$$\frac{5}{4} - \frac{8}{9} + \frac{13}{18} = \frac{45}{36} - \frac{32}{36} + \frac{26}{36}$$

$$= \frac{39}{36} \qquad \textit{Reduce by a factor of 3.}$$

$$= \frac{\overset{13}{\cancel{39}}}{\underset{12}{\cancel{36}}}$$

$$= \boxed{\frac{13}{12}} \qquad \blacksquare$$

EXAMPLE 6 Convert $3\frac{5}{6}$ from a mixed number to a fraction.

6. Convert $5\frac{3}{4}$ to a fraction.

Solution

$$3\frac{5}{6} \quad \text{means} \quad 3 + \frac{5}{6} = \frac{3}{1} + \frac{5}{6} \qquad \textit{The LCD is 6.}$$

$$= \frac{3(6)}{6} + \frac{5}{6}$$

$$= \frac{18}{6} + \frac{5}{6}$$

$$= \boxed{\frac{23}{6}} \qquad \blacksquare$$

The idea of a common denominator is also useful in deciding which of several fractions is largest or smallest.

EXAMPLE 7   Which of the following fractions is the largest?

$$\frac{3}{5}, \quad \frac{5}{9}, \quad \frac{17}{30}$$

Solution   The easiest way to compare the sizes of various fractions is to convert them to equivalent fractions with the same denominator. The LCD for 5, 9, and 30 is 90.

$$\frac{3}{5} = \frac{3(18)}{90} = \frac{54}{90}$$

$$\frac{5}{9} = \frac{5(10)}{90} = \frac{50}{90}$$

$$\frac{17}{30} = \frac{17(3)}{90} = \frac{51}{90}$$

Therefore, we can see that the largest fraction is $\frac{54}{90}$, which is $\boxed{\frac{3}{5}}$. $\blacksquare$

7. Which of the following fractions is the smallest?
$$\frac{2}{3}, \quad \frac{3}{4}, \quad \frac{4}{5}$$

✔ *Answers to Learning Checks in Section 0.4*

1. $\frac{4}{5}$ **2.** $\frac{71}{80}$ **3.** $\frac{13}{30}$ **4.** $\frac{2}{9}$ **5.** $\frac{5}{9}$ **6.** $\frac{23}{4}$ **7.** $\frac{2}{3}$

Exercises 0.4

In Exercises 1–26, perform the indicated operations and simplify as completely as possible.

1. $\dfrac{2}{7} + \dfrac{3}{7}$

2. $\dfrac{5}{9} + \dfrac{2}{9}$

3. $\dfrac{5}{8} - \dfrac{3}{8} + \dfrac{1}{8}$

4. $\dfrac{8}{15} - \dfrac{4}{15} + \dfrac{3}{15}$

5. $\dfrac{5}{12} + \dfrac{1}{12}$

6. $\dfrac{11}{20} + \dfrac{3}{20}$

7. $\dfrac{3}{10} + \dfrac{7}{10}$

8. $\dfrac{23}{18} + \dfrac{13}{18}$

9. $\dfrac{1}{2} + \dfrac{1}{4}$

10. $\dfrac{2}{3} - \dfrac{1}{6}$

11. $\dfrac{5}{6} - \dfrac{3}{8}$

12. $\dfrac{3}{10} - \dfrac{1}{4}$

13. $\dfrac{5}{6} \cdot \dfrac{3}{8}$

14. $\dfrac{3}{10} \div \dfrac{1}{4}$

15. $\dfrac{7}{12} + \dfrac{1}{3}$

16. $\dfrac{8}{15} + \dfrac{2}{5}$

17. $\dfrac{1}{2} + \dfrac{1}{3} + \dfrac{1}{4}$

18. $\dfrac{4}{9} + \dfrac{2}{3} + \dfrac{1}{6}$

19. $\dfrac{4}{15} + \dfrac{7}{20}$

20. $\dfrac{5}{24} + \dfrac{7}{60}$

1. _____
2. _____
3. _____
4. _____
5. _____
6. _____
7. _____
8. _____
9. _____
10. _____
11. _____
12. _____
13. _____
14. _____
15. _____
16. _____
17. _____
18. _____
19. _____
20. _____

21. _____

22. _____

23. _____

24. _____

25. _____

26. _____

27. _____

28. _____

29. _____

30. _____

31. _____

32. _____

33. _____

34. _____

35. _____

36. _____

37. _____

38. _____

21. $\dfrac{3}{28} + \dfrac{2}{35}$ **22.** $\dfrac{11}{18} + \dfrac{3}{20}$ **23.** $5 + \dfrac{3}{5}$

24. $5 \cdot \dfrac{3}{5}$ **25.** $5 \div \dfrac{3}{5}$ **26.** $\dfrac{3}{5} \div 5$

In Exercises 27–30, convert the given mixed number to a fraction.

27. $3\dfrac{3}{4}$ **28.** $4\dfrac{5}{8}$ **29.** $12\dfrac{4}{5}$ **30.** $10\dfrac{1}{9}$

In Exercises 31–34, convert the given fraction to a mixed number.

31. $\dfrac{23}{5}$ **32.** $\dfrac{12}{7}$ **33.** $\dfrac{43}{6}$ **34.** $\dfrac{100}{9}$

In Exercises 35–38, find the smallest of the three fractions.

35. $\dfrac{2}{3},\ \dfrac{3}{4},\ \dfrac{4}{5}$ **36.** $\dfrac{3}{10},\ \dfrac{4}{15},\ \dfrac{7}{20}$ **37.** $\dfrac{5}{12},\ \dfrac{7}{15},\ \dfrac{9}{20}$ **38.** $\dfrac{7}{8},\ \dfrac{8}{9},\ \dfrac{9}{10}$

Decimals

0.5

A *decimal fraction* or *decimal* is another way of writing a fraction. For example, the decimal .4 is another way of writing $\frac{4}{10}$. Similarly, .709 is another way of writing $\frac{709}{1,000}$, and .0037 is another way of writing $\frac{37}{10,000}$.

The process of converting a decimal to a fraction is simply to write the decimal as a fraction and reduce it, if possible. For instance,

$$.458 = \frac{458}{1,000} = \frac{229}{500}$$

To convert a fraction to a decimal, we divide the numerator by the denominator.

For example, to convert $\frac{3}{8}$ to a decimal we divide 3 by 8:

$$
\begin{array}{r}
.375 \\
8\overline{)3.000} \\
2\,4 \\
\overline{60} \\
56 \\
\overline{40} \\
40 \\
\overline{0}
\end{array}
$$

Therefore, $\frac{3}{8} = .375$.

In performing operations with fractions, we often find the decimal form more convenient to use. (If you use a calculator, you may be able to input fractions but the answer which appears in the display is usually a decimal.) However, be aware that if you convert a fraction into a decimal in order to perform some arithmetic operation and the decimal is not exact (because you rounded off), then your answer will not be exact as is generally required by the exercises in this text.

Addition and Subtraction

To add or subtract decimals we arrange the numbers in a column with the decimal points lined up, then add or subtract as usual. The decimal point in the answer is directly beneath the decimal point in the column of numbers.

EXAMPLE 1 *Add.* $2.07 + 81.6 + .084$

Solution   We write

$$
\begin{array}{r}
2.070 \\
81.600 \\
+\ \ .084 \\
\hline
83.754
\end{array}
$$
   $\leftarrow$ *It is helpful to put extra zeros at the end of these numbers so they are all the same length.*

$\uparrow$
Line up the decimal point.

■

✔ **LEARNING CHECKS**

1. *Add.* $143.45 + 8.1 + .019$

2. *Subtract.* $56.2 - 8.47$

EXAMPLE 2 *Subtract.* $21.7 - .095$

Solution   We write

$$\begin{array}{r} 21.700 \\ - \underline{.095} \\ 21.605 \end{array}$$

∎

Multiplication

To multiply two decimals we multiply the numbers as if they were whole numbers. The *number* of decimal places in the answer is the *total* number of decimal places in the numbers being multiplied.

EXAMPLE 3 *Multiply.* 43.6×2.87

Solution

$$\begin{array}{r} 4\,3.6 \quad \leftarrow \; 1 \; decimal \; place \\ \underline{2.8\,7} \quad \leftarrow \; 2 \; decimal \; places \\ 3\,0\,5\,2 \\ 3\,4\,8\,8 \\ \underline{8\,7\,2 } \\ 1\,2\,5.1\,3\,2 \quad \leftarrow \; 3 \; decimal \; places \end{array}$$

∎

3. *Multiply.* 3.52×2.3

Division

EXAMPLE 4 *Divide.* $716.68 \div 16.4$

Solution

Our basic approach is to convert this problem of dividing by a decimal into an equivalent problem of dividing by a whole number. We move the decimal point in 16.4 one place to the right so that it becomes 164 (a whole number). We must then also move the decimal point in 716.68 the same number of places to the right, making it 7,166.8. (In effect what we are doing is viewing $716.68 \div 16.4$ as a fraction $\dfrac{716.68}{16.4}$ and then multiplying both the numerator and denominator of the fraction by 10 to change the denominator 16.4 into a whole number.)
  We write

$$16.4\,)\overline{716.6.8}$$

We now put the decimal point in our answer *directly* above the decimal point in 7,166.8 and divide as we usually do with whole numbers:

$$\begin{array}{r} 43.7 \\ 164\,)\overline{7\,166.8} \\ \underline{6\,56} \\ 606 \\ \underline{492} \\ 114\,8 \\ \underline{114\,8} \\ 0 \end{array}$$

∎

4. *Divide.* $773.15 \div 23.5$

✔ *Answers to Learning Checks in Section 0.5*

1. 151.569 **2.** 47.73 **3.** 8.096
4. 32.9

Exercises 0.5

In Exercises 1–34, perform the indicated operations.

1. $4.7 + 3.5 + 21.7$

2. $2.65 + 8.37 + 11.69$

3. $15.87 - 6.35$

4. $228.7 - 139.8$

5. $21.62 + 4.1 + 57.236$

6. $128.05 + 96.341 + 27.2$

7. $6.5 + .003 + 2.08$

8. $12.86 + 2.015 + 3.1$

9. $9.27 - 7.85$

10. $29.46 - 12.77$

11. $(3.9)(5.8)$

12. $(26.4)(15.7)$

13. $6 - .03$

14. $11 - 5.2$

15. $(2.63)(13.05)$

16. $(31.6)(19.52)$

17. $(.37)(5.16)$

18. $(.0025)(150.3)$

1. _____
2. _____
3. _____
4. _____
5. _____
6. _____
7. _____
8. _____
9. _____
10. _____
11. _____
12. _____
13. _____
14. _____
15. _____
16. _____
17. _____
18. _____

19. $39.6 \div 5$

20. $928.42 \div 6$

21. $630 \div 1.5$

19. _____

20. _____

22. $80 \div 2.5$

23. $3.28 \div .16$

24. $205.63 \div .05$

21. _____

22. _____

23. _____

25. $(20.05)(.004)$

26. $(46.8)(3.06)$

27. $(.032)(.05)$

24. _____

25. _____

26. _____

28. $(56)(.003)$

29. $.28 \div .04$

30. $.096 \div .4$

27. _____

28. _____

31. $(.01)(.02)(.03)$

32. $(1.2)(2.1)(3.2)$

33. $3.9 \div .015$

29. _____

30. _____

31. _____

34. $385.57 \div .285$

32. _____

33. _____

34. _____

Percent

0.6

An important idea that is related to decimals is the idea of percent. The word *percent* means "part of a hundred." The symbol for percent is %. Thus, 8 percent is written 8%.

$$45\% \quad \text{means} \quad \frac{45}{100} = .45$$

$$8\% \quad \text{means} \quad \frac{8}{100} = .08$$

$$137\% \quad \text{means} \quad \frac{137}{100} = 1.37$$

$$.4\% \quad \text{means} \quad \frac{.4}{100} = .004$$

Note the pattern of movement of the decimal point as we change from a percent to a decimal.

RULE FOR CHANGING FROM PERCENT TO DECIMAL

To convert a number from a percent to a decimal, we drop the percent sign and move the decimal point two places to the *left* (we are dividing by 100).

RULE FOR CHANGING FROM DECIMAL TO PERCENT

To convert from a decimal to a percent we move the decimal point two places to the *right* (we are multiplying by 100) and insert the percent sign.

EXAMPLE 1   Convert 28% to a decimal.

Solution   In order to use this rule we must be aware of the location of the decimal point in the given percent. In 28% the decimal point is understood to follow the 8.

$28\% = 28.\% = \boxed{.28}$ *Note that we moved the decimal point two places to the left.*   ■

EXAMPLE 2   Convert .365 to a percent.

Solution

$.365 = \boxed{36.5\%}$ *Note that we moved the decimal point two places to the right.*   ■

Computing Percentages

Since a percent is a fraction, computing a percentage of a number is the same as taking a fraction of a number, and therefore we use the operation of multiplication. However, we must first convert the percent into its decimal equivalent.

LEARNING CHECKS

1. Convert 47% to a decimal.

2. Convert .814 to a percent.

3. Compute 36% of 84.

EXAMPLE 3   Compute 42% of 75.

Solution   We convert 42% into a decimal, 42% = .42, and then we multiply:

$$.42(75) = \boxed{31.5}$$ ■

4. Compute 112% of 56.

EXAMPLE 4   Compute 107.5% of 1,200.

Solution   We convert 107.5% into a decimal, 107.5% = 1.075, and then we multiply:

$$1.075(1,200) = \boxed{1,290}$$ ■

5. Convert $\frac{1}{8}$ to percent.

EXAMPLE 5   Convert $\frac{3}{5}$ to percent.

Solution   We convert $\frac{3}{5}$ to a decimal by dividing 5 into 3, which gives us .6. We then convert .6 into a percentage:

$$.6 = .60 = \boxed{60\%}$$ ■

6. 18 is what percent of 40?

EXAMPLE 6   24 is what percent of 60?

Solution   The problem is asking us to convert the fraction $\frac{24}{60}$ to a percentage. We first convert $\frac{24}{60}$ to a decimal. We can begin by reducing the fraction $\frac{24}{60} = \frac{2}{5}$. We then divide 5 into 2, giving us .4. Finally, we convert .4 into a percentage:

$$.4 = .40 = \boxed{40\%}$$ ■

✔ *Answers to Learning Checks in Section 0.6*

1. .47   **2.** 81.4%   **3.** 30.24   **4.** 62.72   **5.** 12.5%   **6.** 45%

Exercises 0.6

In Exercises 1–24, convert the percents to decimals and the decimals to percents.

1. 25% **2.** .38 **3.** .78 **4.** 43%

5. .05 **6.** 7% **7.** 9% **8.** .03

9. 150% **10.** 2.4 **11.** 28% **12.** .53

13. .67 **14.** 78% **15.** 2 **16.** 40

17. 137% **18.** 5% **19.** .007 **20.** .23%

21. 62.4% **22.** .089 **23.** 8.6% **24.** 10.4%

ANSWERS

1. _____
2. _____
3. _____
4. _____
5. _____
6. _____
7. _____
8. _____
9. _____
10. _____
11. _____
12. _____
13. _____
14. _____
15. _____
16. _____
17. _____
18. _____
19. _____
20. _____
21. _____
22. _____
23. _____
24. _____

In Exercises 25–30, compute the given percentages.

25. 30% of 70

26. 65% of 250

27. 7.2% of 35

25. _____

26. _____

27. _____

28. 26.5% of 900

29. .8% of 5

30. 1% of 314

28. _____

29. _____

30. _____

In Exercises 31–33, convert the given fraction to a percentage.

31. $\frac{1}{4}$

32. $\frac{3}{5}$

33. $\frac{5}{8}$

31. _____

32. _____

33. _____

34. 12 is what percent of 30?

35. 8 is what percent of 20?

36. 25 is what percent of 75?

34. _____

35. _____

36. _____

CHAPTER 0 SUMMARY

After having completed this chapter you should be able to:

1. Use the listing method or set-builder notation to recognize sets (Section 0.1).

   For example:

   Suppose we have the following sets:

   $$A = \{3, 6, 9, 12, 15\}$$
   $$B = \{x \mid x \text{ is an even integer between 8 and 30}\}$$
   $$C = \{17, 19, 21, \ldots, 29\}$$

   Then

   (a) $6 \in A, \quad 10 \notin A$

   (b) $8 \notin B, \quad 28 \in B$

   (c) $\{x \mid x \in A \text{ and } x \in C\} = \varnothing$ (the empty set)

2. Reduce fractions to lowest terms and build fractions to higher terms (Section 0.2).

   For example:

   (a) $\dfrac{21}{36} = \dfrac{\cancel{3} \cdot 7}{\cancel{3} \cdot 12} = \boxed{\dfrac{7}{12}}$

   (b) $\dfrac{4}{9} = \dfrac{4 \cdot 5}{9 \cdot 5} = \boxed{\dfrac{20}{45}}$

3. Multiply and divide fractions (Section 0.3).

   For example:

   (a) $\dfrac{20}{27} \cdot \dfrac{18}{30} = \dfrac{\overset{2}{\cancel{20}}}{\underset{3}{\cancel{27}}} \cdot \dfrac{\overset{2}{\cancel{18}}}{\underset{3}{\cancel{30}}} = \boxed{\dfrac{4}{9}}$

   (b) $\dfrac{4}{5} \div 6 = \dfrac{4}{5} \cdot \dfrac{1}{\underset{3}{\cancel{6}}} = \boxed{\dfrac{2}{15}}$

4. Add and subtract fractions (Section 0.4).

   For example:

   $$\dfrac{3}{8} + \dfrac{7}{10} \qquad \textit{The LCD is } 40.$$

   $$= \dfrac{3 \cdot 5}{40} + \dfrac{7 \cdot 4}{40}$$

   $$= \dfrac{15}{40} + \dfrac{28}{40}$$

   $$= \boxed{\dfrac{43}{40}}$$

5. Perform the four arithmetic operations with decimals (Section 0.5).

For example:

(a) Add $2.83 + .015 + 5.6$.

$$
\begin{array}{r}
2.830 \\
.015 \\
\underline{5.600} \\
8.445
\end{array}
$$

(b) Multiply 4.8×1.06.

$$
\begin{array}{r}
1.0\,6 \\
\underline{4.8} \\
8\,4\,8 \\
\underline{4\,2\,4} \\
5.0\,8\,8
\end{array}
$$

(c) Divide $\dfrac{2.8}{.04}$.

$$
\begin{array}{r}
70. \\
.04\,\overline{)2.80}
\end{array}
$$

6. Compute with percents (Section 0.6).

For example:

(a) $27\% = \boxed{.27}$

(b) $.45 = \boxed{45\%}$

(c) 24% of $50 = (.24)(50) = \boxed{12}$

ANSWERS TO QUESTIONS FOR THOUGHT

1. *W* contains the number 0; *N* does not.

2. In a sum, the numbers to be added are called *terms*; in a product, the numbers to be multiplied are called *factors*. In $2 + 5$, 2 is a term; in $2 \cdot 5$, 2 is a factor.

3. A *factor* of *n* is a number which divides exactly into *n*; a *multiple* of *n* is a number which is exactly divisible by *n*. 3 is a factor of 12; 12 is a multiple of 3.

CHAPTER 0 REVIEW EXERCISES

In Exercises 1–6, list the members of each set.

1. $\{x \mid x$ is a positive even integer less than 20$\}$

2. $\{y \mid y$ is a positive odd integer less than 20$\}$

3. $\{p \mid p$ is a prime number less than 20$\}$

4. $\{c \mid c$ is a positive composite number less than 20$\}$

5. $\{p \mid p$ is a prime number between 20 and 30$\}$

6. $\{x \mid x$ is a prime number and x is a composite number$\}$

In Exercises 7–12, decompose the given number into its prime factors. If the number is prime, say so.

7. 30　　　　　**8.** 28　　　　　**9.** 47

10. 72　　　　　**11.** 100　　　　　**12.** 57

In Exercises 13–26, perform the indicated operations.

13. $\dfrac{2}{5}+\dfrac{3}{4}$　　　**14.** $\dfrac{2}{5}\cdot\dfrac{3}{4}$　　　**15.** $\dfrac{2}{5}\div\dfrac{3}{4}$

16. _____

17. _____

18. _____

19. _____

20. _____

21. _____

22. _____

23. _____

24. _____

25. _____

26. _____

27. _____

28. _____

29. _____

30. _____

31. _____

32. _____

16. $\dfrac{3}{4} \div \dfrac{2}{5}$ **17.** $8.43 + 27.018 + .67$ **18.** $(3.08)(4.9)$

19. $84 \div .021$ **20.** $234.23 - 60.18$ **21.** $.6(.06)(.006)$

22. $52.01 \div .35$ **23.** $\dfrac{5}{12} + \dfrac{7}{8}$ **24.** $\dfrac{2}{3} + \dfrac{4}{9} - \dfrac{1}{12}$

25. $\dfrac{3}{8} \cdot 12$ **26.** $\dfrac{3}{8} \div 12$

In Exercises 27–32, convert fractions and percents to decimals, and decimals to percents.

27. $\dfrac{4}{5}$ **28.** 92% **29.** $.47$

30. $\dfrac{5}{12}$ **31.** 6.3% **32.** $.005$

CHAPTER 0 PRACTICE TEST

1. Let $A = \{0, 4, 8, 12, \ldots, 28\}$ and $B = \{x \mid x$ is an odd integer between 3 and 20$\}$. Answer parts **(a)** through **(d)** True or False:

   (a) $20 \in A$

   (b) $20 \in B$

   (c) $3 \in B$

   (d) Both A and B have the same number of elements.

   (e) List the set $C = \{x \mid x \in A$ and $x \in B\}$.

2. Express 84 as a product of prime factors.

In each of the following, perform the indicated operations.

3. $\dfrac{5}{6} + \dfrac{2}{15}$

4. $\dfrac{5}{6} \cdot \dfrac{2}{15}$

5. $\dfrac{5}{6} \div \dfrac{2}{15}$

6. $27.43 + 102.1 + .052$

7. $12 - .43$

8. $(2.4)(25.6)$

9. $\dfrac{174.838}{2.15}$

10. Convert .841 to a percentage.

11. Find 32% of 150.

12. 12 is what percent of 40?

1. a. _____

   b. _____

   c. _____

   d. _____

   e. _____

2. _____

3. _____

4. _____

5. _____

6. _____

7. _____

8. _____

9. _____

10. _____

11. _____

12. _____

NOTE TO THE STUDENT

Questions frequently come up while you are doing homework exercises. Use the space on this page to write down any questions you have or points you want to review with your instructor.

The Integers

Integers

1.1

In this chapter we will introduce the set of integers and learn how to perform the various arithmetic operations with them. Throughout this chapter we will use many of the ideas and much of the terminology introduced in the first section of Chapter 0. If you have skipped Chapter 0, you may find it a good idea to at least read over Section 0.1 to reacquaint yourself with this very basic material.

Our everyday usage of numbers often requires us to indicate direction as well as size. For example, a checking account in which we have $180 will show a balance of $180, while a checking account in which we are overdrawn by $180 will show a balance of −$180.

Signed numbers tell us not only quantity or size, but also direction. Perhaps the most common everyday use of signed numbers is in describing temperatures: −10°F is 10 degrees *below* 0, while +20°F means 20 degrees *above* 0.

Mathematically, we want to extend our number system beyond the whole numbers to include these signed numbers as well. We do this as follows. We begin with our number line for the whole numbers:

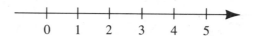

Now we extend the line to the left of 0 and mark off unit lengths going off to the *left*:

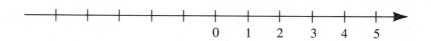

What notation shall we use to designate the units to the left of zero? How shall we name these points?

Rather than trying to memorize a new set of symbols, we can use the same symbols that we use to name units to the right of 0, except we place the symbol "−" before these numbers to indicate that they are to the *left* of 0. Our number line now looks like this:

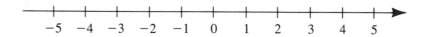

Thus, +2 is 2 units to the *right* of 0, while −2 is 2 units to the *left* of 0. The set of numbers $\{\ldots, -3, -2, -1, 0, 1, 2, 3, \ldots\}$ is called the set of ***integers*** and is usually denoted with the letter Z (you will see later in this chapter that we are saving the letter I for a different set):

$$Z = \{\ldots, -3, -2, -1, 0, 1, 2, 3, \ldots\}$$

The set $\{1, 2, 3, \ldots\}$, which we have called the natural numbers, is also often called the set of ***positive integers***. The set $\{-1, -2, -3, \ldots\}$ is called the set of ***negative integers***. Note that the number 0, although it is an integer, is neither positive nor negative.

Very frequently when we write positive integers we do not write the "+" sign. That is, 5 is understood to mean +5. However, in this chapter, to make things clearer, we will usually write the "extra" + sign.

You might wonder about choosing the minus sign to indicate a negative number, when the minus sign already means subtraction. Will this lead to confusion? As we shall soon see, our various uses and interpretations of the minus sign will be consistent, and we will be free to think of the minus sign in any of several ways.

Opposites

Notice on the number line that −4 is the same distance from 0 as +4, but −4 is in the opposite direction from 0 than is +4 (see Figure 1.1). Similarly for +2 and −2, and +14 and −14. For that reason we call the pair of numbers x and $−x$ *opposites*.

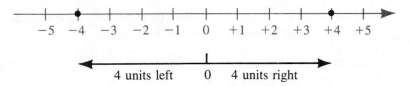

Figure 1.1   *Opposites*

The number +4 is 4 units to the right of 0 and −4 is 4 units from 0 in the *opposite* direction. We can generalize this and say that if x is *any* number then $−x$ will be the same distance from 0 in the opposite direction.

Thus, in addition to naming the points on the left side of the number line, the minus (or negative) sign is also used to change the direction of a number.

> Putting a minus sign in front of a number changes a number into its opposite.

What we have just said is true for both positive and negative integers. That is, we have

−(+5) means the opposite of +5, which is −5

and in exactly the same way

−(−5) means the opposite of −5, which is +5

In general, we have "the opposite of $−x$ is x," regardless of whether x *itself* is positive or negative. That is,

$$-(-x) = x$$

regardless of whether x is positive or negative.

Our ideas about order on the number line continue to be true. That is, a is smaller than b means a is to the left of b, regardless of whether a and b are positive or negative. Thus, the following are true statements:

$2 < 7$ (2 is less than 7)   because 2 is to the left of 7 on the number line.

$-2 > -7$ (-2 is greater than -7)   because -2 is to the right of -7 on the number line.

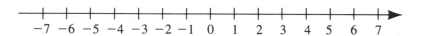

Even though -7 is further away from 0 than 2 is, as integers -7 is smaller because it is further to the left. To use a more familiar example:

40 feet *below* sea level (written -40 feet) is *higher* than 80 feet below sea level (written -80 feet).

As we shall see in the next section, there are times when we are not interested in the direction of a number, but only in its distance from 0.

DEFINITION   The *absolute value* of a number is its distance from 0. The absolute value of x is written $|x|$.

Thus, for example, we have:

$|5| = 5$ $|-5| = 5$ $|234| = 234$ $|-399| = 399$ $|0| = 0$

✔ **LEARNING CHECKS**

1. **(a)** $|9 - 2|$

  (b) $-|-3|$

EXAMPLE 1   Compute each of the following:

(a) $|7 - 3|$ **(b)** $-|-6|$

Solution

(a) Do not interpret the absolute value sign to mean "erase all negative signs." That is not what it means. We must first see what number is *inside* the absolute value sign, and then we take its distance from 0.

$$|7 - 3| = |4| = \boxed{4}$$

(b) Watch the minus signs very carefully. We compute the absolute value first.

$-|-6| = -(6)$ *Note that the first minus sign gets copied.*

$= \boxed{-6}$ ∎

Properties of the Integers

The foundation for much of the work that we do in algebra is the properties of the integers. For example, we are familiar with the fact that we can add or multiply two numbers in any order and get the same answer. That is,

$$6 + 2 = 2 + 6 \quad \text{and} \quad 6 \cdot 2 = 2 \cdot 6$$

These facts are called the *Commutative Laws* of the integers. We can state them algebraically as follows:

COMMUTATIVE LAWS

$a + b = b + a$ Commutative Law of Addition

$ab = ba$ Commutative Law of Multiplication

The emphasis here should be placed on the fact that this law applies to addition and multiplication *only*. A similar property is *not* true for subtraction and division. That is, we cannot change the order of the numbers in a subtraction or division example and expect to get the same answer. For example,

$$6 - 2 \neq 2 - 6 \quad \text{and} \quad 6 \div 2 \neq 2 \div 6$$

Frequently we want to indicate that certain numbers or expressions are to be grouped together. We usually indicate these groupings by using parentheses like ( ) or brackets like [ ] or braces like { }. For example,

$4(3 \cdot 7)$ means multiply 3 by 7 and then multiply the result by 4

while

$(4 \cdot 3)7$ means multiply 4 by 3 and then multiply the result by 7

It is no surprise that in both cases we get the same answer of 84, because we know that we can group multiplications in any way we please. The same holds true for addition. These facts are called the *Associative Laws.* We can state them algebraically as follows:

ASSOCIATIVE LAWS

$a + (b + c) = (a + b) + c$ Associative Law of Addition

$a(bc) = (ab)c$ Associative Law of Multiplication

As with the Commutative Laws, the Associative Laws do not pertain to subtraction or division. For example,

$$10 - (5 - 2) \neq (10 - 5) - 2 \qquad 36 \div (6 \div 3) \neq (36 \div 6) \div 3$$
$$10 - 3 \neq 5 - 2 \qquad\qquad 36 \div 2 \neq 6 \div 3$$
$$7 \neq 3 \qquad\qquad\qquad 18 \neq 2$$

In words, the Commutative Law says that we can *reorder* a sum or a product, while the Associative Law says that we can *regroup* a sum or a product.

There is another very important property called the *Distributive Law,* but we will postpone its discussion until Chapter 2.

Order of Operations

If we look at the example $5 + 4 \cdot 3$, we can interpret the problem in two ways:

$$5 + 4 \cdot 3 = 9 \cdot 3 = 27 \quad \text{or} \quad 5 + 4 \cdot 3 = 5 + 12 = 17$$

Both are valid ways of working out the example. Thus, we must come to some agreement as to what our *order of operations* is going to be, so that we will all understand a given example to mean the same thing.

The accepted order of priority for our operations is given in the accompanying box.

ORDER OF OPERATIONS

1. Evaluate expressions within grouping symbols first. If there are grouping symbols within grouping symbols then work from the innermost grouping symbol outward.
2. Next, perform multiplications and divisions, working from *left to right*.
3. Next, perform additions and subtractions, working from *left to right*.

---

2. (a) $9 - 2(5 - 3)$

(b) $(9 - 2)(5 - 3)$

EXAMPLE 2   Evaluate the following expressions:

(a) $8 - 3(6 - 4)$ **(b)** $(8 - 3)(6 - 4)$

Solution

(a) We begin by following the order of operations and starting inside the parentheses:

$$8 - 3(6 - 4) = 8 - 3(2) \quad \textit{Next we must do the multiplication.}$$
$$= 8 - 6$$
$$= \boxed{2}$$

(b) Again we follow the order of operations. Since we have two sets of parentheses, our first step is to evaluate within each set:

$$(8 - 3)(6 - 4) = (5)(2)$$
$$= \boxed{10} \quad\blacksquare$$

3. $6 + 3[18 - 3(3 + 1)]$

EXAMPLE 3   *Evaluate.* $5 + 2[15 - 4(2 + 1)]$

Solution   Following the order of operations, we work from the innermost grouping symbol outward. In this case, we first evaluate the expression within parentheses and then within the brackets:

$$5 + 2[15 - 4(2 + 1)] = 5 + 2[15 - 4(3)] \quad \textit{Next we continue inside the [ ] and do multiplication before subtraction.}$$

$$= 5 + 2[15 - 12] \quad \textit{Now we evaluate within the [ ].}$$
$$= 5 + 2[3] \quad \textit{Again, we perform multiplication before addition.}$$
$$= 5 + 6$$
$$= \boxed{11} \quad\blacksquare$$

Most of the time, when we want to indicate division we will use the fraction bar. That is, we will usually write

$$a \div b \quad \text{as} \quad \frac{a}{b}$$

EXAMPLE 4 *Evaluate.* $\dfrac{20 - 4(3)}{10 - 2(3)}$

4. $\dfrac{30 - 6(4)}{10 - 2(4)}$

Solution A fraction bar acts as an understood grouping symbol. We must evaluate the numerator (top) and the denominator (bottom), and then divide the results. In both numerator and denominator remember to do the multiplication before the subtraction.

$$\frac{20 - 4(3)}{10 - 2(3)} = \frac{20 - 12}{10 - 6}$$
$$= \frac{8}{4}$$
$$= \boxed{2} \qquad \blacksquare$$

In the next section we will begin discussing how we perform the various arithmetic operations with integers.

STUDY SKILLS 1.1

Studying Algebra—How Often?

In most college courses, you are typically expected to spend 2–4 hours studying outside of class for every hour spent in class.

It is especially important that you spend this amount of time studying algebra since you must both *acquire* and *perfect* skills; and, as most of you who play a musical instrument or participate seriously in athletics already should know, it takes time and lots of practice to develop and perfect a skill.

It is also important that you distribute your studying over time. That is, do not try to do all your studying in 1, 2, or even 3 days, and then skip studying the other days. You will find that understanding algebra and acquiring the necessary skills are much easier if you spread your studying out over the week, doing a little each day. If you study in this way, you will need less time to study just before exams.

In addition, if your study sessions are more than an hour long, it is a good idea to take a 10-minute break within every hour you spend reading math or working exercises. This "break" helps to clear your mind, and allows you to think more clearly.

Answers to Learning Checks in Section 1.1

1. (a) 7 **(b)** −3 **2. (a)** 5 **(b)** 14 **3.** 24 **4.** 3

NOTE TO THE STUDENT

Use the space on this page to write down any questions you have or points you want to review with your instructor.

Exercises 1.1

In Exercises 1–26, determine whether the given statement is true in general. If the statement is true, indicate which property of the integers it illustrates.

1. $9 + 3 = 3 + 9$

2. $10 + 5 = 5 + 10$

3. $x + 3 = 3 + x$

4. $a + 5 = 5 + a$

5. $9 - 3 = 3 - 9$

6. $10 - 5 = 5 - 10$

7. $x - 3 = 3 - x$

8. $a - 5 = 5 - a$

9. $9 \cdot 3 = 3 \cdot 9$

10. $10 \cdot 5 = 5 \cdot 10$

11. $9 \cdot x = x \cdot 9$

12. $5 \cdot x = x \cdot 5$

13. $9 \div 3 = 3 \div 9$

14. $10 \div 5 = 5 \div 10$

15. $10 - (6 - 3) = (10 - 6) - 3$

16. $12 - (7 + 3) = (12 - 7) + 3$

17. $a - (b - c) = (a - b) - c$

18. $a - (b + c) = (a - b) + c$

In Exercises 19–26, name the property used in each step.

19. $(x + 7) + 5 = x + (7 + 5)$
$\quad\quad\quad\quad\ = x + 12$

20. $(a + 9) + 4 = a + (9 + 4)$
$\quad\quad\quad\quad\ = a + 13$

21. $3(4y) = (3 \cdot 4)y$
$\quad\quad\ = 12y$

22. $5(7s) = (5 \cdot 7)s$
$\quad\quad\ = 35s$

23. $3 + (a + 7) = 3 + (7 + a)$
$\quad\quad\quad\quad\quad\ = (3 + 7) + a$
$\quad\quad\quad\quad\quad\ = 10 + a$

24. $2 + (z + 6) = (2 + z) + 6$
$\quad\quad\quad\quad\quad\ = (z + 2) + 6$
$\quad\quad\quad\quad\quad\ = z + (2 + 6)$
$\quad\quad\quad\quad\quad\ = z + 8$

25. $3(a \cdot 7) = 3(7a)$
$\quad\quad\quad\ = (3 \cdot 7)a$
$\quad\quad\quad\ = 21a$

26. $2(z \cdot 6) = 2(6z)$
$\quad\quad\quad\ = (2 \cdot 6)z$
$\quad\quad\quad\ = 12z$

In Exercises 27–42, evaluate each expression.

27. $-(+4)$

28. $-(+7)$

29. $-(-4)$

30. $-(-7)$

31. $|4|$

32. $|7|$

ANSWERS

1. _____
2. _____
3. _____
4. _____
5. _____
6. _____
7. _____
8. _____
9. _____
10. _____
11. _____
12. _____
13. _____
14. _____
15. _____
16. _____
17. _____
18. _____
19. _____
20. _____
21. _____
22. _____
23. _____
24. _____
25. _____
26. _____
27. _____
28. _____
29. _____
30. _____
31. _____
32. _____

ANSWERS

33. _____

34. _____

35. _____

36. _____

37. _____

38. _____

39. _____

40. _____

41. _____

42. _____

43. _____

44. _____

45. _____

46. _____

47. _____

48. _____

49. _____

50. _____

51. _____

52. _____

53. _____

54. _____

55. _____

56. _____

57. _____

58. _____

59. _____

60. _____

61. _____

62. _____

63. _____

64. _____

33. $|-4|$

34. $|-7|$

35. $-|4|$

36. $-|7|$

37. $-|-4|$

38. $-|-7|$

39. $|10 - 5|$

40. $|20 - 12|$

41. $|10| - |-5|$

42. $|20| - |-12|$

In Exercises 43–64, evaluate the given expression. Remember to follow the order of operations.

43. $7 + 3 \cdot 4$

44. $5 + 4 \cdot 6$

45. $15 - 5 \cdot 3$

46. $12 - 2 \cdot 5$

47. $6 + 4(3 + 2)$

48. $30 - 5(4 - 2)$

49. $6 + (4 \cdot 3 + 2)$

50. $30 - (5 \cdot 4 - 2)$

51. $6 + (4 \cdot 3) + 2$

52. $30 - (5 \cdot 4) - 2$

53. $(6 + 4)(3 + 2)$

54. $(30 - 5)(4 - 2)$

55. $\dfrac{15 + 2 \cdot 3}{3 + 2 \cdot 2}$

56. $\dfrac{12 + 6 \cdot 3}{6 + 3 \cdot 3}$

57. $3 + 2[3 + 2(3 + 2)]$

58. $5 + 4[5 + 4(5 + 4)]$

59. $9 - 4[6 - 2(3 - 1)]$

60. $8 - 2[9 - 3(5 - 3)]$

61. $4 + \{3 + 5[2 + 2(3 + 1)]\}$

62. $5 + \{2 + 4[3 + 3(4 + 2)]\}$

63. $\dfrac{10 + 2(5 + 3)}{2 \cdot 5 + 3}$

64. $\dfrac{18 - 3(4 - 1)}{2 \cdot 3 + 3}$

Adding Integers

1.2

We would like to extend our arithmetic procedures to the set of integers. In this section we will focus our attention on determining how we are going to add integers.

One important thing to keep in mind is that whatever rule we develop for adding integers, we will insist that it allow us to add positive integers just as we always have. In other words, $5 + 2$ had better still be equal to 7.

In fact, this is exactly the approach we are going to take. We are going to analyze the addition process we already know, and that will help us see how to reasonably define addition for integers.

Let's consider two similar examples:

$$5 + 2 = ? \quad \text{and} \quad 5 + (-2) = ?$$

To make our work even clearer, let's write these examples as

$$+5 + (+2) = ? \quad \text{and} \quad +5 + (-2) = ?$$

Of course, we all know that the answer to the first example is 7, but let's try to analyze what is happening in $+5 + (+2)$, on the number line.

We can visualize that $+5 + (+2)$ means "start at $+5$ on the number line and move 2 units to the right." The following picture represents $+5 + (+2) = 7$:

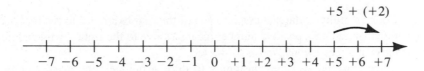

In other words, *adding* $+2$ means "move 2 units to the right."

In view of this, how should we visualize $5 + (-2)$? We are still starting at $+5$, but this time we are *adding* -2 instead of $+2$. It seems reasonable that if adding $+2$ means "move 2 units to the right," then adding -2 should mean "move 2 units to the *left*."

The following picture represents $+5 + (-2)$:

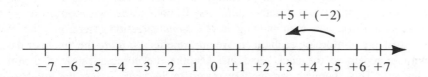

Therefore, we have $+5 + (-2) = +3$.

Let's now consider two more closely related examples:

$$-5 + (+2) = ? \quad \text{and} \quad -5 + (-2) = ?$$

Accepting our understanding of the first two examples, we really do not have much choice as to how we interpret these two. After all, the only difference in these two is that we are starting at -5 instead of at $+5$.

If adding +2 means moving two units to the right, then it should not matter whether we start at +5, or at −5, or at any other number. Similarly, if adding −2 means moving two units to the left, then again it should not matter whether we start at +5 or at −5 or anywhere else.

Let's draw the picture for each of these examples:

The picture for −5 + (+2) is:

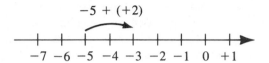

Therefore, we have −5 + (+2) = −3.

The picture for −5 + (−2) is:

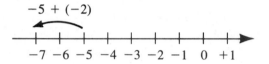

Therefore, we have −5 + (−2) = −7.

Look carefully at the pictures for each of the four examples to see that they illustrate that adding a positive number means "move to the right," while adding a negative number means "move to the left."

Since it would be time-consuming to draw a number line every time we want to add integers, let's summarize these four examples and see if we can extract from them a general rule for adding integers. We have

$$+5 + (+2) = +7$$
$$+5 + (-2) = +3$$
$$-5 + (+2) = -3$$
$$-5 + (-2) = -7$$

In the first and last of these examples we are either adding two positive numbers or adding two negative numbers. In both cases, we are starting in a certain direction and then moving in the *same* direction; therefore, the two numbers are reinforcing each other. We are, in effect, *adding* the "number parts without regard to the sign," and just keeping the common sign of the two numbers. This "number part without regard to the sign" is just the number of units the number is away from 0, which is exactly what we called the absolute value of a number.

RULE FOR ADDING INTEGERS

Part 1.   When adding two integers with the *same sign*, add their absolute values and keep their common sign.

On the other hand, in the middle two examples we are adding numbers that have opposite signs. In these examples, we are starting in one direction and then moving in the *opposite* direction. Therefore, the two numbers are opposing each other. We are in effect *subtracting* the absolute values. However, in such a case we cannot, in general, tell whether the answer will be positive or negative until we know which number is "stronger." Again we make use of the idea of the absolute value of a number.

In the example $+5 + (-2)$ we get an answer of $+3$ because we have more positive strength $(+5)$ than negative strength (-2). In other words, since $+5$ has a larger absolute value than -2, the final answer is positive.

Similarly, in the example $-5 + (+2)$ we get an answer of -3 because -5 has a larger absolute value than $+2$, and therefore we have more negative strength than positive strength. Be careful here. We are *not* saying that -5 is greater than $+2$. In fact, the contrary is true; because $+2$ is to the right of -5, $+2$ is greater than -5. However, when we are adding integers with opposite signs, it is the larger *absolute value* that we are concerned with, not the larger number.

RULE FOR ADDING INTEGERS

Part 2. When adding two integers with *opposite signs*, subtract the smaller absolute value from the larger, and keep the sign of the number with the larger absolute value.

Our formal rule for adding integers is actually one rule with two parts. Do not be intimidated by the formal wording of this rule. In the next several examples we will see that using this rule is quite straightforward.

EXAMPLE 1 Compute each of the following:
(a) $-15 + (-8)$ (b) $-15 + (+8)$
(c) $+15 + (-8)$ (d) $+15 + (+8)$

Solution

(a) Since -15 and -8 have the same sign, we add their absolute values, and keep the common sign, which is negative. Thus, we have

$$-15 + (-8) = \boxed{-(15 + 8)} = \boxed{-23}^*$$

(b) Since -15 and $+8$ have opposite signs, we subtract their absolute values as we would whole numbers (larger absolute value minus smaller absolute value) and keep the sign of the number with the larger absolute value (in this case it is -15), which makes the answer negative. Thus, we have

$$-15 + (+8) = \boxed{-(15 - 8)} = \boxed{-7}$$

---

*Throughout the text we will use a color panel to indicate steps that are normally done *mentally*, but that we include to help clarify the procedure.

LEARNING CHECKS

1. (a) $-9 + (-3)$
   (b) $-9 + (+3)$
   (c) $+9 + (-3)$
   (d) $+9 + (+3)$

(c) Since $+15$ and -8 have opposite signs we proceed as we did in part **(b)**. This time, however, the answer will be positive since the number with the larger absolute value, $+15$, is positive. Thus, we have

$$+15 + (-8) = \boxed{+(15 - 8)} = \boxed{+7}$$

(d) Do not get carried away with the rule. This example is just $15 + 8$ and so we can get the answer without need of the rule, although we could use it if we wanted to. Thus, we have

$$+15 + (+8) = \boxed{23} \qquad \textit{Remember that } 23 \textit{ means } +23. \qquad \blacksquare$$

Once we know how to add two integers we can add more than two as well.

2. **(a)** $-4 + (-7) + (-5)$
   (b) $8 + (-6) + 4 + (-11)$

EXAMPLE 2 Compute each of the following:
(a) $-5 + (-9) + (-6)$ **(b)** $6 + (-4) + 7 + (-10)$

Solution

(a) Since we are adding three numbers whose signs are the same (all negative), we add their absolute values and keep the negative sign:

$$-5 + (-9) + (-6) = \boxed{-(5 + 9 + 6)} = \boxed{-20}$$

(b) We have two choices. The first is to work from left to right as we have been. The second is to reorder and regroup the numbers so that we add all the positive and negative numbers separately first, and then add the results. We illustrate both methods.

Solution 1.   Working from left to right, we have

$$6 + (-4) + 7 + (-10) \qquad \textit{We begin by adding } 6 \textit{ and } -4 \textit{ to get } 2.$$
$$= 2 + 7 + (-10) \qquad \textit{Next we add } 2 \textit{ and } 7.$$
$$= 9 + (-10) \qquad \textit{Now we add } 9 \textit{ and } -10.$$
$$= \boxed{-1}$$

Solution 2.   This time we compute the result by grouping positives and negatives first. Since we are *adding* the numbers, the Commutative and Associative Laws allow us to rearrange and regroup the numbers any way we please.

$$6 + (-4) + 7 + (-10) \qquad \textit{We first group the positive and negative numbers separately.}$$
$$= (6 + 7) + [-4 + (-10)] \qquad \textit{Next we add the positive and negatives separately.}$$
$$= 13 + [-14] \qquad \textit{Now we add } 13 \textit{ and } -14.$$
$$= \boxed{-1} \qquad\qquad\qquad\qquad\qquad\qquad\qquad \blacksquare$$

EXAMPLE 3 *Evaluate.* $-|8 - 3| + |3 + (-8)|$

3. $|7 - 4| - |4 - 7|$

Solution We begin by evaluating the expressions *within* the absolute value signs:

$$-|8 - 3| + |3 + (-8)| = -|5| + |-5|$$
$$= -(5) + 5$$

−5 and 5 have equal absolute values so that when we add we get 0.
Remember that 0 *is neither positive nor negative so we do not write* +0 *or* −0.

$$= \boxed{0} \quad \blacksquare$$

We continue to work within the framework of the same order of operations we outlined in the last section.

EXAMPLE 4 *Evaluate.* $-8 + 3[5 + (-12) + 9]$

4. $-6 + 4[6 + (-3) + 1]$

Solution We begin by working inside the brackets.

$$-8 + 3[5 + (-12) + 9] = -8 + 3[5 + 9 + (-12)]$$
$$= -8 + 3[14 + (-12)]$$
$$= -8 + 3[2]$$
$$= -8 + 6$$
$$= \boxed{-2} \quad \blacksquare$$

It is most important that you be able to add integers as automatically as you do whole numbers. The only way to acquire this skill is to *practice*. That is exactly what the exercise sets are designed for—to give you lots and lots of practice.

STUDY SKILLS 1.2

Previewing Material

Before you attend your next class, preview the material to be covered beforehand. First, skim the section to be covered, look at the headings, and try to guess what the sections will be about. Then read the material carefully.

You will find that when you read the material before you go to class, you will be able to follow the instructor more easily, things will make more sense, and you will learn the material more quickly. Now, if there was something you did not understand when you previewed the material, the teacher will be able to answer your questions *before* you work your assignment at home.

Answers to Learning Checks in Section 1.2

1. (a) −12   **(b)** −6   **(c)** +6   **(d)** 12   **2. (a)** −16   **(b)** −5   **3.** 0
4. 10

NOTE TO THE STUDENT

Use the space on this page to write down any questions you have or points you want to review with your instructor.

Exercises 1.2

In Exercises 1–48, compute the value of each expression.

1. $+5 + (-7)$ **2.** $+9 + (-11)$ **3.** $-9 + (-3)$

4. $-10 + (-4)$ **5.** $-4 + (+11)$ **6.** $-7 + (+12)$

7. $+8 + (-3)$ **8.** $+6 + (-5)$ **9.** $+6 + (-2)$

10. $+5 + (-4)$ **11.** $-6 + (-2)$ **12.** $-5 + (-4)$

13. $-6 + (+2)$ **14.** $-5 + (+4)$ **15.** $+6 + (+2)$

16. $+5 + (+4)$ **17.** $-7 + (-8)$ **18.** $-9 + (-3)$

19. $10 + (-14)$ **20.** $14 + (-10)$ **21.** $8 + (-8)$

22. $-12 + 12$ **23.** $9 + (-16)$ **24.** $-20 + (-5)$

25. $-30 + (-14)$ **26.** $43 + (-28)$ **27.** $-6 + 12$

28. $-10 + 18$ **29.** $-5 + (-4) + (-6)$ **30.** $-8 + (-3) + (-2)$

31. $5 + (-4) + (-6)$ **32.** $8 + (-3) + (-2)$ **33.** $-5 + 4 + (-6)$

34. $8 + (-3) + 2$ **35.** $-5 + (-4) + 6$ **36.** $-8 + 3 + (-2)$

37. $16 + (-5) + (-7) + 2$ **38.** $-10 + 6 + (-4) + (-3)$

39. $-8 + 6 + (-5) + 1$ **40.** $-7 + (-3) + (-5) + 1$

41. $2 + (-9) + (-3) + (-1) + 6$ **42.** $-2 + 9 + 3 + 1 + (-6)$

ANSWERS

1. _____
2. _____
3. _____
4. _____
5. _____
6. _____
7. _____
8. _____
9. _____
10. _____
11. _____
12. _____
13. _____
14. _____
15. _____
16. _____
17. _____
18. _____
19. _____
20. _____
21. _____
22. _____
23. _____
24. _____
25. _____
26. _____
27. _____
28. _____
29. _____
30. _____
31. _____
32. _____
33. _____
34. _____
35. _____
36. _____
37. _____
38. _____
39. _____
40. _____
41. _____
42. _____

ANSWERS

43. _____

44. _____

45. _____

46. _____

47. _____

48. _____

49. _____

50. _____

51. _____

52. _____

53. _____

54. _____

55. _____

56. _____

57. _____

58. _____

59. _____

60. _____

43. $27 + (-56)$

44. $-39 + (-55)$

45. $-22 + (-45)$

46. $-19 + (-72)$

47. $-31 + (-26) + 48$

48. $27 + (-41) + (-9)$

In Exercises 49–56, evaluate the given expression.

49. $-5 + [7 + (-3)]$

50. $-4 + 4[6 + (-3)]$

51. $-3 + 5 \cdot 2 + (-6)$

52. $-8 + 4 \cdot 3 + (-10)$

53. $|2 + (-6)| + |2| + |-6|$

54. $|3 + (-9)| + |3| + |-9|$

55. $|-7| + (-7)$

56. $-|-7| + (-7)$

57. Carla has a balance of $48 in her checking account. She writes checks for $18, $22, and $15; then she deposits $50; then she writes checks for $28 and $12; then she deposits $20; then she writes another check for $17; and finally she deposits $27. What is the final balance of her checking account?

58. Repeat Exercise 57 if Carla begins with a balance of $-$48.

59. A football team takes possession of the ball on their own 25-yard line. On first down they gain 8 yards; on second down they lose 14 yards; and on third down they lose 5 more yards. What is their location for fourth down?

60. On a certain morning the temperature at 6:00 A.M. is 4°C. Two hours later the temperature has fallen 9 degrees, and three hours after that it has risen 2 degrees. What is the temperature at 11:00 A.M.?

QUESTIONS FOR THOUGHT

1. What result should we get when we add an integer to its opposite? Why?

2. If the sum of two integers is 0, what can we conclude about the two integers? Why?

3. $7 - 4 = 3$ because $3 + 4 = 7$. Subtraction is defined in terms of addition. What should the answer to $4 - 7$ be and why?

4. What should the answer to $4 - (-7)$ be and why?

Subtracting Integers

1.3

Even though you probably do not think about it now when you do subtraction, if you can think back to when you first learned how to subtract, subtraction was defined in terms of addition. That is, if you wanted to answer $7 - 4 = ?$, you asked yourself $? + 4 = 7$.

We will continue the approach we took in the last section with addition, in that we will begin with a familiar subtraction problem, analyze it, and see what it tells us about how to subtract integers in general.

As we did with addition, let's consider two examples:

$$5 - 2 = ? \quad \text{and} \quad 5 - (-2) = ?$$

Again, to make our work even clearer let's write these examples as

$$+5 - (+2) = ? \quad \text{and} \quad +5 - (-2) = ?$$

Of course, we all know the answer to the first example is 3, but let's try to analyze what is happening in $+5 - (+2)$, on the number line. We can visualize that $+5 - (+2)$ means "start at $+5$ on the number line and move 2 units to the *left*." This gives us the following picture:

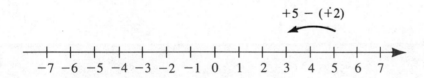

In other words, *subtracting* $+2$ means "move 2 units to the left."

In view of this, how should we view $+5 - (-2)$? We are still starting at $+5$, but this time we are *subtracting* -2. If *subtracting* $+2$ means "move 2 units to the left," then *subtracting* -2 must mean do the opposite—that is, "move 2 units to the right." (What other choice is there?) We visualize this on the number line as follows:

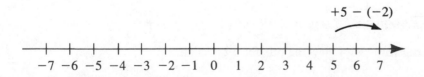

As with addition, if subtracting $+2$ means move 2 units to the left and subtracting -2 means move 2 units to the right, then it should not make any difference where we start. Thus, for the two closely related examples

$$-5 - (+2) = ? \quad \text{and} \quad -5 - (-2) = ?$$

we have the following pictures and resulting answers:

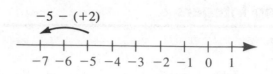

Therefore, $-5 - (+2) = -7$.

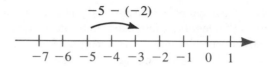

Therefore, $-5 - (-2) = -3$.

Keeping in mind the work we did in the last section on adding integers, we are familiar with the phrases "moving left" and "moving right." We have just seen that *subtracting* $+2$ means moving 2 units to the left, while in the last section we saw that *adding* -2 means exactly the same thing. Similarly, we have just seen that *subtracting* -2 means moving 2 units to the right, while in the last section we saw that *adding* $+2$ means exactly the same thing.

Thus, whenever we are faced with subtracting a number we can accomplish the same result by adding the opposite number, and then following the rules for addition. We state this fact algebraically (symbolically) as follows:

RULE FOR SUBTRACTING INTEGERS
$$a - b = a + (-b)$$

In words, this rule says that in order to subtract an integer we add its opposite. That is, when we are subtracting two integers, we change the subtraction to addition *and* change the sign of the number being subtracted.

✔ **LEARNING CHECKS**

1. $-7 - (+5)$

2. $4 - 9$

EXAMPLE 1   *Compute.* $-9 - (+6)$

Solution   We follow the rule for subtraction.

$-9 - (+6) = -9 + (-6)$ *We have changed the subtraction to addition, and*
$\qquad\qquad = \boxed{-15}$ *we have changed the sign of the second number.*
                                  Now we follow the rule for addition.   ∎

EXAMPLE 2   *Compute.* $7 - 12$

Solution   This example means "$+7$ minus $+12$." At first, you may find it helpful to put in the "understood" plus signs in such an example.

$$7 - 12 = +7 - (+12)$$

We follow the rule for subtraction.
Change the sign of the number being subtracted, and add.

$$= +7 + (-12)$$

Now we follow the rule for addition.

$$= \boxed{-5}$$ ∎

We can apply this rule to examples involving more than one subtraction, or involving several subtractions and additions.

EXAMPLE 3 Compute each of the following:

(a) $7 - 9 - (-4)$ **(b)** $-10 + (-3) - 5 + 11$

Solution

(a) We begin by inserting the understood plus signs.

$$7 - 9 - (-4) = +7 - (+9) - (-4)$$

Change each subtraction to addition, and change the sign of the number following each subtraction.

$$= +7 + (-9) + (+4)$$

Now we follow the rule for addition.

$$= -2 + (+4)$$
$$= \boxed{2}$$

(b) We begin by inserting the understood plus signs. (If *you* feel this step is not necessary then skip it.)

$$-10 + (-3) - 5 + 11 = -10 + (-3) - (+5) + (+11)$$

We follow the subtraction rule.

$$= -10 + (-3) + (-5) + (+11)$$

We add the negative numbers.

$$= -18 + (+11)$$
$$= \boxed{-7}$$

Now we use the addition rule. ∎

As the examples get more complicated we must keep the order of operations in mind.

EXAMPLE 4 Evaluate each of the following:

(a) $4 - 5 - 7$ **(b)** $4 - (5 - 7)$

Solution

(a) $4 - 5 - 7 = 4 - (+5) - (+7)$
$$= 4 + (-5) + (-7)$$
$$= 4 + (-12)$$
$$= \boxed{-8}$$

3. (a) $5 - 6 - (-2)$
(b) $-8 + (-5) - 4 + 6$

4. (a) $3 - 4 - 8$
(b) $3 - (4 - 8)$

(b) The order of operations requires us to work inside the parentheses first.

$$4 - (5 - 7) = 4 - (5 - (+7))$$

$$= 4 - (5 + (-7))$$
$$= 4 - (-2)$$

$$= 4 + (+2)$$
$$= \boxed{6}$$

Note that the numbers in parts **(a)** and **(b)** are the same, but that the parentheses in part **(b)** change the meaning of the example a great deal.   ∎

Up to this point we have been very careful to write each step in both our addition and subtraction examples. However, virtually every example and exercise we do from now on involves adding and subtracting integers in some w ⌄ or another, and it is impossible for us to continue writing each step. In fact, as quickly as possible you must try to reach the point where you can subtract integers as easily and naturally as you subtract whole numbers. That is, in the same way you "just know" that $7 - 4 = 3$, so too you need to "just know" that $4 - 7 = -3$. This means that you must develop the ability to use the addition and subtraction rules mentally (perhaps not at first, of course, but eventually). The only way to acquire this ability is by doing many exercises.

STUDY SKILLS 1.3

What to Do First

Before you attempt any exercises, either for homework or for practicing your skills, it is important to review the relevant portions of your notes and text.

As we mentioned in the introduction, memorizing a bunch of seemingly unrelated algebraic steps to follow in an example may serve you initially, but in the long run (most likely before Chapter 3), your memory will be overburdened—you will tend to confuse examples and/or forget steps.

Reviewing the material before doing exercises makes each solution you go through more meaningful. The better you understand the concepts underlying the exercise, the easier the material becomes, and the less likely you are to confuse examples or forget steps.

When reviewing the material, take the time to *think about what you are reading*. Try not to get frustrated if it takes you an hour or so to read and understand a few pages of a math text—that time will be well spent. As you read your text and your notes, think about the concepts being discussed: **(a)** how they relate to previous concepts covered, and **(b)** how the examples illustrate the concepts being discussed. More than likely, worked-out examples will follow verbal material, so look carefully at these examples and try to understand why each step in the solution is taken. When you finish reading, take a few minutes and think about what you have just read.

    ✔   *Answers to Learning Checks in Section 1.3*

1. -12 **2.** -5 **3. (a)** 1 **(b)** -11 **4. (a)** -9 **(b)** 7

Exercises 1.3

In Exercises 1–50, compute the value of each expression.

1. $6 - (+10)$ **2.** $4 - (+8)$

3. $-7 - (+4)$ **4.** $-9 - (+3)$

5. $3 - (-6)$ **6.** $5 - (-7)$

7. $-8 - (-2)$ **8.** $-6 - (-4)$

9. $-5 + (+8)$ **10.** $-7 + (+5)$

11. $5 - (+8)$ **12.** $4 - (+9)$

13. $5 - (-8)$ **14.** $4 - (-9)$

15. $-5 - (+8)$ **16.** $-4 - (+9)$

17. $-5 - (-8)$ **18.** $-4 - (-9)$

19. $5 + (-8)$ **20.** $4 + (-9)$

21. $-5 + (-8)$ **22.** $-4 + (-9)$

23. $6 - (-7)$ **24.** $2 - (-3)$

25. $-6 - (-7)$ **26.** $-2 - (-3)$

27. $2 + (-6) - (+7)$ **28.** $5 + (-8) - (+3)$

29. $2 - 6 - 7$ **30.** $5 - 8 - 3$

31. $2 - (6 - 7)$ **32.** $5 - (8 - 3)$

33. $7 - 9 - 3 + 2$

34. $2 - 3 + 1 - 8$

35. $11 - 5 + 4 - 7$

36. $4 - 7 - 2 + 3$

37. $4 - 8 - 6 + 3$

38. $9 - 10 - 2 + 5$

39. $4 - 8 - (6 + 3)$

40. $9 - (10 - 2) + 5$

41. $4 - (8 - 6) + 3$

42. $9 - 10 - (2 + 5)$

43. $4 - (8 - 6 + 3)$

44. $9 - (10 - 2 + 5)$

45. $-8 + 8$

46. $-6 + 6$

47. $-8 - 8$

48. $-6 - 6$

49. $-8 - (-8)$

50. $-6 - (-6)$

In Exercises 51–62, evaluate the given expression.

51. $9 - 5 \cdot 4 - 2$

52. $7 - 3 \cdot 5 - 1$

53. $9 - 5(4 - 2)$

54. $7 - 3(5 - 1)$

55. $9 - (5 \cdot 4 - 2)$

56. $7 - (3 \cdot 5 - 1)$

57. $|6 - 2| - |2 - 6|$

58. $|10 - 7| - |7 - 10|$

59. $|2 - 6| - (2 - 6)$

60. $|7 - 10| - (7 - 10)$

61. $|-4 - 3 + 2| - 4 - 3 + 2$

62. $|-8 + 6 - 5| - 8 + 6 - 5$

QUESTIONS FOR THOUGHT

5. What is wrong with $|4 - 8| = |4| - |8|$?

6. We know that multiplication means repeated addition. That is, 4 times 3 means add 3 four times. In view of this and our rule for adding negative integers, what should 4 times -3 be equal to?

7. When we multiply 4 times 3 we are multiplying by *positive* 4 and so we interpret the multiplication as repetitive *addition*. The 4 tells us how many times to *add* 3. How might we interpret -4 times 3 and -4 times -3? Can we interpret multiplying by a negative as repetitive *subtraction*? If so, what should we get as answers?

Multiplying and Dividing Integers

1.4

In order to develop rules for multiplying and dividing integers, we must keep in mind that multiplication is just a shorthand way of writing addition. That is, when we multiply two positive integers—say, for example, 4 times 3—we usually understand it to mean that the 4 tells us how many times we are going to add a number, while the 3 tells us what number to add each time:

$$4(3) = 3 + 3 + 3 + 3 = 12$$

Thus, multiplication is just repetitive addition. In view of this, what result should we get for 4 times -3?

4 times -3 should mean "add -3 four times"

$$4(-3) = (-3) + (-3) + (-3) + (-3) = -12$$

In fact, whenever we multiply a positive number times a negative, we will be repeatedly adding the same *negative* number, and by our rule for addition the answer will always be negative.

It does not matter whether the product is a positive times a negative or a negative times a positive. For example:

Positive times negative

$$5(-2) = (-2) + (-2) + (-2) + (-2) + (-2)$$
$$= -10$$

Negative times positive

$(-2)5 = 5(-2)$ *By the Commutative Law*
$\quad\quad = -10$ *As we just saw*

Thus, we see that the *product* of two numbers which have *opposite* signs should be *negative*.

Deciding what we want a negative times a negative to be requires a slightly more delicate analysis. When we first began talking about the integers in Section 1.2, we pointed out that a negative sign in front of a number can be thought of as meaning the "opposite of." Thus, -4 can be thought of as the opposite of $+4$. On the other hand, now that we have seen that a negative times a positive is negative, we can also think of -4 as -1 times 4 [because $-1(4) = -4$]. Thus, multiplying by -1 changes a number into its opposite. Thus, we can analyze -4 times -3 as follows:

$(-4)(-3) = -1(4)(-3)$ *Because $-4 = -1(4)$*
$\quad\quad\quad = -1(-12)$ *Because 4 times -3 is equal to -12*
$\quad\quad\quad = +12$ *Because multiplying by -1 changes a number into its opposite*

We have come to the conclusion that the product of two negative numbers is positive. Since the product of two positives is also a positive, we can say that the *product* of two numbers which have the *same* sign is *positive*.

Now that we see how multiplication of signed numbers works, what about division? Keep in mind that division is always defined in terms of multiplication.

That is,

$$\frac{12}{3} = 4 \qquad \text{because} \quad (+4) \cdot (+3) = +12 \qquad \textit{Remember that } \frac{a}{b} \textit{ means } a \div b.$$

Therefore, if we want to compute 12 divided by -3, we know the answer must be -4 because -4 times -3 equals 12:

$$\frac{+12}{-3} = -4 \qquad \text{because} \quad (-4)(-3) = +12$$

Similarly, we have

$$\frac{-12}{+3} = -4 \qquad \text{because} \quad (-4)(+3) = -12$$

$$\frac{-12}{-3} = +4 \qquad \text{because} \quad (+4)(-3) = -12$$

We can see that division behaves in the same way that multiplication does. The *quotient* of two numbers will be *positive* if the numbers have the *same* sign and it will be *negative* if the numbers have *opposite* signs.

Basically, we have seen that when we multiply or divide two signed numbers, we simply multiply or divide the numbers ignoring their signs (that is, we use just their absolute values). Then we put down a plus sign if the signs are the same, and a minus sign if the signs are opposite.

Let's formalize this into a rule.

RULE FOR MULTIPLYING AND DIVIDING INTEGERS

When multiplying (or dividing) two integers, multiply (or divide) their absolute values. The sign of the answer is:

Positive if both numbers have the *same* sign.

Negative if the numbers have *opposite* signs.

✔ LEARNING CHECKS

1. **(a)** $(-5)(-7)$

(b) $4(-8)$

(c) $\dfrac{-20}{-5}$

EXAMPLE 1 Compute each of the following:

(a) $(-3)(-4)$ **(b)** $5(-6)$ **(c)** $\dfrac{12}{-6}$

Solution

(a) $(-3)(-4) = \boxed{12}$

(b) $5(-6) = \boxed{-30}$

(c) $\dfrac{12}{-6} = \boxed{-2}$ ■

EXAMPLE 2   *Compute.* $(-3)(-4)(-2)$

2. $(-5)(-2)(-6)$

Solution

$$(-3)(-4)(-2) = (+12)(-2) \qquad \textit{Because } -3 \textit{ times } -4 \textit{ is equal to } +12$$

$$= \boxed{-24} \qquad \textit{Because } +12 \textit{ times } -2 \textit{ is equal to } -24 \quad \blacksquare$$

We must be extremely careful reading examples, particularly when they involve parentheses. For example, both

$$-4(-6) \qquad \text{and} \qquad -(4 - 6)$$

contain parentheses, but they serve different functions in the two examples. In $-4(-6)$ the parentheses serve to indicate that -4 is multiplying -6, and therefore the answer is

$$-4(-6) = \boxed{24}$$

However, in $-(4 - 6)$, the parentheses indicate that we are to take the negative of the *result* of $4 - 6$, and therefore the answer is

$$-(4 - 6) = -(-2) = \boxed{2}$$

We mentioned earlier that even though we have chosen the same symbol (the minus sign) to indicate both subtraction and a negative number, there should not be any confusion because all our uses of the minus sign are consistent. We have seen the minus sign used as subtraction, to indicate a negative number, and to indicate the opposite (which we saw is the same as thinking of it as multiplying by -1). For example:

If we think of $-(-2)$ as the *opposite of negative* 2, we get as our answer $+2$.

If we think of $-(-2)$ as *subtracting negative* 2, then the rule for subtraction gives $+(+2)$, and our answer is again $+2$.

If we think of $-(-2)$ as $-1(-2)$, then following our multiplication rule we also get $+2$.

We are free to think of the minus sign in the way we find most convenient provided we do not change the meaning of the expression.

EXAMPLE 3   Evaluate each of the following:
(a) $3 - 8(2)$ **(b)** $3(-8)(2)$

3. (a) $6 - 5(3)$
   (b) $6(-5)(3)$

Solution
(a) Following the order of operations, we do the multiplication first:

$$3 - 8(2) = 3 - 16$$
$$= \boxed{-13}$$

(b) This example involves multiplication only:

$$3(-8)(2) = -24(2)$$
$$= \boxed{-48} \qquad \blacksquare$$

4. $7 - 3(-2)$

EXAMPLE 4 *Compute.* $8 - 5(-4)$

Solution We will use this example to illustrate a slight shortcut which also offers the advantage of being less prone to making a sign error.

One way of evaluating this expression is as follows: Following the order of operations, we multiply 5 times -4 first.

$$8 - 5(-4) = 8 - (-20) \qquad \textit{Now we follow the rule for subtraction.}$$
$$= 8 + (+20)$$
$$= \boxed{28}$$

Instead of having to follow the subtraction rule, let's anticipate it by *thinking* addition at the outset. In order to do that we think of the minus sign in front of the 5 as part of the 5, so we say "-5 is multiplying -4." Keep in mind that we are thinking addition so we are going to *add* the result of the product. Thus, the alternate solution is as follows:

$$8 - 5(-4) = 8 + 20 \qquad \textit{We added the result of multiplying } -5 \textit{ times } -4.$$
$$= \boxed{28}$$

This approach not only saves a step in the solution, but it also often removes the necessity for "changing signs." The fewer times we have to change signs, the fewer opportunities for making a careless error.

Overall, this approach offers significant advantages, and so we will use it from now on wherever it applies. ∎

5. $8 - 2[4 - 5(-2)]$

EXAMPLE 5 *Evaluate.* $6 - 4[3 - 7(2)]$

Solution Following the order of operations, we begin by working inside the brackets first.

$$6 - 4[3 - 7(2)] = 6 - 4[3 - 14]$$
$$= 6 - 4[-11] \qquad \textit{We think of this as } -4 \textit{ times } -11, \textit{ and}$$
$$= 6 + 44 \qquad\qquad \textit{add the result to 6.}$$
$$= \boxed{50} \qquad\qquad\qquad\qquad\qquad\qquad ∎$$

6. $12 - 3[7 - 4(2 - 5)]$

EXAMPLE 6 *Evaluate.* $10 - 2[8 - 3(4 - 9)]$

Solution Again following the order of operations, we begin with the innermost grouping symbol. Within each grouping symbol we do multiplications before additions and subtractions.

$$10 - 2[8 - 3(4 - 9)] = 10 - 2[8 - 3(-5)] \qquad \textit{We computed } 4 - 9 = -5.$$
$$= 10 - 2[8 + 15] \qquad\quad \textit{We computed } -3(-5) = 15.$$
$$= 10 - 2[23]$$
$$= 10 - 46$$
$$= \boxed{-36} \qquad\qquad\qquad\qquad\qquad ∎$$

EXAMPLE 7 *Evaluate.* $\dfrac{-3 + 9}{1 - 3}$

7. $\dfrac{-6 + 10}{3 - 5}$

Solution Remember that a fraction bar is treated as if the numerator and denominator were in parentheses. Therefore, we must compute the numerator and denominator first and then the resulting quotient.

$$\frac{-3 + 9}{1 - 3} = \frac{6}{-2} \qquad \textit{Since the signs are opposite, the answer is negative.}$$

$$= \boxed{-3} \qquad\qquad\qquad\qquad\qquad\qquad \blacksquare$$

EXAMPLE 8 *Evaluate.* $\dfrac{-2(5)(-10)}{-2(5) - 10}$

8. $\dfrac{2(3)(-4)}{2(3) - 4}$

Solution Look at the example carefully! The numerator involves multiplication only, while the denominator involves multiplication and subtraction.

$$\frac{-2(5)(-10)}{-2(5) - 10} = \frac{-10(-10)}{-10 - 10} \qquad -10 - 10 \textit{ means} -10 + (-10).$$

$$= \frac{100}{-20}$$

$$= \boxed{-5} \qquad\qquad\qquad\qquad\qquad\qquad \blacksquare$$

The location of a parenthesis can make a great deal of difference in what an example *means*. We will frequently be emphasizing the importance of reading an example carefully so that you clearly understand what it is saying, and what it is asking.

Finally, we want to mention in this section some of the special properties of the number 0.

We are familiar with the numerical facts that $5 \cdot 0 = 0$ and $11 \cdot 0 = 0$. In fact, 0 times any whole number is equal to 0, and this fact extends to the negative integers as well. We know the product of any integer and 0 is equal to 0:

$$n \cdot 0 = 0 \quad \text{for all integers } n$$

What about dividing *into* 0? As long as we are not dividing *by* 0 (we will see why in a moment), dividing into 0 is perfectly legitimate. That is,

$$\frac{0}{8} = 0 \quad \text{because} \quad 0 \cdot 8 = 0 \qquad \text{and} \qquad \frac{0}{-4} = 0 \quad \text{because} \quad 0 \cdot (-4) = 0$$

In general, we have

$$\frac{0}{n} = 0 \quad \text{for all integers } n \text{ not equal to } 0$$

What happens when we try to divide *by* 0? There are two cases to consider. If we try to divide 0 into a nonzero number—say, for example, 5—we cannot get an answer. If $\frac{5}{0}$ is going to be equal to some number, say m, then it must follow that $m \cdot 0 = 5$. But we know that *any* number times 0 is equal to 0, and so it is impossible to get 5. Therefore, dividing 0 into a nonzero number is *undefined*. That is, we have no available answer.

If we try to divide 0 into 0, we have a different problem. If $\frac{0}{0}$ is going to be equal to some number, say r, then it must follow that $r \cdot 0 = 0$. But this is true for *all* numbers r. This means that *any* number will work and so $\frac{0}{0}$ is not a unique number. Dividing 0 into 0 is said to be *indeterminate*. That is, we cannot determine a unique answer.

The terminology here is not particularly important for our purposes. What is important is that we realize that division by 0 does not make any sense, and therefore *is not allowed*.

REMEMBER

Division *by* 0 is not allowed.

Division *into* 0 (by a *nonzero* number) is fine. The answer is 0.

STUDY SKILLS 1.4

Doing Exercises

After you have finished reviewing the appropriate material as discussed previously, you should be ready to do the relevant exercises. Although your ultimate goal is to be able to work out the exercises accurately *and* quickly, when you are working out exercises on a topic that is new to you it is a good idea to take your time and think about what you are doing while you are doing it.

Think about how the exercises you are doing illustrate the concepts you have reviewed. Think about the steps you are taking and ask yourself why you are proceeding in this particular way and not some other: Why this technique or step and not a different one?

Do not worry about speed now. If you take the time at home to think about what you are doing, the material becomes more understandable and easier to remember. You will then be less likely to "do the wrong thing" in an exercise. The more complex-looking exercises are less likely to throw you. In addition, if you think about these things in advance, you will need much less time to think about them during an exam, and so you will have more time to work out the problems.

Once you believe you thoroughly understand what you are doing and why, you may work on increasing your speed.

Answers to Learning Checks in Section 1.4

1. (a) 35   (b) −32   (c) 4   **2.** −60   **3.** (a) −9   (b) −90   **4.** 13

5. −20   **6.** −45   **7.** −2   **8.** −12

ANSWERS
1. _____
2. _____
3. _____
4. _____
5. _____
6. _____
7. _____
8. _____
9. _____
10. _____
11. _____
12. _____
13. _____
14. _____
15. _____
16. _____
17. _____
18. _____
19. _____
20. _____
21. _____
22. _____
23. _____
24. _____
25. _____
26. _____
27. _____
28. _____
29. _____
30. _____
31. _____
32. _____
33. _____
34. _____
35. _____
36. _____
37. _____
38. _____
39. _____

Exercises 1.4

Evaluate each of the following expressions, if possible.

1. $(+5)(-3)$

2. $(+7)(-2)$

3. $(-5)(+3)$

4. $(-7)(+2)$

5. $(-5)(-3)$

6. $(-7)(-2)$

7. $-5 - 3$

8. $-7 - 2$

9. $-(5 - 3)$

10. $-(7 - 2)$

11. $\dfrac{20}{-5}$

12. $\dfrac{18}{-3}$

13. $\dfrac{-20}{5}$

14. $\dfrac{-18}{3}$

15. $\dfrac{-20}{-5}$

16. $\dfrac{-18}{-3}$

17. $\dfrac{0}{7}$

18. $\dfrac{0}{-4}$

19. $\dfrac{7}{0}$

20. $\dfrac{-4}{0}$

21. $4(-2) - 6$

22. $3(-1) - 9$

23. $4 - 2 - 6$

24. $3 - 1 - 9$

25. $4 - (2 - 6)$

26. $3 - (1 - 9)$

27. $4(-2 - 6)$

28. $3(-1 - 9)$

29. $-6(2)(-5)(-1)$

30. $8(-4)(-1)(-2)$

31. $9 - 3(1 - 3)$

32. $5 - 2(4 - 8)$

33. $8 - 3(2 - 5)$

34. $7 - 4(3 - 6)$

35. $8 - 3 \cdot 2 - 5$

36. $7 - 4 \cdot 3 - 6$

37. $8 - (3 \cdot 2 - 5)$

38. $7 - (4 \cdot 3 - 6)$

39. $(8 - 3)(2 - 5)$

40. $(7 - 4)(3 - 6)$

41. $\dfrac{-12 - 3 + 1}{-7}$

42. $\dfrac{-11 - 9 + 2}{6}$

43. $\dfrac{-8 - 2 - 4}{-2}$

44. $\dfrac{-18 - 6 - 3}{-3}$

45. $\dfrac{-8 - (2 - 4)}{-2}$

46. $\dfrac{-18 - (6 - 3)}{-3}$

47. $\dfrac{-8 - 2(-4)}{-2}$

48. $\dfrac{-18 - 6(-3)}{-3}$

49. $\dfrac{-4(-5)(-4)}{-4(-5) - 4}$

50. $\dfrac{-3(-2)(-4)}{-3(-2) - 4}$

51. $\dfrac{3(-2) - 4}{-4 - 1}$

52. $\dfrac{6(-3) - 8}{-9 - 4}$

53. $\dfrac{6 - 4}{4 - 4}$

54. $\dfrac{-3 + 9}{-3 + 3}$

55. $-6[-3 - 2(5 - 3)]$

56. $-2 - [5 - 2(4 - 8)]$

57. $5 - 2[3 - (4 + 6)]$

58. $6 - 7[2 - 3(4 - 5)]$

59. $5 - \dfrac{3 - 5}{-2}$

60. $4 + \dfrac{-8 - 6}{7}$

61. $6 - \dfrac{8 - 2(-3)}{-7}$

62. $10 + \dfrac{4 - 3(-2)}{-2}$

CALCULATOR EXERCISES

Compute each of the following:

63. $.831 - .746 - .294$

64. $28.7 - 32.56 + 18.61$

65. $.53(21) - .42(85)$

66. $.28(56) - .36(63)$

67. $12.4 - 20(.8) + 4.7$

68. $15.7 - 35(.6) - 4.6$

69. $.02(28.6 - 13.5)$

70. $.05(120.6 - 73.8)$

71. $5.2 - 1.4(2.8 - .7)$

72. $12.8 - 9.2(3.4 - 1.6)$

QUESTIONS FOR THOUGHT

8. Discuss the similarities and differences between $5 - 2$ and $5(-2)$.

9. Discuss the similarities and differences between $6(-3) - 2$ and $6 - 3(-2)$.

10. Explain what is *wrong* (if anything) with each of the following:

    (a) $-7 - 8 = -56$ **(b)** $-7 - 8 = +56$

    (c) $-3 - (2 - 4) = -3 - 2 = -5$ **(d)** $-4 - 2(5 - 6) = -6(-1) = +6$

11. Taking into account the operations of addition, subtraction, multiplication, and division, is it accurate to say "two negatives make a positive"?

The Real Number System

1.5

Throughout this chapter we have been working within the framework of the set of integers. However, we all recognize that the set of integers is not sufficient to supply us with all the numbers we need to describe various situations. For example, if we have to divide a 3-foot piece of wood into two *equal* pieces, then the length of each piece is 1.5 ft. The number 1.5 (or, if you prefer, the number $\frac{3}{2}$) is not an integer.

The set of "fractions" is called the set of **rational numbers** and is usually designated by the letter Q. The set of rational numbers is difficult to list, primarily because no matter where you start, there is no *next* rational number. (Think about what the "first" fraction after 0 is.) Instead, we use set-builder notation to describe the set of rational numbers:

$$Q = \left\{ \frac{p}{q} \,\middle|\, p, q \in Z \quad \text{and} \quad q \neq 0 \right\}$$

In words, this says that the set of rational numbers, which we are calling Q, is the set of all fractions whose numerators and denominators are integers (Z), provided the denominator is not equal to 0. Thus, the following are all rational numbers:

$$8 \left(= \frac{8}{1} \right) \qquad \frac{-3}{7} \qquad 0 \left(= \frac{0}{3} \right) \qquad .25 \left(= \frac{1}{4} \right)$$

We can easily locate a rational number on the number line. For example, if we want to locate $\frac{7}{4} = 1\frac{3}{4}$, we would simply divide the unit interval into four equal parts, as shown here:

$$0 \quad \tfrac{1}{4} \quad \tfrac{2}{4} \quad \tfrac{3}{4} \quad 1 \quad \tfrac{5}{4} \quad \tfrac{6}{4} \quad \tfrac{7}{4} \quad 2$$

In order to convert a fraction into its decimal form (often called a **decimal fraction**), we divide the numerator by the denominator (see Section 0.5). However, converting from the decimal form to a fraction is not quite so straightforward.

In the example above, we recognized that the decimal .25 is equal to $\frac{1}{4}$. Similarly, we recognize that the decimal $.3333\overline{3}$ (where the dash above the last 3 indicates that the 3 repeats forever) is equal to the fraction $\frac{1}{3}$. On the other hand, it is highly unlikely that we would recognize the decimal $.481481\overline{481}$ as being equal to the fraction $\frac{13}{27}$. (Divide 27 into 13 and verify that you get $.481481\overline{481}$.)

In fact, not all decimals represent rational numbers. It turns out that if a decimal is nonterminating (it does not stop and give 0's after awhile) and nonrepeating, then this decimal is *not* a rational number. In other words, such a decimal cannot be represented as the quotient of two integers.

The set of numbers on the number line which are not rational numbers is called the set of **irrational numbers** and is usually designated with the letter I. It is necessary for us to consider irrational numbers because just as the integers were insufficient to fill all our needs, so too the rational numbers do not quite do the job, either.

If we look for a number which when multiplied by itself gives a product of 9, we will fairly quickly come up with two answers:

$$3 \cdot 3 = 9 \quad \text{and} \quad (-3)(-3) = 9$$

A number x such that $x \cdot x = 9$ is called a **square root** of 9. (Square roots will be discussed in detail in Chapter 9.) Thus, we see that both 3 and -3 are square roots of 9.

Similarly, we can try to find a number which when multiplied by itself gives a product of 2 (such a number is called a *square root* of 2.) It turns out that, if we try to find the answer by trial and error (using a calculator to do the multiplication would help), we can get closer and closer to 2 but we will *never* get 2 exactly. For example, if we try $(1.4)(1.4)$, we get 1.96, so we see that 1.4 is too small. If we try $(1.5)(1.5)$, we get 2.25 and we see that 1.5 is too big. If we continue in this way we can get better and better approximations to a square root of 2. We might reach the approximate answer 1.414235, but $(1.414235)(1.414235) = 2.0000606$ (rounded off to 7 places). The point is that no matter how many places we get to, the decimal will never stop (because we never hit 2 exactly), and it never repeats. Thus, the square root of 2 (written $\sqrt{2}$) is an irrational number.

The important thing for us to recognize is that the irrational numbers also represent points on the number line. If we take all the rational numbers together with all the irrational numbers (both positive and negative), we get *all* the points on the number line. This set is called the set of **real numbers,** and is usually designated by the letter R:

$$R = \{x \mid x \text{ corresponds to a point on the number line}\}$$

We discuss the set of rational numbers in detail in Chapter 4, and the irrational numbers in detail in Chapter 9. Nevertheless, from now on unless we are told otherwise, we will assume that the set of real numbers serves as our basic frame of reference.

We should point out here that very often we describe a set of numbers by using the **double inequality** notation. For instance, we may write

$$\{x \mid -1 < x \leq 4\}$$

This means the set of numbers between -1 and 4, *but* including 4 and excluding -1—that is, the set of numbers x such that $-1 < x$ *and* $x \leq 4$. This set is illustrated on the number line as follows:

Notice that we put an empty circle around the point -1 to indicate that it is *excluded*, and a solid circle around 4 to indicate that it is *included*.

EXAMPLE 1   Sketch the following sets on a number line:

(a) $\{x \mid x > -3\}$ **(b)** $\{x \mid 1 \leq x < 5\}$

Solution

(a) The set $\{x \mid x > -3\}$ is the set of numbers to the right of -3:

✔ **LEARNING CHECKS**

1. **(a)** $\{x \mid x < -2\}$
    (b) $\{x \mid -1 < x \leq 4\}$

(b) The set $\{x \mid 1 \le x < 5\}$ is the set of numbers which are at or to the right of 1 *and* to the left of 5:

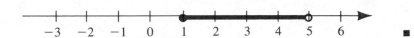

It is important to mention that the properties of the integers which we have discussed, such as the Commutative and Associative Laws, as well as all the rules we have developed for adding, subtracting, multiplying, and dividing integers carry over to the real number system. Thus, if we want to compute $\frac{-5.8688}{1.4} = -4.192$ we would divide 5.8688 by 1.4 and then make our answer negative because we are dividing opposite signs.

In the next chapter, we begin to build and manipulate algebraic expressions. The material on working with signed numbers is absolutely essential for our work ahead, so be sure you can add, subtract, multiply, and divide signed numbers before you go ahead to the next chapter.

STUDY SKILLS 1.5

Reading Directions

One important, frequently overlooked aspect of an algebraic problem is the verbal instructions. Sometimes the instructions are given in a single word, such as "simplify" or "solve" (occasionally it takes more time to understand the instructions than it takes to do the exercise). The verbal instructions tell us what we are expected to do, so make sure you read the instructions carefully and understand what is being asked.

Two examples may look the same, but the information may be asking you to do two different things. For example:

Identify the following property: $a + (b + c) = (a + b) + c$

vs.

Verify the following property by replacing the variables with numbers:

$$a + (b + c) = (a + b) + c$$

On the other hand, two different examples may have the same instructions but require you to do different things. For example:

Evaluate $2(3 + 8)$ vs.   Evaluate $2 + (3 + 8)$

You are asked to evaluate both expressions, but the solutions require different steps.

It is a good idea to familiarize yourself with the various ways the same basic instructions can be worded. In any case, always look at an example carefully and ask yourself what is being asked and what needs to be done, *before you do it.*

Answers to Learning Checks in Section 1.5

1. (a)

NOTE TO THE STUDENT

Use the space on this page to write down any questions you have or points you want to review with your instructor.

Exercises 1.5

In Exercises 1–20, compute the given expression.

1. $4.2 + 5.9$ **2.** $3.4 + 8.8$

3. $4(5.1)$ **4.** $6(3.9)$

5. $\dfrac{12.8}{3.2}$ **6.** $\dfrac{21.9}{7.3}$

7. $\dfrac{36.8}{1.5}$ **8.** $\dfrac{42.6}{2.4}$

9. $\dfrac{8}{.2} + \dfrac{12}{.4}$ **10.** $\dfrac{10}{.25} + \dfrac{5}{.1}$

11. $-2 - 5.3$ **12.** $-6 - 3.7$

13. $-2(5.3)$ **14.** $-6(3.7)$

15. $-2(-5.3)$ **16.** $-6(-3.7)$

17. $\dfrac{-8.4}{1.2}$ **18.** $\dfrac{-12.5}{2.5}$

19. $\dfrac{-6}{1.5} + \dfrac{21.6}{-1.2}$ **20.** $\dfrac{-12}{-1.2} - \dfrac{-10}{2.5}$

In Exercises 21–26, locate the given number between two successive integers on the number line. For example, if the given number is 2.6, it is located between 2 and 3 on the number line.

21. 4.8 **22.** -4.8

23. $2\dfrac{1}{3}$ **24.** $-2\dfrac{1}{3}$

25. $\dfrac{15}{2}$ **26.** $-\dfrac{15}{2}$

1. _____
2. _____
3. _____
4. _____
5. _____
6. _____
7. _____
8. _____
9. _____
10. _____
11. _____
12. _____
13. _____
14. _____
15. _____
16. _____
17. _____
18. _____
19. _____
20. _____
21. _____
22. _____
23. _____
24. _____
25. _____
26. _____

In Exercises 27–36, sketch the given set on a number line.

27. $\{x \mid x < 2\}$

28. $\{x \mid x > -3\}$

29. $\{x \mid x > 4\}$

30. $\{x \mid x < -2\}$

31. $\{x \mid -3 \leq x \leq 2\}$

32. $\{x \mid 2 < x < 7\}$

33. $\{x \mid 3 < x < 8\}$

34. $\{x \mid -2 \leq x \leq 2\}$

35. $\{x \mid 1 \leq x < 3\}$

36. $\{x \mid -1 < x \leq 3\}$

QUESTIONS FOR THOUGHT

12. Explain how you could use the Commutative and/or Associative Laws to make the following computations easier to do:

   (a) $-1.7 + 12.9 - 8.3$

   (b) $10(7)(-2.3)$

   (c) $\dfrac{-10(3.2)}{-1.6}$

   (d) $-30\left(\dfrac{1.6}{3}\right)(-10)$

13. Try to find an approximate value for a square root of 3. That is, try to find a number which when multiplied by itself gives a product of 3. Give answers correct to the nearest whole number, nearest tenth, nearest hundredth, and nearest thousandth. Use a calculator to help with the multiplication, but not to compute the square root itself.

CHAPTER 1 SUMMARY

After having completed this chapter you should be able to:

1. Evaluate a numerical expression using the order of operations (Section 1.1).

   For example:

$$8 + 4(7 - 2) = 8 + 4(5)$$
$$= 8 + 20$$
$$= \boxed{28}$$

2. Use the rules developed in this chapter to add, subtract, multiply, and divide signed numbers (Sections 1.2–1.5).

   For example:

   (a) $-6 + 3 = \boxed{-3}$

   (b) $-6 + (-3) = \boxed{-9}$

   (c) $-6 - (-3) = -6 + 3 = \boxed{-3}$

   (d) $-6 - 3 = \boxed{-9}$

   (e) $-6(-3) = \boxed{+18}$

   (f) $\dfrac{-6}{-3} = \boxed{2}$

   (g) $8 - 5(-3) = 8 + 15 = \boxed{23}$

   (h) $\dfrac{-6(-3)}{-6 - 3} = \dfrac{18}{-9} = \boxed{-2}$

   (i) $6 - 2[4 - 3(2 - 5)] = 6 - 2[4 - 3(-3)]$
   $$= 6 - 2[4 + 9]$$
   $$= 6 - 2[13]$$
   $$= 6 - 26$$
   $$= \boxed{-20}$$

ANSWERS TO QUESTIONS FOR THOUGHT

1. When we add an integer to its opposite, we get 0 as our result. This occurs because an integer and its opposite have the same absolute value.

2. If the sum of two integers is 0, the integers must be opposites. In all other cases, the sum of the integers would either be positive or negative.

3. $4 - 7 = -3$, because $-3 + 7 = 4$.

4. $4 - (-7) = 11$, because $11 + (-7) = 4$.

5. The left side is equal to 4, while the right side is equal to -4. In general, the absolute value of a difference of two integers is not the same as the difference of the absolute values of those integers.

6. 4 times -3 should mean "add -3 four times," giving $(-3) + (-3) + (-3) + (-3) = -12$.

7. We can interpret -4 times 3 as "subtract 3 four times": $-3 - 3 - 3 - 3 = -12$. Similarly, we can interpret -4 times -3 to mean "subtract -3 four times," giving $-(-3) - (-3) - (-3) - (-3) = +3 + 3 + 3 + 3 = +12$.

8. Even though the symbols are the same in both examples, the parentheses lead to different results. Without the parentheses, we have a subtraction example: $5 - 2 = 3$; with the parentheses, we have a multiplication example: $5(-2) = -10$.

9. Again, the same symbols are used in both examples, but the location of the parentheses causes the difference this time. Our order of operations requires that we multiply before we subtract. Therefore, $6(-3) - 2 = -18 - 2 = -20$, while $6 - 3(-2) = 6 + 6 = 12$.

10. When we see $-7 - 8$, it must be interpreted as a subtraction, since no parentheses are present. Therefore, (a) and (b) are both wrong, since $-7 - 8 = -15$. [It would be correct to write $-7(-8) = +56$ or $(-7)(-8) = +56$.] Part (c) is wrong also. Since $2 - 4 = -2$, it follows that $-3 - (2 - 4) = -3 - (-2) = -3 + 2 = -1$. Here, $-4 - 2(5 - 6) = -4 - 2(-1) = -4 + 2 = -2$. ($-4 - 2 = -6$ never legitimately enters into the calculation.)

11. The statement "two negatives make a positive" is accurate for multiplication and division but *not* for addition and subtraction. For instance, $(-3) + (-2) = -5$ and $(-3) - (-2) = -1$, and neither of these answers is positive.

12. (a) Use the Commutative Law of Addition to rewrite the given problem as $12.9 - 1.7 - 8.3$. Then this equals $12.9 - 10 = 2.9$.

    (b) Use the Commutative Law of Multiplication to rewrite the problem as $7(10)(-2.3)$. Then this becomes $7(-23) = -161$.

    (c) Use the Associative Law of Multiplication to rewrite the problem as $-10\left(\dfrac{3.2}{-1.6}\right)$. Then this equals $-10(-2) = 20$.

    (d) First, use the Commutative Law of Multiplication to rewrite the problem as $-30(-10)\left(\dfrac{1.6}{3}\right)$, which becomes $300\left(\dfrac{1.6}{3}\right)$. Then use the Associative Law of Multiplication to write this as $\dfrac{300}{3}(1.6) = 100(1.6) = 160$.

13. Since $(1)(1) = 1$ and $(2)(2) = 4$, the square root of 3, to the nearest whole number, is 2.

    Since $(1.7)(1.7) = 2.89$ and $(1.8)(1.8) = 3.24$, the square root of 3, to the nearest tenth, is 1.7.

    Since $(1.73)(1.73) = 2.9929$ and $(1.74)(1.74) = 3.0276$, the square root of 3, to the nearest hundredth, is 1.73.

    Since $(1.732)(1.732) = 2.999824$ and $(1.7333)(1.7333) = 3.00432889$, the square root of 3, to the nearest thousandth, is 1.732.

CHAPTER 1 REVIEW EXERCISES

In Exercises 1–44, evaluate the given expression.

1. $3 - 7$ **2.** $4 - 9$ **3.** $-7 - 5$

4. $-8 - 2$ **5.** $-2 - (-6)$ **6.** $-6 - (-2)$

7. $-4 - 5 - 6$ **8.** $-2 - 3 - 4$ **9.** $-7 + 12 - 5$

10. $-6 + 8 - 12$ **11.** $7 - 4 + 3 - 9$ **12.** $2 - 5 - 3 + 1$

13. $8 - 5 - 6$ **14.** $4 - 9 - 3$ **15.** $8 - (5 - 6)$

16. $4 - (9 - 3)$ **17.** $8(5 - 6)$ **18.** $4(9 - 3)$

19. $8 - 3 - 6$ **20.** $4 - 9 - 2$ **21.** $8 - 3(-6)$

22. $4 - 9(-2)$ **23.** $8 - (3 - 6)$ **24.** $4 - (9 - 2)$

25. $8(-3)(-6)$ **26.** $4(-9)(-2)$ **27.** $9 - 4(3 - 7)$

28. $7 - 5(2 - 8)$ **29.** $9 - 4 \cdot 3 - 7$ **30.** $7 - 5 \cdot 2 - 8$

31. $9 - (4 \cdot 3 - 7)$ **32.** $7 - (5 \cdot 2 - 8)$ **33.** $(9 - 4)(3 - 7)$

34. $(7 - 5)(2 - 8)$ **35.** $|4 - 9| - |3 - 7|$ **36.** $|2 - 5| - |1 - 6|$

37. $\dfrac{-7 - 3}{-2(-5)}$ **38.** $\dfrac{-4 - 15 - 2}{1 - 4}$ **39.** $\dfrac{-4(-2)(-8)}{-4(2) - 8}$

40. $\dfrac{-3(2)(-6)}{3(-2) - 6}$ **41.** $\dfrac{6 - 4(3 - 1)}{-2(-3) - 4}$ **42.** $\dfrac{7 + 2(5 - 10)}{-3(-3) - 6}$

43. $8 + 2[3 - 4(1 - 6)]$ **44.** $6 - 4[7 + 2(1 - 5)]$

ANSWERS

1. _____
2. _____
3. _____
4. _____
5. _____
6. _____
7. _____
8. _____
9. _____
10. _____
11. _____
12. _____
13. _____
14. _____
15. _____
16. _____
17. _____
18. _____
19. _____
20. _____
21. _____
22. _____
23. _____
24. _____
25. _____
26. _____
27. _____
28. _____
29. _____
30. _____
31. _____
32. _____
33. _____
34. _____
35. _____
36. _____
37. _____
38. _____
39. _____
40. _____
41. _____
42. _____
43. _____
44. _____

In Exercises 45–50, sketch the given set on a number line.

45. $\{x \mid x > 3\}$

46. $\{x \mid x \le 5\}$

47. $\{x \mid x \le -2\}$

48. $\{x \mid x > -4\}$

49. $\{x \mid 0 < x < 4\}$

50. $\{x \mid -3 \le x < -1\}$

CHAPTER 1 PRACTICE TEST

ANSWERS

In each of the following problems, compute the given expression.

1. $-9 - 4 + 3 - 6 + 5$

2. $|3 - 8| - |1 - 6|$

3. $4 - 7 - 3 - (-2)$

4. $\dfrac{5(-4)(-3)}{1 - 7}$

5. $\dfrac{(-2)(-3)(-4)}{(-2)(-3) - 4}$

6. $8 - 5(4 - 7)$

7. $8 - 5 \cdot 4 - 7$

8. $-7[5 + 4(3 - 7) - 2]$

9. $8 - 3[8 - 3(8 - 3)]$

10. Sketch the given set on a number line.
$\{x \mid -4 < x \le 3\}$

1. _____

2. _____

3. _____

4. _____

5. _____

6. _____

7. _____

8. _____

9. _____

10. _____

Algebraic Expressions

Exponents

2.1

In the last chapter we discussed various number systems and some of their basic properties. When we wanted to make a general statement about numbers, we used letters to represent numbers. This is in fact the essence of algebra—it is the generalization of arithmetic.

When we want to describe even fairly simple properties of numbers in words it can be quite cumbersome. It is usually much more convenient to use letters when we want to talk about "any numbers." The key idea here is the *variable,* a symbol which stands for a number (or numbers). In this book a variable will usually be represented as an italic Latin letter such as x, y, A, M, etc.

In algebra, there are two ways variables are primarily used. One use of a variable is as a place holder. That is, the variable is holding the place of a particular number (or numbers) which has not yet been identified but which needs to be found. The equation

$$x + 7 = 5$$

is an example of this type of use of a variable. Here we would like to figure out the number which when added to 7 gives 5. Variables are used in this way when we are solving equations.

A second use of a variable is to describe a general relationship between numbers and/or arithmetic operations. In the statement of the Associative Law when we write

$$a + (b + c) = (a + b) + c$$

we mean that a, b, and c can be any real numbers.

While variables represent unknown quantities, a *constant,* on the other hand, is a symbol whose value is fixed. The numbers 8, -5, 2.43, and π are examples of constants.

Our goal in this chapter is to learn how to take algebraic expressions and, given a basic set of guidelines and properties, change them into simpler expressions. This process is called *simplifying algebraic expressions.*

We have used the phrase *algebraic expression* without having actually defined it. For the time being we will accept the following definition of an algebraic expression. (We will extend this definition a bit further in Chapter 9.)

> **DEFINITION** An *algebraic expression* is made up of a finite number of additions, subtractions, multiplications, and divisions of constants and/or variables.

Do not be intimidated by this formal definition. It merely makes our terminology precise. For example, each of the following is an algebraic expression:

$$5 \qquad -8xy \qquad 3x + 4y - \frac{2}{w}$$

As we proceed through this course we will develop more specific terminology for different types of algebraic expressions.

To a great extent, the meaning of the word *simplify* will depend on the type of algebraic expression we are working with.

In this chapter we are going to deal with a certain type of expression which involves some new notation. We are often going to encounter expressions which consist of repeated multiplication. For example:

$$2 \cdot 2 \cdot 2 \cdot 2 \cdot 2 \qquad \textit{Remember that we use } \cdot \textit{ to indicate a product.}$$

$$x \cdot x \cdot x \cdot x \cdot x$$

If we multiply out the expression $2 \cdot 2 \cdot 2 \cdot 2 \cdot 2$, we see that it has a numerical value of 32. On the other hand, the expression $x \cdot x \cdot x \cdot x \cdot x$ has no numerical value until we know the value of x. However, our focus now is not so much on the value of such expressions, but rather on their *form*. Consequently, we introduce what is called *exponent notation*:

$$2 \cdot 2 \cdot 2 \cdot 2 \cdot 2 = 2^5$$

In the expression 2^5,

2 is called the *base*

5 is called the *exponent*

We read 2^5 as "2 raised to the fifth power." Similarly, $5^4 = 5 \cdot 5 \cdot 5 \cdot 5$. The exponent 4 tells us how many factors of 5 to *multiply*.

If, in addition, we want to *evaluate* or compute the value of 5^4, we would proceed as follows:

$$
\begin{aligned}
5^4 &= 5 \cdot 5 \cdot 5 \cdot 5 \\
&= 25 \cdot 5 \cdot 5 \\
&= 125 \cdot 5 \\
&= \boxed{625}
\end{aligned}
$$

DEFINITION *Exponent notation:*

$$\underbrace{x^n = x \cdot x \cdot x \cdot \cdots \cdot x}_{n \text{ times}}$$

where x appears as a factor n times. Of course, n is a positive integer.

In general, x^n is read "x to the nth power." However, exponents 2 and 3 are given special names:

x^2 is read "x squared"

x^3 is read "x cubed"

Note: $x = x^1$ but we usually do not write the exponent 1, even though we do often think of it.

EXAMPLE 1 Evaluate each of the following:

(a) 4^3 **(b)** $(-2)^4$ **(c)** -2^4 **(d)** $5 \cdot 3^2$

Solution

(a) $4^3 = 4 \cdot 4 \cdot 4$ *4 taken as a factor 3 times*

$\qquad = 16 \cdot 4$

$\qquad = \boxed{64}$

✔ **LEARNING CHECKS**

1. (a) 3^4

(b) $(-6)^2$

(c) -6^2

(d) $7 \cdot 2^3$

(b) $(-2)^4 = (-2)(-2)(-2)(-2)$

$= 4(-2)(-2)$

$= -8(-2)$

$= \boxed{16}$

(c) $-2^4 = -2 \cdot 2 \cdot 2 \cdot 2$

$= -4 \cdot 2 \cdot 2$

$= -8 \cdot 2$

$= \boxed{-16}$

Note the difference between parts **(b)** and **(c)**. In $(-2)^4$ the exponent 4 refers to the number -2; therefore, -2 appears as a factor 4 times. But in -2^4 the exponent 4 refers only to the number 2; therefore, 2 appears as a factor 4 times with only *one* minus sign in front. An exponent applies only to what is to its immediate left. If the intention is for the exponent 4 to apply to the -2 then parentheses are necessary as in part **(b)** above.

(d) $5 \cdot 3^2 = 5 \cdot 3 \cdot 3$

$= 5 \cdot 9$

$= \boxed{45}$

Here again, do not make the mistake of first multiplying 5 times 3 and then squaring the result. The exponent 2 applies *only* to the 3. As long as we clearly keep this in mind we will carry out the order of operations correctly.   ∎

2. (a) *aaabbcccc*

(b) 5*aab*

(c) *aaaa · aa*

(d) *aaaa + aa*

EXAMPLE 2   Write each of the following with exponents.

(a) *xxyzzzz*   **(b)** 3*xyyy*   **(c)** *xxx · xxxx*   **(d)** *xxx + xxxx*

Solution

(a) $xxyzzzz = \boxed{x^2yz^4}$

(b) $3xyyy = \boxed{3xy^3}$

(c) $xxx \cdot xxxx = \boxed{x^7}$

(d) $xxx + xxxx = \boxed{x^3 + x^4}$ ∎

3. (a) uv^4

(b) $(uv)^4$

(c) u^3u^5

(d) $u^3 + u^5$

EXAMPLE 3   Write each of the following without exponents.

(a) xy^3 **(b)** $(xy)^3$ **(c)** $x^2 \cdot x^3$ **(d)** $x^2 + x^3$

Solution

(a) $xy^3 = \boxed{xyyy}$

(b) $(xy)^3 = \boxed{(xy)(xy)(xy)}$

(c) $x^2 \cdot x^3 = \boxed{xx \cdot xxx}$

(d) $x^2 + x^3 = \boxed{xx + xxx}$ ∎

Let's look at Example 3, parts (c) and (d), more carefully:

$$x^2 \cdot x^3 = xx \cdot xxx = x^5 \quad \text{but} \quad x^2 + x^3 = xx + xxx \neq x^5$$

There is sometimes a tendency to look at $x^2 + x^3$ and to think something like

"There are 5 x's so it must be x^5."

Algebraic notation is very precise. x^5 means x taken as a *factor* 5 times, which is not at all the same as x taken as a factor 2 times *added to* x taken as a factor 3 times. If you are still not sure of the difference, let's evaluate both expressions for a particular value of x, say $x = 4$:

$$4^5 = 4 \cdot 4 \cdot 4 \cdot 4 \cdot 4 = 1{,}024$$
$$\text{while} \quad 4^2 + 4^3 = 4 \cdot 4 + 4 \cdot 4 \cdot 4 = 16 + 64 = 80$$

We get very different values because the expressions are really very different even though they may superficially look the same.

When asked to simplify an expression involving exponents, we are expected to write an equivalent expression with bases and exponents occurring as few times as possible.

EXAMPLE 4   Simplify each of the following as completely as possible.

(a) x^2x^5 (b) a^4a^5 (c) yy^7

Solution   We can compute each expression by writing out all the factors and simply counting them up.

(a) $x^2x^5 = xx \cdot xxxxx = \boxed{x^7}$ *Note the trend in exponents: $2 + 5 = 7$*

(b) $a^4a^5 = aaaa \cdot aaaaa = \boxed{a^9}$ *Note: $4 + 5 = 9$*

(c) $yy^7 = y \cdot yyyyyyy = \boxed{y^8}$ *Note: $y = y^1$ and $1 + 7 = 8$*   ∎

From these examples we see that if we are multiplying powers of the same base, we keep the base and add the exponents. This follows from simply understanding what an exponent means, and being able to count. This fact is called the first rule for exponents.

EXPONENT RULE 1

$$a^m \cdot a^n = a^{m+n}$$

This same rule extends to a product of more than two powers of the same base.

4. (a) m^3m^7

   (b) n^2n^6

   (c) zz^4

5. (a) $x^3x^4x^2$

(b) yy^4y

(c) $2^3 \cdot 2^6$

(d) $3^2 \cdot 5^3$

(e) u^3v^7

6. $aa^2a^4 + a^3a^6$

7. (a) $(3x^4)(5x^8)$

(b) $(-4x^2)(3x^3)(-2x)$

EXAMPLE 5 Simplify each of the following as completely as possible.

(a) $r^2r^4r^5$ **(b)** a^6aa **(c)** 3^43^6 **(d)** 2^35^2 **(e)** x^5y^4

Solution We use Exponent Rule 1:

(a) $r^2r^4r^5 = \boxed{r^{2+4+5}} = \boxed{r^{11}}$ *Remember that the shaded steps are usually done mentally.*

(b) $a^6aa = \boxed{a^{6+1+1}} = \boxed{a^8}$

(c) $3^43^6 = \boxed{3^{4+6}} = \boxed{3^{10}}$ *Note that the answer is **not** 9^{10}.*

We are not multiplying the 3's. Rule 1 says that we count up the factors of 3. There are 10 factors of 3, not 10 factors of 9!

(d) We cannot simplify this expression in the way we just did part **(c)**, because the bases are not the same; we can, however, compute its value.

$$2^35^2 = (2 \cdot 2 \cdot 2)(5 \cdot 5) = (8)(25) = \boxed{200}$$

(e) x^5y^4 remains unchanged, since the bases, x and y, are not identical. ∎

EXAMPLE 6 *Simplify as completely as possible.* $x^3x^5 + xx^2x^4$

Solution

$$x^3x^5 + xx^2x^4 = \boxed{x^{3+5} + x^{1+2+4}}$$

$$= \boxed{x^8 + x^7}$$

This is the complete answer. You do not get x^{15} as an answer. Rule 1 says that you add the exponents when you are *multiplying* the powers, not when you are adding them. ∎

The Commutative and Associative Laws of Multiplication which we discussed in the last chapter allow us to multiply simple algebraic expressions by rearranging and regrouping the factors. Consider the following illustration:

$$\begin{aligned}(3x^2)(5x^4) &= 3(x^2 \cdot 5)x^4 &&\text{\textit{Associative Law of Multiplication}}\\ &= 3(5 \cdot x^2)x^4 &&\text{\textit{Commutative Law of Multiplication}}\\ &= (3 \cdot 5)(x^2 \cdot x^4) &&\text{\textit{Associative Law of Multiplication}}\\ &= \boxed{15x^6} &&\text{\textit{Exponent Rule} 1}\end{aligned}$$

Go back and reread the last illustration, making sure you *understand* each step completely. Essentially, the Commutative and Associative Laws allow us to ignore the original order and grouping of the factors so that we may multiply all constants together, and multiply identical variables together using Exponent Rule 1.

EXAMPLE 7 Simplify each of the following as completely as possible.

(a) $(4x^3)(7x^6)$ **(b)** $(-2x)(5x^4)(3x^3)$ **(c)** $(-2x^3y)(-6x^4y^5)$ **(d)** $(2x^2)^3$

Solution Since these expressions all consist entirely of multiplication, we can use the Commutative and Associative Laws to rearrange and regroup the factors.

(a) $(4x^3)(7x^6) = \boxed{4 \cdot x^3 \cdot 7 \cdot x^6 = (4 \cdot 7)(x^3 \cdot x^6)} = \boxed{28x^9}$

Notice how we use the Commutative and Associative Laws to rearrange and regroup the factors.

(b) $(-2x)(5x^4)(3x^3) = \boxed{(-2)(5)(3)(xx^4x^3)} = \boxed{-30x^8}$

(c) $(-2x^3y)(-6x^4y^5) = \boxed{(-2)(-6)(x^3x^4yy^5)} = \boxed{12x^7y^6}$

(d) $(2x^2)^3 = (2x^2)(2x^2)(2x^2) = \boxed{(2 \cdot 2 \cdot 2)(x^2x^2x^2)} = \boxed{8x^6}$

Notice how the coefficient (8) and exponent (6) of the final answer were computed. ∎

While we have only scratched the surface as far as simplifying expressions is concerned, understanding the examples we have just worked out is basic to being able to do the more complex problems in the sections ahead.

(c) $(-5xy^3)(8x^5y^4)$

(d) $(3x^5)^4$

STUDY SKILLS 2.1

Comparing and Contrasting Examples

When learning most things for the first time, it is very easy to get confused and to treat things which are different as though they were the same because they "look" similar. Algebraic notation can be especially confusing because of the detail involved. Move or change one symbol in an expression and the entire example is different; change one word in a verbal problem and the whole problem has a new meaning.

It is important that you be capable of making these distinctions. The best way to do this is by comparing and contrasting examples and concepts which look almost identical, but are not. It is also important that you ask yourself in what ways these things are similar and in what ways they differ. For example, the Associative Law of Addition is similar in some respects to the Associative Law of Multiplication, but different from it in others. Also, the expressions $3 + 2 \cdot 4$ and $3 \cdot 2 + 4$ look similar, but are actually very different.

When you are working out exercises (or reading a concept), ask yourself, "What examples or concepts are similar to those which I am now doing? In what ways are they similar? How do I recognize the differences?" Doing this while you are working the exercises will help prevent you from making careless errors later on.

✔ *Answers to Learning Checks in Section 2.1*

1. (a) 81   **(b)** 36   **(c)** -36 **(d)** 56

2. (a) $a^3b^2c^4$ **(b)** $5a^2b$ **(c)** a^6 **(d)** $a^4 + a^2$

3. (a) $uvvvv$ **(b)** $(uv)(uv)(uv)(uv)$ **(c)** $uuu \cdot uuuuu$ **(d)** $uuu + uuuuu$

4. (a) m^{10} **(b)** n^8 **(c)** z^5

5. (a) x^9 **(b)** y^6 **(c)** $2^9 = 512$ **(d)** 1,125   **(e)** u^3v^7

6. $a^7 + a^9$ **7. (a)** $15x^{12}$ **(b)** $24x^6$ **(c)** $-40x^6y^7$ **(d)** $81x^{20}$

NOTE TO THE STUDENT

Use the space on this page to write down any questions you have or points you want to review with your instructor.

Exercises 2.1

In Exercises 1–12, write the given expression without exponents.

1. x^6

2. y^5

3. $(-x)^4$

4. $(-y)^5$

5. $-x^4$

6. $-y^6$

7. x^2y^3

8. x^4y^6

9. $x^2 + y^3$

10. $x^4 + y^6$

11. xy^3

12. $(xy)^3$

In Exercises 13–22, write the given expression with exponents.

13. $aaaa$

14. $sssss$

15. $xxyyy$

16. $xx + yyy$

17. $-rrsss$

18. $(-r)(-r)sss$

19. $-xx(-y)(-y)(-y)$

20. $-xx - yyy$

21. $(xxx)(xxxxx)$

22. $(xxx) + (xxxxx)$

In Exercises 23–38, evaluate the given expression.

23. 3^5

24. 5^3

25. -2^3

26. $(-2)^3$

27. $(-2)^4$

28. -2^4

29. $3^2 + 3^3 - 3^4$

30. $4^2 + 4^3 - 4^4$

31. $4 \cdot 3^2 - 2 \cdot 5^2$

32. $7 \cdot 2^3 - 5 \cdot 3^2$

33. $(3 - 7)^2 - (4 - 5)^3$

34. $(6 - 8)^2 - (1 - 3)^3$

35. $3^2 4^3$

36. $2^3 2^4$

37. $3^2 3^3$

38. $2^3 3^4$

ANSWERS

1. _____
2. _____
3. _____
4. _____
5. _____
6. _____
7. _____
8. _____
9. _____
10. _____
11. _____
12. _____
13. _____
14. _____
15. _____
16. _____
17. _____
18. _____
19. _____
20. _____
21. _____
22. _____
23. _____
24. _____
25. _____
26. _____
27. _____
28. _____
29. _____
30. _____
31. _____
32. _____
33. _____
34. _____
35. _____
36. _____
37. _____
38. _____

39. _____

40. _____

41. _____

42. _____

43. _____

44. _____

45. _____

46. _____

47. _____

48. _____

49. _____

50. _____

51. _____

52. _____

53. _____

54. _____

55. _____

56. _____

57. _____

58. _____

59. _____

60. _____

61. _____

62. _____

63. _____

64. _____

65. _____

66. _____

67. _____

68. _____

In Exercises 39–60, simplify the given expression as completely as possible.

39. x^3x^5

40. y^4y^6

41. aa^2a^4

42. m^3mm^5

43. $(3x)(5x)(4x)$

44. $(2a)(3a)(6a)$

45. $(3r^2)(2r^3)$

46. $(4w^5)(5w^4)$

47. $(-3x^3)(5x^2)$

48. $(-4x)(6x^5)$

49. $(-c^4)(2^3)(-5c)$

50. $(-3p)(-2p^5)(p^2)$

51. $(8x^3y^2)(4xy^5)$

52. $(7x^2y^3)(5xy^4)$

53. $(-2xy)(x^2y^2)(-3xy)$

54. $(-8x)(-3xy^2)(5x^3y^2)$

55. $(3a^4)^2$

56. $(4a^3)^2$

57. $(-4n^2)^3$

58. $(-5n^4)^3$

59. $(x^2)^3(x^4)^2$

60. $(x^3)^2(x^2)^4$

CALCULATOR EXERCISES

Evaluate each of the following expressions. Round off your answer to the nearest thousandth.

61. $(.52)^4$

62. $(1.83)^5$

63. $(1.4)^8$

64. $(-2.7)^9$

65. $5.1(4.6)^2$

66. $2.3(7.1)^3$

67. $(3.81)^3 - (2.64)^2$

68. $(-5.4)^2 - (4.1)^2$

QUESTIONS FOR THOUGHT

1. State in *words* the difference between $3 \cdot 2^4$ and $(3 \cdot 2)^4$.

2. Explain what is *wrong* with each of the following:

(a) $3 \cdot 2^2 \overset{?}{=} 6^2 = 36$

(b) $3x^4 \overset{?}{=} 3 \cdot 3 \cdot 3 \cdot 3 \cdot x \cdot x \cdot x \cdot x$

(c) $5^4 \cdot 2 \overset{?}{=} 20 \cdot 2 = 40$

(d) $(3x^3)^2 \overset{?}{=} 6x^6$

(e) $-3^4 \overset{?}{=} (-3) \cdot (-3) \cdot (-3) \cdot (-3)$
$$= 81$$

(f) $x^4 + x^5 \overset{?}{=} x^9$

(g) $x^2 \cdot x^7 \overset{?}{=} x^{14}$

(h) $(x^2)^4 \overset{?}{=} x^6$

Algebraic Substitution

2.2

A complaint often heard from students is that the expressions encountered in algebra all "look alike." If you do not happen to play the piano you may think the piano presents a similar problem—all the keys look alike. It is only through practice and study that you can learn to distinguish the keys. The same practice and study are required in algebra.

As we have begun to see in the last section, algebraic notation is very precise. It means exactly what it says. One of the most basic skills necessary for success in algebra is the ability to read carefully so that you can see what a particular problem says. For example, the two expressions

$$(-5)(-3) \quad \text{and} \quad (-5) - (3)$$

may look quite similar at first glance, but they are in fact quite different.

$$(-5)(-3) = 15 \quad \text{and} \quad (-5) - (3) = -8$$

The first expression involves multiplication; the second involves subtraction. You must sensitize yourself to recognize these differences.

Algebraic substitution refers to the process of replacing the variables in an expression with specific values and then evaluating the result.

As you will see in the examples that follow, we must be careful to follow the order of operations and *not* change the arithmetic steps which appear in the example.

EXAMPLE 1   Evaluate each of the following for $x = 2$ and $y = -3$:
(a) $5x - 3$ **(b)** $-y$ **(c)** $x - y$ **(d)** $|x - y|$ **(e)** $|x| - |y|$

Solution   In each case we begin by replacing every occurrence of a variable with its assigned value, and then we perform the indicated operations. In order to make this process even clearer you may find it is a good idea to put parentheses in wherever you see a variable, and then substitute the assigned value into the parentheses. However, be careful because putting in too many parentheses can clutter up an example.

(a) Evaluate $5x - 3$ for $x = 2$:

$$5x - 3 = 5() - 3 \qquad \textit{We rewrite with parentheses.}$$
$$= 5(2) - 3 \qquad \textit{We put the value for x into the parentheses.}$$
$$= \boxed{7}$$

(b) Evaluate $-y$ for $y = -3$:

$$-y = -()$$
$$= -(-3)$$
$$= \boxed{3}$$

(c) Evaluate $x - y$ for $x = 2$ and $y = -3$:

$$x - y = () - ()$$
$$= (2) - (-3) \qquad \textit{Now we follow the rule for subtraction.}$$
$$= (2) + (+3)$$
$$= \boxed{5}$$

✔ LEARNING CHECKS

1. Evaluate for $x = -2$ and $y = 4$:
   (a) $4x - 7$
   (b) $-x$
   (c) $y - x$
   (d) $|x + y|$
   (e) $|y| - |x|$

ANSWERS TO PROGRESS CHECK

1. -2 **2.** -320 **3.** 2 **4.** 5
5. -36 **6.** 36 **7.** 9 **8.** 17

(d) Evaluate $|x - y|$ for $x = 2$ and $y = -3$:

$$\begin{aligned} |x - y| &= |(\ \ ) - (\ \ )| \\ &= |(2) - (-3)| \\ &= |2 + 3| \\ &= |5| \\ &= \boxed{5} \end{aligned}$$

(e) Evaluate $|x| - |y|$ for $x = 2$ and $y = -3$:

$$\begin{aligned} |x| - |y| &= |(\ \ )| - |(\ \ )| \\ &= |(2)| - |(-3)| \\ &= 2 - (+3) \\ &= \boxed{-1} \end{aligned}$$

Note the difference here as compared to part (d) above.

∎

2. Evaluate each of the following:

(a) y^4 for $y = -1$

(b) $-y^4$ for $y = -1$

(c) $(5a + 1)(3a - 4)$ for $a = 2$

(d) $2t^2 - 3t + 1$ for $t = 4$

EXAMPLE 2 Evaluate each of the following expressions for the given value of the variable:

(a) x^2 for $x = -3$ **(b)** $-x^2$ for $x = -3$

(c) $(3t + 2)(2t + 1)$ for $t = 6$ **(d)** $3w^2 + 4w - 2$ for $w = 5$

Solution Again we begin by replacing each occurrence of the variable with parentheses and inserting the given value of the variable into the parentheses.

(a) For $x = -3$:

$$x^2 = (-3)^2 = (-3)(-3) = \boxed{9}$$

(b) For $x = -3$:

$$-x^2 = -(-3)^2 = -(-3)(-3) = \boxed{-9}$$

Note that the minus sign in front of the x^2 does not alter the fact that each x is replaced by -3.

(c) For $t = 6$:

$$\begin{aligned} (3t + 2)(2t + 1) &= [3(6) + 2][2(6) + 1] \\ &= [18 + 2][12 + 1] \\ &= [20][13] \\ &= \boxed{260} \end{aligned}$$

We could have written $(3(6) + 2)(2(6) + 1)$. However, brackets can be used instead and using brackets makes it easier to read.

(d) For $w = 5$:

$$\begin{aligned} 3w^2 + 4w - 2 &= 3(\ \ )^2 + 4(\ \ ) - 2 \\ &= 3(5)^2 + 4(5) - 2 \\ &= 3(25) + 4(5) - 2 \\ &= 75 + 20 - 2 \\ &= \boxed{93} \end{aligned}$$

Be sure to square the 5 first; the exponent 2 is only on the 5.

∎

As we mentioned earlier, be careful not to change the arithmetic operations in the example when you substitute values. We will continue to insert parentheses whenever we think it makes the substitution clearer.

EXAMPLE 3 Evaluate each of the following for $x = 3$, $y = -4$, and $z = 5$:

(a) $x + yz$ (b) $(x + y)z$ (c) $x - y - z$ (d) $x - (y - z)$

(e) $x - 7(y - 2)$ (f) $(x - 7)(y - 2)$ (g) $(x + y)^2 - (xy)^2$

Solution

(a) $x + yz = (\ ) + (\ )(\ )$

$\qquad = (3) + (-4)(5)$ *Watch your order of operations.*

$\qquad = 3 + (-20)$

$\qquad = \boxed{-17}$

(b) $(x + y)z = [3 + (-4)]5$ *Note the difference the given parentheses make as*

$\qquad = (-1)5$ *compared with part (a).*

$\qquad = \boxed{-5}$

(c) $x - y - z = 3 - (-4) - 5$

$\qquad = 3 + 4 - 5$

$\qquad = \boxed{2}$

(d) $x - (y - z) = 3 - (-4 - 5)$ *Note the difference the given parentheses make*

$\qquad = 3 - (-9)$ *as compared with part (c).*

$\qquad = \boxed{12}$

(e) $x - 7(y - 2) = 3 - 7(-4 - 2)$

$\qquad = 3 - 7(-6)$ *Watch your order of operations.*

$\qquad = 3 + 42$

$\qquad = \boxed{45}$

(f) $(x - 7)(y - 2) = (3 - 7)(-4 - 2)$ *Note the difference the extra paren-*

$\qquad = (-4)(-6)$ *theses make as compared with part*
(e).

$\qquad = \boxed{24}$

(g) $(x + y)^2 - (xy)^2 = [3 + (-4)]^2 - [3(-4)]^2$ *Note that the first bracket*

$\qquad = [-1]^2 - [-12]^2$ *has addition while the*
second bracket has

$\qquad = 1 - 144$ *multiplication.*

$\qquad = \boxed{-143}$ ■

This seems to be an opportune place to review some terminology which we will find both very important and extremely useful.

Informally we have said that a *term* is a member of a sum while a *factor* is a member of a product. It would be a bit more precise to say that a **term** is an algebraic expression which is connected by multiplication (and/or division). For example, the expression $2x^3y^2$ is a term because

$$2x^3y^2 = 2xxxyy$$

is a *product* of a constant and variables.

3. Evaluate for $x = 3$, $y = -4$, and $z = 5$.

(a) $xy + z$

(b) $x(y + z)$

(c) $z - x - y$

(d) $z - (x - y)$

(e) $z - 5(x - 3)$

(f) $(z - 5)(x - 3)$

(g) $(y - z)^2 - (yz)^2$

Similarly,

$$-4 \quad \text{and} \quad x \quad \text{and} \quad -7x^3y^4z^8$$

are examples of terms.

As we shall soon see, there are many situations in algebra where it is crucial to be able to distinguish between terms and factors.

$3x + y + z$ consists of three terms, the first of which, $(3x)$, has two factors (3 and x).

$3(x + y)$ is one term which is made up of two factors (3 and $x + y$).

$3xy$ is one term made up of three factors (3, x, and y).

In other words, a term is all *connected* by multiplication, while terms are *separated* by addition or subtraction. (Those terms involving division of constants and variables will be discussed in Chapter 4.)

Here is some more terminology we will often use.

> **DEFINITION** The *constant* multiplier of a term is called the **numerical coefficient**, or usually just the **coefficient**. The variable part of a term is often called the **literal** part of the term.

We will always include the sign as part of the coefficient:

The coefficient of $5x$ is 5; its literal or variable part is x.

The coefficient of $-2y^3$ is -2; its literal part is y^3.

The expression $4x^2 - 3y$ is thought of as $4x^2 + (-3y)$ and consists of two terms. The first term, $4x^2$, has coefficient 4, while the second term, $-3y$, has coefficient -3.

If no constant appears, then the coefficient is understood to be 1. Thus, the term x^2yz has a coefficient of 1. Similarly, the term $-z^4$ has a coefficient of -1.

Thus, the process of algebraic substitution that we have described in this section can also be called *evaluating literal expressions*.

One final point: If we look at the expression $(3x)(-4x^3)$ carefully, we can see that it is one term. What is its coefficient? Normally we simplify an expression before we determine the coefficient:

$$(3x)(-4x^3) = -12x^4$$

and so the coefficient is -12.

4. (a) $7x + 4y$
    (b) $7x(4y)$

EXAMPLE 4 In each of the following expressions determine the number of terms, and the coefficient of each term.

(a) $8m - 3n$ **(b)** $8m(-3n)$

Solution

(a) $8m - 3n$ is made up of two terms. The coefficient of the first term is 8 and the coefficient of the second term is -3.

(b) $8m(-3n)$ is one term since it is all connected by multiplication. First we multiply:

$$8m(-3n) = -24mn$$

and we see that the coefficient is -24.   ∎

STUDY SKILLS 2.2

Coping with Getting Stuck

All of us have had the frustrating experience of getting stuck on a problem; sometimes even the simple problems can give us difficulty.

Perhaps you do not know how to begin; or, you are stuck halfway through an exercise and are at a loss as to how to continue; or, your answer and the book's answer do not seem to match. (Do not assume the book's solutions are 100% correct—we are only human even if we are math teachers. But do be sure to check that you have copied the problem accurately.)

Assuming you have reviewed all the relevant material beforehand, be sure you have spent enough time on the problem. Some people take one look at a problem and simply give up without giving the problem much thought. This is not what we regard as "getting stuck," since it is giving up before having even gotten started.

If you find after a reasonable amount of time, effort, and *thought*, that you are still not getting anywhere, if you have looked back through your notes and textbook and still have no clue as to what to do, try to find exercises similar to the one you are stuck on (with answers in the back) that you can do. Analyze what you did to arrive at the solution and try to apply those principles to the problem you are finding difficult. If you have difficulty with those similar problems as well, you may have missed something in your notes or in the textbook. Reread the material and try again. If you are still not successful, go on to different problems or take a break and come back to it later.

If you are still stuck, wait until the next day. Sometimes a good night's rest is helpful. Finally, if you are still stuck after rereading the material, see your teacher (or tutor) as soon as possible.

Answers to Learning Checks in Section 2.2

1. (a) -15 (b) 2   (c) 6   (d) 2   (e) 2
2. (a) 1   (b) -1 (c) 22   (d) 21
3. (a) -7 (b) 3   (c) 6   (d) -2 (e) 5   (f) 0   (g) -319
4. (a) 2 terms; coefficients 7 and 4   (b) 1 term; coefficient 28

NOTE TO THE STUDENT

Use the space on this page to write down any questions you have or points you want to review with your instructor.

Exercises 2.2

In Exercises 1–50, evaluate the given expression for $x = 2$, $y = -3$, and $z = -4$.

1. $-y$

2. $-z$

3. $-|y|$

4. $-|z|$

5. $x + y$

6. $x + z$

7. $x - y$

8. $x - z$

9. $|x - y|$

10. $|x - z|$

11. $|x| - y$

12. $|y| - z$

13. $|x| - |y|$

14. $|y| - |z|$

15. $x + y + z$

16. $x - y - z$

17. $x + y - z$

18. $x - y + z$

19. $xy - z$

20. $x - yz$

21. $x(y - z)$

22. $(x - y)z$

23. $x - (y - z)$

24. $x - (y + z)$

25. xy^2

26. $(xy)^2$

27. $x + y^2$

28. $(x + y)^2$

29. $x^2 + y^2$

30. $x^2 + 2xy + y^2$

31. $(x + y + z)^2$

32. $x^2 + y^2 + z^2$

1. _____
2. _____
3. _____
4. _____
5. _____
6. _____
7. _____
8. _____
9. _____
10. _____
11. _____
12. _____
13. _____
14. _____
15. _____
16. _____
17. _____
18. _____
19. _____
20. _____
21. _____
22. _____
23. _____
24. _____
25. _____
26. _____
27. _____
28. _____
29. _____
30. _____
31. _____
32. _____

ANSWERS

33. _____

34. _____

35. _____

36. _____

37. _____

38. _____

39. _____

40. _____

41. _____

42. _____

43. _____

44. _____

45. _____

46. _____

47. _____

48. _____

49. _____

50. _____

51. _____

52. _____

53. _____

54. _____

55. _____

56. _____

57. _____

58. _____

59. _____

60. _____

61. _____

62. _____

63. _____

64. _____

65. _____

66. _____

67. _____

68. _____

69. _____

70. _____

33. xyz^2

34. $(xyz)^2$

35. $x(y - z)^2$

36. $y(x - z)^2$

37. $xy^2 - (xy)^2$

38. $(xy)^2 - (x - y)^2$

39. $(z - 3x)^2$

40. $z - 3x^2$

41. $(5x + y)(3x - y)$

42. $(y - 2z)(y + 3z)$

43. $y^2 - 3y + 2$

44. $2z^2 - 5z - 4$

45. $3x^2 + 4x + 1$

46. $5x^2 - 8x + 5$

47. $x^2 + 3x^2y - 3xy^2 + y^3$

48. $5x^2 - 2xy^2 + zx^2y$

49. $5x + 2[3 + 2(x + 1)]$

50. $3y + 5[2y + 3(y - 20)]$

*In Exercises 51–62, look at each of the following expressions, **as given**, and indicate the number of terms in the expression, the coefficient of each term, and the literal part of each term.*

51. $3x - 4y$

52. $4a - 5b$

53. $3x(-4y)$

54. $4a(-5b)$

55. $3x(z - y)$

56. $4a(c - b)$

57. $4x^2 - 3x + 2$

58. $2a^2 + 7a - 3$

59. $-x^2 + y - 13$

60. $-a^2 - a - 1$

61. $3x(2x) + 4y(5y)$

62. $7a(2a) - 3c(6c)$

CALCULATOR EXERCISES

Evaluate the given expression for $x = .24$ and $y = -.5$. Round off to the nearest thousandth.

63. $8.4x - .03y$

64. $12.7x + 8y$

65. $10x^3 + 20y$

66. $.25x^2 - .43y$

67. $xy^2 - (xy)^2$

68. $(x + y)^2$

69. $\dfrac{10}{x} + \dfrac{8}{y}$

70. $\dfrac{5}{xy}$

QUESTIONS FOR THOUGHT

3. What is the difference between a *term* and a *factor*?

4. Make up an example of an expression that consists of three terms, one of which has one factor, one of which has two factors, and one of which has three factors.

The Distributive Law and Combining Like Terms

2.3

During the course of our earlier discussion of the real number system in Chapter 1, we talked about the special properties that the real numbers exhibit under addition and multiplication. Namely, we can reorder the terms of a sum and the factors of a product by the Commutative Law, and we can regroup those terms or factors by the Associative Law.

A third, and equally important, property which was only briefly mentioned earlier, is one that describes how the operations of multiplication and addition interact. Let's begin by considering the numerical example $3(5 + 2)$. We would normally evaluate this as

$$3(5 + 2) = 3(7) = 21$$

However, we know that multiplication is really just a shorthand way of writing repeated addition. That is,

$$3(5 + 2) \quad \text{means} \quad \text{add } (5 + 2) \text{ 3 times.}$$

We can write this as

$3(5 + 2) = (5 + 2) + (5 + 2) + (5 + 2)$ *Let's regroup and reorder the addition.*

$\quad\quad = (5 + 5 + 5) + (2 + 2 + 2)$ *Now let's rewrite the addition back to multiplication.*

$\quad\quad = 3(5) + 3(2)$

$\quad\quad = 15 + 6$

$\quad\quad = 21$

Thus, at the third step we see that $3(5 + 2) = 3(5) + 3(2)$.

Now you may be asking yourself why anyone would bother doing the computation the second way. The first way was easier and also followed our agreement on the order of operations. Of course you are right! The point of this example was not to suggest another method for doing this particular problem, but rather to illustrate a property of the real numbers. This property is called the **Distributive Law of Multiplication over Addition** (usually just the **Distributive Law** for short). It can be expressed algebraically as indicated in the box.

DISTRIBUTIVE LAW

$$a(b + c) = ab + ac$$

In words, it says that the *factor* outside the parentheses multiplies each *term* inside the parentheses.

Before we proceed to look at several examples, a few comments are in order.

First, in our analysis of how $3(5 + 2)$ became $3 \cdot 5 + 3 \cdot 2$, you will see that it does not make any difference whether we are adding two numbers or twenty numbers in the parentheses. Each term inside still gets multiplied by the factor outside.

Second, since we have defined subtraction in terms of addition, the Distributive Law holds equally well for multiplication over subtraction or over addition *and* subtraction. That is,

$$a(b - c) = a \cdot b - a \cdot c$$

and

$$a(b + c - d) = a \cdot b + a \cdot c - a \cdot d$$

Third, because multiplication is commutative, we can also write the Distributive Law as

$$(b + c)a = b \cdot a + c \cdot a$$

When we use the Distributive Law this way to remove a set of parentheses, we say that we are *multiplying out* the expression.

✔ **LEARNING CHECKS**

1. **(a)** $6(a - 4)$
   (b) $4(u^2 + v - 3)$
   (c) $a^2(a^4 + 5a)$
   (d) $(7 - 3x)x$

EXAMPLE 1   Use the Distributive Law to multiply out each of the following, and simplify as completely as possible.

(a) $5(x + 3)$ **(b)** $4(x^2 - y + 7)$ **(c)** $x(x^3 + 3x)$ **(d)** $(3 + 2a)a$

Solution

(a) $5(x + 3) = \boxed{5} \cdot x + \boxed{5} \cdot 3 = \boxed{5x + 15}$

(b) $4(x^2 - y + 7) = \boxed{4} \cdot x^2 - \boxed{4} \cdot y + \boxed{4} \cdot 7 = \boxed{4x^2 - 4y + 28}$

(c) $x(x^3 + 3x) = \boxed{x} \cdot x^3 + \boxed{x} \cdot 3x = \boxed{x^4 + 3x^2}$

(d) $(3 + 2a)a = 3 \cdot \boxed{a} + 2a \cdot \boxed{a} = \boxed{3a + 2a^2}$ ■

Recall that when we first described the Distributive Law as applied to the example $3(5 + 2) = 3 \cdot 5 + 3 \cdot 2$, we mentioned that normally we would add $5 + 2$ and then multiply by 3 rather than use the Distributive Law. If you look at the example we have just completed, you will notice that we could not work inside the parentheses first. We had no choice here; we had to use the Distributive Law.

2. **(a)** $5(x - y)$
   (b) $5x(-y)$

EXAMPLE 2   Multiply out each expression.

(a) $3(x + y)$ **(b)** $3(xy)$

Solution   It is very important to see that the Distributive Law applies to part **(a)** but *not* to part **(b)**.

(a) $3(x + y) = 3x + 3y = \boxed{3x + 3y}$

(b) $3(xy)$ is not an expression to which we can apply the Distributive Law, because there is no addition or subtraction within the parentheses. Do not make the *mistake* of thinking that $3(xy)$ is the same as $3x \cdot 3y$, which is $9xy$. This is *wrong*.

$$3(xy) = \boxed{3xy} \qquad \text{We simply "erase" the parentheses.}$$ ■

Since we will often be simplifying expressions by removing parentheses, knowing when a set of parentheses makes a difference is obviously important.

Let's agree to call parentheses *essential* if their erasure changes the arithmetic meaning of the problem. In other words, if we mentally erase the parentheses and that changes the meaning of the problem, then those parentheses are essential. If erasing the parentheses does not make any difference, then those parentheses are called *nonessential*.

In Example 2 we have both types of parentheses. In part **(a)**, the parentheses in $3(x + y)$ are essential because *if* we erase them, we get $3x + y$ and the 3 will be multiplying only the x instead of the sum of x and y, which is not the same thing. On the other hand, in part **(b)** if we erase the parentheses in $3(xy)$ we get $3xy$, which is exactly the same thing because multiplication is associative. Thus, the parentheses in part **(b)** are nonessential.

EXAMPLE 3 In each of the following, determine whether each set of parentheses is essential or nonessential.

(a) $3x + (2y + z)$ **(b)** $3x + 2(y + z)$ **(c)** $x(y + z) + x(yz)$

3. (a) $(5x + 4y) + 2z$
    (b) $5(x + 4y) + 2z$
    (c) $r(s - t) + r(st)$

Solution

(a) In $3x + (2y + z)$ the parentheses are nonessential since if we erase them we will still be adding the same three quantities. In other words, since addition is associative, the parentheses make no difference.

(b) In $3x + 2(y + z)$ the parentheses are essential since if we erase them we get $3x + 2y + z$ and then the 2 will be multiplying only the y instead of $y + z$. The parentheses do make a difference.

(c) In $x(y + z) + x(yz)$ the first parentheses are essential, the second are not. If we erase the first parentheses we get $xy + z$ and so x will be multiplying only y instead of $y + z$. If we erase the second parentheses, x will still be multiplying the product yz. ∎

While the vocabulary of essential and nonessential parentheses is convenient, it is not—pardon the pun—essential. However, being able to recognize when parentheses make a difference in an expression is an extremely important skill.

As with any equality, the Distributive Law can be read two ways. If we read it from left to right, we are removing the parentheses and this is called *multiplying out*. If we read from right to left, we are creating parentheses and this is called *factoring*.

$$a(b + c) = \boxed{a}\,(b + c) = a \cdot b + a \cdot c \qquad \textit{Multiplying out}$$

$$a \cdot b + a \cdot c = \boxed{a}\cdot b + \boxed{a}\cdot c = a(b + c) \qquad \textit{Factoring}$$

Notice that when we multiply out and remove parentheses, we are changing a product into a sum. When we factor and create parentheses, we are changing a sum into a product.

Thus far we have been using the Distributive Law to remove parentheses. Now let's turn our attention to the other side of the coin—using the Distributive Law to create parentheses, a process called *factoring*.

Looking at the Distributive Law as saying

$$ab + ac = a(b + c)$$

4. (a) $5x - 5y$

(b) $6a + 6b - 6c$

(c) $7x + 14$

(d) $ab + bc$

we see that both terms on the left-hand side have a *common factor* of a. The Distributive Law says that we can "factor out" the common factor of a by inserting the parentheses.

EXAMPLE 4   Use the Distributive Law to factor each of the following:

(a) $3x + 3y$ **(b)** $4x - 4y + 4z$ **(c)** $2x + 6$ **(d)** $yz + xz$

Solution

(a) Both $3x$ and $3y$ contain a common factor of 3. So we have

$$3x + 3y = \boxed{3}\,x + \boxed{3}\,y = \boxed{3}\,(x + y) = \boxed{3(x + y)}$$

(b) $4x - 4y + 4z = \boxed{4}\,x - \boxed{4}\,y + \boxed{4}\,z$

$$= \boxed{4}\,(x - y + z) = \boxed{4(x - y + z)}$$

(c) In $2x + 6$ it helps to think of 6 as $2 \cdot 3$ so we can "see" the common factor of 2.

$$2x + 6 = \boxed{2}\,x + \boxed{2} \cdot 3 = \boxed{2}\,(x + 3) = \boxed{2(x + 3)}$$

(d) $yz + xz = y \cdot \boxed{z} + x \cdot \boxed{z} = (y + z)\,\boxed{z} = \boxed{(y + x)z}$

Note that in each case we can immediately check our answer by multiplying it out. ∎

While there is much more to say about factoring (which we will do in Chapter 6), we are already in a position to put this idea to use.

Many students can look at the expression $5x + 3x$ and "see" that the answer is $8x$. We can now justify this statement mathematically as follows:

$$5x + 3x = 5x + 3x \qquad \textit{The common factor is x.}$$
$$= (5 + 3)x \qquad \textit{We factor out the common factor of x.}$$
$$= 8 \cdot x$$

We are able to combine $5x$ and $3x$ because they are *like terms*.

DEFINITION   *Like terms* are terms whose variable parts are identical.

Thus, we similarly have

$$4a^2 + 5a^2 - 2a^2 = 4\,\boxed{a^2} + 5\,\boxed{a^2} - 2\,\boxed{a^2}$$

$$= (4 + 5 - 2)\,\boxed{a^2}$$

$$= (7)a^2$$

$$= \boxed{7a^2}$$

These are again like terms (they are all a^2 terms) and so we combine them by simply adding their coefficients. Note we said *add* the coefficients. Recall that we have defined the coefficient of a term to include the sign. Thus, in the expression $4a^2 + 5a^2 - 2a^2$ the coefficients are 4, 5, and -2, and we are adding them.

This point of view makes it easy to state the following rule.

RULE FOR COMBINING LIKE TERMS

To combine like terms simply add their coefficients.

We must take care to combine terms properly.

EXAMPLE 5 Separate the following list into groups of like terms and name the coefficient of each term.

$$3x \quad x^3 \quad -2x \quad -x^2y \quad 5x \quad xy^2 \quad 2y^2x \quad 5x^3$$

Solution As our definition of like terms says, like terms must have variable parts that are *identical*. Do not think that just because all the above terms contain the variable x, they are all like terms. Remember that exponents are just shorthand for repetitive multiplication.

$$x^2 = x \cdot x \text{ while } x^3 = x \cdot x \cdot x \text{ and therefore they are not like terms.}$$

The groups of like terms are:

Group 1. $3x$, $-2x$, $5x$ with coefficients 3, -2, and 5, respectively
Group 2. x^3 and $5x^3$ with coefficients 1 and 5, respectively
            (Remember that the coefficient 1 is usually not written.)
Group 3. $-x^2y$ with coefficient -1
Group 4. xy^2 and $2y^2x$ with coefficients 1 and 2, respectively
            Note: By the Commutative Law, y^2x is the same as xy^2. ∎

Important: From now on we are automatically expected to combine like terms whenever possible. The process of combining like terms is a basic step in simplifying an expression.

EXAMPLE 6 *Simplify as completely as possible.* $3x + 5y - 7x + 2y$

Solution Keep in mind that we are "thinking addition" with the sign as part of the coefficient.

$$
\begin{aligned}
3x + 5y - 7x + 2y &= 3x - 7x + 5y + 2y && \text{\textit{Reorder by the}} \\
 & && \text{\textit{Commutative Law.}} \\
&= 3\,x - 7\,x + 5\,y + 2\,y \\
&= (3 - 7)x + (5 + 2)y && \text{\textit{Combine like terms.}} \\
&= \boxed{-4x + 7y}
\end{aligned}
$$

The answer could also have been written $7y - 4x$. ∎

5. $5a^2$, $4a$, $-a^2b$, $6a$, $3ba^2$, $-2a^2$

6. $6a + 3b + 5a - 8b$

7. $4y^2 - 7y - 8 - 3y^2 - y + 4$

EXAMPLE 7 *Simplify as completely as possible.*

$$5x^2 + 9x - 7 - x^2 + x - 2$$

Solution

$$5x^2 + 9x - 7 - x^2 + x - 2 = (5 - 1)x^2 + (9 + 1)x - 7 - 2$$

$$= 4x^2 + 10x - 9$$

Do not forget that the understood coefficient of x is 1 and of $-x^2$ is -1. ∎

8. (a) $3(a^2 + 5b) + 5(a^2 - 4b)$
   (b) $2x^2(x - 3y) + 4(7 - x^2y)$

EXAMPLE 8 *Multiply and simplify.*

(a) $2(x^2 + 4y) + 3(x^2 - 2y)$ **(b)** $3x(x^3 + 2y) + 5(2 - xy)$

Solution

(a) $2(x^2 + 4y) + 3(x^2 - 2y) = 2x^2 + 8y + 3x^2 - 6y$ *Multiply out by Distributive Law.*

$$= (2 + 3)x^2 + (8 - 6)y$$ *Combine like terms.*

$$= 5x^2 + 2y$$

(b) $3x(x^3 + 2y) + 5(2 - xy) = 3x \cdot x^3 + 3x \cdot 2y + 5 \cdot 2 - 5xy$

By the Distributive Law

$$= 3x^4 + 6xy + 10 - 5xy$$ *Combine like terms.*

$$= 3x^4 + xy + 10$$

You probably noticed that in this example we used the Distributive Law both ways—multiplying out to remove parentheses, and factoring in order to combine like terms. ∎

   In the next section we will look at examples that require several steps before we reach their simplest form.

✔ *Answers to Learning Checks in Section 2.3*

1. (a) $6a - 24$ **(b)** $4u^2 + 4v - 12$ **(c)** $a^6 + 5a^3$ **(d)** $7x - 3x^2$

2. (a) $5x - 5y$ **(b)** $-5xy$

3. (a) Nonessential   **(b)** Essential   **(c)** First is essential; second is nonessential

4. (a) $5(x - y)$ **(b)** $6(a + b - c)$ **(c)** $7(x + 2)$ **(d)** $b(a + c)$

5. Group 1: $5a^2$, $-2a^2$; coefficients 5, -2. Group 2: $4a$, $6a$; coefficients 4, 6.
   Group 3: $-a^2b$, $3ba^2$; coefficients -1, 3

6. $11a - 5b$ **7.** $y^2 - 8y - 4$ **8. (a)** $8a^2 - 5b$ **(b)** $2x^3 - 10x^2y + 28$

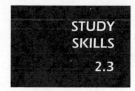

STUDY
SKILLS
2.3

Reviewing Old Material

One of the most difficult aspects of learning algebra is that each skill and concept are dependent on those previously learned. If you have not acquired a certain skill or learned a particular concept well enough, this will, more than likely, affect your ability to learn the next skill or concept.

Thus, even though you have finished a topic that was particularly difficult for you, you should not breathe too big a sigh of relief. Eventually you will have to learn that topic well in order to understand subsequent topics. It is important that you try to master all skills and understand all concepts.

Whether or not you have had difficulty with a topic, you should be constantly reviewing previous material as you continue to learn new subject matter. Reviewing helps to give you a perspective of the material you have covered. It helps you tie the different topics together and makes them *all* more meaningful.

Some statement you read 3 weeks ago, and which may have seemed very abstract then, is suddenly simple and obvious in the light of all you now know.

Since many problems require you to draw on many of the skills you have developed previously, it is important for you to review so that you will not forget or confuse them. You will be surprised to find how much constant reviewing aids in the learning of new material.

When working the exercises, always try to work out some exercises from earlier chapters or sections. Try to include some review exercises at every study session, or at least every other session. Take the time to reread the text material in previous chapters. When you review, think about how the material you are reviewing relates to the topic you are presently learning.

NOTE TO THE STUDENT

Use the space on this page to write down any questions you have or points you want to review with your instructor.

Exercises 2.3

In Exercises 1–6, identify each set of parentheses as essential or nonessential.

1. $2x + 3(y + z)$

2. $5x + 4(y + z)$

3. $2x + (3y + z)$

4. $5x + (4y + z)$

5. $(2x + y) + 2(x + y)$

6. $5(x + y) + (5x + y)$

In Exercises 7–12, separate each list into groups of like terms and name the coefficient and literal part of each term.

7. $x, \quad y, \quad 2x, \quad 3y$

8. $3u, \quad v, \quad 5v, \quad 7u$

9. $2x^2, \quad -3x, \quad 4x^3, \quad -x, \quad -x^2$

10. $5u^3, \quad 4u^2, \quad -u, \quad u^2, \quad u^3$

11. $-x^2y, \quad 2xy^2, \quad x^2y^2, \quad 3xy^2, \quad -2x^2y$

12. $-4s^2t, \quad 2st^2, \quad 3s^2t, \quad -s^2t^2, \quad s^2t^2$

In Exercises 13–38, simplify by combining like terms wherever possible.

13. $2x + 5x$

14. $3a + 7a$

15. $2x^2 + 5x^2$

16. $3a^2 + 7a^2$

17. $3a - 8a + 2a$

18. $4m - 9m + 3m$

19. $-3y + y - 2y$

20. $-2z - 5z + z$

21. $-x - 2x - 3x$

22. $-a - 4a - 2a$

23. $2x - 3y - 7x + 5y$

24. $3s - 4t - 8s + t$

25. $3x + 5y + 2z$

26. $4a - 2b - 5c$

27. $3x^2 + 7x + x^2 + 3x$

28. $5m^2 + 7m + m^2 + 2m$

29. $x^2 - 2x + x^2 - x$

30. $y^2 - 6y + y^2 - y$

31. $5x^2y - 3x^2 + x^2y - x^2$

32. $8s^2t - 4s^2 - s^2 - s^2t$

33. $2x + 5x - 3y + y - 7x$

34. $m + 9m - 3n - n - 10m$

35. $-5s^2 + 3st - s^2 + 6s^2$

36. $-3x^2 + 5xy - xy - x^2 - 2x^2$

37. $3a^2b + ab^2 - ab^2 - 2a^2b - ab^2$

38. $2x^2y + xy^2 - x^2y - x^2y - xy^2$

1. _____
2. _____
3. _____
4. _____
5. _____
6. _____
7. _____
8. _____
9. _____
10. _____
11. _____
12. _____
13. _____
14. _____
15. _____
16. _____
17. _____
18. _____
19. _____
20. _____
21. _____
22. _____
23. _____
24. _____
25. _____
26. _____
27. _____
28. _____
29. _____
30. _____
31. _____
32. _____
33. _____
34. _____
35. _____
36. _____
37. _____
38. _____

In Exercises 39–46, factor the given expression by taking out the common factor.

39. $2x + 10$

40. $3x + 12$

41. $5y - 20$

42. $7x - 14$

43. $9x + 3y - 6$

44. $10m - 2n - 8$

45. $x^2 + xy$

46. $3x^3 + x$

In Exercises 47–72, multiply out and simplify as completely as possible.

47. $3(x + 4)$

48. $5(a + 3)$

49. $5(y - 2)$

50. $7(b - 4)$

51. $-2(x + 7)$

52. $-3(y + 2)$

53. $3(5x + 2)$

54. $5(3x + 4)$

55. $-4(3x + 1)$

56. $-6(2x + 3)$

57. $x(x + 3)$

58. $a(a + 4)$

59. $x(x^2 + 3x)$

60. $a(a^2 + 4a)$

61. $5x(2x - 4)$

62. $2y(5y - 4)$

63. $3(x + y) + 4x - y$

64. $4(a + b) + 4a - b$

65. $3(x + y) + 4(x - y)$

66. $4(a + b) + 4(a - b)$

67. $3(x + y) + 4x(-y)$

68. $4(a + b) + 4a(-b)$

69. $5x(x^2 + 3) + 2x(3 + x^2)$

70. $7y^2(y + 2) + 2y(5 + y^2)$

71. $5x(x^2 + 3) + 2x(3x^2)$

72. $7y^2(y + 2) + 2y(5y^2)$

CALCULATOR EXERCISES

Simplify the following expressions as completely as possible.

73. $8.3x - .2x - 2.6x$

74. $5.6t^2 - 10.1t^2 + .9t^2$

75. $3.56y^3 - .08x^2 - 5.7y^3 + 4.7x^2$

76. $21.6a^2 - 34.7t^5 - 18.62t^5 - 7.85a^2$

QUESTIONS FOR THOUGHT

5. Which property of the real numbers allows us to combine like terms?

6. Discuss what is *wrong* with each of the following (if anything):

   (a) $5 + 3(x - 4) \stackrel{?}{=} 8(x - 4) = 8x - 32$

   (b) $5 + 3(x - 4) \stackrel{?}{=} 5 + 3x - 4 = 3x + 1$

   (c) $5 + 3(x - 4) \stackrel{?}{=} 5 + 3x - 12 \stackrel{?}{=} 8x - 12$

   (d) $5 + 3(x - 4) \stackrel{?}{=} 5 + 3x - 12 \stackrel{?}{=} 3x - 7$

Simplifying Algebraic Expressions

2.4

Up to this point we have been developing a variety of concepts and procedures which we shall now integrate into one unified approach to simplifying algebraic expressions. Let's begin by looking at a few examples which offer additional practice and will help us appraise the type of examples we can currently handle.

LEARNING CHECKS

EXAMPLE 1 *Multiply and simplify.* $2(3x - 4) + 5(2x - 1)$

Solution   We remove each set of parentheses by using the Distributive Law.

$$2(3x - 4) + 5(2x - 1) = 2 \cdot 3x - 2 \cdot 4 + 5 \cdot 2x + 5(-1)$$
$$= 6x - 8 + 10x - 5$$
$$= (6 + 10)x - 8 - 5 \qquad \textit{Combine like terms.}$$

$$= \boxed{16x - 13} \qquad \blacksquare$$

1. $3(2a - 5) + 4(3a - 2)$

EXAMPLE 2 *Multiply and simplify.* $3(4a - 7) + (2a - 3) + 5(a + 2)$

Solution   We note that the first and third sets of parentheses are essential while the second is not. We remove the first and third sets by the Distributive Law.

$$3(4a - 7) + (2a - 3) + 5(a + 2) = 12a - 21 + 2a - 3 + 5a + 10$$
$$= (12 + 2 + 5)a - 21 - 3 + 10$$

$$\textit{Combine like terms.}$$

$$= \boxed{19a - 14} \qquad \blacksquare$$

2. $5(6u - 5) + 2(u + 3) + (4 - 7u)$

EXAMPLE 3 *Multiply and simplify.* $2x^2(x - 3) + x(3x^2 - 5x)$

Solution   We multiply out using the Distributive Law.

$$2x^2(x - 3) + x(3x^2 - 5x) = 2x^2 \cdot x - 2x^2 \cdot 3 + x \cdot 3x^2 - x \cdot 5x$$
$$= 2x^3 - 6x^2 + 3x^3 - 5x^2$$
$$= (2 + 3)x^3 + (-6 - 5)x^2 \qquad \textit{Combine like terms.}$$
$$= \boxed{5x^3 - 11x^2} \qquad \blacksquare$$

3. $7z^2(z^3 - 2) + 4z(z^4 + 3z)$

EXAMPLE 4 *Multiply and simplify.* $2x^2y(x + 3y) + 4y(x^3 + 2x^2y)$

Solution   While our basic outline remains the same, we do have to exercise more care as the examples become more complex.

$$2x^2y(x + 3y) + 4y(x^3 + 2x^2y) = 2x^2y \cdot x + 2x^2y \cdot 3y + 4y \cdot x^3 + 4y \cdot 2x^2y$$
$$= 2x^3y + 6x^2y^2 + 4yx^3 + 8x^2y^2$$

Notice that $2x^3y$ and $4yx^3$ are like terms. The order of the factors does not matter.

$$= \boxed{6x^3y + 14x^2y^2} \qquad \blacksquare$$

4. $4ab^3(3a + b) + 6ab(ab^2 - b^3)$

In order to make it easier to recognize like terms, it helps to have the variables in the same order in all the terms. From now on, as we obtain our products when we multiply out, we will use the same order for the variables throughout a particular example. In this text we will normally use alphabetical order.

Next let us examine how our understanding of a minus sign, our definition of a numerical coefficient, and the Distributive Law work together.

If we use the Distributive Law on the next two expressions and write out *all* the steps, we get the following:

$$-3(x + 4) = -3x + (-3)(4) = -3x + (-12) = \boxed{-3x - 12}$$

$$-3(x - 4) = -3x - (-3)(4) = -3x - (-12) = \boxed{-3x + 12}$$

Since any subtraction can be converted to addition, it is much easier to do these examples by performing the addition mentally as we do with integers. We simply multiply the coefficients according to the Distributive Law. In other words, we would do the above examples as follows:

$$-3(x + 4) = -3x - 12 \qquad \textit{Keep in mind that the addition is understood.}$$
$$\textit{-3x - 12 means -3x + (-12).}$$
$$-3(x - 4) = -3x + 12 \qquad \textit{-3 times -4 equals +12.}$$

This way of looking at these examples becomes even more useful as the examples become more complex.

5. $7y - 2(y - 8)$

EXAMPLE 5   *Simplify.* $2x - 5(x - 2)$

Solution   We think of -5 as the coefficient of $(x - 2)$, and so we distribute -5 into the parentheses.

$$2x - 5(x - 2) = 2x - 5x + 10 \qquad \textit{-5 times x and -5 times -2}$$
$$= \boxed{-3x + 10} \qquad \textit{After combining like terms} \qquad \blacksquare$$

6. $3(4u + v) - (9u + 4v)$

EXAMPLE 6   *Simplify as completely as possible.* $2(3x - y) - (10x - 3y)$

Solution   In using the Dsitributive Law, we multiply out each set of parentheses by the coefficient that precedes it. The coefficient of the first set of parentheses is 2. What is the coefficient of the second set? Anytime we have a negative sign we can think of it as -1. For example, $-5 = -1 \cdot 5$. In applying the Distributive Law it is helpful to have some number to use as the coefficient of the parentheses. Thus, the answer to the question is that the coefficient of the second set of parentheses is -1.

$$2(3x - y) - (10x - 3y) = 2(3x - y) - \boxed{1}(10x - 3y)$$

We distribute the -1.

$$= 6x - 2y - 10x + 3y$$
$$= \boxed{-4x + y}$$

Note that this answer can also be written as $y - 4x$. Since we are "thinking" addition, the terms in an expression can always be rearranged as long as the coefficient of each term is unchanged. ∎

EXAMPLE 7 *Simplify as completely as possible.* $8 - 3[x - 4(x - 3)]$

7. $5 - 2[a - 3(a - 4)]$

Solution Remember that both square brackets and parentheses are types of grouping symbols. Whenever one grouping symbol occurs within another it is usually easiest and most efficient to work from the innermost grouping symbol outward. We begin by removing the parentheses around $x - 3$ by distributing the -4.

$8 - 3[x - 4(x - 3)] = 8 - 3[x - 4x + 12]$ *Watch the order of operations. We do not perform the subtraction $8 - 3 = 5$. Remember that multiplication precedes addition and subtraction. Next combine like terms within the brackets.*

$= 8 - 3[-3x + 12]$ *Now remove the brackets by distributing the -3.*
$= 8 + 9x - 36$
$= \boxed{9x - 28}$

Note that in the second step we could have distributed the -3 to remove the bracket, but we chose to combine like terms within the bracket first. This is usually the better procedure to follow. As soon as like terms present themselves combine them. ∎

Having looked at these examples let's pause to consolidate our ideas into a procedural outline.

PROCEDURE FOR SIMPLIFYING ALGEBRAIC EXPRESSIONS

1. Remove any essential grouping symbols by using the Distributive Law. Multiply each term inside the grouping symbol by its coefficient. Remember that the coefficient includes the sign that precedes it.
2. If there are grouping symbols within grouping symbols, always work from the innermost one outward.
3. When multiplying terms, keep the same order of the variables to make it easier to recognize like terms. The best method is to write terms with their numerical coefficients first, followed by the variables in alphabetical order.
4. In each step of the solution look for any like terms that can be combined.

In each of the remaining examples in this section, simplify the given expression as completely as possible.

8. $6a^2b(a - 2)$
$- (4a^2b - ab^2) - 2b(ab + 3a^3)$

EXAMPLE 8 $4xy(x - 3) - (xy^2 - 12xy) - 3y(xy + x^2)$

Solution Following the outline given in the box, we proceed as follows:

$$4xy(x - 3) \quad - (xy^2 - 12xy) \quad - 3y(xy + x^2)$$

Use the Distributive Law to remove each set of parentheses.
The first parentheses get multiplied by 4xy.
The second parentheses get multiplied by −1.
The third parentheses get multiplied by −3y.

$$= 4x^2y - 12xy - xy^2 + 12xy - 3xy^2 - 3x^2y$$

Note that since the very first term came out with x appearing before y, we kept that same order throughout. Now we are ready to combine like terms.

$$= \boxed{x^2y - 4xy^2} \qquad \blacksquare$$

9. $3x^2(y^2 - 8xy) - 2x(4x)(-3xy)$

EXAMPLE 9 $4x(3x^2 - y) + (x^3 - 4xy) + 2x(3x)(-4y)$

Solution First of all, we should recognize that the second set of parentheses in this example is nonessential, because its coefficient is 1. Second, it is important to see that the last term in the example does *not* call for the Distributive Law. Why not?

$$4x(3x^2 - y) + (x^3 - 4xy) + 2x(3x)(-4y) = 12x^3 - 4xy + x^3 - 4xy - 24x^2y$$

Combine like terms.

$$= \boxed{13x^3 - 8xy - 24x^2y} \qquad \blacksquare$$

10. $5 - [4 - 3(2 - a)]$

EXAMPLE 10 $4 - [4 - 4(4 - a)]$

Solution We begin by working on the innermost parentheses first by distributing the -4.

$$\begin{aligned} 4 - [4 - 4(4 - a)] &= 4 - [4 - 16 + 4a] \qquad \textit{Combine like terms within [ ].} \\ &= 4 - [-12 + 4a] \qquad \textit{Remove [ ] by distributing −1.} \\ &= 4 + 12 - 4a \\ &= \boxed{16 - 4a} \qquad \blacksquare \end{aligned}$$

11. $a\{a + 2[a - 3(a - 4)]\}$

EXAMPLE 11 $x\{2x^2 + x[x - 3(x - 1)]\}$

Solution We begin by distributing the -3 to remove parentheses.

$$\begin{aligned} x\{2x^2 + x[x - 3(x - 1)]\} &= x\{2x^2 + x[x - 3x + 3]\} \qquad \textit{Combine like terms within [ ].} \\ &= x\{2x^2 + x[-2x + 3]\} \qquad \textit{Remove [ ] by distributing x.} \\ &= x\{2x^2 - 2x^2 + 3x\} \qquad \textit{Combine like terms within \{ \}.} \\ &= x\{3x\} \\ &= \boxed{3x^2} \qquad \blacksquare \end{aligned}$$

STUDY
SKILLS
2.4

Reflecting

When you have finished reading or doing examples, it is always a good idea to take a few minutes to think about what you have just covered. Think about how the examples relate to the verbal material, and how the material just covered relates to what you have learned previously. How are the examples and concepts you have just covered similar to or different from those you have already learned?

✔ *Answers to Learning Checks in Section 2.4*

1. $18a - 23$ **2.** $25u - 15$ **3.** $11z^5 - 2z^2$ **4.** $18a^2b^3 - 2ab^4$

5. $5y + 16$ **6.** $3u - v$ **7.** $4a - 19$ **8.** $-16a^2b - ab^2$ **9.** $3x^2y^2$

10. $7 - 3a$ **11.** $-3a^2 + 24a$

NOTE TO THE STUDENT

Use the space on this page to write down any questions you have or points you want to review with your instructor.

Exercises 2.4

In Exercises 1–60, simplify each of the expressions as completely as possible.

1. $4x + y + 4(x + y)$

2. $3r + s + 3(r + s)$

3. $5(m + 2n) + 3(m - n)$

4. $5(a - 2b) + 4(2a + b)$

5. $-2(x - 3y) + 5(y - x)$

6. $-4(2r - 3s) + 6(s - r)$

7. $5 + 3(x - 2)$

8. $7 + 2(x - 5)$

9. $5 - 3(x - 2)$

10. $7 - 2(x - 5)$

11. $(5 - 3)(x - 2)$

12. $(7 - 2)(x - 5)$

13. $8 - (3x - 4)$

14. $10 - (6x - 5)$

15. $5y - (1 - 2y)$

16. $7a - (2 - 3a)$

17. $5(x - 3y) - x - 3y$

18. $7(a - 5b) - (a - 5b)$

19. $5(x - 3y) - (x - 3y)$

20. $7(a - 5b) - a - 5b$

21. $5(x - 3y) - x(-3y)$

22. $7(a - 5b) - a(-5b)$

23. $5x(-3y) - x(-3y)$

24. $7a(-5b) - a(-5b)$

25. $5x(-3y)(-x)(-3y)$

26. $7a(-5b)(-a)(-5b)$

27. $2x^2(x - 2) + x(3x^2 - 4x)$

28. $5y(y^2 - 3) + y^2(y - 3)$

29. $3a(4a - 1) - a(4 - a)$

30. $6z(z - 2) - z(4z - 12)$

ANSWERS

1. _____
2. _____
3. _____
4. _____
5. _____
6. _____
7. _____
8. _____
9. _____
10. _____
11. _____
12. _____
13. _____
14. _____
15. _____
16. _____
17. _____
18. _____
19. _____
20. _____
21. _____
22. _____
23. _____
24. _____
25. _____
26. _____
27. _____
28. _____
29. _____
30. _____

31. _____

32. _____

33. _____

34. _____

35. _____

36. _____

37. _____

38. _____

39. _____

40. _____

41. _____

42. _____

43. _____

44. _____

45. _____

46. _____

47. _____

48. _____

49. _____

50. _____

51. _____

52. _____

53. _____

54. _____

55. _____

56. _____

57. _____

58. _____

59. _____

60. _____

31. $4(x^2 + 7x) - (x^2 + 7x)$

32. $9(m^3 - 4m) - (m^3 - 4m)$

33. $3a(a^2 + 3b) + 4b^2(a^2 - b)$

34. $7z(z^2 - 4y) + 3z^2(z - 2)$

35. $3x^2 - 7x + 4 - 8x^2 - 3 - x$

36. $4y^3 - y^2 + 7 - 2y - y^2 + y$

37. $x^2y(xy - x) - 5xy(x^2y - x^2)$

38. $2a^2b(ab - a) - 3ab(a^2b - a^2)$

39. $4u^2v(u - v) - (uv^3 + u^2v^2)$

40. $6st^2(s - t) + (s^2t^2 + 3s^3t)$

41. $4(x + 3y) + (4x + 3y) + 4x(3y)$

42. $5(a + 2b) + (5a + 2b) + 5a(2b)$

43. $6(m - 2n) + (6m - 2n) + 6m(-2n)$

44. $3(u - 4v) + (3u - 4v) + 3u(-4v)$

45. $3t^5(t^4 - 4) - (t^5 + t^4) - 2t^3(3t)(-t^5)$

46. $u^3(u^3 - 5) - (u^6 + u^3) - 3u^2(u^3)(-u)$

47. $-3(-x + 2) + (8 - 5x) - (2 - 2x)$

48. $-5(-t + 3) - (3t - 6) + (9 - 2t)$

49. $a - 2[a - 2(a - 2)]$

50. $x - 3[x - 3(x - 3)]$

51. $x\{x - 4[x - (x - 4)]\}$

52. $m^2\{2m - 5[m - (m - 5)]\}$

53. $3(x + 2) + 4[x - 3(2 - x)]$

54. $5(y - 3) + 2[y - 5(3 - y)]$

55. $4(y - 3) - 2[3y - 5(y - 1)]$

56. $7(a - 1) - 6[2a - 4(a - 3)]$

CALCULATOR EXERCISES

57. $8.64x - (2.3y - 4.7x)$

58. $4.6a^2 - 2.1(3a^2 - 5.2)$

59. $3.4(4.8x^2 - 5.6xy) - 3x(2.9x - 9.1y)$

60. $27.351t^3 - t^2(7.83t - 6)$

QUESTIONS FOR THOUGHT

7. In words describe the error (or errors) in each of the following "solutions," and explain how to correct the errors.

(a) $3 + 2[x + 4(x + 3)]$
$= 3 + 2[x + 4x + 12]$
$= 3 + 2[5x + 12]$
$= 5[5x + 12]$
$= 25x + 60$

(b) $3x + 5[x + (x + 3)]$
$= 3x + 5[x^2 + 3x]$
$= 3x + 5x^2 + 15x$
$= 5x^2 + 18x$

8. Discuss what is *wrong* (if anything) with each of the following examples:

(a) $5 - 3(x - 4) \overset{?}{=} 2(x - 4) \overset{?}{=} 2x - 8$

(b) $5 - 3(x - 4) \overset{?}{=} 5 - 3x - 4 \overset{?}{=} 1 - 3x$

(c) $5 - 3(x - 4) \overset{?}{=} 5 - 3x - 12 \overset{?}{=} -7 - 3x$

(d) $5 - 3(x - 4) \overset{?}{=} 5 - 3x + 12 \overset{?}{=} 17 - 3x$

Translating Phrases and Sentences Algebraically

2.5

Even though it is still quite early in our study of algebra, we have already developed *algebraic* skills adequate enough for us to begin solving "real-life problems." This we will do in the next chapter. However, in order to be able to apply these skills to such problems, we need to be able to formulate and translate a problem stated in words into its equivalent algebraic form.

In this section we will focus our attention on translating phrases and sentences into their algebraic form, and leave the formulation and solution of entire problems to the next chapter. Let us begin by considering some common phrases and how they are translated from English into algebra.

EXAMPLE 1   In each of the following, let x represent the *number*, and write an algebraic expression which translates the given phrase.

(a) 5 more than a number   (b) 5 less than a number

(c) 5 times a number   (d) the product of 5 and a number

Solution   Translating a phrase algebraically does not mean simply reading the phrase from left to right and replacing each word with an equivalent algebraic expression. We must translate the *meaning* of the phrase as well as the words. Keep in mind that we are letting x represent the number.

(a) 5 more than a number is translated as

$$\boxed{x + 5}$$

We recognize that the phrase "5 more than" conveys addition. We are starting with something (a number, which in this example we are calling x) and adding 5 to it. This becomes $x + 5$. While $5 + x$ is also correct, $x + 5$ is usually considered preferable. The next part of this example will clarify why.

(b) 5 less than a number is translated as

$$\boxed{x - 5}$$

We recognize that the phrase "5 less than" conveys subtraction. We are starting with something (again, a number which we are calling x) and subtracting 5 *from it*. This becomes $x - 5$, *not* $5 - x$, which is what we would get if we ignored the *meaning* of the statement and simply translated from left to right.

Remember: $5 - x$ is not the same as $x - 5$. For example, if we translate "5 less than 8" correctly we should get $8 - 5$, not $5 - 8$.

(c) 5 times a number is translated as

$$\boxed{5x}$$

The word "times" conveys multiplication.

(d) The product of 5 and a number is also translated as

$$\boxed{5x}$$

The word "product" also conveys multiplication.   ■

✔ **LEARNING CHECKS**

1. (a) 8 more than a number

   (b) 8 less than a number

   (c) 8 times a number

   (d) the product of 8 and a number

Many students look for *key words* to help them quickly determine which algebraic symbols to use in translating an English phrase or sentence. For instance, in Example 1 we recognized the key words "more than" to indicate addition, "less than" to indicate subtraction, and "product" to indicate multiplication.

However, the key words do not usually indicate *how* the symbols should be put together to yield an accurate translation of the words. In order to be able to put the symbols together in a meaningful way we must understand how the key words are being used in the context of the given verbal statement.

Let's look at two verbal expressions which use the same key words and yet do not have the same meaning:

1. Three times the sum of five and four
2. The sum of three times five and four

Both expressions contain the following key words:

"three"	the number 3	(3)
"times"	meaning multiply	($\cdot$)
"sum"	meaning addition	($+$)
"five"	the number 5	(5)
"four"	the number 4	(4)

How should the symbols be put together?

Expression 1 says: Three times . . . .

Three times what? . . . The sum.

Thus, we must *first* find the sum of 5 and 4, before multiplying by 3.

Expression 2 says: The sum of . . . .

The sum of what *and* what? . . . The sum of 3 times 5 *and* 4.

Thus, we must *first* compute 3 times 5 before we can determine what the sum is.

Translating the two verbal expressions algebraically, we get:

Expression 1 is $3(5 + 4) = 3 \cdot 9 = 27$

Expression 2 is $3 \cdot 5 + 4 = 15 + 4 = 19$

Thus, in comparing expressions 1 and 2, we can see that the simple change of word positions in a phrase can change the meaning quite a bit.

2. (a) 6 more than 4 times a number

(b) 4 less than 7 times a number

(c) The product of two numbers is 10 less than their sum.

EXAMPLE 2   Translate each of the following algebraically:

(a) 8 more than 3 times a number

(b) 2 less than 5 times a number is 18.

(c) The sum of two numbers is 4 less than their product.

Solution

(a) "8 more than" means we are going to *add* 8 to something. To what? To "3 times a number." If we represent the number by n (*you* are free to choose any letter you like), we get

8 more than   3 times a number

$$3n + 8$$

Thus, the answer is

$$\boxed{3n + 8}$$

Of course, $8 + 3n$ is also correct.

(b) "2 less than" means we are going to *subtract* 2 from something. From what? From "5 times a number."

The word "is" translates to " $=$." If we represent the number by s we get

$$\underbrace{\text{2 less than}} \quad \underbrace{\text{5 times a number}} \quad \underbrace{\text{is}} \quad \underbrace{\text{18}}$$
$$\qquad \qquad \qquad 5s - 2 \qquad = \quad 18$$

Thus, the final answer is

$$\boxed{5s - 2 = 18}$$

It is very important to note here that we must subtract 2 *from* $5s$ and therefore $2 - 5s = 18$ is *not* a correct translation.

(c) The sentence mentions two numbers so let's call them x and y.

$$\underbrace{\text{The sum of two numbers}} \quad \underbrace{\text{is}} \quad \underbrace{\text{4 less than}} \quad \underbrace{\text{their product}}$$
$$\qquad \quad x + y \qquad \qquad = \qquad \qquad \qquad x \cdot y - 4$$

Thus, our translation is

$$\boxed{x + y = xy - 4} \qquad \qquad \blacksquare$$

EXAMPLE 3 Translate each of the following algebraically:

(a) The sum of two consecutive integers

(b) The sum of two consecutive even integers

(c) The sum of two consecutive odd integers

Solution This example illustrates how important it is to state clearly what the variable you are using represents. Let's begin by looking at some numerical examples for guidance.

Examples of two consecutive integers	*Examples of two consecutive even integers*	*Examples of two consecutive odd integers*
2 and 3	4 and 6	7 and 9
19 and 20	10 and 12	15 and 17
25 and 26	28 and 30	31 and 33
x and $x + 1$	x and $x + 2$	x and $x + 2$

(a) The case of two consecutive integers is fairly straightforward. If we let x represent the first integer, then the next consecutive integer is $x + 1$. Therefore, the sum of two consecutive integers can be represented as

$$x + (x + 1) \quad \text{or} \quad \boxed{2x + 1}$$

3. (a) The sum of three consecutive integers

(b) The sum of three consecutive odd integers

(c) The sum of three consecutive even integers

Based upon the numerical examples given above, we notice that the two cases of consecutive even or odd integers are basically the same. In order to get from one even integer to the next we have to add 2, and to get from one odd integer to the next we have to add 2. It is not the adding of 2 which makes the numbers even or odd, but rather whether the *first* number is even or odd. If x is even, then so is $x + 2$. If x is odd, then so is $x + 2$.

(b) If we let x represent the first *even* integer, then the sum of two consecutive even integers can be represented as

$$x + (x + 2) \quad \text{or} \quad \boxed{2x + 2}$$

(c) If we let x represent the first *odd* integer, then the sum of two consecutive odd integers can be represented as

$$x + (x + 2) \quad \text{or} \quad \boxed{2x + 2}$$

It all depends on how we designate x.   ∎

The next example again illustrates the need to understand clearly the meaning of the words in an example in order to be able to put them together properly.

While it is certainly necessary to recognize that particular words and phrases imply specific arithmetic operations, and to understand what the problem means taken as a whole, there is still one more important factor necessary to be able to successfully translate from verbal expressions to algebraic ones. This additional factor is the ability to take your basic general knowledge and common sense, and apply it to a particular problem.

Let's look at how these various components blend together in formulating the solution to a problem.

4. Suppose a floppy disk costs $2 and you buy two packs of disks. The first pack contains x disks and the second pack contains four more disks than the first pack.

(a) How many disks are in the first pack?

(b) How much does the first pack cost?

(c) How many disks are in the second pack?

(d) How much does the second pack cost?

(e) How much did the two packs cost all together?

EXAMPLE 4   Suppose one roll of film costs $3 and you buy two multiroll packs of film. The first pack contains a certain number of rolls of film (let's say n rolls), and the second pack contains five less than the number of rolls in the first pack.

(a) How many rolls of film are there in the first pack?

(b) How much does the first pack cost?

(c) How many rolls are there in the second pack?

(d) How much does the second pack cost?

(e) What is the total cost of the two packs of film?

Solution

(a) The example tells us that the number of rolls of film in the first pack is to be represented by

$$\boxed{n}$$

(b) How much do n rolls of film cost? Let's think numerically a moment.

If n were equal to 4 (that is, if there were 4 rolls of film in the first pack), then the cost of the 4 rolls would be $4 \cdot \$3 = \$12$.

If n were equal to 7, the cost would be $7 \cdot \$3 = \$21$.

If n were equal to 11, the cost would be $11 \cdot \$3 = \$33$.

Whatever the number of rolls in the pack, we multiply by 3 to get the cost of the entire pack. Therefore, the cost of the first pack of film is

$$n \cdot \$3 = \boxed{3n \text{ dollars}}$$

(c) In part (a) we followed the suggestion in the example and let n represent the number of rolls in the first pack; how then do we represent the number of rolls in the second pack?

   We are told that the number of rolls in the second pack is "five less than the number in the first pack." But we have already represented the number of rolls in the first pack as n. So the number in the second pack is

$$5 \text{ less than } n$$

We know this translates to

Thus, the answer to part (c) is

$$\boxed{n - 5}$$

(d) To represent the cost of the second pack of film we reason as we did in part (b). Since the number of rolls in the second pack is $n - 5$, the second pack costs

$$(n - 5) \cdot \$3 = \boxed{3(n - 5) \text{ dollars}}$$

(e) To get the total cost of the two packs of film we add the results obtained in parts (b) and (d):

$$\boxed{3n + 3(n - 5) \text{ dollars}}$$

Simplifying, we get

$$3n + 3n - 15 \text{ dollars} \quad \text{or} \quad \boxed{6n - 15 \quad \text{dollars}}$$

   Notice that in order to successfully answer the various parts of this question we must be able to distinguish clearly between the *number* of rolls of film in each pack, and the *cost* of the film in each pack. These are two distinct ideas and this is where our own basic knowledge came into play.

   We have been quite detailed in our discussion of this example in order to emphasize that you must understand what the problem is asking as well as what it is saying. ∎

Answers to Learning Checks in Section 2.5

1. (a) $x + 8$ (b) $x - 8$ (c) $8x$ (d) $8x$

2. (a) $4n + 6$ (b) $7s - 4$ (c) $m \cdot n = m + n - 10$

3. (a) $3x + 3$ (b) $3x + 6$ (c) $3x + 6$

4. (a) n (b) $2n$ dollars   (c) $n + 4$ (d) $2(n + 4) = 2n + 8$ dollars

   (e) $2n + 2n + 8 = 4n + 8$ dollars

STUDY SKILLS 2.5

Checking Your Work

We develop confidence in what we do by knowing that we are right. One way to check to see if we are right is to look at the answers usually provided in the back of the book. However, few algebra texts provide *all* the answers. And of course, answers are not provided during exams, when we need the confidence most.

Isn't it frustrating to find out that you incorrectly worked a problem on an exam, and then to discover that you would easily have seen your error had you just taken the time to check over your work? Therefore, you should know how to check your answers.

The method of checking your work should be different from the method used in the solution. In this way you are more likely to discover any errors you might have made. If you simply rework the problem the same way, you cannot be sure you did not make the same mistake twice.

Ideally, the checking method should be quicker than the method for solving the problem (although this is not always possible).

Learn how to check your answers, and practice checking your homework exercises as you do them.

Exercises 2.5

*In Exercises 1–18, let n represent the **number** and translate each phrase or sentence algebraically.*

1. Four more than a number

2. Four times a number

3. Four less than a number

4. A number increased by four

5. A number decreased by four

6. The product of four and a number

7. Six more than five times a number

8. Six less than five times a number

9. Nine less than twice a number

10. Five more than twice a number

11. The product of a number and seven more than the number

12. The product of a number and seven less than the number

13. The product of two more than a number and six less than the number

14. The product of three less than a number and four more than the number

15. Eight less than twice a number is fourteen.

16. One less than three times a number is seven.

17. Four more than five times a number is two less than the number.

18. Ten less than a number is three more than six times the number.

In Exercises 19–22, let r and s represent the two numbers and translate the sentence algebraically.

19. The sum of two numbers is equal to their product.

20. Five more than the sum of two numbers is equal to their product.

21. Twice the sum of two numbers is three less than their product.

22. Five times the product of two numbers is eight more than their sum.

In Exercises 23–32, use any letter you choose to translate the given phrase or sentence algebraically. Be sure to identify clearly what your variable represents.

23. The sum of two consecutive integers

24. The product of two consecutive integers

25. The sum of two consecutive even integers

26. The sum of two consecutive odd integers

27. The product of three consecutive odd integers

28. The product of four consecutive even integers

29. Eight times an even integer is four less than seven times the next even integer.

30. Five less than twice an odd integer is four more than the next odd integer.

31. The sum of the squares of three consecutive integers is five.

32. The cube of the sum of two consecutive odd integers is sixty-four.

ANSWERS

1. _____
2. _____
3. _____
4. _____
5. _____
6. _____
7. _____
8. _____
9. _____
10. _____
11. _____
12. _____
13. _____
14. _____
15. _____
16. _____
17. _____
18. _____
19. _____
20. _____
21. _____
22. _____
23. _____
24. _____
25. _____
26. _____
27. _____
28. _____
29. _____
30. _____
31. _____
32. _____

ANSWERS

33. a. _____

    b. _____

    c. _____

    d. _____

    e. _____

    f. _____

    g. _____

    h. _____

34. a. _____

    b. _____

    c. _____

    d. _____

    e. _____

    f. _____

    g. _____

    h. _____

35. a. _____

    b. _____

    c. _____

    d. _____

    e. _____

    f. _____

    g. _____

36. a. _____

    b. _____

    c. _____

    d. _____

    e. _____

    f. _____

    g. _____

37. a. _____

    b. _____

    c. _____

    d. _____

    e. _____

    f. _____

    g. _____

33. A box contains 8 nickels, 12 dimes, and 9 quarters.

    (a) How many nickels are in the box?

    (b) What is the *value* of the nickels in the box?

    (c) How many dimes are in the box?

    (d) What is the value of the dimes in the box?

    (e) How many quarters are in the box?

    (f) What is the value of the quarters in the box?

    (g) All together, how many coins are there in the box?

    (h) What is the total value of all the coins in the box?

34. Repeat Exercise 33 for n nickels, d dimes, and q quarters.

35. Electronic card reader A can read 200 cards per minute and reads cards for 15 minutes, while electronic card reader B can read 160 cards per minute and reads cards for 20 minutes.

    (a) How many cards per minute does A read?

    (b) How many minutes does A read cards?

    (c) How many cards does A read?

    (d) How many cards per minute does B read?

    (e) How many minutes does B read cards?

    (f) How many cards does B read?

    (g) How many cards do A and B read all together?

36. Repeat Exercise 35 if card reader A reads x cards per minute for 25 minutes, and card reader B reads 250 cards per minute for t minutes.

37. Ruth walks at the rate of 100 meters per minute and jogs at the rate of 220 meters per minute. She walks for 25 minutes and then jogs for 35 minutes.

    (a) How fast does she walk?

    (b) For how long does she walk?

    (c) How far does she walk?

    (d) How fast does she jog?

    (e) For how long does she jog?

    (f) How far does she jog?

    (g) How much distance has Ruth covered all together?

38. Repeat Exercise 37 if Ruth walks at the rate of r meters per minute, jogs at twice that rate, walks for t minutes, and jogs for 20 minutes longer than she walks.

CHAPTER 2 SUMMARY

After having completed this chapter you should be able to:

1. Understand and evaluate algebraic expressions involving exponents (Section 2.1).

   For example:

   (a) *Write without exponents.* $-x^2y^3 = \boxed{-xxyyy}$

   (b) *Write with exponents.* $2xxx + 3xxxx = \boxed{2x^3 + 3x^4}$

   (c) *Evaluate.*

   $$-6^2 = -6 \cdot 6 = \boxed{-36}$$

   $$(-6)^2 = (-6)(-6) = \boxed{36}$$

2. Substitute numerical values into algebraic expressions and evaluate the results (Section 2.2).

   For example:

   Evaluate. $xy^2 - (x - y)^3$ for $x = 2$ and $y = -1$

   $$\begin{aligned}
   xy^2 - (x - y)^3 &= (2)(-1)^2 - [2 - (-1)]^3 \\
   &= (2)(1) - [3]^3 \\
   &= 2 - 27 \\
   &= \boxed{-25}
   \end{aligned}$$

3. Distinguish between *terms* and *factors* (Section 2.2).

   For example:

   $x + y + z$ consists of three terms, each term consisting of one factor.

   xyz is one term consisting of three factors.

4. Use the Distributive Law to *multiply out*, *factor*, and *combine like terms* (Section 2.3).

   For example:

   (a) *Multiply out.* $3x(x^2 - 5x) = \boxed{3x^3 - 15x^2}$

   (b) *Factor.* $14x + 28y - 7 = \boxed{7(2x + 4y - 1)}$ *Factor out the common factor of 7.*

   (c) *Combine like terms.* $3x^2 - 5x - x^2 + 2x = \boxed{2x^2 - 3x}$

5. Simplify algebraic expressions by removing grouping symbols and combining like terms (Section 2.4).

   For example:

   $3x^2(x - 4) - 2x(x^2 + 3) - (x - 12x^2)$

   Remove each set of parentheses by distributing its coefficient.

   $= 3x^3 - 12x^2 - 2x^3 - 6x - x + 12x^2$ *Now combine like terms.*

   $= \boxed{x^3 - 7x}$

6. Translate phrases and sentences from their English language form into their algebraic form (Section 2.5).

For example:

Four less than three times a number is 26.

If we let *n* represent the number, then the translation becomes:

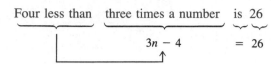

Thus, the answer is

$$3n - 4 = 26$$

1. In $3 \cdot 2^4$, the exponent of 4 applies only to the 2, not to the 3; in $(3 \cdot 2)^4$, the exponent of 4 applies to both the 2 and the 3. Put another way, we compute $3 \cdot 2^4$ by first raising 2 to the fourth power and then multiplying the result by 3. We compute $(3 \cdot 2)^4$ by first multiplying 2 by 3 and then raising the result to the fourth power.

2. **(a)** The exponent of 2 applies only to the 2, not to the 3.

   (b) Only x should be raised to the fourth power.

   (c) 5^4 is not the same as $5 \cdot 4$.

   (d) When we square 3, we get 9, not 6. Addition is performed in the *exponents*.

   (e) -3^4 is the opposite of 3^4, so is equal to -81.

   (f) We add exponents when we multiply powers of x, not when we add powers of x.

   (g) When multiplying powers of x, we add exponents. So $x^2x^7 = x^{2+7} = x^9$.

   (h) Here, we do not add exponents, since $(x^2)^4 = x^2x^2x^2x^2 = x^{2+2+2+2} = x^8$.

3. A term is an algebraic expression which is connected by multiplication (and/or division). If a term is formed by multiplying two or more expressions, each is called a factor of that term.

4. $x + xy + xyz$ is one of many possible answers.

5. The Distributive Law allows us to combine like terms.

6. **(a)** Our order of operations requires that multiplication be done before addition. Therefore, we should never add 5 and 3.

   (b) The Distributive Law requires that we multiply each term of $x - 4$ by 3.

   (c) We cannot add 5 and $3x$ to get $8x$, since these are not like terms.

   (d) This is correct as written.

7. **(a)** The solution is correct up to $3 + 2[5x + 12]$. At this point, it is wrong to add $3 + 2 = 5$. Instead, use the Distributive Law:

$$3 + 2[5x + 12] = 3 + 10x + 24$$
$$= 10x + 3 + 24$$
$$= 10x + 27$$

(b) Within the brackets, we must add x and $x + 3$, not multiply them. As a result, we get

$$3x + 5[x + (x + 3)] = 3x + 5[2x + 3]$$
$$= 3x + 10x + 15$$
$$= 13x + 15$$

8. (a) Do not subtract $5 - 3$; distribute -3 instead.

(b) When -3 is distributed, it must multiply -4 as well as x.

(c) When we multiply -3 by -4, we get 12, not -12.

(d) This is done correctly.

NOTE TO THE STUDENT

Use the space on this page to write down any questions you have or points you want to review with your instructor.

ANSWERS
1. _____
2. _____
3. _____
4. _____
5. _____
6. _____
7. _____
8. _____
9. _____
10. _____
11. _____
12. _____
13. _____
14. _____
15. _____
16. _____
17. _____
18. _____
19. _____
20. _____
21. _____
22. _____
23. _____
24. _____
25. _____
26. _____
27. _____
28. _____
29. _____
30. _____
31. _____
32. _____
33. _____
34. _____
35. _____
36. _____
37. _____
38. _____
39. _____
40. _____

CHAPTER 2 REVIEW EXERCISES

In Exercises 1–6, write the given expression without exponents.

1. xy^3 **2.** $(xy)^3$

3. $-x^4$ **4.** $(-x)^4$

5. $3x^2$ **6.** $(3x)^2$

In Exercises 7–10, write the given expression with exponents.

7. $xxyyy$ **8.** $xx + xxx$

9. $aa - bbb$ **10.** $-aa(bbb)$

In Exercises 11–20, evaluate the given expression for $x = -3$, $y = 4$, and $z = -2$.

11. $-z^4$ **12.** $(-z)^4$

13. xy^2 **14.** $(xy)^2$

15. $xyz - (x + y + z)$ **16.** $xyz - x + y + z$

17. $|xy| + z - |z|$ **18.** $|x + y + z| + |x| - |y| - |z|$

19. $2x^2 - (x + y)^2$ **20.** $3z^3 - (x - z)^2$

In Exercises 21–44, simplify each expression as completely as possible.

21. x^3x^4x **22.** r^5rr^3

23. $a^7a^2 + a^3a^6$ **24.** $2y^4y^2 + 3y^3y^3$

25. $4x^3x^2 + 3x^2x^4$ **26.** $5a^7a + 2a^4a^3$

27. $3x^2 - 7x + 7 - 5x^2 - x - 3$ **28.** $4m^3 - 8 - m^2 - 3m^2 - m^3 - 1$

29. $2a^2b(3ab^4)$ **30.** $-5ab(2a^2b^3)$

31. $2a^2b(3a + b^4)$ **32.** $-5ab(2a^2 + b^3)$

33. $3x(2x + 4) + 5(x^2 - 3)$ **34.** $2z(3z - 5) + 4(z^2 + 1)$

35. $4y^2(y - 2) - y(y^2 - 5y)$ **36.** $c^3(2c - 3) - c^2(c^2 - 8c)$

37. $3xy(x^2 - 2y) + 4xy^2(y - x)$ **38.** $2r^2s(s - rs) - 5s(r^2s + r^2s^2)$

39. $3(2x - 4y) - (x + 2y) - (x - 2y)$ **40.** $4(3a - 2b) - (a - 2b) - (a + 2b)$

41. _____

42. _____

43. _____

44. _____

45. _____

46. _____

47. _____

48. _____

49. _____

50. _____

51. _____

52. _____

53. a. _____

b. _____

c. _____

d. _____

e. _____

f. _____

g. _____

54. a. _____

b. _____

c. _____

d. _____

e. _____

f. _____

g. _____

41. $3x^4(x^3 - 2y^2) - 4x(x^2)(x^4)$

42. $6y^5(x^3 - 5x^4) + 2x^2(3xy^2)(5xy^3)$

43. $3 - [x - 3 - (x - 3)]$

44. $3 - x[3 - x(3 - x)]$

In Exercises 45–48, use the Distributive Law to factor the given expression.

45. $5x^2 + 10$

46. $2a^5 + 16$

47. $3y - 6z + 9$

48. $22x - 33y + 11$

In Exercises 49–52, translate the given statement algebraically. Let n represent the number.

49. The sum of a number and seven is four less than three times the number.

50. Five less than twice a number is four more than three times the number.

51. The sum of two consecutive odd integers is five less than the smaller one.

52. We have three consecutive integers such that the sum of the first integer and four times the second integer is eight more than three times the third integer.

53. John earns a commission of $2 for each newspaper subscription he sells, and $5 for each magazine subscription he sells. He sells 12 newspaper subscriptions and 9 magazine subscriptions.

(a) How many newspaper subscriptions does he sell?

(b) How much does he earn for each newspaper subscription?

(c) How many magazine subscriptions does he sell?

(d) How much does he earn for each magazine subscription?

(e) How much does he earn for the newspaper subscriptions?

(f) How much does he earn for the magazine subscriptions?

(g) How much does he earn all together?

54. Repeat Exercise 53 if John sells *n* newspaper subscriptions and four less than twice that many magazine subscriptions.

CHAPTER 2 PRACTICE TEST

ANSWERS

1. _____

2. _____

3. _____

4. _____

5. _____

6. _____

7. _____

8. _____

9. _____

10. _____

11. _____

12. _____

13. _____

14. a. _____

     b. _____

15. a. _____

     b. _____

     c. _____

     d. _____

     e. _____

     f. _____

     g. _____

In Problems 1–4, evaluate each of the given expressions.

1. -3^4

2. $(-3)^4$

3. $(-3 - 4 + 6)^5$

4. $7 - 3(2 - 6)^2$

In Problems 5–8, evaluate the given expression for $x = -2$, $y = 3$, and $z = -4$.

5. $x - y - z$

6. $xy - z$

7. $x^3 - z^2$

8. $2yz^2 - |x|$

In Problems 9–13, perform the indicated operations and simplify as completely as possible.

9. $4x^2y - 5xy + y^2 - 3xy - 2y^2 - x^2y$

10. $-3xy^2(-4x^2y)(-2x^3)$

11. $2x(x^2 - y) - 3(x - xy) - (2x^3 - 3x)$

12. $3x^3(x^2 - 4xy) - 2x(xy)(-6x^2)$

13. $4 - [x - 4(x - 4)]$

14. If we let n stand for the *number*, translate each of the following phrases:

   (a) Four more than twice a number

   (b) Twenty less than five times a number is equal to the number.

15. A blank 60-minute audio cassette costs $2, and a blank 90-minute audio cassette costs $3. Someone buys a certain number of 60-minute cassettes (let's say x) and five less than twice that many 90-minute cassettes.

   (a) How much does a 90-minute cassette cost?

   (b) How many 60-minute cassettes were bought?

   (c) How much does a 60-minute cassette cost?

   (d) How many 90-minute cassettes were bought?

   (e) How much was spent on the 90-minute cassettes?

   (f) How much was spent on the 60-minute cassettes?

   (g) How much was spent all together on the cassettes?

NOTE TO THE STUDENT

Use the space on this page to write down any questions you have or points you want to review with your instructor.

First-Degree Equations and Inequalities

3

- 3.1 Types of Equations and Basic Properties of Equalities
- 3.2 Solving First-Degree Equations in One Variable
- 3.3 Verbal Problems
- 3.4 Types of Inequalities and Basic Properties of Inequalities
- 3.5 Solving First-Degree Inequalities in One Variable

STUDY SKILLS

3.1 Preparing for Exams: Is Doing Homework Enough?

3.2 Preparing for Exams: When to Study

3.3 Preparing for Exams: Study Activities

3.4 Preparing for Exams: Making Study Cards

3.5 Preparing for Exams: Using Study Cards

3.6 Preparing for Exams: Reviewing Your Notes and Text; Reflecting

3.7 Preparing for Exams: Using Quiz Cards

In Chapter 2 we mentioned that variables are used in algebra primarily in two ways: **(a)** To describe a general relationship between numbers and/or arithmetic operations, as for example in the statement of the Distributive Law; and **(b)** as a place holder where the variable represents a number which is as yet not identified, but which needs to be found. It is this second use of variables that we are going to discuss in this chapter.

Types of Equations and Basic Properties of Equalities

3.1

When we read the mathematical sentences $5 + 2 = 7$ and $4 + 7 = 12$, we see clearly that the first is true and the second is false. The first is true because the left-hand side of the equation, $5 + 2$, represents the same number as the right-hand side of the equation, 7. The second is false because the two sides of the equation do not represent the same number.

On the other hand, the sentence $x + 3 = 5$ is neither true nor false. Since the letter x represents some number, we cannot tell whether the sentence is true until we know the value of x. Such an equation, whose truth or falsehood depends on the value of the variable, is called a ***conditional equation***.

The value or values of the variable which make a conditional equation true are called the ***solutions*** of the equation. They are the values which, when substituted for the variable into the equation, make both sides of the equation equal. Any such value is said to ***satisfy*** the equation.

In the equation mentioned above, $x + 3 = 5$, we can see that $x = 2$ is a solution to the equation because $2 + 3 = 5$ is true, and so $x = 2$ satisfies the equation. Another number such as $x = 8$ does not satisfy the equation because $8 + 3 \neq 5$.

Not every equation which involves a variable is necessarily conditional. For example, the equation

$$5(x + 2) - 4 = 5x + 6$$

is not conditional. If we simplify the left-hand side of this equation we get

$$5(x + 2) - 4 = 5x + 6$$
$$5x + 10 - 4 = 5x + 6$$
$$5x + 6 = 5x + 6$$

which is *always true*, no matter what the value of the variable is. Such an equation, which is always true regardless of the value of the variable, is called an ***identity***.

On the other hand, the equation

$$2(x + 3) - 2x = 5$$

is also not conditional. If we simplify the left-hand side of this equation we get

$$2(x + 3) - 2x = 5$$
$$2x + 6 - 2x = 5$$
$$6 = 5$$

which is *always false*, no matter what the value of the variable is. Such an equation, which is always false regardless of the value of the variable, is called a ***contradiction***.

Often, if we simplify both sides of an equation as completely as possible, we can recognize it as an identity or a contradiction. (Here is where we begin to use the skills we developed in the last chapter.)

EXAMPLE 1 Determine the type of each of the following equations:

(a) $4x - 3(x - 2) = x + 6$

(b) $5y(y - 4) - y(5y - 20) = 3(2y + 1) - (6y - 7)$

Solution Looking at each equation in the form that it is given, it is not clear what type of equation we have. We will simplify both sides of each equation as completely as possible according to the methods we learned in Chapter 2. Hopefully, we will then be able to recognize what type of equation we have.

(a) $4x - 3(x - 2) = x + 6$ *Remove the parentheses by distributing the -3.*

    $4x - 3x + 6 = x + 6$ *Combine like terms.*

         $x + 6 = x + 6$

Since $x + 6 = x + 6$ is always true, we see that the original equation is an identity.

(b) $5y(y - 4) - y(5y - 20) = 3(2y + 1) - (6y - 7)$

Remove the parentheses by applying the Distributive Law.

Remember: $-(6y - 7) = -1(6y - 7) = -6y + 7$

    $5y^2 - 20y - 5y^2 + 20y = 6y + 3 - 6y + 7$

                $0 = 10$

Since $0 = 10$ is always false we see that the original equation is a contradiction. ∎

As you might expect, the most interesting situation is that of a conditional equation. Since a conditional equation is neither always true nor always false, we would like to be able to determine exactly which values satisfy the equation. The process of finding the values that satisfy an equation is called *solving the equation*.

Once we find the values that we think are the solutions to an equation we would, of course, like to be able to check our answers. This consists of substituting the proposed value into the original equation and verifying that it satisfies the equation.

Several examples will illustrate this idea.

EXAMPLE 2 In each of the following determine which of the listed values satisfies the given equation.

(a) $2x + 11 = 2 - x$; $x = 4, -3$

(b) $5 - 2(a - 4) = 3a + 4(a - 9) + 13$; $a = 0, 4$

(c) $2z^2 - 12 = z^2 - z$; $z = 3, -4, 4$

Solution We replace each occurrence of the variable with the value we are checking.

(a) CHECK $x = 4$: $2x + 11 = 2 - x$

              $2(4) + 11 \stackrel{?}{=} 2 - (4)$

                  $8 + 11 \stackrel{?}{=} -2$

                    $19 \neq -2$ *Therefore $x = 4$ does not satisfy the equation.*

✔ **LEARNING CHECKS**

1. (a) $3(x - 5) - (5 + 3x) = 8$

   (b) $4x(x + 1) - 3(x^2 - 2)$
       $= x(x + 4) + 6$

2. (a) $3a + 10 = 6 - a$;
     $a = 2, -1$

   (b) $5 + 2(t - 3) = 9 - (t + 1)$;
     $t = 3, -2$

   (c) $u^2 - 5u = 2u^2 + u$;
     $u = 0, -6$

CHECK $x = -3$:

$$2x + 11 = 2 - x$$
$$2(-3) + 11 \stackrel{?}{=} 2 - (-3)$$
$$-6 + 11 \stackrel{?}{=} 2 + 3$$
$$5 \stackrel{\checkmark}{=} 5 \qquad \text{Therefore } x = -3 \text{ } does \text{ satisfy the equation.}$$

(b) CHECK $a = 0$:

$$5 - 2(a - 4) = 3a + 4(a - 9) + 13$$
$$5 - 2(0 - 4) \stackrel{?}{=} 3(0) + 4(0 - 9) + 13$$
$$5 - 2(-4) \stackrel{?}{=} 0 + 4(-9) + 13$$
$$5 + 8 \stackrel{?}{=} -36 + 13$$
$$13 \ne -23 \qquad \text{Therefore } a = 0 \text{ does } not \text{ satisfy the equation.}$$

CHECK $a = 4$:

$$5 - 2(a - 4) = 3a + 4(a - 9) + 13$$
$$5 - 2(4 - 4) \stackrel{?}{=} 3(4) + 4(4 - 9) + 13$$
$$5 - 2(0) \stackrel{?}{=} 12 + 4(-5) + 13$$
$$5 - 0 \stackrel{?}{=} 12 - 20 + 13$$
$$5 \stackrel{\checkmark}{=} 5 \qquad \text{Therefore } a = 4 \text{ } does \text{ satisfy the equation.}$$

(c) CHECK $z = 3$:

$$2z^2 - 12 = z^2 - z$$
$$2(3)^2 - 12 \stackrel{?}{=} (3)^2 - (3) \qquad \textit{Evaluate the powers first.}$$
$$2(9) - 12 \stackrel{?}{=} 9 - 3$$
$$18 - 12 \stackrel{?}{=} 9 - 3$$
$$6 \stackrel{\checkmark}{=} 6 \qquad \text{Therefore } z = 3 \text{ } is \text{ a solution to the equation.}$$

CHECK $z = -4$:

$$2z^2 - 12 = z^2 - z$$
$$2(-4)^2 - 12 \stackrel{?}{=} (-4)^2 - (-4)$$
$$2(16) - 12 \stackrel{?}{=} 16 + 4$$
$$32 - 12 \stackrel{?}{=} 20$$
$$20 \stackrel{\checkmark}{=} 20 \qquad \text{Therefore } z = -4 \text{ } is \text{ a solution to the equation.}$$

CHECK $z = 4$:

$$2z^2 - 12 = z^2 - z$$
$$2(4)^2 - 12 \stackrel{?}{=} (4)^2 - (4)$$
$$2(16) - 12 \stackrel{?}{=} 16 - 4$$
$$32 - 12 \stackrel{?}{=} 12$$
$$20 \ne 12 \qquad \text{Therefore } z = 4 \text{ does } not \text{ satisfy the equation.}$$

Note that each of the equations in the example is conditional. It is neither always true nor always false. ∎

While we have just seen how to check whether a particular number is a solution to a specific equation, we have not yet discussed how to actually *find* the solution(s) to conditional equations. In order to develop a systematic method for solving certain kinds of equations, we need to start by discussing the properties of equalities.

If we think of the "=" sign as indicating that the two sides of the equation are "perfectly balanced," then we will not upset this balance by changing each side of the equation in exactly the same way. If we start with an equation of the form

$$a = b$$

and we

add the same quantity to both sides of the equation,

or subtract the same quantity from both sides of the equation,

or multiply both sides of the equation by the same quantity,

or divide both sides of the equation by the same (nonzero) quantity,

then the two sides of the equation remain equal. Algebraically we can write this as shown in the accompanying box.

PROPERTIES OF EQUALITY

1. If $a = b$ then $a + c = b + c$

   Addition Property of Equality

2. If $a = b$ then $a - c = b - c$

   Subtraction Property of Equality

3. If $a = b$ then $a \cdot c = b \cdot c$

   Multiplication Property of Equality

4. If $a = b$ then $\dfrac{a}{c} = \dfrac{b}{c}; \quad c \neq 0$

   Division Property of Equality

When we write $a = b$ we mean that in any expression where a and b occur they are interchangeable. This idea is known as the **substitution principle**.

Looking carefully at property 1 of equality and applying the substitution principle, we can see that what property 1 actually says is

If $a = b$ then $a + c = b + c$ *Now substitute a for b.*

giving us $a + c = a + c$

Sometimes this idea is expressed as "If equals are added to equals the results are equal."

Consider the following four equations:

$$3x - (x - 1) = x + 3$$
$$2x + 1 = x + 3$$
$$x + 1 = 3$$
$$x = 2$$

We can easily check that $x = 2$ is a solution to each of them (verify this). As we shall discover in the next section, $x = 2$ is the only solution to these four equations. Equations which have exactly the same set of solutions are called *equivalent equations*. Of the four equations, the solution to the last one was the most obvious. An equation of this kind (that is, $x = 2$) is often called an *obvious equation*.

In the next section we will develop a systematic method for using the properties of equality to take a certain kind of equation and transform it into an obvious equation.

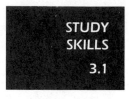

STUDY SKILLS 3.1

Preparing for Exams

Is Doing Homework Enough?

At some point in time you will probably take an algebra exam. More than likely, your time will be limited and you will not be allowed to refer to any books or notes during the test.

Working problems at home may help you to develop your skills and to better understand the material, but there is still no guarantee that you can demonstrate the same high level of performance during a test as you may be showing on your homework. There are several reasons for this:

1. Unlike homework, exams must be completed during a limited time.
2. The fact that you are being assigned a grade may make you anxious and therefore more prone to careless errors.
3. Homework usually covers a limited amount of material while exams usually cover much more material. This increases the chance of confusing or forgetting skills, concepts, and rules.
4. Your books and notes *are* available as a guide while working homework exercises.

Even if you do not deliberately go through your textbook or notes while working on homework exercises, the fact that you know what sections the exercises are from, and what your last lecture was about, cues you in on how the exercises are to be solved. You may not realize how much you depend on these cues and may be at a loss when an exam does not provide them for you.

If you believe that you understand the material and/or you do well on your homework, but your exam grades just do not seem to be as high as you think they should be, then the study skills discussed in this chapter should be helpful.

✔ *Answers to Learning Checks in Section 3.1*

1. (a) Contradiction   **(b)** Identity   **2. (a)** $a = -1$ **(b)** $t = 3$
(c) $u = 0, -6$

Exercises 3.1

In Exercises 1–16, determine whether the given equation is an identity or a contradiction.

1. $2(x - 3) = 2x - 6$

2. $3(x + 4) = 3x + 12$

3. $5(x + 2) = 5x + 2$

4. $4x - 1 = 4(x - 1)$

5. $2a + 4 + 3a = 6a + 4 - a$

6. $3a - 5 + a = 6a + 5 - 2a$

7. $2z^2 + 3z - 2z^2 - z = z + z + 1$

8. $2z + z - 5 = 3z^2 + z - 3z^2 + 2z$

9. $5u - 4(u - 1) - u = u - 4 - (u - 2)$

10. $7u - 5(u - 1) - 2u = 2u - 1 - (2u - 3)$

11. $7 - 3(y - 2) = y + 4(5 - y) - 7$

12. $8 - 2(y - 4) = y + 3(4 - y) + 4$

13. $w(w - 2) - w^2 + 2w = 3(w + 1) - (3w - 1)$

14. $2w(w + 1) - (2w^2 + 2w) = 4(w - 3) - 4w + 12$

15. $2(x^2 - 3) - x(2x - 1) + x = 2 - x - (x + 8) + 4x$

16. $6(x^2 - x) + 2x(3 - 3x) + 1 = 4 + x - (x + 4) - 1$

1. _____

2. _____

3. _____

4. _____

5. _____

6. _____

7. _____

8. _____

9. _____

10. _____

11. _____

12. _____

13. _____

14. _____

15. _____

16. _____

ANSWERS

17. _____

18. _____

19. _____

20. _____

21. _____

22. _____

23. _____

24. _____

25. _____

26. _____

27. _____

28. _____

29. _____

30. _____

31. _____

32. _____

33. _____

34. _____

In Exercises 17–34, determine whether the given equation is satisfied by the values listed following it.

17. $x + 5 = -2$; $x = -3, -7$ **18.** $x - 5 = 2$; $x = -3, 3$

19. $2 - a = 3$; $a = -5, 5, 1$ **20.** $4 - a = -2$; $a = -6, 2, 6$

21. $3y + 7 = y - 3$; $y = 2, -5$ **22.** $5y - 2 = y - 10$; $y = 2, -2$

23. $6 - 2w = 10 - 3w$; $w = -4, 1$ **24.** $11 - 5w = -1 - w$; $w = 3, -3$

25. $4(x - 7) - (x + 1) = 15 - x$; $x = 0, 7$

26. $3(x - 6) - (x - 2) = 10 - x$; $x = 0, 6$

27. $3z + 2(z - 1) = 4(z + 2) - (z + 5)$; $z = -2, 1$

28. $5z + 4(z - 2) = 2(z + 6) - (z + 4)$; $z = -3, 2$

29. $x^2 - 3x = 2x - 6$; $x = -2, 2$ **30.** $x^2 - 5x = 15 - 3x$; $x = 0, -3$

31. $a^2 - 4a = 4 - a$; $a = -1, 4$ **32.** $a^2 - 5a = 10 - 2a$; $a = -2, 5$

33. $y(y + 6) = (y + 2)^2$; $y = -2, 2$ **34.** $y(y + 5) = (y + 2)^2$; $y = -4, 4$

QUESTIONS FOR THOUGHT

1. What does it mean when we say that a value satisfies or is a solution to an equation?

2. What is the difference between a conditional equation, an identity, and a contradiction?

3. Summarize the properties of equality in one sentence.

4. When are two equations equivalent?

Solving First-Degree Equations in One Variable

3.2

When we are faced with the task of solving an equation, perhaps the most important thing to do first is to ask, "What kind of an equation am I trying to solve?" As we shall see time and again, the method of solution will very much depend on the kind of equation we are trying to solve.

For the time being, we are going to restrict ourselves to solving what we refer to as *first-degree equations in one variable*, that is, equations which involve only one variable and in which that variable appears to the first power only. For example, the equation $3x - 5 = x + 3$ is a first-degree equation, while $3x^2 = x + 5$ is called a *second-degree equation*. (The degree of an equation is not always obvious and we will discuss this in a bit more detail later in this section.)

Given an equation to solve, we can apply the four properties of equality that were listed in the last section to transform the given equation into one for which the solution may be more obvious. Since we are interested in the solution to a given equation, we want to make sure that each time we apply one of the properties of equality we obtain an equivalent equation—that is, an equation with the same solution set.

Applying the four properties of equality appropriately does yield an equivalent equation with one important exception. Property 3 says "If $a = b$ then $ac = bc$." While this *is* always true, if $c = 0$ we do not necessarily obtain an equivalent equation. For example, the equation $x = 3$ is a conditional equation whose only solution is 3. If we multiply both sides of this equation by 0, we get $0 \cdot x = 0 \cdot 3$ or $0 = 0$, which is an identity, and so all numbers are solutions. Consequently, if we want to be sure that we will always obtain an equivalent equation, we must restrict property 3 to have $c \neq 0$.

We can summarize this as shown in the box.

An equation can be transformed into an equivalent equation by adding or subtracting the same quantity to both sides of the equation, and by multiplying or dividing both sides of the equation by the same *nonzero* quantity.

Recalling that the simplest equation is one of the form $x =$ "a number" (which we called the *obvious* equation), we will solve a first-degree equation in one variable by applying the properties of equality to transform the original equation into progressively simpler *equivalent* equations until we eventually end up with an obvious equation. Let's illustrate the method with several examples.

EXAMPLE 1   *Solve for x.* $x - 5 = 2$

Solution   Since the obvious equation is always of the form $x =$ "number," our goal in solving an equation is to "isolate x" (or whatever the variable happens to be) on one side of the equation. We proceed as follows:

$$
\begin{array}{rl}
x - 5 = & 2 \\
\underline{+5 \quad +5} & \\
x = & 7
\end{array}
$$

We add 5 to both sides of the equation using the Addition Property of Equality.

 **LEARNING CHECKS**

1. $t - 7 = 1$

CHECK $x = 7$: $x - 5 = 2$

$(7) - 5 \overset{\checkmark}{=} 2$ ∎

2. $t + 8 = 5$

EXAMPLE 2 *Solve for x.* $x + 7 = 2$

Solution Before we proceed to the solution a comment is in order. Solving equations involves not only using the properties of equality, but also using the properties in a way which leads to the solution. For instance, if we want to we can write

$$
\begin{array}{rr}
x + 7 = & 2 \\
+3 & +3 \\
\hline
x + 10 = & 5
\end{array}
$$ *By the Addition Property of Equality*

The equation we have obtained, $x + 10 = 5$, is equivalent to the given equation. Adding 3 to both sides of the equation is mathematically correct, but it is *not* particularly useful, since we are no closer to a solution.

Learning to solve equations means learning how to use the properties of equality in a way which leads to a solution. The proper way to solve the equation is

$$
\begin{array}{rr}
x + 7 = & 2 \\
-7 & -7 \\
\hline
\boxed{x = -5}
\end{array}
$$ *We can think of this as the Subtraction Property (subtract 7 from both sides) or as the Addition Property (add −7 to both sides).*

CHECK $x = -5$: $x + 7 = 2$

$-5 + 7 \overset{\checkmark}{=} 2$ ∎

3. $5t = 20$

EXAMPLE 3 *Solve for x.* $3x = 24$

Solution In order to isolate x we want to eliminate the 3 which is multiplying the x, so we *divide both* sides of the equation by 3.

$$3x = 24$$

$$\frac{3x}{3} = \frac{24}{3}$$ *By the Division Property of Equality*

$$\frac{\cancel{3}x}{\cancel{3}} = \frac{24}{3}$$

$$\boxed{x = 8}$$

CHECK $x = 8$: $3x = 24$

$3(8) \overset{\checkmark}{=} 24$ ∎

Let's analyze the three examples given above to see if we can determine why a particular property was needed in each example in order to achieve our goal of isolating x (the variable) and obtaining the obvious equation.

If we examine	what is being done to x,	then we can determine how to isolate x.

In $x - 5 = 2$ 5 is *subtracted* from x,   so *add* to both sides in order to isolate x.

In $x + 7 = 2$ 7 is *added* to x,   so *subtract* 7 from both sides in order to isolate x.

In $3x = 24$ 3 is *multiplying* x,   so *divide* both sides by 3 in order to isolate x.

Note the pattern: We use the *inverse* operation to isolate x.

What if the equation involves more than one operation?

EXAMPLE 4   *Solve for a.* $2a - 7 = -9$.

4. $3t - 1 = -7$

Solution   In order to isolate a we must deal with the 2 multiplying a and the 7 being subtracted. There is nothing wrong with using the Division Property first to divide both sides of the equation by 2. However, keep in mind that we must divide each *entire* side of the equation by 2. Specifically, the entire left-hand side, $2a - 7$, must be divided by 2, *not* just the $2a$. Hence, the equation would look like this:

$$2a - 7 = -9$$
$$\frac{2a - 7}{2} = \frac{-9}{2} \quad \textit{Dividing both entire sides of the equation by 2}$$

This equation is equivalent to the given equation but, because it involves fractions, is messy to solve (even if done properly). Whenever we use the Division Property we run the risk of introducing fractions into the equation. Therefore, as a general procedure, it is best to use the Addition and Subtraction Properties first whenever possible.

We proceed as follows:

$$2a - 7 = -9$$
$$\underline{+7 \quad\quad +7} \quad \textit{Add 7 to both sides (Addition Property).}$$
$$2a = -2$$

$$\frac{2a}{2} = \frac{-2}{2} \quad \textit{Divide both sides by 2 (Division Property).}$$

$$\frac{\cancel{2}a}{\cancel{2}} = \frac{-2}{2}$$

$$\boxed{a = -1}$$

CHECK $a = -1$: $2a - 7 = -9$
$$2(-1) - 7 \overset{?}{=} -9$$
$$-2 - 7 \overset{\checkmark}{=} -9 \qquad\qquad \blacksquare$$

Sometimes we have to simplify the equation before we proceed to solve it.

5. $2(12 - t) + 8 = 14$

EXAMPLE 5 *Solve for y.* $3(12 - y) - 16 = 2$

Solution Before we actually isolate the variable we want to make sure that each side of the equation is simplified as completely as possible. In this example, the left-hand side is not.

$3(12 - y) - 16 = 2$	*Multiply out using the Distributive Law.*
$36 - 3y - 16 = 2$	*Combine like terms.*
$20 - 3y = 2$	*Now we can solve this equation.*
$\underline{-20 \qquad\qquad -20}$	*Subtract 20 from both sides.*
$-3y = -18$	*In order to isolate y, we divide both sides by -3.*
$\dfrac{-3y}{-3} = \dfrac{-18}{-3}$	
$\boxed{y = 6}$	

CHECK $y = 6$:
$$3(12 - y) - 16 = 2$$
$$3(12 - 6) - 16 \overset{?}{=} 2$$
$$3(6) - 16 \overset{?}{=} 2$$
$$18 - 16 \overset{\checkmark}{=} 2 \qquad\qquad\blacksquare$$

Sometimes the variable appears on both sides of the equation.

6. $11 - t = -1 - 4t$

EXAMPLE 6 *Solve for x.* $17 - 6x = 2 - x$

[We want to isolate x on one side of the equation. As illustrated in the two solutions that follow, which side we decide to isolate x on does not matter.]

Solution 1

$17 - 6x = 2 - x$	*We decide to isolate x on the right-hand side.*
$\underline{+6x \qquad\qquad +6x}$	*We add 6x to both sides, which eliminates x from the left-hand side.*
$17 = 2 + 5x$	
$\underline{-2 \quad -2}$	
$15 = 5x$	
$\dfrac{15}{5} = \dfrac{5x}{5}$	
$\boxed{3 = x}$	

Solution 2

$17 - 6x = 2 - x$	*We decide to isolate x on the left-hand side.*
$\underline{+x \qquad\qquad +x}$	*We add x to both sides, which eliminates x from the right-hand side.*
$17 - 5x = 2$	
$\underline{-17 \qquad\qquad -17}$	
$-5x = -15$	
$\dfrac{-5x}{-5} = \dfrac{-15}{-5}$	
$\boxed{x = 3}$	Thus, both approaches yield the same solution.

CHECK $x = 3$: $17 - 6x = 2 - x$

$$17 - 6(3) \stackrel{?}{=} 2 - (3)$$

$$17 - 18 \stackrel{?}{=} -1$$

$$-1 \stackrel{\checkmark}{=} -1 \qquad\blacksquare$$

Before proceeding to several more examples, let's pause to organize the approach we have used in the last few examples into a basic strategy for solving first-degree equations.

STRATEGY FOR SOLVING FIRST-DEGREE EQUATIONS

1. Remove any grouping symbols by using the Distributive Law.
2. Simplify both sides of the equation as completely as possible.
3. By using the Addition and/or Subtraction Properties of Equality, isolate the variable term on one side of the equation and the numerical term on the other side.
4. By using the Division Property of Equality, make the coefficient of the variable 1, and in so doing obtain an obvious equation.
5. Check the solution in the original equation.

EXAMPLE 7 *Solve for t.* $2t - 5(t - 2) = 20 - (2t + 6)$

7. $5x - 4(3 - x) = 16 - (x - 2)$

Solution Following the outline given in the box, we first remove the parentheses in the equation by using the Distributive Law.

$2t - 5(t - 2) = 20 - (2t + 6)$	*Remove parentheses.*
$2t - 5t + 10 = 20 - 2t - 6$	*Combine like terms on both sides.*
$-3t + 10 = 14 - 2t$	
$\underline{+2t \qquad\qquad +2t}$	*Eliminate t from the right-hand side.*
$-t + 10 = \quad 14$	
$\underline{\quad -10 \quad -10}$	
$-t = 4$	*We are not finished yet. We want t alone, so*
$\dfrac{-t}{-1} = \dfrac{4}{-1}$	*we divide both sides of the equation by the coefficient of t, which is -1.*
$\boxed{t = -4}$	

CHECK $t = -4$:

$$2t - 5(t - 2) = 20 - (2t + 6)$$

$$2(-4) - 5(-4 - 2) \stackrel{?}{=} 20 - (2(-4) + 6)$$

$$-8 - 5(-6) \stackrel{?}{=} 20 - (-8 + 6)$$

$$-8 + 30 \stackrel{?}{=} 20 - (-2)$$

$$22 \stackrel{\checkmark}{=} 20 + 2 \qquad\blacksquare$$

As we mentioned earlier, the technique we are using here of isolating x applies to first-degree equations—that is, to equations involving the variable to the first power only. The fact that an equation is of the first degree is not always obvious, as we see in the next example.

8. $t^2 - 3t = t(t - 4) - 6$

EXAMPLE 8 *Solve for x.* $x(x + 5) - 2(x - 1) = x^2 + 2$

Solution This equation may not appear to be a first-degree equation (because of the x^2 term), but in fact we shall see that it is. This example also highlights why it is important to simplify equations first.

$$
\begin{array}{rll}
x(x + 5) - 2(x - 1) = & x^2 + 2 & \textit{Remove parentheses.} \\
x^2 + 5x - 2x + 2 = & x^2 + 2 & \textit{Combine like terms.} \\
x^2 + 3x + 2 = & x^2 + 2 & \\
\underline{-x^2 \qquad\qquad -x^2} & & \\
3x + 2 = & 2 & \textit{Now we see that, in fact, we do have a} \\
\underline{-2 \quad -2} & & \textit{first-degree equation.} \\
3x = & 0 & \textit{Do not treat } 0 \textit{ any differently.} \\
\dfrac{3x}{3} = & \dfrac{0}{3} & \\
\boxed{x = 0} & &
\end{array}
$$

CHECK $x = 0$:
$$
\begin{array}{rcl}
x(x + 5) - 2(x - 1) &=& x^2 + 2 \\
0(0 + 5) - 2(0 - 1) &\overset{?}{=}& 0^2 + 2 \\
0(5) - 2(-1) &\overset{?}{=}& 0 + 2 \\
0 + 2 &\overset{\checkmark}{=}& 2
\end{array}
$$ ∎

In order to determine the degree of an equation, isolate all the variable terms on one side of the equation; the highest exponent of the variable is the degree of the equation. If no variable appears at all, then we have either an *identity* or a *contradiction*. The next two examples illustrate these last two possibilities.

9. $4(t - 1) - 2(t + 1) = 2t - 6$

EXAMPLE 9 *Solve for a.* $2(a + 1) - 3(a - 1) = 5 - a$

Solution

$$
\begin{array}{rll}
2(a + 1) - 3(a - 1) = 5 - a & & \textit{Remove parentheses.} \\
2a + 2 - 3a + 3 = 5 - a & & \textit{Combine like terms.} \\
-a + 5 = 5 - a & & \\
\underline{+a \qquad\qquad +a} & & \\
5 = 5 & &
\end{array}
$$

Since $5 = 5$ is *always* true and is an equivalent equation, the original equation is an identity. In fact, you might have already noticed that we have an identity in the third line of our solution because the expressions $-a + 5$ and $5 - a$ are always equal regardless of the value of a.

In solving an equation which is in fact an identity, the variable drops out entirely, and we get an equation of the form "a number is equal to itself," which is always true. ∎

EXAMPLE 10   *Solve for y.* $2 - (y - 4) + 3y = 2y + 1$

10. $8 - (3 - t) = t - 6$

Solution

$$2 - (y - 4) + 3y = 2y + 1 \qquad \textit{Remove the parentheses.}$$
$$2 - y + 4 + 3y = 2y + 1 \qquad \textit{Combine like terms.}$$
$$6 + 2y = 2y + 1$$
$$\underline{-2y \quad -2y}$$
$$6 = 1$$

Since $6 = 1$ is always false and is an equivalent equation, the original equation is a contradiction. In solving an equation which is in fact a contradiction, the variable drops out entirely, and we get an equation which says that two unequal numbers are equal, which is always false. ∎

Based on the examples we have done we can state that if a first-degree equation is conditional, then it has a unique solution, which we find by the methods outlined above. Otherwise, it is an identity or a contradiction.

STUDY SKILLS 3.2

Preparing for Exams

When to Study

When to start studying and how to distribute your time studying for exams is as important as how to study. To begin with, "pulling all-nighters" (staying up all night to study just prior to an exam) seldom works. As with athletic or musical skills, algebraic skills cannot be developed overnight. In addition, without an adequate amount of rest, you will not have the clear head you need to work on an algebra exam. It is usually best to start studying early—from $1\frac{1}{2}$ to 2 weeks before the exam. In this way you have the time to perfect your skills and, if you run into a problem, you can consult your teacher to get an answer in time to include it as part of your studying.

It is also a good idea to distribute your study sessions over a period of time. That is, instead of putting in 6 hours in one day and none the next two days, put in 2 hours each day over the three days. You will find that not only will your studying be less boring, but also you will retain more with less effort.

As we mentioned before, your study activity should be varied during a study session. It is also a good idea to take short breaks and relax. A study "hour" could consist of about 50 minutes of studying and a 10-minute break.

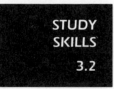 *Answers to Learning Checks in Section 3.2*

1. $t = 8$ **2.** $t = -3$ **3.** $t = 4$ **4.** $t = -2$ **5.** $t = 9$ **6.** $t = -4$

7. $x = 3$ **8.** $t = -6$ **9.** Identity   **10.** Contradiction

NOTE TO THE STUDENT

Use the space on this page to write down any questions you have or points you want to review with your instructor.

Exercises 3.2

Solve each of the following equations. If the equation is an identity or a contradiction, indicate this.

1. $x + 3 = 8$ **2.** $x + 5 = 11$ **3.** $y - 4 = 7$

4. $y - 6 = 3$ **5.** $a + 3 = 1$ **6.** $a + 5 = 2$

7. $a - 5 = -8$ **8.** $a - 3 = -7$ **9.** $3x = 21$

10. $8x = 48$ **11.** $-4x = -12$ **12.** $-9x = -36$

13. $4x - x = 2 - 8$ **14.** $6x - x = 4 - 14$ **15.** $2z - 3z - 11z = -4(6)$

16. $4z - z - 9z = -8(3)$ **17.** $2x + 1 = 7$ **18.** $5x - 2 = 8$

19. $2t = 3t + 5$ **20.** $3z = 4z - 6$ **21.** $9 = 6 - 3a$

22. $15 = 7 - 4b$ **23.** $20 = 3w - 1$ **24.** $37 = 13r - 2$

1. _____
2. _____
3. _____
4. _____
5. _____
6. _____
7. _____
8. _____
9. _____
10. _____
11. _____
12. _____
13. _____
14. _____
15. _____
16. _____
17. _____
18. _____
19. _____
20. _____
21. _____
22. _____
23. _____
24. _____

25. _____

26. _____

27. _____

28. _____

29. _____

30. _____

31. _____

32. _____

33. _____

34. _____

35. _____

36. _____

37. _____

38. _____

39. _____

40. _____

41. _____

42. _____

43. _____

44. _____

45. _____

46. _____

47. _____

48. _____

25. $8y + 4 = 5y + 19$

26. $7y + 5 = 3y + 17$

27. $2a + 5 = 4a + 13$

28. $3a + 7 = 10a + 14$

29. $5r - 8 = 3r - 20$

30. $9r - 7 = 6r - 19$

31. $10 - x = 4 - 3x$

32. $9 - 5x = 1 - x$

33. $-4 - 3u = -2 - u$

34. $-13 - u = -5u - 1$

35. $x + 7 = 7 - x$

36. $x + 5 = 5 - x$

37. $x + 7 = 7 + x$

38. $x + 5 = 5 + x$

39. $x - 7 = 7 + x$

40. $x - 5 = 5 + x$

41. $x - 7 = 7 - x$

42. $x - 5 = 5 - x$

43. $2(t + 1) + 3t = 27$

44. $3(t + 2) + t = 30$

45. $2(y + 3) + 4(y - 2) = 22$

46. $6(y - 3) + 3(y + 5) = 51$

47. $4 + 3(3y - 5) = 2y - 11 + y$

48. $6 + 4(y - 2) = 5y - 8 - y$

49. $3(a - 2) + 4(2 - a) = a + 2(a + 1)$

50. $5(a - 4) + 3(5 - a) = a + 4(a - 1) - 1$

51. $8z - 3(z - 2) = -9$ **52.** $12z - 8(z - 1) = -8$

53. $3t - 5(t - 1) = 23$ **54.** $5t - 7(t - 2) = 40$

55. $3 - 5(t - 1) = 23$ **56.** $5 - 7(t - 2) = 40$

57. $2(y - 3) - 3(y - 5) = 5y - 5(y - 2)$

58. $4(y - 2) - 5(y - 3) = 7y - 7(y - 1)$

59. $4x - 3(x + 8) = 5x - 2(x - 12) - 2x$

60. $7a - 5(a - 2) - a = 4a - 2(a - 5) - a$

61. $a - (5 - 3a) = 7a - (a - 3) - 8$

62. $2y - (7 - 4y) = 10y - (y - 2)$

ANSWERS

49. _____

50. _____

51. _____

52. _____

53. _____

54. _____

55. _____

56. _____

57. _____

58. _____

59. _____

60. _____

61. _____

62. _____

ANSWERS

63. _____

64. _____

65. _____

66. _____

67. _____

68. _____

69. _____

70. _____

71. _____

72. _____

63. $x(x + 2) + 3x = x(x - 1) - 12$

64. $a(a - 2) - a = a(a + 1) - 8$

65. $2z(z + 1) + 3(z + 2) = 3z(z + 2) - z^2$

66. $3y(y - 1) = 2y(y - 2) - (3 - y^2)$

CALCULATOR EXERCISES

Solve each of the following equations. Round off your answers to the nearest hundredth.

67. $.3x - .82 = 1.13$ **68.** $6.7a - 13.4 = 2.8a + 110.23$

69. $2.3t - 1.6(t + .1) = -.139$ **70.** $1.4 + .5(8 - t) = .7t$

71. $3.4(t - 8) = 10.6(t + 3)$ **72.** $.03w - .8(w - .62) = 40$

QUESTIONS FOR THOUGHT

5. How do you check your answer after you have solved an equation?

6. What is *wrong* (if anything) with each of the following and why?
Solve for x.

(a)
$$2x + 4 = 8$$
$$\underline{-4 \quad -4}$$
$$2x = 12$$
$$\frac{2x}{2} = \frac{12}{2}$$
$$x = 6$$

(b)
$$3x - 2 = 19$$
$$\underline{-3 \qquad -3}$$
$$x - 2 = 16$$
$$x - 2 = 16$$
$$\underline{+2 \quad +2}$$
$$x = 18$$

(c)
$$3x - 6 = 12$$
$$\frac{3x - 6}{3} = \frac{12}{3}$$
$$x - 6 = 4$$
$$x - 6 = 4$$
$$\underline{+6 \quad +6}$$
$$x = 10$$

(d)
$$5x = 2x$$
$$\frac{5x}{x} = \frac{2x}{x}$$
$$5 = 2$$
Contradiction

(e)
$$2x = 5x - 6$$
$$\underline{-5x \quad -5x}$$
$$-3x = -6$$
$$\frac{-3x}{-3} = \frac{-6}{3}$$
$$x = -2$$

(f)
$$2x = 5x + 6$$
$$\underline{-5x \quad -5x}$$
$$-3x = 6$$
$$\frac{-3x}{3} = \frac{6}{3}$$
$$-x = 2$$

(g)
$$7x = 4x$$
$$\underline{-4x \quad -4x}$$
$$3x = 0$$
$$\frac{3x}{3} = \frac{0}{3}$$
$$x = 0$$

Verbal Problems

3.3

Most applications of mathematics in the "real" world involve translating a particular problem into mathematical language, getting an equation out of this translation, and solving it.

While some of the problems in this book may strike you as artificial and having nothing to do with the real world, they do afford us an opportunity to practice those skills necessary to do a wide variety of problems.

Before we proceed any further, it is worthwhile mentioning that you may be able to solve some of the problems that follow by "playing around with the numbers." This may be a valid way of getting the answer to a specific problem, *but* it is a very risky strategy for problem-solving. What if the numbers are complicated, the trial-and-error procedure very long, or the problem has no answer? Consequently, we will not accept the trial-and-error method as an acceptable method of solution for our purposes. Instead, we are going to develop an outline of our strategy for solving verbal problems which, hopefully, we will be able to apply to just about every problem from the simplest to the most complicated.

Let's begin by looking at two examples and their solutions, and then generalize the procedure into a strategy outline for solving verbal problems.

EXAMPLE 1 If one number is three more than twice another number and their sum is 30, find the numbers.

Solution Since the algebraic skills we have learned thus far limit us to solving equations involving only one variable, we shall try to translate all the relevant information in the problem in terms of one variable.

We are seeking two numbers with the properties described in the example. To make it easier to talk about the two numbers, let's refer to them as the "first number" and the "second number."

Let x = the first number.

The problem tells us that the "second number" is "three more than twice" the first number. Therefore, we can represent the second number as

$$\text{"second number"} = \underbrace{\text{three more than}}\ \underbrace{\text{twice the first}}$$
$$2x \quad + 3$$

The problem also tells us that the sum of the numbers is 30. Our equation is

"first number" + "second number"   is equal to   30

or written algebraically,

$$x \quad + \quad 2x + 3 \quad = \quad 30 \ *$$

---

*We will indicate the equation we obtain from each problem by enclosing it in a rectangular shaded panel in this way.

Now we proceed to solve the equation.

$$x + 2x + 3 = 30 \qquad \textit{Combine like terms.}$$
$$3x + 3 = 30$$
$$\underline{-3 \qquad -3}$$
$$3x = 27$$
$$\frac{3x}{3} = \frac{27}{3}$$
$$x = 9$$

Thus, the first number is 9. The second number is

$$2x + 3 = 2(9) + 3 = 18 + 3 = 21$$

The *answer* to the problem is the pair of numbers

$$\boxed{9 \text{ and } 21}$$

In order to check this answer it is not sufficient to check the answer in the equation since we may have formulated our equation incorrectly. In other words, we may have the right answer to a wrong equation. Instead, we must check our answer in the original words of the problem.

CHECK: Is 21 equal to 3 more than twice 9? Yes. Is the sum of 9 and 21 equal to 30? Yes. Our answer checks. ∎

2. The length of a rectangle is 5 more than twice the width. If the perimeter is 58 inches, find the dimensions.

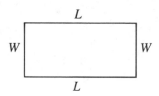

Figure 3.1 *Rectangle*

EXAMPLE 2   The length of a rectangle is 2 less than 3 times the width. If the perimeter of the rectangle is 76 cm., find the dimensions of the rectangle.

Solution   In order to solve this problem we need to know exactly what a rectangle looks like, what perimeter means, and how to compute it.

A diagram is particularly useful in a problem of this kind. A rectangle is a figure in which the opposite sides are equal and the sides meet at right angles (see Figure 3.1). The perimeter, P, of a geometric figure means the length around it. Thus, for a rectangle,

$$P = L + W + L + W$$

or in simplified form,

$$P = 2L + 2W$$

In our problem we are told that the length is 2 less than 3 times the width. Since the length is described in terms of the width we can let

$$W = \text{Width of the rectangle}$$

Then

$$3W - 2 = \text{Length of the rectangle} \quad \text{(length is 2 less than 3 times the width)}$$

Alternatively, we could simply draw a diagram and label it as shown in Figure 3.2.

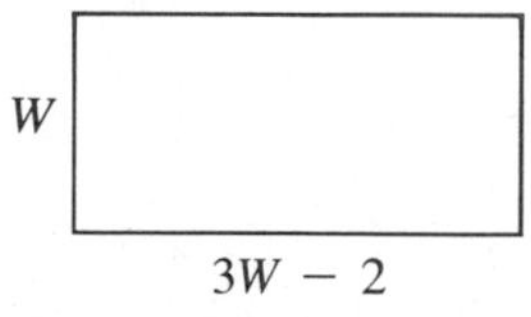

The perimeter being 76 cm. gives us the equation

$$2W \quad + \quad 2(3W - 2) \quad = \quad 76$$

Twice the width + Twice the length = Perimeter

Figure 3.2 *Length described in terms of width*

Now we proceed to solve this equation:

$$2W + 2(3W - 2) = 76$$
$$2W + 6W - 4 = 76$$
$$8W - 4 = 76$$
$$\underline{+4 \quad +4}$$
$$8W = 80$$
$$\frac{8W}{8} = \frac{80}{8}$$
$$W = 10$$

If the width is 10 cm., then the length is

$$3W - 2 = 3(10) - 2 = 30 - 2 = 28 \text{ cm.}$$

CHECK: Is the length 2 less than 3 times the width? Is 28 equal to 2 less than 3 times 10? Yes. Is the perimeter 76 cm.? Is $2(28) + 2(10) \stackrel{?}{=} 76$? Yes.

Thus, the answer to the problem is: The dimensions of the rectangle are

$$\boxed{W = 10 \text{ cm.}; \quad L = 28 \text{ cm.}}$$ ■

In analyzing the solutions to these two examples we can generalize our method to obtain the outline (given in the next box) for solving verbal problems.

OUTLINE OF STRATEGY FOR SOLVING VERBAL PROBLEMS

1. Read the problem carefully, as many times as is necessary to understand what the problem is saying and what it is asking.

2. Use diagrams whenever you think it will make the given information clearer.

3. Ask whether there is some underlying relationship or formula you need to know. If not, then the words of the problem themselves give the required relationship.

4. Clearly identify the unknown quantity (or quantities) in the problem, and label it (them) using one variable.

   Step 4 is very important, and not always easy.

5. By using the underlying formula or relationship in the problem, write an equation involving the unknown quantity (or quantities).

   Step 5 is the *crucial step*.

6. Solve the equation.

7. Make sure you have answered the question that was asked.

8. Check the answer(s) in the original words of the problem.

Being asked to solve a problem *algebraically* means to apply this outline (or some variation of it) to a problem, as opposed to a trial-and-error procedure. Do not be surprised if it takes a while to solve a problem, especially if it is a type you have not seen before.

3. A rock weighing 100 pounds is broken into three parts. The heaviest piece is 6 times the weight of the lightest piece, and the medium piece weighs 30 pounds less than the heaviest piece. Find the weight of each piece.

One other word of advice: Take the time to look over your work and think about what you have done. The more time you spend thinking about the problem and the relationships you have uncovered, the easier subsequent problems will be. If you can learn to apply this outline to relatively simple problems it is more likely that you will be able to apply it to more complicated ones as well.

EXAMPLE 3 A wooden board, 20 meters long, is cut into three pieces. The medium piece is twice as long as the shortest piece, and the longest piece is 5 meters longer than the medium piece. Find the length of each piece.

Solution Since the medium piece is described in terms of the shortest piece, it seems reasonable to begin as follows. Let

x = Length of the shortest piece

$2x$ = Length of the medium piece   (The problem tells us that the medium piece is twice as long as the shortest piece.)

$2x + 5$ = Length of the longest piece   (The problem tells us that the longest piece is 5 meters longer than the medium piece.)

If we had preferred, we could simply have drawn Figure 3.3 instead.

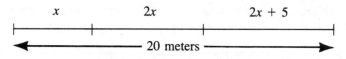

Figure 3.3 *Diagram for Example* 3

Our common sense tells us that the three pieces together add up to the total length of 20 meters. Therefore, the equation is

$$x + 2x + 2x + 5 = 20$$

Now we solve this equation.

$$x + 2x + 2x + 5 = 20 \qquad \textit{Combine like terms.}$$
$$5x + 5 = 20 \qquad \textit{Subtract 5 from both sides.}$$
$$5x = 15 \qquad \textit{Divide both sides by 5.}$$
$$x = 3$$

Thus,

The length of the shortest piece is $x =$ $\boxed{3 \text{ meters}}$

The length of the medium piece is $2x =$ $\boxed{6 \text{ meters}}$

The length of the longest piece is $2x + 5 = 2(3) + 5 =$ $\boxed{11 \text{ meters}}$

The check is left to the student. ∎

4. A collection of forty coins consists of dimes and quarters and has a total value of $5.80. How many coins of each type are there?

EXAMPLE 4 A collection of twenty coins consists of dimes and nickels. If the total value of the coins is $1.65, how many dimes and how many nickels are there?

Solution In formulating a solution to this problem it is important to recognize the difference between how many coins we have and how much they are worth or their *value*. For example, if we have 16 dimes and 4 nickels, then we have 20 coins. But to compute how much money we have (the total value of the coins), we must consider the value of each type of coin:

The value of 16 dimes is 16(10)   =   160 cents

The value of 4 nickels is 4(5)   =   20 cents

Total *number* of coins = 20   BUT   Total *value* of the coins = 180 cents or $1.80

We multiply the number of dimes by the value of one dime (10) to get the value of all the dimes. Similarly, we multiply the number of nickels by the value of one nickel (5) to get the value of all the nickels. The total value of all the coins is the sum of the value of the nickels plus the value of the dimes.

In exactly the same way, the *value* of 20 coins consisting of 8 dimes and 12 nickels is

$$8(10) + 12(5) = 80 + 60 = 140 \text{ cents} \quad \text{or} \quad \$1.40$$

In this problem we are asked to find out how many nickels and how many dimes there are. It may seem that there are two unknown quantities, but in reality there is only one. If it should turn out that there are 15 dimes, then, since there are 20 coins all together, there must be $20 - 15 = 5$ nickels. Similarly, if it turns out that there are 9 dimes, then there must be $20 - 9 = 11$ nickels. Consequently, if we let

$$x = \text{Number of dimes}$$

then

$$20 - x = \text{Number of nickels (since there are 20 coins all together)}$$

Now that we have expressed the unknown quantities in terms of one variable, we will write an equation which describes the *value* relationship in the problem. The equation is somewhat simpler if we express everything in cents.

Value of the dimes + Value of the nickels = Total value

(*# of dimes*) · (*Value of* 1 *dime*) + (*# of nickels*) · (*Value of* 1 *nickel*) = *Total value*

$$(x) \cdot (10) + 5 \cdot (20 - x) \qquad = 165$$

Now we solve this equation.

$$10x + 5(20 - x) = 165$$
$$10x + 100 - 5x = 165$$
$$5x + 100 = 165$$
$$5x = 65$$
$$x = 13$$

Thus, there are

13 dimes

and $20 - 13$ or

7 nickels

CHECK: 13 dimes and 7 nickels give us 20 coins as required

$$13 \text{ dimes are worth } 13(10) = 130 \text{ cents}$$
$$7 \text{ nickels are worth } 7(5) = 35 \text{ cents}$$
$$\text{Total value} \stackrel{\checkmark}{=} 165 \text{ cents} \quad \text{or} \quad \$1.65$$

Two points are worth mentioning. First, we did not have to let x be the number of dimes. We could just as well have let x be the number of nickels. Then x would have worked out to be 7 instead of 13, but the answer to the problem would still have been the same: 13 dimes and 7 nickels. In some cases the choice of what the variable represents may make the problem a bit easier (see Example 3 above). Only experience will teach you how to make the best choice.

Second, since we knew that there were 20 coins all together, if we labeled the number of one type of coin x, then the number of the other type had to be $20 - x$. This idea comes up frequently. If we have two quantities and we know their total and we represent one of the two unknown quantities by x, then the other quantity must be *total* $- x$.   ∎

5. A discount book store sells old textbooks for \$4 each and old novels for \$5 each. If Jim bought four more novels than textbooks for a total of \$65, how many of each did he buy?

EXAMPLE 5   The amount \$1,025 was collected on the sale of tickets to a school play. The ticket price was \$5 for adults and \$2 for students. If 75 more students than adults purchased tickets, how many tickets were purchased all together?

Solution   Let x be the number of adult tickets sold. Then

$$x + 75 = \text{Number of student tickets sold} \quad (75 \text{ more students than adults})$$

Our equation involves the amount of money collected on the sale of the tickets—that is, the *value* of the tickets sold. We compute this amount as follows:

$$\begin{bmatrix} \text{\# of adult} \\ \text{tickets sold} \end{bmatrix} \cdot \begin{bmatrix} \text{Price of an} \\ \text{adult ticket} \end{bmatrix} + \begin{bmatrix} \text{\# of student} \\ \text{tickets sold} \end{bmatrix} \cdot \begin{bmatrix} \text{Price of a} \\ \text{student ticket} \end{bmatrix} = \begin{matrix} \text{Total} \\ \text{collected} \end{matrix}$$

$$x \quad\quad \cdot \quad\quad 5 \quad\quad + \quad\quad (x + 75) \quad\quad \cdot \quad\quad 2 \quad\quad = \quad 1{,}025$$

Thus, our equation is

$$5x + 2(x + 75) = 1{,}025$$

We now solve this equation.

$$5x + 2(x + 75) = 1{,}025$$
$$5x + 2x + 150 = 1{,}025$$
$$7x + 150 = 1{,}025$$
$$7x = 875$$
$$x = \frac{875}{7}$$
$$x = 125$$

There were $x = 125$ adult tickets sold and $x + 75 = 200$ student tickets sold.

Do not forget to answer the question. The problem asked "How many tickets were purchased all together?" Therefore, the answer is

$$125 + 200 = \boxed{325 \text{ tickets}}$$

CHECK: 200 is 75 more than 125

125 tickets at $5 each = 5(125) = $625

200 tickets at $2 each = 2(200) = $400

Total collected for all the tickets = $625 + $400 $\overset{\checkmark}{=}$ $1,025 ∎

At first glance the last two examples might seem unrelated. However, if you look them over you will see that the structure of their solutions is very similar.

EXAMPLE 6 Susan jogs to the post office at the rate of 12 kilometers per hour (kph) and walks home at the rate of 6 kilometers per hour. If her total time, jogging and walking, is 3 hours, how far is the post office from her home?

Solution Let's begin by drawing a little diagram to help us visualize the problem.

$$\text{HOME} \overset{\text{Jogging}}{\underset{\text{Walking}}{\rightleftarrows}} \text{POST OFFICE}$$

The diagram emphasizes for us the fact that the jogging distance is the same as the walking distance.

The underlying relationship or formula needed in this problem is that if you are travelling at a constant rate of speed then

Distance covered = [Rate (or speed) at which you travel] · [Time travelling]

In short, we write

$$d = rt$$

For example, if you travel at 50 miles per hour for 3 hours you have covered 150 miles:

$$d = rt$$

$$d = 50\frac{\text{miles}}{\text{hr.}} \cdot 3 \text{ hr.}$$

$$d = 150 \text{ miles}$$

We have already noted the fact that the distances covered jogging to and walking from the post office are the same. That is, we have

$d_{\text{jogging}} = d_{\text{walking}}$ d_{jogging} is called a **subscripted variable**, and is used to indicate the distance covered jogging. d_{jogging} and d_{walking} are two different variables just like x and y.

Using the formula $d = rt$ we can rewrite this as

$$r_{\text{jogging}} \cdot t_{\text{jogging}} = r_{\text{walking}} \cdot t_{\text{walking}}$$

We are given the information that $r_{\text{jogging}} = 12$ kph and that $r_{\text{walking}} = 6$ kph. So our equation now looks like

$$12 \cdot t_{\text{jogging}} = 6 \cdot t_{\text{walking}}$$

Thus, we need to figure out how the different times are related.

Since we are told that Susan's *total* time is 3 hours, let

$t = t_{\text{jogging}}$,   the amount of time for Susan to jog to the post office

6. Marge walks to a store at the rate of 100 meters per minute, and then walks home more quickly at the rate of 150 meters per minute in a total of 15 minutes. How far away is the store?

Then

$$3 - t = t_{\text{walking}}, \quad \text{the amount of time for Susan to walk home}$$

(The total time was 3 hours, so if $t = $ time jogging, then the time walking is what is left over or $3 - t$.) We can now fill in this information and our equation above becomes

$$12 \cdot t = 6 \cdot (3 - t)$$

Now we solve this equation.

$$12t = 6(3 - t)$$
$$12t = 18 - 6t$$
$$18t = 18$$
$$t = 1$$

Therefore, the time jogging was 1 hour, but we are not finished yet. The example asks for the distance of the post office from her home:

$$\text{Distance jogging} = \left(12\frac{\text{km.}}{\text{hr.}}\right) \cdot (1 \text{ hr.}) = 12 \text{ kilometers}$$

CHECK: If Susan jogged for 1 hour then she walked for $3 - 1 = 2$ hours.

$$\text{Distance walking} = \left(6\frac{\text{km.}}{\text{hr.}} \cdot 2 \text{ hr.}\right) \stackrel{\checkmark}{=} 12 \text{ km.}$$

Note that this problem was a bit different in that the variable t did not represent the quantity that was asked for (distance). However, once we found t we were easily able to compute the distance we were asked to find. ∎

    In doing the exercises do not be discouraged if you do not get a complete solution to every problem. Make an honest effort to solve the problems and keep a written record of your work so that when you go over the problems in class you can see how far you got and exactly where you got stuck.

✔ *Answers to Learning Checks in Section 3.3*

1. 10 and 28   **2.** $W = 8$ in., $L = 21$ in.   **3.** 60 lb., 30 lb., and 10 lb.

4. 28 dimes and 12 quarters   **5.** 5 textbooks and 9 novels   **6.** 900 meters

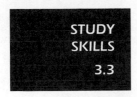

**STUDY
SKILLS

3.3**

**Preparing
for Exams**

Study Activities

If you are going to learn algebra well enough to be able to demonstrate high levels of performance on exams, then you must concern yourself with both developing your skills in algebraic manipulation and understanding what you are doing and why you are doing it.

Many students concentrate only on skills and resort to memorizing the procedures for algebraic manipulations. This may work for quizzes or a test covering just a few topics. For exams covering a chapter's worth of material or more this can be quite a burden on the memory. Eventually interference occurs and problems and procedures get confused. If you find yourself doing well on quizzes but not on longer exams, this may be your problem.

Concentrating on understanding what a method is and why it works is important. Neither the teacher nor the textbook can cover every possible way in which a particular concept may present itself in a problem. If you understand the concept, you should be able to recognize it in a problem. But again, if you concentrate only on understanding concepts and not on developing skills, you may find yourself prone to making careless and costly errors under the pressure of an exam.

In order to achieve the goal of both skill development and understanding, your studying should include four activities: **(1)** practicing problems, **(2)** reviewing your notes and textbook, **(3)** drilling with study cards (to be discussed in the next section), and **(4)** reflecting on the material being reviewed and the exercises being done.

Rather than doing any one of these activities over a long period of time, it is best to do a little of the first three activities during a study session and save some time for reflection at the end of the session.

NOTE TO THE STUDENT

Use the space on this page to write down any questions you have or points you want to review with your instructor.

Exercises 3.3

Solve each of the following problems algebraically. That is, set up an equation and solve it. Be sure to clearly label what the variable represents.

1. One number is 4 more than 3 times another. If the sum of the two numbers is 24, find the numbers.

2. One number is 8 more than twice another. If the sum of the two numbers is 38, find the numbers.

3. One number is 5 less than 4 times another. If the sum of the two numbers is 10, find the numbers.

4. One number is 9 less than twice another. If the sum of the two numbers is 6, find the numbers.

5. If a number is added to 3 more than 5 times itself the result is 27. Find the number.

6. If a number is added to 3 less than 5 times itself the result is 27. Find the number.

7. If a number is subtracted from 4 more than twice itself the result is 12. Find the number.

8. If a number is subtracted from 6 less than 3 times itself the result is 18. Find the number.

9. The sum of three numbers is 80. The largest number is 10 more than twice the smallest, and the middle number is 5 less than twice the smallest. Find the three numbers.

10. The sum of three numbers is 68. The largest number is 6 less than twice the smallest, and the middle number is 10 less than the largest. Find the three numbers.

11. The length of a rectangle is 1 more than twice the width. If the perimeter is 26 centimeters, find the dimensions of the rectangle.

11. _____

12. If the length of a rectangle is 4 more than 5 times the width, and the perimeter is 32 meters, what are its dimensions?

13. The first side of a triangle is 10 inches more than the second side. If the third side is 3 times as long as the second side and the perimeter is 45 inches, find the lengths of the three sides.

12. _____

13. _____

14. The length of a rectangle is 6 more than the width. If the width is increased by 10 while the length is tripled, the new rectangle has a perimeter which is 56 more than the original perimeter. Find the original dimensions of the rectangle.

14. _____

15. The length of a rectangle is 2 less than 3 times the width. If the width is tripled while the length is decreased by 2, the new perimeter is 12 more than the original perimeter. Find the original dimensions.

15. _____

16. A collection of 20 coins consisting of dimes and quarters has a total value of $4.25. How many of each type of coin are there?

16. _____

17. Jack bought 29 stamps at the post office. Some were 12¢ stamps and the rest were 15¢ stamps. If the total cost of the stamps was $3.99, how many of each type did he buy?

17. _____

18. Advanced-purchase tickets to an art exhibition cost $4, while tickets purchased at the door cost $6. If a total of 150 tickets were sold and $680 was collected, how many advanced-purchase tickets were sold?

18. _____

19. In a [   ] wice as many dimes as nickels, and [   ] e coins is $4.50, how many of each [   ]

20. In a collection of dimes, quarters, and half-dollars there are 45 coins in all. There are 11 more quarters than half-dollars and the remaining coins are dimes. If the total value of the coins is $11.10, how many of each type of coin are there?

21. Susan works for a publisher and earns a commission of $1.50 for each book she sells and $2.25 for each magazine subscription she sells. During a certain week she made a total of 80 sales and earned $157.50. How many books did she sell?

22. An electrician charges $20 per hour for her time and $12 per hour for her assistant's time. On a certain job the assistant worked alone for 3 hours preparing the site, and then the electrician and her assistant completed the job together. If the total bill for the job was $164, how many hours did the electrician work?

23. Two trains leave cities 300 miles apart at 10:00 A.M. travelling toward each other. One train travels at 20 mph and the other train travels at 40 mph. At what time do they pass by each other?

24. Repeat Problem 23, if the 40-mph train leaves at 8:00 A.M. and the 20-mph train leaves at 10:00 A.M.

ANSWERS

19. ——————

20. ——————

21. ——————

22. ——————

23. ——————

24. ——————

25. Two people leave by car from the same location travelling in opposite directions. One leaves at 2:00 P.M. driving at 55 kph, while the other leaves at 3:00 P.M. driving at 45 kph. At what time will they be 355 kilometers apart?

25. _____

26. How long would it take someone driving at 90 kph to overtake someone driving at 75 kph with a 1-hour head start?

26. _____

27. A relay race requires each team of two contestants to complete a 172-kilometer course. The first person runs part of the course then the second person completes the remainder of the course by bicycle. One team covers the running section at 18 kph and the bicycle section at 50 kph in a total of 6 hours. How long did it take to complete the running section? How long is the running section of the course?

27. _____

28. A person can drive from town A to town B at a certain rate of speed in 5 hours. If he increases his speed by 15 kph, he can make the trip in 4 hours. How far is it from town A to town B?

28. _____

29. A secretary and a trainee are processing a pile of 124 forms. The secretary processes 15 forms per hour while the trainee processes 7 forms per hour. If the trainee begins working at 9:00 A.M. and is then joined by the secretary 2 hours later, at what time will they finish the pile of forms?

29. _____

30. A stack of 9,500 computer cards is to be read by an electronic card reader. A card reader which reads 200 cards per minute begins reading the cards but breaks down before completing the job. A new card reader, which reads 300 cards per minute, then completes the job. If it took a total of 40 minutes to read the stack of cards, how long did the first machine work before it broke down?

30. _____

Types of Inequalities and Basic Properties of Inequalities

3.4

In Section 1.1 we discussed the meaning of the inequality symbols $<$, $>$, $\leq$, $\geq$. In this section we are going to discuss inequalities in much greater detail. Much of our discussion will follow along the same lines as our previous one on equalities earlier in this chapter.

Inequalities such as $-2 < 5$ and $3 > 8$ can be categorized as being true or false, while an equality such as $x + 3 < 5$ is neither true nor false since its truth depends on the value of x. An inequality whose truth depends on the value of the variable is called a **conditional inequality**.

As we saw with equations, the presence of a variable does not necessarily mean that we have a conditional inequality. If an inequality is always true regardless of the value of the variable it is called an **identity**, while if it is always false it is called a **contradiction**. We will examine these possibilities and learn how to recognize them in the next section.

The values of the variable which make a conditional inequality true are called the **solutions** of the inequality.

EXAMPLE 1 In each of the following inequalities, determine whether the given value of the variable is a solution.

(a) $x < 5$; $x = 3$ **(b)** $x > -4$; $x = -3$ **(c)** $a \leq 7$; $a = 7$

(d) $4x - 10 < -20$; $x = -2$

Solution

(a) $x < 5$ *Substitute 3 for x.*

$3 \overset{?}{<} 5$

$3 \overset{\checkmark}{<} 5$ *Therefore, 3 is a solution.*

(b) $x > -4$ *Substitute -3 for x.*

$-3 \overset{?}{>} -4$

$-3 \overset{\checkmark}{>} -4$ *Remember that $a > b$ means that a is to the right of b on the number line. Since -3 is to the right of -4, we have $-3 > -4$. Therefore, -3 is a solution.*

(c) $a \leq 7$ *Substitute 7 for a.*

$7 \overset{?}{\leq} 7$

$7 \overset{\checkmark}{\leq} 7$ *Therefore, $a = 7$ is a solution.*

Remember that $a \leq b$ means $a < b$ or $a = b$ so that $7 \leq 7$ means $7 < 7$ or $7 = 7$, which is true.

(d) $4x - 10 < -20$ *Substitute -2 for x.*

$4(-2) - 10 \overset{?}{<} -20$

$-8 - 10 \overset{?}{<} -20$

$-18 \not< -20$ *Therefore, -2 is not a solution. On the number line, -18 is not to the left of -20.* ∎

EXAMPLE 2 Determine whether the specified value satisfies the given inequality.

(a) $8 - 3(y - 5) \geq 17$; $y = 1$ **(b)** $2 < 3x + 8 < 7$; $x = 0$

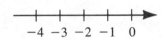

Solution

(a) $8 - 3(y - 5) \geq 17$ *Substitute $y = 1$.*

$8 - 3(1 - 5) \overset{?}{\geq} 17$ *Remember the order of operations.*

$8 - 3(-4) \overset{?}{\geq} 17$

$8 + 12 \overset{?}{\geq} 17$

$20 \overset{?}{\geq} 17$

$20 \overset{\checkmark}{\geq} 17$ *Therefore, $y = 1$ does satisfy this inequality.*

(b) Remember that a double inequality like $2 < 3x + 8 < 7$ means that $2 < 3x + 8$ *and* $3x + 8 < 7$. In other words, $3x + 8$ must be *between* 2 and 7.

$2 < 3x + 8 < 7$ *Substitute $x = 0$.*

$2 \overset{?}{<} 3(0) + 8 \overset{?}{<} 7$

$2 \overset{?}{<} 0 + 8 \overset{?}{<} 7$

$2 \overset{?}{<} 8 \overset{?}{<} 7$

$2 < 8 \not< 7$ *Therefore, $x = 0$ does not satisfy the inequality.* ∎

In solving first-degree equations our goal was to isolate x. In order to do this we used the properties of equality to transform the original equation into simpler equivalent equations until we ended up with an obvious equation. ***First-degree inequalities***, meaning those which involve the variable to the first power only, are going to be handled in a very similar way.

Keep in mind, however, that while a first-degree equality such as $x = 4$ has only one solution, a first-degree inequality such as $x > 4$ can have infinitely many solutions. Any number greater than 4 satisfies this inequality.

In discussing inequalities, those of the form

$$x < 3 \qquad a \geq -2 \qquad 1 < y < 5$$

(that is, those in which the variable is isolated) are called ***obvious inequalities***. Given our experience with solving equations, it seems natural to ask what are the properties of inequalities that we can use to obtain equivalent inequalities (that is, inequalities with the same solution set)?

Let's start with a true inequality such as $-6 < 8$, perform the same kinds of operations on this inequality as we did when we were solving equations, and see what happens.

$-6 < 8$ Add 2 to both sides. $-6 + 2 \; ? \; 8 + 2$

$-4 \; ? \; 10$

$-4 < 10$ *The inequality symbol remains the same.*

$-6 < 8$ Subtract 2 from both sides. $-6 - 2 \; ? \; 8 - 2$

$-8 \; ? \; 6$

$-8 < 6$ *The inequality symbol remains the same.*

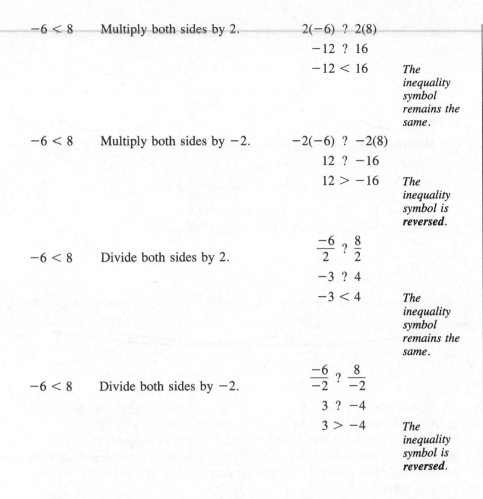

$-6 < 8$ Multiply both sides by 2. $2(-6)$? $2(8)$

-12 ? 16

$-12 < 16$ *The inequality symbol remains the same.*

$-6 < 8$ Multiply both sides by -2. $-2(-6)$? $-2(8)$

12 ? -16

$12 > -16$ *The inequality symbol is **reversed**.*

$-6 < 8$ Divide both sides by 2. $\dfrac{-6}{2}$? $\dfrac{8}{2}$

-3 ? 4

$-3 < 4$ *The inequality symbol remains the same.*

$-6 < 8$ Divide both sides by -2. $\dfrac{-6}{-2}$? $\dfrac{8}{-2}$

3 ? -4

$3 > -4$ *The inequality symbol is **reversed**.*

If the reversal of the inequality symbol were simply a random occurrence, we would not be able to formulate the properties of inequalities. Fortunately, the examples we have just looked at illustrate what happens in general. If we illustrate a few of the above examples on the number line, the mechanics of what is going on should become clear.

On the number line, the original inequality $-6 < 8$ is true because -6 is to the left of 8:

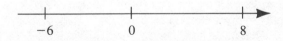

Adding or subtracting the same quantity from both sides of the inequality simply shifts both numbers the same number of units either to the left or right. In either case, the number further to the left remains further to the left, and hence *the inequality symbol remains the same.*

Multiplying both sides by 2 looks like this:

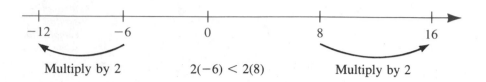

Multiply by 2 $2(-6) < 2(8)$ Multiply by 2

and again *the inequality symbol remains the same.*

However, multiplying both sides by -2 looks like this:

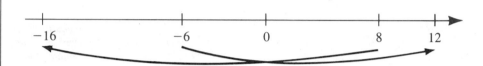

We can see that after multiplying by -2, the positions of the resulting numbers are reversed, and therefore *the inequality symbol must be reversed.*

(You might find it interesting to consider some examples starting with an inequality $a < b$ where a and b are both positive or both negative and see what happens when you multiply or divide the inequality by first a positive and then a negative number.)

While we have only analyzed some numerical evidence, inequalities behave this way all the time. We formulate these properties as indicated in the box.

PROPERTIES OF INEQUALITIES

1. If we add or subtract the same quantity to each side of an inequality, the inequality symbol remains the same.

2. If we multiply or divide each side of an inequality by a *positive* quantity, the inequality symbol remains the same.

3. If we multiply or divide each side of an inequality by a *negative* quantity, the inequality symbol is *reversed.*

These properties tell us what operations we can perform on an inequality and still obtain an equivalent inequality. Algebraically, we can write these properties as shown in the next box. (In each case, the $<$ symbol can be replaced by $>$, $\leq$, or $\geq$.)

PROPERTIES OF INEQUALITIES

1. If $a < b$ then $a + c < b + c$.

2. If $a < b$ then $a - c < b - c$.

3. If $a < b$ then $\begin{cases} ac < bc & \text{when } c \text{ is positive.} \\ ac > bc & \text{when } c \text{ is negative.} \end{cases}$

4. If $a < b$ then $\begin{cases} \dfrac{a}{c} < \dfrac{b}{c} & \text{when } c \text{ is positive.} \\ \dfrac{a}{c} > \dfrac{b}{c} & \text{when } c \text{ is negative.} \end{cases}$

EXAMPLE 3   In each of the following use the properties of inequalities to perform the indicated operation on the given inequality. Sketch the resulting equivalent inequality on a number line.

(a) $x - 5 < 4$;   add 5 to each side

(b) $a + 3 \geq -2$;   subtract 3 from each side

(c) $4y > 12$;   divide each side by 4

(d) $-3x \geq 6$;   divide each side by -3

(e) $3 < x + 2 \leq 7$;   subtract 2 from each member

3. (a) $t - 6 > 1$;   add 6 to both sides

(b) $t + 2 \leq 5$;   subtract 2 from both sides

(c) $5t < 10$;   divide both sides by 5

(d) $-7t \geq 21$;   divide both sides by -7

(e) $1 \leq t - 4 < 3$;   add 4 to each member

Solution

(a) $\begin{array}{rr} x - 5 < & 4 \\ +5 & +5 \\ \hline \boxed{x < \quad 9} \end{array}$ *The inequality symbol remains the same under addition (see property 1).*

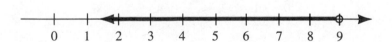

The open circle at 9 means that 9 is *excluded*.

(b) $\begin{array}{rr} a + 3 \geq & -2 \\ -3 & -3 \\ \hline \boxed{a \geq -5} \end{array}$ *The inequality symbol remains the same under subtraction (see property 2).*

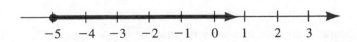

The solid circle at -5 means that -5 is *included*.

(c) $4y > 12$

$$\frac{4y}{4} > \frac{12}{4}$$

The inequality symbol remains the same when we divide by a **positive** *number (see property 4).*

$$\boxed{y > 3}$$

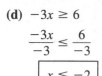

(d) $-3x \geq 6$

$$\frac{-3x}{-3} \leq \frac{6}{-3}$$

The inequality symbol **reverses** *when we divide by a* **negative** *number (see property 4).*

$$\boxed{x \leq -2}$$

(e) The double inequality $3 < x + 2 \leq 7$ has three members to it. They are 3, $x + 2$, and 7. In order to produce an equivalent inequality we must perform our operations on each member of the inequality. In this example we are asked to subtract 2 from each member.

$$\begin{array}{ccc} 3 < & x + 2 \leq & 7 \\ -2 & -2 & -2 \\ \hline \boxed{1 < x \leq 5} \end{array}$$

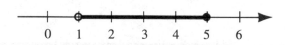

STUDY SKILLS

3.4

Preparing for Exams

Making Study Cards

Study cards are $3'' \times 5''$ or $5'' \times 8''$ index cards which contain summary information needed for convenient review. We will discuss three types of cards: the definition/principle card, the warning card, and the quiz card.

The *definition/principle* (**D/P**) *cards* are cards which contain a single definition, concept, or rule for a particular topic. The front of each D/P card should contain

1. A heading of a few words

2. The definition, concept, or rule accurately recorded

3. If possible, a restatement of the definition, concept, or rule in your own words

The back of the card should contain examples illustrating the idea on the front of the card.

Here is an example of a D/P card.

FRONT

The <u>Distributive</u> <u>Law</u>
$a(b+c) = ab + ac$
    or
$(b+c)a = ba + ca$
Multiply each <u>term</u> by a.

BACK

(1) $3x(x+2y) = 3x(x) + 3x(2y)$
                   $= 3x^2 + 6xy$
(2) $-2x(3x-y) = -2x(3x) - 2x(-y)$
                   $= -6x^2 + 2xy$
(3) $(2x+y)(x+y) = 2x(x+y) + y(x+y)$

Warning (**W**) *cards* are cards which contain errors that you may be consistently making on homework, quizzes, or exams, or those common errors pointed out by your teacher or your text. The front of the warning card should contain the word WARNING; the back of the card should contain an example of both the correct and incorrect way an example should be done. Be sure to label clearly which solution is correct and which is not. For example:

FRONT

WARNING
<u>EXPONENTS</u>
    An exponent refers only to the factors immediately to the left of the exponent.

BACK

EXAMPLES
$2 \cdot 3^2 = 2 \cdot 3 \cdot 3$ *NOT* $(2 \cdot 3)(2 \cdot 3)$
$(-3)^2 = (-3)(-3) = 9$
↑ Parentheses mean -3 is the factor to be squared.
BUT
$-3^2 = -3 \cdot 3 = -9$
↑ The factor being squared here is 3, not -3.

Quiz cards are another type of study card. We will discuss how to use them later. For now, go through your text and pick out a few of the odd-numbered exercises (just the problem) from each section, putting one or two problems on one side of each card. Make sure that you copy the *instructions* as well as the problem accurately. On the back of the card write down the exercise number and section of the book where the problem was found. For example:

FRONT

Translate the following algebraically using n as the <u>number.</u>

Four more than five times a number is two less than the number.

BACK

Exercise 17
Section 2.5

✔ *Answers to Learning Checks in Section 3.4*

1. (a) Yes   **(b)** Yes   **(c)** Yes   **(d)** No   **2. (a)** No   **(b)** Yes

3. (a) $t > 7$

(b) $t \leq 3$

(c) $t < 2$

(d) $t \leq -3$

(e) $5 \leq t < 7$

Exercises 3.4

In Exercises 1–18, determine whether the given value of the variable satisfies the inequality.

1. $x + 4 < 3$; $x = -2$

2. $x + 7 < 5$; $x = -1$

3. $a - 2 > -1$; $a = -3$

4. $a - 4 > 2$; $a = -4$

5. $-y + 3 \leq 5$; $y = -2$

6. $-y + 1 \leq 7$; $y = -6$

7. $2z - 5 < -3$; $z = 1$

8. $3z - 7 \leq 5$; $z = 4$

9. $5 + 2u > 12$; $u = 3$

10. $6 + 4u > 15$; $u = 2$

11. $7 - 4x < 8$; $x = -4$

12. $8 - 3x > -5$; $x = 5$

13. $-2 < 8 - x < 3$; $x = 6$

14. $-4 \leq 9 - 2x < 7$; $x = -1$

15. $6 + 2(a - 3) < 1$; $a = -2$

16. $5 + 3(a - 7) > -12$; $a = 2$

17. $-12 < 9 - 5(x + 1) < -5$; $x = 3$

18. $-5 < 8 - 2(x + 3) \leq 0$; $x = 4$

In Exercises 19–46, perform the indicated operations on the given inequality. Sketch the resulting inequality on a number line.

19. $x - 3 < 2$;   add 3 to each side

20. $x - 5 < 1$;   add 5 to each side

21. $a + 7 > 4$;   subtract 7 from each side

22. $a + 6 \geq 3$;   subtract 6 from each side

23. $3x \leq 12$;   divide each side by 3

24. $5x \geq 10$;   divide each side by 5

25. $4y > -8$;   divide each side by 4

26. $2y < -10$;   divide each side by 2

ANSWERS

1. _____
2. _____
3. _____
4. _____
5. _____
6. _____
7. _____
8. _____
9. _____
10. _____
11. _____
12. _____
13. _____
14. _____
15. _____
16. _____
17. _____
18. _____
19. _____
20. _____
21. _____
22. _____
23. _____
24. _____
25. _____
26. _____

27. _____

28. _____

29. _____

30. _____

31. _____

32. _____

33. _____

34. _____

35. _____

36. _____

37. _____

38. _____

39. _____

40. _____

41. _____

42. _____

43. _____

44. _____

45. _____

46. _____

27. $-3x < 6$; divide each side by -3

28. $-6x > -12$; divide each side by -6

29. $-3x < -6$; divide each side by -3

30. $-6x > 12$; divide each side by -6

31. $7a > 0$; divide each side by 7

32. $2a < 0$; divide each side by 2

33. $-7a \geq 0$; divide each side by -7

34. $-2a < 0$; divide each side by -2

35. $-x < 3$; multiply each side by -1

36. $-x < 3$; divide each side by -1

37. $-x < -3$; multiply each side by -1

38. $-x < -3$; divide each side by -1

39. $-5 < a - 4 \leq 2$; add 4 to each member

40. $-2 < a - 5 < 1$; add 5 to each member

41. $1 \leq x + 3 \leq 5$; subtract 3 from each member

42. $2 < x + 4 \leq 6$; subtract 4 from each member

43. $-6 < 3y < 3$; divide each member by 3

44. $0 \leq -2x < 2$; divide each member by -2

45. $-5 < -x < -1$; multiply each member by -1

46. $-5 < -x < -1$; divide each member by -1

QUESTIONS FOR THOUGHT

7. What does it mean for one number to be less than another?

8. Summarize the properties of inequalities in two sentences.

9. What is the basic difference between the properties of equalities and those of inequalities?

10. Explain in words what the inequality $2 < x \leq 5$ means.

Solving First-Degree Inequalities in One Variable

3.5

If we put the method we developed in Section 3.2 for solving first-degree equations together with the properties of inequalities discussed in the last section, we can formulate the procedure described in the accompanying box.

> **TO SOLVE A FIRST-DEGREE INEQUALITY**
>
> Use the same procedure as in solving a first-degree equation, *except* that when multiplying or dividing the inequality by a *negative* quantity the inequality symbol must be *reversed*.

Let's apply this outline to several examples.

EXAMPLE 1 *Solve for x.* $x + 4 < 6$

Solution Our goal is to isolate x and obtain an obvious inequality. If the example were an equation, that is, $x + 4 = 6$, we would subtract 4 from each side. Therefore, according to our outline we proceed here in exactly the same way:

$$
\begin{array}{r}
x + 4 < 6 \\
-4 \quad -4 \\
\hline
\boxed{x < 2}
\end{array}
$$
   The inequality symbol remains the same under subtraction.

The solution $x < 2$ means that *any number* less than 2 makes the original inequality true. It is impossible to check every number less than 2 in the inequality. Instead, we can check a number less than 2 to see that it does satisfy the inequality, *and* a number greater than 2 to see that it does not satisfy the inequality.

CHECK:

We can choose any number less than 2, say $x = 1$ because $1 < 2$. This *should* satisfy the inequality.

$$
\begin{array}{c}
x + 4 < 6 \\
1 + 4 \overset{?}{<} 6 \\
5 \overset{\checkmark}{<} 6
\end{array}
$$

We can choose any number greater than or equal to 2, say $x = 3$ because $3 > 2$. This *should not* satisfy the inequality.

$$
\begin{array}{c}
x + 4 < 6 \\
3 + 4 \overset{?}{<} 6 \\
7 \not< 6
\end{array}
$$

While this is not a conclusive check, as in the case of an equation, it does make us feel much more confident about our solution. ∎

EXAMPLE 2 *Solve for t.* $-4t < 8$

✔ LEARNING CHECKS

1. $y + 5 < 8$

2. $-6y < 12$

Solution

$$-4t < 8 \qquad \textit{We divide both sides by } -4 \textit{ to isolate } t.$$

$$\frac{-4t}{-4} > \frac{8}{-4} \qquad \textit{Since we are dividing by a \textbf{negative} number, we \textbf{reverse} the inequality symbol.}$$

$$\boxed{t > -2}$$

CHECK:

We can choose any number greater than -2, say $t = -1$. This should satisfy the inequality.	We can choose any number less than -2, say $t = -3$. This should not satisfy the inequality.
$-4t < 8$	$-4t < 8$
$-4(-1) \overset{?}{<} 8$	$-4(-3) \overset{?}{<} 8$
$4 \overset{\checkmark}{<} 8$	$12 \not< 8$ ∎

3. $4y - 3 \le 17$

EXAMPLE 3 *Solve for a.* $\quad 3a + 5 \ge 2$

Solution

$$\begin{array}{r} 3a + 5 \ge \quad 2 \\ \underline{-5 \quad -5} \\ 3a \ge -3 \end{array} \qquad \textit{The inequality symbol remains the same under subtraction.}$$

$$\frac{3a}{3} \ge \frac{-3}{3} \qquad \textit{The inequality symbol remains the same when we divide by a positive number. The fact that we are dividing \textbf{into} } -3 \textit{ is irrelevant.}$$

$$\boxed{a \ge -1}$$

CHECK:

We can choose any number greater than or equal to -1, say $a = 0$. This should satisfy the inequality.	We can choose any number less than -1, say $a = -2$. This should not satisfy the inequality.
$3a + 5 \ge 2$	$3a + 5 \ge 2$
$3(0) + 5 \overset{?}{\ge} 2$	$3(-2) + 5 \overset{?}{\ge} 2$
$5 \overset{\checkmark}{\ge} 2$	$-6 + 5 \overset{?}{\ge} 2$
	$-1 \not\ge 2$ ∎

4. $3y + 2 \le 7y + 14$

EXAMPLE 4 *Solve for x and sketch the solution set on a number line.*

$$5x - 1 < 7x + 9$$

Solution Since x appears on both sides of the inequality, we can choose to isolate x on either side. Let's do it both ways and compare the results.

Solution 1

$$\begin{array}{r} 5x - 1 < 7x + 9 \\ \underline{-7x \qquad -7x} \\ -2x - 1 < \quad 9 \\ \underline{+1 \quad +1} \\ -2x < \quad 10 \end{array}$$

$$\frac{-2x}{-2} > \frac{10}{-2}$$

Since we are dividing by a negative number the inequality symbol is reversed.

$$\boxed{x > -5}$$

Solution 2

$$\begin{array}{r} 5x - 1 < 7x + 9 \\ \underline{-5x \qquad -5x} \\ -1 < 2x + 9 \\ \underline{-9 \qquad -9} \\ -10 < 2x \end{array}$$

$$\frac{-10}{2} < \frac{2x}{2}$$

Since we are dividing by a positive number the inequality symbol remains the same.

$$\boxed{-5 < x}$$

It is important to look at these two answers and see that they say the same thing. They are both saying that x is greater than -5. Both methods of solution are correct.

The sketch of the solution set is shown in Figure 3.4.

Figure 3.4 *Solution set for Example 4*   ■

EXAMPLE 5   *Solve for x and sketch the graph of the solution set.*   **5.** $1 < y - 2 < 3$

$$2 < x + 4 < 5$$

Solution   First let's review what $2 < x + 4 < 5$ means. Essentially we are trying to find numbers which when added to 4 will yield a result which is both greater than 2 and less than 5. In other words, we are really solving two inequalities:

$$2 < x + 4 \qquad \text{and} \qquad x + 4 < 5$$

Both inequalities can be solved by applying the same properties.

$$
\begin{array}{ccc}
2 < x + 4 & & x + 4 < 5 \\
\underline{-4 \qquad -4} & & \underline{-4 \quad -4} \\
-2 < x & \text{and} & x < 1
\end{array}
$$

Combining these two answers and using the double inequality ("between") notation, we get

$$-2 < x < 1$$

Since we are doing the same thing to both inequalities we can think of the solution in the following simpler way. Solving a double inequality means that we want to isolate x in the middle.

$$
\begin{array}{l}
2 < x + 4 < 5 \\
\underline{-4 \qquad -4 \ -4} \\
\boxed{-2 < x < 1}
\end{array}
$$
We subtract 4 from each member of the inequality. The inequality symbol remains the same.

The sketch of the solution set is shown in Figure 3.5.

Figure 3.5   *Solution set for Example 5*

The check here requires us to choose three numbers: one number less than or equal to -2, which should not satisfy the inequality; one number between -2 and 1, which should satisfy the inequality; and one number greater than or equal to 1, which should not satisfy the inequality. It is left to the student to carry out this check.   ■

6. $7 \le 1 - 3y < 13$

EXAMPLE 6 *Solve for a.* $5 < 5 - 2a \le 11$

Solution We want to isolate a in the middle.

$$5 < 5 - 2a \le 11$$
$$\underline{-5 \quad -5 \qquad\quad -5}$$
$$0 < \quad -2a \le \quad 6$$
$$\frac{0}{-2} > \frac{-2a}{-2} \ge \frac{6}{-2}$$

We divide all three members of the inequality by -2. Both inequality symbols must be reversed.

$$\boxed{0 > a \ge -3}$$

The same answer could be written as

$$\boxed{-3 \le a < 0}$$

This second form of the inequality is much easier to visualize, although both forms are correct.

The check is left to the student. ∎

As was mentioned in the last section, not all inequalities involving variables are conditional. The last two examples in this section illustrate these possibilities.

7. $y - (7 - 2y) \le 3y + 2$

EXAMPLE 7 *Solve for x.* $5x - 3(x - 2) < 2x + 9$

Solution

$$5x - 3(x - 2) < 2x + 9$$
$$5x - 3x + 6 < 2x + 9$$
$$2x + 6 < 2x + 9$$
$$\underline{-2x \qquad\quad -2x}$$
$$6 < 9$$

The variable has been eliminated entirely and the resulting inequality is always true. Therefore, the original inequality is an *identity*. All values of x are solutions to this inequality. ∎

8. $2(y + 4) - (2y - 3) < 5$

EXAMPLE 8 *Solve for z.* $4z - (z - 7) < 3z + 4$

Solution

$$4z - (z - 7) < 3z + 4$$
$$4z - z + 7 < 3z + 4$$
$$3z + 7 < 3z + 4$$
$$\underline{-3z \qquad\quad -3z}$$
$$7 < 4$$

This statement is always false.

Therefore, the original inequality is a *contradiction*. No values of z are solutions to this inequality. ∎

Using Study Cards

The very process of making up study cards is a learning experience in itself. Study cards are convenient to use—you can carry them along with you and use them for review in between classes or as you wait for a bus.

Use the (D/P and W) cards as follows:

1. Look at the beginning of a card and, covering the rest of the card, see if you can remember what the rest of the card says.

2. Continue this process with the remaining cards. Pull out those cards you know well and put them aside, but do review them from time to time. Study those cards you do not know.

3. Shuffle the cards so that they are in random order and repeat the process again from the beginning.

4. As you go through the cards, ask yourself the following questions (where appropriate):
   (a) When do I use this rule, method, or principle?
   (b) What are the differences and similarities between problems?
   (c) What are some examples of the definitions or concepts?
   (d) What concept is illustrated by the problem?
   (e) Why does this process work?
   (f) Is there a way to check this problem?

✔ *Answers to Learning Checks in Section 3.5*

1. $y < 3$ **2.** $y > -2$ **3.** $y \le 5$ **4.** $y \ge -3$

5. $3 < y < 5$

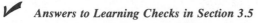

6. $-4 < y \le -2$ **7.** Identity   **8.** Contradiction

NOTE TO THE STUDENT

Use the space on this page to write down any questions you have or points you want to review with your instructor.

Exercises 3.5

Solve each of the following inequalities:

1. $x + 5 < 3$ **2.** $x + 7 < 4$

3. $a - 2 > -3$ **4.** $a - 5 > -2$

5. $2y < 8$ **6.** $4y < 12$

7. $2y > -8$ **8.** $4y > -12$

9. $-2y < 8$ **10.** $-4y > 12$

11. $-2y > -8$ **12.** $-4y < -12$

13. $-x < 4$ **14.** $-t > 1$

15. $-1 > -y$ **16.** $-w > -6$

1. _____

2. _____

3. _____

4. _____

5. _____

6. _____

7. _____

8. _____

9. _____

10. _____

11. _____

12. _____

13. _____

14. _____

15. _____

16. _____

17. _____

18. _____

19. _____

20. _____

21. _____

22. _____

23. _____

24. _____

25. _____

26. _____

27. _____

28. _____

29. _____

30. _____

17. $5x + 3 \leq 8$

18. $3x + 7 \geq 13$

19. $2x - 9 \geq 15$

20. $4x - 5 \geq -9$

21. $2(z - 3) + 4 \geq -6$

22. $4(z - 1) + 3z < -4$

23. $5(w + 3) - 7w \leq 7$

24. $2(w + 4) - 5w < 2$

25. $3(a + 4) - 4(a - 1) < 10$

26. $5(a - 2) - 6(a + 1) > -5$

27. $4(y - 3) - (3y - 12) \geq 2$

28. $7(y + 1) - (6y + 7) \leq 4$

29. $2(u + 2) - 2(u - 1) < 5$

30. $3(u - 1) - 3(u + 1) < 2$

31. $4(x - 2) - (4x - 3) < 6$ **32.** $5(x + 1) - (5x - 1) > 12$

33. $x + 3 < 2x + 7$ **34.** $2x + 5 < 3x + 8$

35. $2(a - 5) + 3a > 6a - 6$ **36.** $3(a - 4) + 5a \leq 9a - 8$

37. $2y - 4(y + 1) \leq 8 - (y + 2)$ **38.** $4y - 6(y - 2) \geq 10(y - 6)$

39. $2 < x + 7 < 10$ **40.** $1 < x + 5 < 9$

41. $3 < 2a + 5 < 7$ **42.** $2 < 3a + 2 \leq 8$

43. $1 \leq 6 - x < 3$ **44.** $0 < 9 - x \leq 5$

ANSWERS

31. _____

32. _____

33. _____

34. _____

35. _____

36. _____

37. _____

38. _____

39. _____

40. _____

41. _____

42. _____

43. _____

44. _____

ANSWERS

45. _____

46. _____

47. _____

48. _____

49. _____

50. _____

51. _____

52. _____

53. _____

54. _____

55. _____

56. _____

57. _____

58. _____

In Exercises 45–54, solve the inequality and sketch the solution set on a number line.

45. $x + 4 < 2x - 1$

46. $x + 5 < 2x - 3$

47. $3(a + 2) - 5a \geq 2 - a$

48. $5(a - 1) - 8a \geq 3 - a$

49. $-1 < x + 3 < 2$

50. $-3 < x + 5 < 4$

51. $-3 \leq 4t + 5 < 9$

52. $-6 < 5t - 1 \leq -1$

53. $-5 < 3 - 2x \leq 9$

54. $-1 \leq 2 - 3x < 11$

CALCULATOR EXERCISES

Solve each of the following equations. Round off your answers to the nearest hundredth.

55. $.8 - .45(x - 2) \leq .26$

56. $.5(x - .3) > .25(x + 3)$

57. $5.468 < 2.9t - 12.86 \leq 20.519$

58. $-15.45 \leq 53.67 - .45t < 36.93$

QUESTIONS FOR THOUGHT

11. What is *wrong* with each of the following "solutions"?

(a)
$$
\begin{array}{r}
2 + 7x \leq \quad 5x \\
\underline{-7x \quad\quad -7x} \\
2 \leq -2x \\
-2 \leq x
\end{array}
$$

(b)
$$
\begin{array}{r}
2x + 4 < \quad 2 \\
\underline{-4 \quad -4} \\
2x > -2 \\
x > -1
\end{array}
$$

(c)
$$
\begin{array}{r}
3x - 9 > \quad 6x \\
\underline{-3x \quad\quad -3x} \\
-9 > 3x \\
-3 < x
\end{array}
$$

12. Look at each of the following inequality statements and determine whether they make sense. Explain your answers.

(a) $-3 < x < 2$

(b) $-5 < x < -8$

(c) $7 < x < 4$

(d) $6 > x < 3$

(e) $3 < x < -2$

(f) $-5 > x > 4$

CHAPTER 3 SUMMARY

After having completed this chapter you should be able to:

1. Understand and recognize the basic types of equations (*conditional*, *identity*, and *contradiction*), and the properties of equality (Section 3.1).

2. Determine whether or not a particular value is a solution to a given equation or inequality (Sections 3.1, 3.4).

   For example:

   Does $x = -3$ satisfy the equation $3x - 4(2 - x) = -5$?

   $$3x - 4(2 - x) = -5 \qquad \textit{Substitute } x = -3.$$
   $$3(-3) - 4(2 - (-3)) \overset{?}{=} -5$$
   $$-9 - 4(5) \overset{?}{=} -5$$
   $$-9 - 20 \overset{?}{=} -5$$
   $$-29 \neq -5$$

3. Use the properties of equations and inequalities to solve first-degree equations and inequalities (Sections 3.2, 3.5).

   For example:

   (a) *Solve for t.* $3t - 4(2 - t) = 48$

   SOLUTION: $3t - 4(2 - t) = 48$ *Simplify.*
   $$3t - 8 + 4t = 48$$
   $$7t - 8 = 48$$
   $$\underline{+8 \quad +8} \qquad \textit{Add 8 to both sides.}$$
   $$7t = 56 \qquad \textit{Divide both sides by 7.}$$
   $$\frac{7t}{7} = \frac{56}{7}$$
   $$\boxed{t = 8}$$

   (b) *Solve for y.* $5 - 3(y + 1) < 8$

   SOLUTION: $5 - 3(y + 1) < 8$ *Simplify.*
   $$5 - 3y - 3 < 8$$
   $$2 - 3y < 8$$
   $$\underline{-2 \qquad\quad -2}$$
   $$-3y < 6$$
   $$\frac{-3y}{-3} > \frac{6}{-3} \qquad \textit{Inequality sign reverses.}$$
   $$\boxed{y > -2}$$

4. Solve verbal problems that give rise to first-degree equations in one variable (Section 3.3).

   For example:

   Six less than three times a number is 12 more than the number. Find the number.

SOLUTION: Let $n =$ the number. We translate the given relationship as follows:

Six less than | three times a number | is | 12 more than | the number
$3n - 6 =$ | | | $n + 12$

Thus, the equation is

$$3n - 6 = n + 12$$

$$
\begin{aligned}
3n - 6 &= n + 12 \\
-n \qquad &\quad -n \\
\hline
2n - 6 &= 12 \\
+6 \quad &\quad +6 \\
\hline
2n &= 18 \\
\frac{2n}{2} &= \frac{18}{2} \\
\boxed{n = 9}
\end{aligned}
$$

ANSWERS TO QUESTIONS FOR THOUGHT

1. A value satisfies or is a solution to an equation if both sides of the equation are equal when the value is substituted for the variable.

2. An identity is an equation that is always true, no matter what value is chosen for the variable. A contradiction is an equation that is never true, no matter what value is chosen for the variable. A conditional equation is one that is true for some values of the variable and false for the others.

3. If two quantities are equal, then we will not disturb this equality if we change each quantity in exactly the same way.

4. Two equations are called equivalent if their solution sets are exactly the same.

5. To check your answer after you have solved an equation, write your answer in place of the variable in the original equation. If the result is a true statement, then your answer satisfies the equation.

6. (a) After subtracting 4 from both sides of the equation, we should get $2x = 4$, not $2x = 12$.

   (b) Subtracting 3 from $3x$ will not give x. (These are not like terms.)

   (c) When we divide both sides of the equation by 3, we must remember to divide both terms on the left. We would then get $x - 2 = 4$, not $x - 6 = 4$.

   (d) We cannot divide both sides by x, since x might be equal to 0. (In this case, it is!)

   (e) In the last step, we must divide both sides by the same amount. If we divide by -3, we get $\dfrac{-3x}{-3} = \dfrac{-6}{-3}$, which gives $x = 2$.

   (f) We need to divide the final equation by -1 to obtain $x = -2$.

   (g) This is correct as written.

7. One number is less than a second number if the first number is located to the left of the second on the number line. (Equivalently, we could say that the second number is located to the right of the first on the number line.)

8. If we begin with an inequality, we may add the same quantity to both sides, subtract the same quantity from both sides, or multiply or divide both sides by the same positive quantity and get another inequality with the same symbol. If we multiply or divide both sides by the same negative quantity, we get another inequality with the reversed symbol.

9. When we multiply or divide both sides of an equality by the same quantity, it does not matter whether the quantity is positive or negative. This is not the case for inequalities (see previous question).

10. $2 < x \le 5$ means that x is between 2 and 5 on the number line, possibly equal to 5, but not equal to 2.

11. **(a)** Starting with $2 \le -2x$, we must divide both sides of this inequality by -2, obtaining $\dfrac{2}{-2} \ge \dfrac{-2x}{-2}$ or $-1 \ge x$.

   (b) When we subtract 4 from both sides of the original inequality, the resulting inequality should read $2x < -2$, not $2x > -2$. (We reverse the inequality sign only when we multiply or divide by a negative quantity.)

   (c) We must divide both sides of the inequality $-9 > 3x$ by 3, giving $\dfrac{-9}{3} > \dfrac{\cancel{3}x}{\cancel{3}}$ or $-3 > x$.

12. Only part **(a)** makes sense, since it requires that x must be a number between -3 and 2. Parts **(b)**, **(c)**, **(e)**, and **(f)** all lead to contradictions. For instance, the inequality $-5 < x < -8$ implies that $-5 < -8$, which is false. (The others are similar.) Part **(d)** makes no sense, since we cannot write double inequalities in which the inequality symbols point in opposite directions.

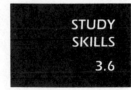

STUDY SKILLS 3.6

Reviewing Your Notes and Text; Reflecting

Another activity we suggested as an important facet of studying for exams is to review your notes and text. Your notes are a summary of the information you feel is important at the time you write them down. In the process of reviewing your notes and text you may turn up something you missed: some gap in your understanding may get filled which may give more meaning to (and make easier to remember) some of the definitions, rules, and concepts on your study cards. Perhaps you will understand a shortcut that you missed the first time around.

Reviewing the explanations or problems in the text *and* your notes gives you a better perspective and helps tie the material together. Concepts will begin to make more sense when you review and think about how they are interrelated. It is also important to practice review problems so that you will not forget those skills you have already learned. Do not forget to review old homework exercises, quizzes, and exams—especially those problems which were incorrectly done. Review problems also offer an excellent opportunity to work on your speed as well as your accuracy.

We discussed reflecting on the material you are reading and the exercises you are doing. Your thinking time is usually limited during an exam, and you want to anticipate variations in problems and make sure that your careless errors will be minimized at that time. For this reason it is a good idea to try to think about possible problems ahead of time. Make as clear as possible the distinctions that exist in those areas where you tend to get confused.

As you review material, ask yourself the study questions given in Study Skill 2.1. Also look at the Questions for Thought at the end of most of the exercise sets and ask yourself those questions as well.

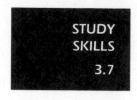

STUDY SKILLS 3.7

Preparing for Exams

Using Quiz Cards

A few days before the exam, select an appropriate number of problems from the quiz cards or old exams or quizzes, and make up a practice test for yourself. You may need the advice of your teacher as to the number of problems, and the amount of time to allow yourself for the test. If available, old quizzes and exams may help guide you.

Now find a quiet, well-lit place with no distractions, set your clock for the appropriate time limit (the same as your class exam will be) and take the test. Pretend it is a real test; that is, do not leave your seat or look at your notes, books, or answers until your time is up. (Before giving yourself a test you may want to refer to next chapter's discussion on taking exams.)

When your time is up, stop; you may now look up the answers and grade yourself. If you are making errors, check over what you are doing wrong. Find the section where those problem types are covered, review the material, and try more problems of that type.

If you do not finish your practice test on time, you should definitely work on your speed. Remember speed as well as accuracy are important on most exams.

Think about what you were doing as you took your test. You may want to change your test-taking strategy or reread the next chapter's discussion on taking exams. If you were not satisfied with your performance and you have the time after the review, make up and give yourself another practice test.

CHAPTER 3 REVIEW EXERCISES

In Exercises 1–4, determine whether the given equation or inequality is conditional, an identity, or a contradiction.

1. $5(x - 4) - 3(x - 3) = 3 - (14 - 2x)$

2. $3x - 4(x - 3) < 3 - (x - 4)$

3. $3a(a + 3) - a(2a + 4) = a^2 + 10$

4. $a(a - 5) - 2(a - 5) = a^2 - 7a + 10$

In Exercises 5–14, determine whether the given values of the variable satisfy the equation or inequality.

5. $2x - 5 = -7$; $x = -6, -1$

6. $3x - 7 = 5$; $x = -4, 4$

7. $4y + 3 \leq 10 - 3y$; $y = -1, 1$

8. $6w + 11 \geq 31 - 4w$; $w = -2, 2$

9. $3t + 2(t - 5) = 3t - 14$; $t = -2, 2$

10. $5z + 3(z - 6) = 10z - 12$; $z = -3, 3$

11. $8 - 3(x - 2) > x - 4$; $x = -5, 5$

12. $7 - 5(x - 4) \leq x + 2$; $x = -4, 4$

13. $a^2 + (a - 2)^2 = 20$; $a = -2, 2$

14. $u^2 - (u - 5)^2 = 4$; $u = -3, 3$

ANSWERS

1. _____
2. _____
3. _____
4. _____
5. _____
6. _____
7. _____
8. _____
9. _____
10. _____
11. _____
12. _____
13. _____
14. _____

ANSWERS

15. _____

16. _____

17. _____

18. _____

19. _____

20. _____

21. _____

22. _____

23. _____

24. _____

25. _____

26. _____

In Exercises 15–26, solve the equation or inequality.

15. $5x + 8 = 2x - 7$

16. $3x - 11 = 7x + 5$

17. $2(y + 4) - 2y = 8$

18. $3r + 2(r - 4) = r + 4(r - 1)$

19. $2(3a + 4) + 4 = 3(a - 1)$

20. $5(2t - 3) + 2(t - 2) = 5$

21. $8x - 3(x - 4) = 4(x + 3) + 28$

22. $4(x - 2) - 7x = 2x - 3(x + 2)$

23. $a(a + 3) - 2(a - 1) = a(a - 1) + 6$

24. $2w(w + 1) - 3(w - 2) = w^2 - 1 + w(w - 8)$

25. $8 - 3(x - 1) < 2$

26. $9 - 5(x - 2) \geq 4$

In Exercises 27–30, solve the inequality and sketch the solution set on a number line.

27. $2(x - 3) - 4(x - 1) \geq 7 - x$ **28.** $3(2x + 1) - 4(3x - 1) < 17 - 4x$

29. $2 \leq 3a + 8 < 20$ **30.** $-2 < 4 - 2t < 12$

*Solve each of the following problems **algebraically**.*

31. One number is 3 less than twice another. If their sum is 18, find the numbers.

32. The larger of two numbers is 7 less than 3 times the smaller. If their sum is 8 more than the smaller number, find the numbers.

33. The length of a rectangle is 4 more than 5 times the width. If the perimeter is 80 cm., find the dimensions of the rectangle.

34. The width of a rectangle is 8 less than twice the length. If the perimeter is 5 times the width, find the dimensions of the rectangle.

35. A manufacturer sells first-quality skirts for $12 each, and irregulars for $7 each. If a wholesaler spends $1,500 on 150 skirts, how many of each type did she buy?

36. A round trip by car takes 7 hours. If the rate going was 45 kph and the rate returning was 60 kph, how far was the round trip?

37. A laborer earns $6/hour regular wages and $9/hour for overtime (overtime being computed for working more than 40 hours per week). If a laborer wants to earn at least $348 during a certain week, what is the minimum number of overtime hours he must work?

ANSWERS

27. _____

28. _____

29. _____

30. _____

31. _____

32. _____

33. _____

34. _____

35. _____

36. _____

37. _____

ANSWERS

1. a. _____

   b. _____

   c. _____

2. a. _____

   b. _____

   c. _____

3. a. _____

   b. _____

   c. _____

   d. _____

   e. _____

4. _____

5. _____

6. _____

CHAPTER 3 PRACTICE TEST

1. Determine whether the given equation is conditional, an identity, or a contradiction.

   (a) $3x - 5(x - 2) = -2x + 8$

   (b) $3x - 5(x - 2) = -2x + 10$

   (c) $3x - 5(x - 2) = 2x - 10$

2. Determine whether the given value is a solution to the equation or inequality.

   (a) $2(x + 3) - (x - 1) = 9 - x;$ $x = 1$

   (b) $a^2 - 3a = a(a + 1) + 5;$ $a = -2$

   (c) $8 - 3(t - 4) < 10;$ $t = 3$

3. Solve each of the following equations or inequalities:

   (a) $8 - 3x = 3x - 10$

   (b) $2(3y - 5) - 4y = 2 - (y + 12)$

   (c) $2a^2 - 3(a - 4) = 2a(a - 6) + 3a$

   (d) Solve and sketch the solution set on a number line: $9 - 5(x - 2) \geq 4$

   (e) Solve and sketch the solution set on a number line: $1 < 3 - x \leq 5$

4. The length of a rectangle is 5 less than 4 times its width. If the perimeter is 11 more than 7 times the width, find the dimensions of the rectangle.

5. At a flea market used audio cassettes sell for $1 each while new ones sell for $3 each. If a person spends $46 on 20 cassettes, how many of each type did he buy?

6. If 8 less than twice a number is 3 less than the number, what is the number?

CUMULATIVE REVIEW
Chapters 1–3

ANSWERS

1. _____
2. _____
3. _____
4. _____
5. _____
6. _____
7. _____
8. _____
9. _____
10. _____
11. _____
12. _____
13. _____
14. _____
15. _____
16. _____
17. _____
18. _____
19. _____
20. _____
21. _____
22. _____
23. _____
24. _____
25. _____
26. _____
27. _____
28. _____
29. _____
30. _____
31. _____
32. _____

In Exercises 1–24, perform the indicated operations and simplify as completely as possible.

1. $-8 - 5 - 7$

2. $-8 - 5(-7)$

3. $12 - 4(3 - 5)$

4. $12 - (4 \cdot 3 - 5)$

5. -5^2

6. $(-5)^2$

7. xx^2x^3

8. $x^2x^3x^4 + x^5x^3x$

9. $x^2y - 2xy^2 - xy^2 - 3x^2y$

10. $3z^2 - 5z - 7 - 2z^2 - z - 1$

11. $2x(3x^2 - 4y)$

12. $2x(3x^2)(-4y)$

13. $-3u^2(u^3)(-5v)$

14. $-3u^2(u^3 - 5v)$

15. $4(m - 3n) + 3(2m - n)$

16. $7(2t^2 - 3r^3) + 5(r^3 - 3t^2)$

17. $2ab(a^2 - ab) - 4a^2(ab - b^2)$

18. $3x^2yz(xz - 6y^2) - 9y^2z(x^2 - 2x^2y)$

19. $x^2y - xy^2 - (xy^2 - x^2y)$

20. $8(x - y) - (8x - y)$

21. $x - 3[x - 4(x - 5)]$

22. $a - [b - a(b - a)]$

23. $3xy(4x^3y - 2y) - 2x(3y^2)(2x^3)$

24. $x(x - y) - y(y - x) - (x^2 - y^2)$

In Exercises 25–32, evaluate the given expression for $x = -2$, $y = -3$, and $z = 5$.

25. x^2

26. $-x^2$

27. $xy^2 - (xy)^2$

28. $(x - z)^2$

29. $|x - y - z|$

30. $xyz + xy - z$

31. $2x - 4y^2$

32. $|3x - z|$

ANSWERS	*In Exercises 33–44, solve the equation or inequality. In the case of an inequality, sketch the solution set on a number line.*

ANSWERS

33. _____

34. _____

35. _____

36. _____

37. _____

38. _____

39. _____

40. _____

41. _____

42. _____

43. _____

44. _____

45. _____

46. _____

47. _____

48. _____

In Exercises 33–44, solve the equation or inequality. In the case of an inequality, sketch the solution set on a number line.

33. $2x - 11 = 5x + 10$

34. $9 - 5t = 13 - 4t$

35. $9(a + 1) - 3(2a - 2) = 12$

36. $3(4w - 5) - (w + 2) = 16$

37. $4(5 - x) - 2(6 - 2x) = 8$

38. $8 - 3(a - 7) \leq 2$

39. $2(s + 4) + 3(2s + 2) > 6s$

40. $6(t - 3) + 4(6 - t) = 8 + 2(t - 5)$

41. $1 < 2y - 5 \leq 3$

42. $1 \leq 4 - z \leq 6$

43. $4(3d - 2) + 6(8 - d) = 9d - 4(d - 10)$

44. $2[x + 2(x + 2)] = 8$

Solve each of the following problems algebraically. Be sure to label clearly what the variable represents.

45. One number is 7 less than 3 times another. If their sum is 41 find the numbers.

46. The length of a rectangle is 8 more than 3 times the width. If the perimeter is 24 cm., find the dimensions of the rectangle.

47. A bakery charges 40¢ for "danishes" and 55¢ for "pastries." Louise pays $8.25 for an assortment of 18 danishes and pastries. How many of each type were in the assortment?

48. An office has an old copier which makes 20 copies per minute, and a newer model which makes 25 copies per minute. If the older machine begins making copies at 10:00 A.M. and is joined by the newer machine at 10:15 A.M., at what time will they have made a total of 885 copies?

CUMULATIVE PRACTICE TEST
Chapters 1–3

1. a. _____

1. Evaluate each of the following:

    (a) $-3^4 + 3(-2)^3$

    (b) $(3 - 7 - 2)^2$

    b. _____

2. Evaluate each of the following for $u = -4$, $v = -1$.

    (a) $(u - v)^2 - uv^2$

    (b) $|2u - 3v|$

    2. a. _____

3. Perform the indicated operations and simplify as completely as possible.

    (a) $5x^2 - 4x - 8 - 7x^2 - x + 11$

    b. _____

    (b) $2(x - 3y) + 5(y - x)$

    3. a. _____

    (c) $3a(a^2 - 2b) - 5(a^3 - ab)$

    b. _____

    (d) $2(u - 4v) - 3(v - 2u) - (8u - 11v)$

    c. _____

    (e) $4x^2y^3(x - 5y) - 2xy(5y^2)(-2xy)$

    d. _____

    (f) $6 - a[6 - a(6 - a)]$

    e. _____

4. Solve each of the following equations or inequalities.

    (a) $8 - 5x = 14 - 3x$

    f. _____

    (b) $2(w - 4) - 3(w - 2) = 7$

    4. a. _____

    (c) $9 - 5(x - 2) \leq 4$

    b. _____

    c. _____

ANSWERS

(d) $3(a + 5) - 2(1 - a) = 1 - (8 - a)$

(e) $6(2 - z) - 5(3 - z) = 4 - z$

4. d. _____

(f) $5t + 2(t - 7) - (t - 3) = 6t - 11$

e. _____

5. Solve the following inequalities and sketch the solution set on a number line.

(a) $7 - 3a \geq 13$ (b) $1 \leq 4x - 3 < 17$

f. _____

6. Solve each of the following problems algebraically.

(a) One number is 5 less than 3 times another. If their sum is 27, find the numbers.

5. a. _____

b. _____

(b) A tire retailer pays $1,124 for 40 tires. Some were new tires costing $32 each, while the rest were retreads costing $19 each. How many of each type were there?

6. a. _____

(c) Judy completes a bike race at an average speed of 20 kph. If she had averaged 25 kph she would have finished the race 1 hour faster. What was the distance that the race covered?

b. _____

c. _____

Rational Expressions

- 4.1 Fundamental Principle of Fractions
- 4.2 Multiplying and Dividing Rational Expressions
- 4.3 Adding and Subtracting Rational Expressions
- 4.4 Solving Fractional Equations and Inequalities
- 4.5 Ratio and Proportion
- 4.6 Verbal Problems

STUDY SKILLS

Now that we have examined addition, subtraction, and multiplication of simple algebraic expressions, we are going to turn our attention to division. This will lead us to a new type of expression called an *algebraic fraction*—that is, a fractional expression such as $\dfrac{x+1}{3x}$, in which variables appear in the *numerator* (top) and/or *denominator* (bottom). For our purposes the phrases *algebraic fractions* and *rational expressions* are synonymous. We will then be led quite naturally to see how to perform the four basic arithmetic operations with algebraic fractions.

In the sections that follow we will begin by reviewing a procedure for arithmetic fractions, and then see how the same procedure carries over to algebraic fractions.

Fundamental Principle of Fractions

4.1

Recall that an arithmetic fraction (rational number) is simply the quotient of two integers, where the denominator is not equal to 0. In Chapter 1 we defined the set of rational numbers as those numbers which *can be* represented as the quotient of integers.

For example, $\frac{2}{3}$, $-\frac{3}{5}$, and 4 are all arithmetic fractions. Remember that every integer is also a fraction since it has an understood denominator of 1 which, while it is seldom written, is very useful to think of in many situations. Thus, $4 = \frac{4}{1}$.

We can locate any rational number, $\dfrac{p}{q}$, on the number line by recalling that $\dfrac{p}{q}$ means "p of q equal parts." For example, $\frac{5}{8}$ is 5 of 8 equal parts. In order to locate the point $\frac{5}{8}$, we divide the interval between 0 and 1 on the number line into 8 equal parts, and count out 5 of these parts to the right of 0:

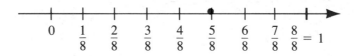

Similarly, $-\frac{5}{4}$ would be 5 of 4 equal parts counted out to the left of 0:

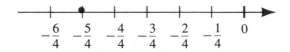

The basic starting point in dealing with and understanding fractions is simply recognizing the fact that two fractions can look different and yet be equal (represent the same amount). It does not make any difference if a certain whole is divided into 3 equal parts and you have 2, or if that same whole is divided into 6 equal parts and you have 4. In both cases you have the same amount, as indicated in the next figure. Thus, the fractions $\frac{2}{3}$ and $\frac{4}{6}$ are equal—they each represent the same amount.

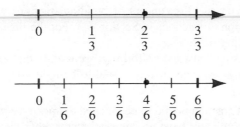

What we have just described is called the *Fundamental Principle of Fractions*. It says that the value of a fraction is unchanged if the numerator and denominator are both multiplied or divided by the same nonzero quantity. (We must specify nonzero for if not we would be dividing by 0 which, we have already seen, does not make any sense.) Algebraically we have the result given in the box.

FUNDAMENTAL PRINCIPLE OF FRACTIONS

$$\frac{a}{b} = \frac{a \cdot k}{b \cdot k} \qquad \text{where} \quad b, k \neq 0$$

If we read this from left to right, we are multiplying both the numerator and the denominator by k—this is called **building fractions to higher terms**. If we read from right to left, we are dividing both the numerator and the denominator by k—this is called **reducing the fraction to lower terms**.

$$\frac{12}{15} = \frac{12 \cdot 4}{15 \cdot 4} = \frac{48}{60} \qquad \textit{This is an example of building fractions.}$$

$$\frac{12}{15} = \frac{4 \cdot 3}{5 \cdot 3} = \frac{4 \cdot \cancel{3}}{5 \cdot \cancel{3}} = \frac{4}{5} \qquad \textit{This is an example of reducing fractions.}$$

The slash indicates that we have divided both numerator and denominator by 3.

When using the Fundamental Principle of Fractions, the key thing to keep in mind is that we are allowed to cancel only *common factors* and not common terms. Thus, it is very important that we be able to distinguish terms from factors. Recall from Chapter 2 that a **term** is an algebraic expression that is connected by multiplication and/or division. Thus, for example, the expression $3at + xyz$ has two terms, each of which consists of three factors, while the expression $3xyz$ is one term consisting of four factors.

The following very common error should be noted:

$$\frac{3 + 2}{1 + 2} \neq \frac{3 + \cancel{2}}{1 + \cancel{2}} \neq 3 \qquad \text{because} \qquad \frac{3 + 2}{1 + 2} = \frac{5}{3} \neq 3$$

The 2's *cannot* be reduced because 2 is not a common factor, but a common term.

REMEMBER

$$\frac{a \cdot k}{b \cdot k} = \frac{a}{b} \qquad \text{but} \qquad \frac{a + k}{b + k} \neq \frac{a}{b} \qquad \text{where } k \neq 0$$

In words, this says **common terms cannot be reduced**.

We will further discuss the concept of building fractions when we get to addition and subtraction of fractions in Section 4.3. For the time being, we will concentrate on reducing fractions.

For the sake of simplicity and uniformity, we require that all final answers be reduced completely. Such an answer is said to be reduced to *lowest terms*.

> **DEFINITION**   A fraction is said to be reduced to *lowest terms* if the numerator and the denominator have no common factor other than 1.

✔ LEARNING CHECKS

1. (a) $\dfrac{45}{80}$

   (b) $\dfrac{16x^9}{8x^3}$

EXAMPLE 1   *Reduce to lowest terms.*

(a) $\dfrac{42}{70}$ (b) $\dfrac{18x^5}{12x^2}$

Solution

(a) Using the Fundamental Principle of Fractions, we get

$$\frac{42}{70} = \frac{3 \cdot 14}{5 \cdot 14} = \frac{3 \cdot \cancel{14}}{5 \cdot \cancel{14}} = \boxed{\frac{3}{5}}$$

This entire process is often written in a shorthand fashion as follows:

$$\frac{\overset{3}{\cancel{42}}}{\underset{5}{\cancel{70}}} = \boxed{\frac{3}{5}}$$

Writing the solution this way "hides" the fact that the common factor that was reduced was 14. Additionally, it is quite possible that we might not see that 14 is a common factor of both 42 and 70. Consequently, we could also have proceeded as follows:

$$\frac{42}{70} = \frac{6 \cdot 7}{10 \cdot 7} = \frac{6 \cdot \cancel{7}}{10 \cdot \cancel{7}} = \frac{6}{10} = \frac{3 \cdot 2}{5 \cdot 2} = \frac{3 \cdot \cancel{2}}{5 \cdot \cancel{2}} = \boxed{\frac{3}{5}}$$

In shorthand, this would be written

$$\frac{\overset{\overset{3}{\cancel{6}}}{\cancel{42}}}{\underset{\underset{5}{\cancel{10}}}{\cancel{70}}} = \boxed{\frac{3}{5}}$$

While $\frac{6}{10}$ is a *partially* reduced form of $\frac{42}{70}$, it is not the final answer because 6 and 10 still have a common factor of 2. The common factor of 2 must also be reduced for the fraction to be in *lowest terms*.

In order to make the reducing process as efficient as possible, start reducing with the *largest* common factor that you see.

(b) Keep in mind that this example could have been written $(18x^5) \div (12x^2)$ because the fraction bar is an alternative way of writing division.

In this example, as in all those that follow, we automatically exclude any value of the variable which makes the denominator equal to 0. Thus, for this expression, $x \neq 0$.

$$\frac{18x^5}{12x^2} = \frac{2 \cdot 3 \cdot 3xxxxx}{2 \cdot 2 \cdot 3xx} = \frac{\cancel{2} \cdot \cancel{3} \cdot 3\cancel{xx}xxx}{\cancel{2} \cdot 2 \cdot \cancel{3xx}} = \boxed{\frac{3x^3}{2}}$$

Or, we could write the solution as follows:

$$\frac{18x^5}{12x^2} = \frac{3 \cdot 6 \cdot x^3 \cdot x^2}{2 \cdot 6 \cdot x^2} = \frac{3 \cdot \cancel{6} \cdot x^3 \cdot \cancel{x^2}}{2 \cdot \cancel{6} \cdot \cancel{x^2}} = \boxed{\frac{3x^3}{2}}$$

The solution is usually written as

$$\frac{18x^5}{12x^2} = \frac{\overset{3}{\cancel{18}}\overset{x^3}{x^5}}{\underset{2}{\cancel{12}}x^2} = \boxed{\frac{3x^3}{2}}$$

∎

EXAMPLE 2   *Reduce to lowest terms.*

(a) $\dfrac{-12x^3y^4}{2x^5y^3}$ **(b)** $\dfrac{x^2}{x+2}$

Solution

(a) $\dfrac{-12x^3y^4}{2x^5y^3} = \dfrac{-6 \cdot 2x^3 \cdot y^3 \cdot y}{2x^3 \cdot x^2 \cdot y^3} = \dfrac{-6 \cdot \cancel{2x^3} \cdot \cancel{y^3} \cdot y}{\cancel{2x^3} \cdot x^2 \cdot \cancel{y^3}} = \boxed{\dfrac{-6y}{x^2}}$

As we indicated in the last example, much of this work is often done mentally, and the solution would look like

$$\frac{-12x^3y^4}{2x^5y^3} = \frac{\overset{-6}{\cancel{-12}}\overset{y}{x^3y^4}}{\underset{x^2}{\cancel{2x^5}}y^3} = \boxed{\frac{-6y}{x^2}}$$

(b) We cannot apply the Fundamental Principle of Fractions to $\dfrac{x^2}{x+2}$.

$\dfrac{x^2}{x+2}$ *cannot* be reduced because there is no common *factor*.

The x in the numerator is a factor but the x in the denominator is *not*—it is a term. The fraction as given is already in lowest terms.   ∎

EXAMPLE 3   *Reduce to lowest terms.*

(a) $\dfrac{(4x^3)(-3xy)}{(6xy)(x^2)}$ **(b)** $\dfrac{(3x^2y)^2}{(2xy^2)^3}$ **(c)** $\dfrac{8x - 7x}{4x^2 - 7x^2}$

Solution

(a) As with many problems of this type there is more than one correct method. We illustrate two possible approaches.

First approach	*Second approach*
We first multiply out, and then reduce.	We first reduce, and then multiply out.

$$\frac{(4x^3)(-3xy)}{(6xy)(x^2)} = \frac{-12x^4y}{6x^3y} \qquad\qquad \frac{(4x^3)(-3xy)}{(6xy)(x^2)} = \frac{(4x^3)(-3xy)}{(6xy)(x^2)}$$

First approach — Now reduce.

$$= \frac{\overset{-2}{\cancel{-12}}\overset{x}{x^4}y}{\cancel{6}x^3y}$$

$$= \frac{-2x}{1}$$

$$= \boxed{-2x}$$

Second approach:

$$= \frac{-12x^3}{6x^2}$$

We can reduce more.

$$= \frac{\overset{-2}{\cancel{-12}}\overset{x}{x^3}}{\cancel{6}x^2}$$

$$= \frac{-2x}{1}$$

$$= \boxed{-2x}$$

2. (a) $\dfrac{-18x^7y}{24x^3y^6}$

(b) $\dfrac{4a^3}{a^3 + 4}$

3. (a) $\dfrac{(3m^2)(-5mn)}{(mn)(10mn^2)}$

(b) $\dfrac{(4uv)^3}{(8u^2v)^2}$

(c) $\dfrac{4a^2 - 5a^2}{2a^4 - 10a^4}$

(b) We *cannot* cancel within the parentheses because of the outside exponents. Instead, we can proceed as follows.

$$\frac{(3x^2y)^2}{(2xy^2)^3} = \frac{(3x^2y)(3x^2y)}{(2xy^2)(2xy^2)(2xy^2)} \qquad \textit{Multiply.}$$

$$= \frac{9x^4y^2}{8x^3y^6} \qquad \textit{Now reduce.}$$

$$= \frac{\overset{x}{\cancel{9x^4y^2}}}{\underset{y^4}{\cancel{8x^3y^6}}}$$

$$= \boxed{\frac{9x}{8y^4}}$$

(c) Our first step must be to simplify the numerator and denominator by combining like terms. Then we can look further to see if we have any common factors that can be reduced.

$$\frac{8x - 7x}{4x^2 - 7x^2} = \frac{x}{-3x^2} \qquad \textit{Now we can reduce.}$$

$$= \frac{\cancel{x}}{\underset{x}{-3\cancel{x^2}}}$$

$$= \frac{1}{-3x}$$

$$= \boxed{-\frac{1}{3x}} \qquad \blacksquare$$

A comment is in order about the location of a minus sign in a fraction. Since dividing quantities with opposite signs yields an answer with a negative sign, it makes no difference whether the minus sign appears in the numerator, denominator, or in front of the fraction. However, the preferred form is to have the minus sign either in front or in the numerator.

$$-\frac{6}{7} = \frac{-6}{7} = \frac{6}{-7} \qquad \text{These fractions are all equal, but the first and second are the preferred forms.}$$

Notice that in part **(c)** of Example 3, we wrote our final answer in the preferred form.

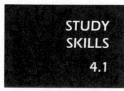

Answers to Learning Checks in Section 4.1

1. (a) $\dfrac{9}{16}$ (b) $2x^6$

2. (a) $\dfrac{-3x^4}{4y^5}$ (b) Cannot be reduced

3. (a) $\dfrac{-3m}{2n^2}$ (b) $\dfrac{v}{u}$ (c) $\dfrac{1}{8a^2}$

STUDY SKILLS

4.1

Taking an Algebra Exam

Just Before the Exam

You will need to concentrate and think clearly during the exam. For this reason it is important that you get plenty of rest the night before the exam, and that you have adequate nourishment.

It is *not* a good idea to study up until the last possible moment. You may find something that you missed and become anxious because there is not enough time to learn it. Then rather than simply missing a problem or two on the exam, the anxiety may affect your performance on the entire exam. It is better to stop studying some time before the exam and do something else. You could, however, review formulas you need to remember and warnings (common errors you want to avoid) just before the exam.

Also, be sure to give yourself plenty of time to get to the exam.

Exercises 4.1

Reduce each of the following rational expressions to lowest terms.

1. $\dfrac{18}{30}$

2. $\dfrac{12}{28}$

3. $\dfrac{-9}{21}$

4. $\dfrac{-10}{24}$

5. $\dfrac{-15}{-6}$

6. $\dfrac{-14}{-8}$

7. $\dfrac{-5 + 8}{10 - 4}$

8. $\dfrac{4 - 6}{4 - 2}$

9. $\dfrac{8 - 5(2)}{6 - 4(3)}$

10. $\dfrac{3 - 6(4)}{-2 - 3(4)}$

11. $\dfrac{x^3}{x}$

12. $\dfrac{y^5}{y^2}$

13. $\dfrac{x}{x^3}$

14. $\dfrac{y^2}{y^5}$

15. $\dfrac{10x}{4x^2}$

16. $\dfrac{8y^2}{6y^3}$

17. $\dfrac{-3z^6}{5z^2}$

18. $\dfrac{-7w^9}{9w^3}$

19. $\dfrac{12t^5}{30t^{10}}$

20. $\dfrac{-28u^8}{16u^4}$

21. $\dfrac{6ab^5}{-2a^3b^2}$

22. $\dfrac{10a^2b^6}{-5a^4b^4}$

23. $\dfrac{(2x^3)(6x^2)}{(4x)(3x^4)}$

24. $\dfrac{(4y^4)(5y^5)}{(8y^3)(y^6)}$

25. $\dfrac{(r^3t^2)(-rt^3)}{2r^2t^7}$

26. $\dfrac{(-a^4b^3)(a^2b^5)}{3a^3b^6}$

27. $\dfrac{3a(5b)(-4ab^3)}{6ab(2a^2b^2)}$

28. $\dfrac{-2s^2(8t^2)(5st)}{6s^3t^2(5st^3)}$

29. $\dfrac{(2x)^5}{(4x)^3}$

30. $\dfrac{(3y)^4}{(6y)^2}$

1. _____
2. _____
3. _____
4. _____
5. _____
6. _____
7. _____
8. _____
9. _____
10. _____
11. _____
12. _____
13. _____
14. _____
15. _____
16. _____
17. _____
18. _____
19. _____
20. _____
21. _____
22. _____
23. _____
24. _____
25. _____
26. _____
27. _____
28. _____
29. _____
30. _____

ANSWERS

31. _____

32. _____

33. _____

34. _____

35. _____

36. _____

37. _____

38. _____

39. _____

40. _____

41. _____

42. _____

43. _____

44. _____

31. $\dfrac{(-4x^3)^2}{(-2x^4)^3}$

32. $\dfrac{(-6y^2)^2}{(-2y^4)^3}$

33. $\dfrac{(2xy^2)^3}{(4x^2y^3)^3}$

34. $\dfrac{(3x^2y)^4}{(9xy^2)^2}$

35. $\dfrac{5x - 2x}{10x - 4x}$

36. $\dfrac{6y - 3y}{12y - 6y}$

37. $\dfrac{5a(2x)}{15a(8x)}$

38. $\dfrac{5a - 2x}{15a - 8x}$

39. $\dfrac{4s - 3t}{8s - 9t}$

40. $\dfrac{4s(3t)}{8s(9t)}$

41. $\dfrac{7a^2 - 5a^2 - 6a^2}{4a - 8a}$

42. $\dfrac{10z^2 - 8z^2 - 4z^2}{5z^3 - 7z^3}$

43. $\dfrac{5x^2 - 3x - x^2 + 2x}{6x^2 - 5x - 2x^2 + 4x}$

44. $\dfrac{2y^2 - 4y - y^2 - 2y}{5y^2 - 5y - 4y^2 - y}$

QUESTIONS FOR THOUGHT

1. In your own words, state the Fundamental Principle of Fractions.

2. If each step of the following were correct, what would you conclude?

$$2 = \frac{6}{3} = \frac{4+2}{1+2} = \frac{4+\cancel{2}}{1+\cancel{2}} = \frac{4}{1} = 4$$

Which step was *incorrect*? Explain why.

3. Discuss what is *wrong* with each of the following:

   (a) $\dfrac{3x^2}{2x} = \dfrac{3}{2x}$

   (b) $\dfrac{5x}{25x^2} = \dfrac{\cancel{5}\cancel{x}}{\cancel{5} \cdot 5kx} = \dfrac{0}{5x} = 0$

   (c) $\dfrac{2xy}{6x^3y^2} = \dfrac{\cancel{2}\cancel{x}\cancel{y}}{\cancel{2} \cdot 3\cancel{x}\cancel{y}x^2y} = 3x^2y$

   (d) $\dfrac{3x + 2x^2}{x} = \dfrac{3\cancel{k} + 2x^2}{\cancel{k}} = 3 + 2x^2$

4. Group the equivalent fractions together:

   (a) $-\dfrac{3}{4}$ (b) $\dfrac{-3}{-4}$ (c) $\dfrac{3}{-4}$ (d) $-\dfrac{-3}{-4}$

   (e) $\dfrac{-3}{4}$ (f) $-\dfrac{-3}{4}$ (g) $-\dfrac{3}{-4}$

Multiplying and Dividing Rational Expressions

4.2

Having begun to handle algebraic fractions in the last section, we now turn our attention to formulating methods for performing the arithmetic operations with them. Since the procedures are simpler for multiplication and division than for addition and subtraction we will begin with the former.

Multiplication

We take as our starting point the multiplication of ordinary arithmetic fractions. For example,

$$\frac{2}{3} \cdot \frac{5}{7} = \frac{2 \cdot 5}{3 \cdot 7} = \boxed{\frac{10}{21}}$$

The multiplication is accomplished by multiplying the numerators and dividing by the product of the denominators. It is quite natural then to extend this rule to the multiplication of any two algebraic fractions.

MULTIPLICATION OF RATIONAL EXPRESSIONS

$$\frac{a}{b} \cdot \frac{c}{d} = \frac{a \cdot c}{b \cdot d} \qquad b, d \neq 0$$

Before we look at an example involving variables, let's look at one more numerical example.

EXAMPLE 1 *Multiply.* $\frac{8}{9} \cdot \frac{3}{10}$

Solution Following the rule for multiplying fractions as stated in the box, we get

$$\frac{8}{9} \cdot \frac{3}{10} = \frac{8 \cdot 3}{9 \cdot 10}$$

$$= \frac{24}{90} \quad \text{\textit{Now we reduce the fraction.}}$$

$$= \frac{4 \cdot 6}{15 \cdot 6}$$

$$= \frac{4 \cdot \cancel{6}}{15 \cdot \cancel{6}}$$

$$= \boxed{\frac{4}{15}}$$

In getting the final answer in Example 1, we followed the ground rules laid down in the last section requiring that our final answer be reduced to lowest terms. Since when the multiplication is carried out, any factor in a numerator ends up in the numerator of the product, and any factor in a denominator ends

✔ **LEARNING CHECKS**

1. $\frac{4}{27} \cdot \frac{15}{16}$

up in the denominator of the product, it is much more efficient to reduce any common factors *before* we actually carry out the multiplication.

In other words, it is much easier to do Example 1 as follows:

$$\frac{8}{9} \cdot \frac{3}{10} = \frac{\overset{4}{\cancel{8}}}{\underset{3}{\cancel{9}}} \cdot \frac{\cancel{3}}{\underset{5}{\cancel{10}}}$$

$$= \frac{4 \cdot 1}{3 \cdot 5} = \boxed{\frac{4}{15}}$$

In the examples that follow we will adopt the approach of trying to reduce any common factors before we multiply.

2. $\dfrac{20}{a^8} \cdot \dfrac{a^4}{-6}$

EXAMPLE 2 *Multiply.* $\dfrac{-14}{x^6} \cdot \dfrac{x^3}{4}$

Solution

$$\frac{-14}{x^6} \cdot \frac{x^3}{4} = \frac{-14}{x^3 x^3} \cdot \frac{x^3}{4}$$

$$= \frac{\overset{-7}{\cancel{-14}}}{x^3 \cancel{x^3}} \cdot \frac{\cancel{x^3}}{\underset{2}{\cancel{4}}}$$

$$= \boxed{\frac{-7}{2x^3}}$$

∎

3. $\dfrac{-8u^2}{3v} \cdot \dfrac{9v^3}{10u^6}$

EXAMPLE 3 *Multiply.* $\dfrac{5x^3}{4y^2} \cdot \dfrac{-6y^8}{25x^4}$

Solution An example such as this one, which has many common factors that can be reduced, can be difficult to follow. Consequently, we will show each cancellation as a separate step. (The order we choose to reduce factors is arbitrary.) When you do such an exercise, however, you will most likely do the reducing all in one step.

$$\frac{5x^3}{4y^2} \cdot \frac{-6y^8}{25x^4} \qquad \textit{Reduce the 5 with the 25 and the 4 with the } -6.$$

$$= \frac{\cancel{5}x^3}{\underset{2}{\cancel{4}}y^2} \cdot \frac{\overset{-3}{\cancel{-6}}y^8}{\underset{5}{\cancel{25}}x^4} \qquad \textit{Now reduce } x^3 \textit{ with } x^4.$$

$$= \frac{\overset{}{\cancel{x^3}}}{2y^2} \cdot \frac{-3y^8}{5\underset{x}{\cancel{x^4}}} \qquad \textit{Finally we reduce } y^2 \textit{ with } y^8.$$

$$= \frac{1}{2\cancel{y^2}} \cdot \frac{-3\overset{y^6}{\cancel{y^8}}}{5x}$$

$$= \frac{-3y^6}{2 \cdot 5x} = \boxed{\frac{-3y^6}{10x}}$$

∎

Now that we have worked a bit with multiplication, let's look at the Fundamental Principle of Fractions again and see how the two ideas are related. We know that

$$\frac{k}{k} = 1, \qquad k \neq 0$$

Since multiplying by 1 does not change the value of an expression, we have the following:

$$\frac{a}{b} \cdot 1 = \frac{a}{b} \qquad \textit{Multiply by } 1.$$

$$\frac{a}{b} \cdot \frac{k}{k} = \frac{a}{b} \qquad \textit{Since } \frac{k}{k} = 1$$

$$\frac{a \cdot k}{b \cdot k} = \frac{a}{b} \qquad \textit{From the definition of multiplication}$$

This last line is exactly what the Fundamental Principle asserts. In other words, the Fundamental Principle simply says that multiplying by 1 does not change the value of a fraction.

When more than two fractions are to be multiplied together, we proceed in exactly the same way.

EXAMPLE 4 *Multiply.* $\dfrac{2}{3y} \cdot \dfrac{x}{5a} \cdot \dfrac{5}{8x^2}$

Solution We reduce the 2 with the 8; the 5 with the 5; the x with the x^2.

$$\frac{2}{3y} \cdot \frac{x}{5a} \cdot \frac{5}{8x^2} = \frac{\cancel{2}}{3y} \cdot \frac{\cancel{x}}{\cancel{5}a} \cdot \frac{\cancel{5}}{\cancel{8}x^2}$$

Note that the final answer

$$= \boxed{\frac{1}{12axy}} \qquad \textit{is } \frac{1}{12axy} \textit{ and not } 12axy.$$

Since all the factors in each numerator have been cancelled, we are left with a factor of 1 each time. ∎

EXAMPLE 5 *Multiply.* $5 \cdot \dfrac{3}{x}$

Solution When doing examples involving expressions some of which have denominators and some of which do not, it is a very good idea to put in the "understood" denominator of 1 for those expressions without a denominator. Thus, in this example we think of (and write) 5 in the fractional form $5 = \frac{5}{1}$.

$$5 \cdot \frac{3}{x} = \frac{5}{1} \cdot \frac{3}{x} = \boxed{\frac{15}{x}}$$

Do not make the mistake of multiplying both the numerator 3 and the denominator x by 5. ∎

Division

You probably remember the "rule" for dividing fractions as "invert and multiply." Before we state the rule explicitly and explain why the division of fractions is carried out in this way, let's keep in mind that division is defined to be the inverse of multiplication.

If someone asks "Why is $35 \div 7 = 5$?", a reasonable reply would be "because $5 \cdot 7 = 35$." In other words, the answer to a division problem is checked by multiplication. So let's state the rule for dividing fractions, and then verify it.

4. $\dfrac{5}{6m^2} \cdot \dfrac{pm}{4n} \cdot \dfrac{3n^2}{25p^2}$

5. $8 \cdot \dfrac{2}{a^2}$

DIVISION OF RATIONAL EXPRESSIONS

$$\frac{a}{b} \div \frac{c}{d} = \frac{a}{b} \cdot \frac{d}{c} \qquad b, c, d \neq 0$$

To see why division is defined in this way remember that we verified

$$35 \div 7 = 5 \qquad \text{because} \qquad 5 \quad \cdot \quad 7 = 35 \qquad \text{is true}$$

$$\frac{a}{b} \div \frac{c}{d} = \frac{a}{b} \cdot \frac{d}{c} \quad \text{because} \quad \left(\frac{a}{b} \cdot \frac{d}{c}\right) \cdot \frac{c}{d} = \frac{a}{b} \quad \text{is true!} \quad \text{since} \quad \frac{a}{b} \cdot \frac{d}{c} \cdot \frac{c}{d} = \frac{a}{b}$$

In other words, the rule for division was formulated in such a way as to make sure that it works out correctly.

Another definition would be appropriate here.

DEFINITION   The *reciprocal* of a nonzero number x is defined to be $\frac{1}{x}$.

Thus, the reciprocal of 3 is $\boxed{\dfrac{1}{3}}$

The reciprocal of $\dfrac{3}{4}$ is $\dfrac{1}{\frac{3}{4}}$. Remember that this means $1 \div \dfrac{3}{4}$. Therefore,

$$\frac{1}{\frac{3}{4}} = 1 \div \frac{3}{4} = 1 \cdot \frac{4}{3} = \boxed{\frac{4}{3}}$$

Thus, we can see that in general,

$$\boxed{\text{Reciprocal of } \frac{a}{b} \quad \text{is} \quad \frac{b}{a} \qquad \text{where} \quad a, b \neq 0}$$

In light of this definition, the rule for dividing fractions can be restated as

"To divide by a fraction, multiply by its reciprocal."

6. $\dfrac{10u^3v^4}{6w^2} \div \dfrac{15uv}{w}$

EXAMPLE 6   *Divide.* $\quad \dfrac{4x^2y^3}{5a^4} \div \dfrac{20ax}{y^3}$

Solution   Following the rule for division we get:

$$\frac{4x^2y^3}{5a^4} \div \frac{20ax}{y^3} = \frac{4x^2y^3}{5a^4} \cdot \frac{y^3}{20ax} \qquad \textit{Reduce the 4 with the 20, the x with the } x^2.$$

$$= \frac{\overset{x}{\cancel{4}x^2y^3}}{5a^4} \cdot \frac{y^3}{\underset{5}{\cancel{20}}a\cancel{x}}$$

$$= \frac{xy^3 \cdot y^3}{5a^4 \cdot 5a}$$

$$= \boxed{\frac{xy^6}{25a^5}}$$

■

EXAMPLE 7 *Divide.* $\dfrac{x}{y} \div (xy)$

Solution As we mentioned previously, it is very helpful to think of xy as $\dfrac{xy}{1}$.

$$\dfrac{x}{y} \div (xy) = \dfrac{x}{y} \div \dfrac{xy}{1} \qquad \textit{Use the rule for division.}$$

$$= \dfrac{x}{y} \cdot \dfrac{1}{xy}$$

$$= \dfrac{\cancel{x}}{y} \cdot \dfrac{1}{\cancel{x}y}$$

$$= \boxed{\dfrac{1}{y^2}}$$

∎

Since a quotient *is* a fraction, another way of expressing a quotient of fractions is by using a large fraction bar rather than the division sign, $\div$. Thus, we can write

$$\dfrac{a}{b} \div \dfrac{c}{d} \quad \text{as} \quad \dfrac{\dfrac{a}{b}}{\dfrac{c}{d}}$$

A fractional expression which contains fractions within it, such as the preceding one, is called a **complex fraction**.

EXAMPLE 8 *Change the following to a simple fraction reduced to lowest terms.* $\dfrac{\dfrac{3x^2}{2y}}{\dfrac{9xy^2}{4x}}$

Solution The given complex fraction means

$$\dfrac{3x^2}{2y} \div \dfrac{9xy^2}{4x}$$

We therefore apply the rule for the division of fractions to change the division to multiplication and to invert the divisor.

$$\dfrac{\dfrac{3x^2}{2y}}{\dfrac{9xy^2}{4x}} = \dfrac{3x^2}{2y} \div \dfrac{9xy^2}{4x}$$

$$= \dfrac{3x^2}{2y} \cdot \dfrac{4x}{9xy^2} \qquad \textit{Reduce.}$$

$$= \dfrac{\cancel{3}x^2}{\cancel{2}y} \cdot \dfrac{\overset{2}{\cancel{4}}x}{\underset{3}{\cancel{9}}xy^2}$$

$$= \boxed{\dfrac{2x^2}{3y^3}}$$

∎

7. $\dfrac{a^2}{b} \div (ab^2)$

8. $\dfrac{\dfrac{2m}{n^2}}{\dfrac{6}{mn}}$

Beginning the Exam

At the exam, make sure that you listen carefully to the instructions given by your instructor or the proctor.

As soon as you are allowed to begin, jot down the formulas you think you might need, and write some key words (warnings) to remind you to avoid common errors or errors you have previously made. Writing down the formulas first will relieve you of the burden of worrying about whether you will remember them when you need to, thus allowing you to concentrate more.

You should refer back to the relevant warnings as you go through the exam to make sure you avoid those errors.

Remember to read the directions carefully.

✔ *Answers to Learning Checks in Section 4.2*

1. $\dfrac{5}{36}$ 2. $\dfrac{-10}{3a^4}$ 3. $\dfrac{-12v^2}{5u^4}$ 4. $\dfrac{n}{40mp}$

5. $\dfrac{16}{a^2}$ 6. $\dfrac{u^2v^3}{9w}$ 7. $\dfrac{a}{b^3}$ 8. $\dfrac{m^2}{3n}$

Exercises 4.2

Perform the indicated operations. Final answers should be reduced to lowest terms.

1. $\dfrac{-4}{9} \cdot \dfrac{-2}{3}$ **2.** $\dfrac{6}{-25} \cdot \dfrac{-3}{5}$ **3.** $\dfrac{-6}{10} \cdot \dfrac{15}{9}$

4. $\dfrac{12}{20} \cdot \dfrac{-15}{8}$ **5.** $\dfrac{2}{3y} \cdot \dfrac{x}{5}$ **6.** $\dfrac{4}{5m} \cdot \dfrac{n}{7}$

7. $\dfrac{x^2}{4y} \cdot \dfrac{5x}{3y}$ **8.** $\dfrac{a^3}{2b} \cdot \dfrac{3a^2}{4b}$ **9.** $\dfrac{4}{5} \div \dfrac{5}{4}$

10. $\dfrac{-7}{10} \div \dfrac{10}{-7}$ **11.** $\dfrac{6x}{y} \div \dfrac{y^2}{2x^2}$ **12.** $\dfrac{8a}{3b} \div \dfrac{b^3}{4a^2}$

13. $\dfrac{3}{2t} \cdot \dfrac{tw}{6}$ **14.** $\dfrac{5}{4} \cdot \dfrac{mn}{20}$ **15.** $4 \cdot \dfrac{x}{12}$

16. $8 \cdot \dfrac{y}{4}$ **17.** $4 \div \dfrac{x}{12}$ **18.** $8 \div \dfrac{y}{4}$

19. $\dfrac{x}{12} \div 4$ **20.** $\dfrac{y}{4} \div 8$ **21.** $\dfrac{-2x}{3y^2} \cdot \dfrac{-9y}{4x}$

22. $\dfrac{3x^2}{-4y} \cdot \dfrac{-16y}{12x^3}$ **23.** $\dfrac{m^3n^2}{2m} \cdot \dfrac{6}{n^3}$ **24.** $\dfrac{3t}{r^2t^3} \cdot \dfrac{r^3}{9}$

25. $\dfrac{3uv^2}{5w} \div \dfrac{6u^2v}{15w}$ **26.** $\dfrac{21y^2z^2}{12uv} \div \dfrac{14yz}{3v}$ **27.** $6xy \cdot \dfrac{2x}{3y}$

28. $(10a^2b) \cdot \dfrac{2a}{5b}$ **29.** $6xy \div \dfrac{2x}{3y}$ **30.** $(10a^2b) \div \dfrac{2a}{5b}$

1. _____
2. _____
3. _____
4. _____
5. _____
6. _____
7. _____
8. _____
9. _____
10. _____
11. _____
12. _____
13. _____
14. _____
15. _____
16. _____
17. _____
18. _____
19. _____
20. _____
21. _____
22. _____
23. _____
24. _____
25. _____
26. _____
27. _____
28. _____
29. _____
30. _____

31. $\dfrac{2x}{3y} \div (6xy)$

32. $\dfrac{2a}{5b} \div (10a^2b)$

33. $\dfrac{-4x}{9y} \cdot \dfrac{x^2}{y^2} \cdot \dfrac{3y}{2x}$

34. $\dfrac{5m}{4n} \cdot \dfrac{m^3}{n^2} \cdot \dfrac{-2}{10m^2n}$

35. $\dfrac{9}{a^2}\left(\dfrac{a}{3} \div \dfrac{3}{a}\right)$

36. $\dfrac{x^2}{10}\left(\dfrac{2}{x} \div \dfrac{x}{5}\right)$

37. $\dfrac{9}{a^2} \div \left(\dfrac{a}{3} \cdot \dfrac{3}{a}\right)$

38. $\dfrac{x^2}{10} \div \left(\dfrac{2}{x} \cdot \dfrac{x}{5}\right)$

39. $\dfrac{9}{a^2} \div \left(\dfrac{a}{3} \div \dfrac{3}{a}\right)$

40. $\dfrac{x^2}{10} \div \left(\dfrac{2}{x} \div \dfrac{x}{5}\right)$

41. $\left(\dfrac{9}{a^2} \div \dfrac{a}{3}\right) \div \dfrac{a}{3}$

42. $\left(\dfrac{x^2}{10} \div \dfrac{x}{2}\right) \div \dfrac{x}{2}$

43. $\dfrac{\dfrac{2x}{3}}{\dfrac{10x}{9}}$

44. $\dfrac{\dfrac{5s}{8}}{\dfrac{15s}{2}}$

45. $\dfrac{\dfrac{x^2}{3}}{\dfrac{x}{6}}$

46. $\dfrac{\dfrac{y^2}{4}}{\dfrac{y}{10}}$

47. $\dfrac{\dfrac{x}{y^2}}{\dfrac{y}{x^2}}$

48. $\dfrac{\dfrac{y}{x^2}}{\dfrac{y^2}{x}}$

49. $\dfrac{\dfrac{2u}{z^2}}{\dfrac{4z}{u}}$

50. $\dfrac{\dfrac{3w}{t^3}}{\dfrac{t^2}{6w}}$

51. $\dfrac{3x^2 - x^2}{4y^2 - y^2} \cdot \dfrac{2y + y}{x^2 + x^2}$

52. $\dfrac{3x^2 \cdot x^2}{4y^2 \cdot y^2} \cdot \dfrac{2y \cdot y}{x^2 \cdot x^2}$

53. $\dfrac{4x^2 - x^2}{2y + y} \div \dfrac{4x - x}{2y^2 + y^2}$

54. $\dfrac{4x^2(x^2)}{2y(y)} \div \dfrac{4x(x)}{2y^2(y^2)}$

QUESTIONS FOR THOUGHT

5. Explain the rule for dividing fractions. Discuss why it works.

6. Explain what is *wrong* with each of the following:

(a) $\dfrac{3x}{2y} \div \dfrac{2y}{3x} = \dfrac{\cancel{3x}}{\cancel{2y}} \div \dfrac{\cancel{2y}}{\cancel{3x}} = 1$

(b) $5 \cdot \dfrac{3x}{2} = \dfrac{5 \cdot 3x}{5 \cdot 2} = \dfrac{15x}{10}$

Adding and Subtracting Rational Expressions

4.3

As was mentioned previously, many of the basic procedures in algebra have as their basis the Distributive Law, and the method for combining fractions is a prime illustration of this.

Recall that when we learned how to combine like terms, we reasoned by the Distributive Law that $5x + 3x = (5 + 3)x = 8x$. In light of our discussion of multiplication of fractions in the last section, we can see that any fraction $\frac{a}{c}$ can be written as $a \cdot \frac{1}{c}$ because

$$a \cdot \frac{1}{c} = \frac{a}{1} \cdot \frac{1}{c} = \frac{a}{c}$$

For example, $\frac{3}{7} = 3 \cdot \frac{1}{7}$. Thus, to add $\frac{3}{7} + \frac{2}{7}$ we can proceed as follows:

$$\frac{3}{7} + \frac{2}{7} = 3 \cdot \frac{1}{7} + 2 \cdot \frac{1}{7}$$

$$= (3 + 2) \cdot \frac{1}{7} \quad \textit{We have factored out the common factor of } \frac{1}{7}.$$

$$= 5 \cdot \frac{1}{7}$$

$$= \boxed{\frac{5}{7}}$$

This process can, of course, be used whenever the fractions have the same denominator. Such fractions are said to have a **common denominator.**

As long as the denominators are exactly the same, we can perform addition, subtraction, or a combination of addition and subtraction of several fractions. We can formulate this idea as follows:

RULE FOR ADDING AND SUBTRACTING RATIONAL EXPRESSIONS

$$\frac{a}{b} + \frac{b}{c} = \frac{a + b}{c} \qquad \frac{a}{c} - \frac{b}{c} = \frac{a - b}{c}, \quad c \neq 0$$

In words, this rule says that we can add or subtract fractions with common denominators by adding or subtracting the numerators and putting the result over the common denominator.

EXAMPLE 1 *Add.* $\dfrac{x}{6} + \dfrac{x + 3}{6}$

Solution Since the denominators are the same we can apply the above rule to add the numerators and keep the common denominator.

$$\frac{x}{6} + \frac{x + 3}{6} = \frac{x + x + 3}{6}$$

$$= \boxed{\frac{2x + 3}{6}} \quad \textit{Note that this answer } \textbf{cannot} \textit{ be reduced any further;}$$
$$\textit{the 2 and the 3 are } \textbf{not} \textit{ factors of the numerator.} \quad \blacksquare$$

LEARNING CHECKS

1. $\dfrac{2a + 5}{10} + \dfrac{a}{10}$

2. $\dfrac{3a^2 + 8}{6a} + \dfrac{3 - a^2}{6a} - \dfrac{11}{6a}$

EXAMPLE 2 *Combine.* $\dfrac{3x + 5}{4x} - \dfrac{7}{4x} + \dfrac{2 - x}{4x}$

Solution Again, since the denominators are the same, we simply "combine the numerators and keep the denominator."

$$\dfrac{3x + 5}{4x} - \dfrac{7}{4x} + \dfrac{2 - x}{4x} = \dfrac{3x + 5 - 7 + 2 - x}{4x} \qquad \textit{Combine like terms.}$$

$$= \dfrac{2x}{4x} \qquad \textit{Reduce.}$$

$$= \dfrac{2\cancel{x}}{\underset{2}{4\cancel{x}}} \qquad \begin{array}{l}\textit{Remember the understood } 1 \\ \textit{in the numerator.}\end{array}$$

$$= \boxed{\dfrac{1}{2}} \qquad\qquad\qquad\qquad \blacksquare$$

Even when the denominators are the same, subtraction problems often require extra care.

3. $\dfrac{5x - 4}{8} - \dfrac{x - 4}{8}$

EXAMPLE 3 *Subtract.* $\dfrac{3a - 7}{10} - \dfrac{a - 7}{10}$

Solution The rule we have stated tells us that we must subtract the *entire* second numerator from the first.

$$\dfrac{3a - 7}{10} - \dfrac{a - 7}{10} = \dfrac{3a - 7 - (a - 7)}{10} \qquad \begin{array}{l}\textit{Notice that the parentheses are} \\ \textit{essential; we must subtract the} \\ \textit{entire second numerator.}\end{array}$$

$$= \dfrac{3a - 7 - a + 7}{10} \qquad \textit{Because } -(a - 7) = -a + 7$$

$$= \dfrac{2a}{10} \qquad \textit{Reduce.}$$

$$= \dfrac{2a}{\underset{5}{10}}$$

$$= \boxed{\dfrac{a}{5}} \qquad\qquad\qquad\qquad \blacksquare$$

The next type of problem to consider is, quite naturally, one in which the denominators are not the same. If the denominators are not the same we will apply the Fundamental Principle of Fractions to build each denominator into a common denominator. In order to keep the fractions as simple as possible, we generally try to use what is called the **least common denominator** (LCD for short), which is the least common multiple of the denominators—that is, the "smallest" expression that is exactly divisible by each of the denominators.

While the LCD can sometimes be found by simply looking at the denominators, it is very useful, particularly for algebraic fractions, to have a mechanical procedure for finding it. We will first state the procedures, and then explain its use with a numerical example.

OUTLINE FOR FINDING THE LCD

Step 1   Factor each denominator as completely as possible.

Step 2   The LCD consists of the product of each *distinct* factor the *maximum* number of times it appears in any one denominator.

Unless it is stated to the contrary, whenever we discuss factoring we mean using integer factors.

A numerical example will help clarify the process.

EXAMPLE 4   *Combine.* $\dfrac{5}{18} + \dfrac{7}{24} - \dfrac{11}{30}$

4. $\dfrac{7}{12} - \dfrac{5}{36} + \dfrac{1}{20}$

Solution   Since the denominators are not the same, we seek the LCD, which we will find by using the outline given in the box. In dealing with a numerical example, to factor as completely as possible means to break each number into its prime factorization.

Step 1 $\left.\begin{array}{l} 18 = 2 \cdot 3 \cdot 3 \\ 24 = 2 \cdot 2 \cdot 2 \cdot 3 \\ 30 = 2 \cdot 3 \cdot 5 \end{array}\right\}$ *Notice that the **distinct** factors are 2, 3, 5.*

Step 2   We must make a decision for each distinct factor.

How many factors of 2 do we need?

2 appears as a factor once in 18, three times in 24, and once in 30. Therefore, according to the outline, we need to take *three* factors of 2. We take the factor 2 the *maximum* number of times it appears in any *one* denominator, *not* the total number of times it appears all together.

How many factors of 3 do we need?

3 appears as a factor twice in 18, once in 24, and once in 30. Therefore, we need to take *two* factors of 3.

How many factors of 5 do we need?

5 appears as a factor zero times in 18, zero times in 24, and once in 30. Therefore, we need to take *one* factor of 5.

Following the instructions in step 2 of the outline, we get

$$\text{LCD} = 2 \cdot 2 \cdot 2 \cdot 3 \cdot 3 \cdot 5 = 360$$

If you look at 360 in its factored form you can see each of the original denominators contained in it, and therefore it is a common denominator. For example, 360 "contains" 18 and 24 and 30.

$2 \cdot 2 \cdot \underbrace{2 \cdot 3 \cdot 3}_{18} \cdot 5$ and $\underbrace{2 \cdot 2 \cdot 2 \cdot 3}_{24} \cdot 3 \cdot 5$ and $2 \cdot 2 \cdot 3 \cdot \underbrace{2 \cdot 3 \cdot 5}_{30}$

On the other hand, because we chose each factor the *maximum* number of times it appears as a factor in any of the denominators, we do not have any extra factors. If we try to delete any of the factors, one of the denominators will not be represented. Thus, 360 is the *least common denominator*.

Now that we have decided on 360 and the LCD, we want to build each of our original denominators into 360 by applying the Fundamental Principle. The easiest way to see how to build a denominator into the LCD (360 in this example) is to look at the LCD in its factored form and fill in the missing factors. In other words, we look at 360 and each denominator in its factored form, and then use the Fundamental Principle to fill in the factors necessary to change the denominator into 360.

Looking at $18 = 2 \cdot 3 \cdot 3$ and $360 = 2 \cdot 2 \cdot 2 \cdot 3 \cdot 3 \cdot 5$, we can see that 18 is missing $2 \cdot 2 \cdot 5$. Thus, we multiply the numerator and denominator by $2 \cdot 2 \cdot 5$.

$$\frac{5}{18} = \frac{5}{2 \cdot 3 \cdot 3} = \frac{5}{2 \cdot 3 \cdot 3} \frac{(2 \cdot 2 \cdot 5)}{(2 \cdot 2 \cdot 5)} = \frac{100}{360}$$

Now we repeat this process for the other two fractions.

$$\frac{7}{24} = \frac{7}{2 \cdot 2 \cdot 2 \cdot 3} = \frac{7}{2 \cdot 2 \cdot 2 \cdot 3} \frac{(3 \cdot 5)}{(3 \cdot 5)} = \frac{105}{360}$$

$$\frac{11}{30} = \frac{11}{2 \cdot 3 \cdot 5} = \frac{11}{2 \cdot 3 \cdot 5} \frac{(2 \cdot 2 \cdot 3)}{(2 \cdot 2 \cdot 3)} = \frac{132}{360}$$

Even though we are focusing our attention on the denominator to tell us what the missing factors are in each case, the Fundamental Principle tells us nevertheless that in order to obtain an equivalent fraction we must multiply both the numerator and denominator by the same quantity.

Therefore, the original example has become

$$\frac{5}{18} + \frac{7}{24} - \frac{11}{30} = \frac{100}{360} + \frac{105}{360} - \frac{132}{360}$$

$$= \frac{100 + 105 - 132}{360}$$

$$= \boxed{\frac{73}{360}} \qquad \blacksquare$$

While it may seem to have taken a very long time to complete Example 4 due to all the explanations along the way, the effort was worthwhile because this same process can be used in every problem where we need to combine fractions with unlike denominators.

You may have previously learned to build arithmetic fractions by asking:

$\dfrac{5}{18} = \dfrac{?}{360}$, dividing 18 into 360, which gives 20, and then multiplying 5 times 20 to get $\dfrac{100}{360}$

While this procedure for building fractions works well for arithmetic fractions, it can become very messy with algebraic fractions. On the other hand, the process of building fractions by analyzing and filling in the missing factors works well for both types of fractions.

5. $\dfrac{4}{3a^3} + \dfrac{3}{5a^4}$

EXAMPLE 5 *Add.* $\dfrac{5}{2x^3} + \dfrac{7}{3x^2}$

Solution Since the denominators are not the same, we want to find the LCD. You may be able to "see" that the LCD is $6x^3$. If so, fine. If not, we can find it by following the outline.

Step 1 We factor each denominator completely.

$$\left.\begin{array}{l} 2x^3 = 2 \cdot x \cdot x \cdot x \\ 3x^2 = 3 \cdot x \cdot x \end{array}\right\} \quad \textit{The distinct factors are 2, 3, and x.}$$

While this step is not absolutely necessary here ($2x^3$ and $3x^2$ are already in factored form), it probably helps to make the distinct factors clearer.

Step 2 We take each distinct factor the maximum number of times it appears in any one denominator. Therefore, we take one factor of 2, one factor of 3, and three factors of x. We have the LCD $= 2 \cdot 3 \cdot x \cdot x \cdot x = 6x^3$.

Now we want to build each of our original fractions into an equivalent fraction having $6x^3$ as its denominator.

$$\frac{5}{2x^3} = \frac{5}{2 \cdot x \cdot x \cdot x} = \frac{5 \,(3)}{2 \cdot x \cdot x \cdot x \cdot (3)} = \frac{15}{6x^3}$$

$$\frac{7}{3x^2} = \frac{7}{3 \cdot x \cdot x} = \frac{7 \,(2x)}{3 \cdot x \cdot x \,(2x)} = \frac{14x}{6x^3}$$

Thus, the original example has become

$$\frac{5}{2x^3} + \frac{7}{3x^2} = \frac{15}{6x^3} + \frac{14x}{6x^3}$$
$$= \boxed{\frac{15 + 14x}{6x^3}}$$

It is important to note that this final answer cannot be reduced any further. The x which appears in the numerator is *not a factor* of the numerator. ∎

EXAMPLE 6 *Multiply.* $\dfrac{5}{2x^3} \cdot \dfrac{7}{3x^2}$

6. $\dfrac{4}{3a^3} \cdot \dfrac{3}{5a^4}$

Solution Read examples carefully! This is a multiplication problem, not an addition problem, so no common denominator is needed.

$$\frac{5}{2x^3} \cdot \frac{7}{3x^2} = \boxed{\frac{35}{6x^5}}$$ ∎

EXAMPLE 7 *Subtract.* $\dfrac{7}{4xy^2} - \dfrac{1}{6y^4}$

7. $\dfrac{5}{6u^2v} - \dfrac{4}{9v^3}$

Solution We begin by finding the LCD.

Step 1 $\left.\begin{array}{l} 4xy^2 = 2 \cdot 2 \cdot xyy \\ 6y^4 = 2 \cdot 3 \cdot yyyy \end{array}\right\}$ *The distinct factors are 2, 3, x, y.*

Step 2 LCD $= 2 \cdot 2 \cdot 3 \cdot xyyyy = 12xy^4$

Next we build the original fractions into equivalent fractions which have the LCD as denominator.

$$\frac{7}{4xy^2} = \frac{7\,(3y^2)}{4xy^2\,(3y^2)} = \frac{21y^2}{12xy^4}$$

$$\frac{1}{6y^4} = \frac{1\,(2x)}{6y^4\,(2x)} = \frac{2x}{12xy^4}$$

Thus, the original problem has become

$$\frac{7}{4xy^2} - \frac{1}{6y^4} = \frac{21y^2}{12xy^4} - \frac{2x}{12xy^4}$$

$$= \boxed{\frac{21y^2 - 2x}{12xy^4}}$$

∎

The procedure used in Example 7 can be applied to examples involving more than two fractions.

8. $\dfrac{5}{8uv^2} - \dfrac{1}{10u^2v} + \dfrac{3}{4uv}$

EXAMPLE 8 *Combine.* $\dfrac{9}{10x^2y^3} + \dfrac{7}{6xy^2} - \dfrac{2}{15x^3y}$

Solution Here is the solution without numbering all the steps and in a more concise form.

$$\left. \begin{array}{l} 10x^2y^3 = 2 \cdot 5 \cdot xxyyy \\ 6xy^2 = 2 \cdot 3 \cdot xyy \\ 15x^3y = 3 \cdot 5 \cdot xxxy \end{array} \right\} \quad \textit{The distinct factors are 2, 3, 5, x, and y.}$$

The LCD is $2 \cdot 3 \cdot 5 \cdot xxxyyy = 30x^3y^3$. For each fraction, look at the denominator to determine the missing factors, then apply the Fundamental Principle.

$$\frac{9}{10x^2y^3} + \frac{7}{6xy^2} - \frac{2}{15x^3y} = \frac{9\,(3x)}{10x^2y^3\,(3x)} + \frac{7\,(5x^2y)}{6xy^2\,(5x^2y)} - \frac{2\,(2y^2)}{15x^3y\,(2y^2)}$$

$$= \frac{27x}{30x^3y^3} + \frac{35x^2y}{30x^3y^3} - \frac{4y^2}{30x^3y^3}$$

$$= \frac{27x + 35x^2y - 4y^2}{30x^3y^3}$$

∎

9. $\dfrac{7b}{6a^3} - \dfrac{1}{2a} + a$

EXAMPLE 9 *Combine.* $\dfrac{5}{4x} + \dfrac{1}{2x^2} + 3y$

Solution It helps to think of $3y$ as $\dfrac{3y}{1}$.

$$\left. \begin{array}{l} 4x = 2 \cdot 2 \cdot x \\ 2x^2 = 2 \cdot xx \\ 1 = 1 \end{array} \right\} \quad \textit{The distinct factor are 2 and x.}$$

The LCD is $2 \cdot 2 \cdot x \cdot x = 4x^2$.

$$\frac{5}{4x} + \frac{1}{2x^2} + \frac{3y}{1} = \frac{5\,(x)}{4x\,(x)} + \frac{1\,(2)}{2x^2\,(2)} + \frac{3y\,(4x^2)}{1\,(4x^2)}$$

$$= \frac{5x}{4x^2} + \frac{2}{4x^2} + \frac{12x^2y}{4x^2}$$

$$= \boxed{\frac{5x + 2 + 12x^2y}{4x^2}}$$

STUDY SKILLS 4.3

Taking an Algebra Exam

What to Do First

Not all exams are arranged in ascending order of difficulty (from easiest to most difficult). Since time is usually an important factor, you do not want to spend so much time working on a few problems that you find difficult and then find that you do not have enough time to solve the problems that are easier for you. Therefore, it is strongly recommended that you first look over the exam and then follow the order given below:

1. Start with the problems which you know how to solve quickly.
2. Then go back and work on problems which you know how to solve but take longer.
3. Then work on those problems which you find more difficult, but for which you have a general idea of how to proceed.
4. Finally, divide the remaining time between the problems you find most difficult and checking your solutions. Do not forget to check the warnings you wrote down at the beginning of the exam.

You probably should not be spending a lot of time on any single problem. To determine the average amount of time you should be spending on a problem, divide the amount of time given for the exam by the number of problems on the exam. For example, if the exam lasts 50 minutes and there are 20 problems, you should spend an average of $\frac{50}{20} = 2\frac{1}{2}$ minutes per problem. Remember, this is just an estimate. You should spend less time on "quick" problems (or those worth fewer points), and more time on the more difficult problems (or those worth more points). As you work the problems be aware of the time; if half the time is gone you should have completed about half of the exam.

Answers to Learning Checks in Section 4.3

1. $\frac{3a + 5}{10}$ 2. $\frac{a}{3}$ 3. $\frac{x}{2}$ 4. $\frac{89}{180}$ 5. $\frac{20a + 9}{15a^4}$ 6. $\frac{4}{5a^7}$

7. $\frac{15v^2 - 8u^2}{18u^2v^3}$ 8. $\frac{25u - 4v + 30uv}{40u^2v^2}$ 9. $\frac{7b - 3a^2 + 6a^4}{6a^3}$

NOTE TO THE STUDENT

Use the space on this page to write down any questions you have or points you want to review with your instructor.

Exercises 4.3

In each of the following exercises perform the indicated operations. Express your answer as a single fraction reduced to lowest terms.

1. $\dfrac{5}{3} + \dfrac{4}{3}$

2. $\dfrac{6}{5} + \dfrac{4}{5}$

3. $\dfrac{3}{5} - \dfrac{7}{5}$

4. $\dfrac{5}{7} - \dfrac{8}{7}$

5. $\dfrac{7}{9} - \dfrac{5}{9} - \dfrac{8}{9}$

6. $\dfrac{4}{15} - \dfrac{11}{15} - \dfrac{2}{15}$

7. $\dfrac{2}{3} + \dfrac{4}{5}$

8. $\dfrac{3}{2} \cdot \dfrac{5}{7}$

9. $\dfrac{2}{3} \cdot \dfrac{4}{5}$

10. $\dfrac{3}{2} + \dfrac{5}{7}$

11. $\dfrac{2}{3} - \dfrac{5}{6}$

12. $\dfrac{1}{2} - \dfrac{7}{10}$

13. $3 + \dfrac{3}{4} - \dfrac{3}{8}$

14. $5 + \dfrac{5}{3} - \dfrac{5}{6}$

15. $\dfrac{8}{3x} + \dfrac{4}{3x}$

16. $\dfrac{5}{4y} + \dfrac{3}{4y}$

17. $\dfrac{8}{3x} \cdot \dfrac{4}{3x}$

18. $\dfrac{5}{4y} \cdot \dfrac{3}{4y}$

19. $\dfrac{3y}{7x} - \dfrac{5y}{7x} + \dfrac{4y}{7x}$

20. $\dfrac{4a}{5b} - \dfrac{7a}{5b} + \dfrac{6a}{5b}$

21. $\dfrac{w}{9z} - \dfrac{5w}{9z} + \dfrac{4w}{9z}$

22. $\dfrac{3m}{5n} - \dfrac{7m}{5n} + \dfrac{4m}{5n}$

23. $\dfrac{x+3}{3x} + \dfrac{x-6}{3x}$

24. $\dfrac{x+5}{2x} + \dfrac{2x-3}{2x}$

25. $\dfrac{3y^2-5}{4y} + \dfrac{5-4y^2}{4y}$

26. $\dfrac{2y^2-1}{6y} + \dfrac{y^2+1}{6y}$

1. _____
2. _____
3. _____
4. _____
5. _____
6. _____
7. _____
8. _____
9. _____
10. _____
11. _____
12. _____
13. _____
14. _____
15. _____
16. _____
17. _____
18. _____
19. _____
20. _____
21. _____
22. _____
23. _____
24. _____
25. _____
26. _____

ANSWERS

27. _____

28. _____

29. _____

30. _____

31. _____

32. _____

33. _____

34. _____

35. _____

36. _____

37. _____

38. _____

39. _____

40. _____

41. _____

42. _____

27. $\dfrac{5x + 2}{10x} - \dfrac{x + 2}{10x}$

28. $\dfrac{3x + 7}{4x} - \dfrac{x + 7}{4x}$

29. $\dfrac{w - 4}{6w} - \dfrac{w - 3}{6w} + \dfrac{5}{6w}$

30. $\dfrac{w - 3}{8w} - \dfrac{2w - 3}{8w} + \dfrac{w + 2}{8w}$

31. $\dfrac{3}{x} + \dfrac{2}{y}$

32. $\dfrac{4}{x} \cdot \dfrac{7}{y}$

33. $\dfrac{3}{x} \cdot \dfrac{2}{y}$

34. $\dfrac{4}{x} + \dfrac{7}{y}$

35. $\dfrac{5}{3x} - \dfrac{7}{2}$

36. $\dfrac{4}{5x} - \dfrac{5}{3}$

37. $\dfrac{5}{4x} + \dfrac{3}{2y}$

38. $\dfrac{7}{6y} + \dfrac{2}{3x}$

39. $\dfrac{4}{x^2} - \dfrac{3}{2x}$

40. $\dfrac{2}{x^3} - \dfrac{4}{3x}$

41. $\dfrac{4}{x^2} \cdot \dfrac{3}{2x}$

42. $\dfrac{2}{x^3} \cdot \dfrac{4}{3x}$

43. $\dfrac{7}{4a^2} - \dfrac{9}{20a}$

44. $\dfrac{5}{6a^3} - \dfrac{1}{18a^2}$

45. $\dfrac{1}{x} + 2$

46. $\dfrac{2}{y} + 1$

47. $\dfrac{5}{3xy} + \dfrac{1}{6y^2}$

48. $\dfrac{2}{5xy} + \dfrac{1}{10x^2}$

49. $\dfrac{7}{6a^2b} + \dfrac{3}{4ab^3}$

50. $\dfrac{5}{4ab^2} + \dfrac{9}{10a^3b}$

51. $\dfrac{7}{6a^2b} \cdot \dfrac{3}{4ab^3}$

52. $\dfrac{5}{4ab^2} \cdot \dfrac{9}{10a^3b}$

53. $\dfrac{3}{4m^2n} - \dfrac{5}{6mn^3} + \dfrac{1}{8n^2}$

54. $\dfrac{1}{6rt^2} - \dfrac{7}{4r^3t} + \dfrac{4}{9r^2}$

43. _____

44. _____

45. _____

46. _____

47. _____

48. _____

49. _____

50. _____

51. _____

52. _____

53. _____

54. _____

ANSWERS

55. _____

56. _____

57. _____

58. _____

59. _____

60. _____

55. $\dfrac{x}{y} + \dfrac{y}{x} + \dfrac{3x}{2y}$

56. $\dfrac{3y}{x} - \dfrac{x}{4y} + \dfrac{2y}{5x}$

57. $t - \dfrac{3}{t}$

58. $u^2 + \dfrac{2}{u}$

59. $\dfrac{a-5}{2} + \dfrac{3}{a}$

60. $\dfrac{a-4}{3} + \dfrac{5}{a}$

QUESTIONS FOR THOUGHT

7. What is the least common denominator? Why is it needed?

8. Explain why the procedure outlined in this section produces the LCD for a given set of denominators.

9. Discuss in detail what is *wrong* with each of the following.

(a) $\dfrac{x+3}{x} - \dfrac{5-x}{x} = \dfrac{x+3-5-x}{x} = \dfrac{-2}{x}$

(b) $\dfrac{2x}{y} + \dfrac{3y}{x} = \dfrac{2\cancel{x}}{\cancel{y}} + \dfrac{3\cancel{y}}{\cancel{x}} = 2 + 3 = 5$

(c) $\dfrac{5x}{2y} + \dfrac{7y}{6x} = \dfrac{5x}{6xy} + \dfrac{7y}{6xy} = \dfrac{5x+7y}{6xy}$

(d) $\dfrac{5x}{2y} + \dfrac{7y}{6x} = \dfrac{15x^2}{6xy} + \dfrac{7y^2}{6xy} = \dfrac{\overset{5\ x}{\cancel{15x^2}}}{\underset{2}{\cancel{6xy}}} + \dfrac{\overset{y}{\cancel{7y^2}}}{\cancel{6xy}} = \dfrac{5x}{2y} + \dfrac{7y}{6x} = \dfrac{5x+7y}{2y+6x}$

(e) $\dfrac{5x}{2y} + \dfrac{7y}{6x} = \dfrac{15x^2}{6xy} + \dfrac{7y^2}{6xy} = \dfrac{15x^2+7y^2}{6xy} = \dfrac{\overset{5\ x}{\cancel{15x^2}} + \overset{y}{\cancel{7y^2}}}{\underset{2}{\cancel{6xy}}} = \dfrac{5x+7y}{2}$

10. Look back at the solution to Example 8 of this section. The next to the last line of the solution contains three fractions which *can* be reduced. Why were they *not* reduced?

Solving Fractional Equations and Inequalities

4.4

Solving equations which contain fractional expressions involves combining the ideas we learned in Chapter 3 together with the material in the last section. By using the idea of the least common denominator and the Multiplication Property of equality, we can convert an equation involving fractional expressions into an equivalent equation without fractions. We can then solve it by the methods we have already learned.

Several examples will illustrate the process.

__EXAMPLE 1__ *Solve for x.* $\dfrac{x}{4} - \dfrac{2}{3} = \dfrac{7}{12}$

✔ **LEARNING CHECKS**

1. $\dfrac{a}{4} - \dfrac{1}{5} = \dfrac{11}{20}$

__Solution__ Before we attempt to isolate x on one side of the equation, we first ask if there is a way of eliminating the denominators. We need to multiply both sides of the equation by a number which is exactly divisible by each of the denominators, and so will cancel out each of the denominators. The LCD is exactly the smallest number which will do the job, so we multiply both sides of the equation by 12, which is the LCD for 3, 4, and 12.

$\dfrac{x}{4} - \dfrac{2}{3} = \dfrac{7}{12}$ *Multiply both sides by 12. According to the Multiplication Property of equality, multiplying both sides by 12 yields an equivalent equation.*

$12\left(\dfrac{x}{4} - \dfrac{2}{3}\right) = 12 \cdot \dfrac{7}{12}$ *We must use the Distributive Law on the left side.*

$\dfrac{12}{1} \cdot \dfrac{x}{4} - \dfrac{12}{1} \cdot \dfrac{2}{3} = \dfrac{12}{1} \cdot \dfrac{7}{12}$ *Note that each **term** gets multiplied by 12.*

$\dfrac{\overset{3}{\cancel{12}}}{1} \cdot \dfrac{x}{\cancel{4}} - \dfrac{\overset{4}{\cancel{12}}}{1} \cdot \dfrac{2}{\cancel{3}} = \dfrac{\cancel{12}}{1} \cdot \dfrac{7}{\cancel{12}}$

In this way we have obtained the following *equivalent* equation *without* fractions:

$$3x - 8 = 7$$
$$\underline{+8 \quad +8}$$
$$3x = 15$$

$$\dfrac{\cancel{3}x}{\cancel{3}} = \dfrac{15}{3}$$

$$\boxed{x = 5}$$

Note that we did not use the LCD to convert each of the original fractions into an equivalent one with the LCD as denominator. Rather, we used the LCD to multiply both sides of the equation to "clear" the denominators.

CHECK $x = 5$: $\dfrac{x}{4} - \dfrac{2}{3} = \dfrac{7}{12}$

$\dfrac{5}{4} - \dfrac{2}{3} \overset{?}{=} \dfrac{7}{12}$

$\dfrac{15}{12} - \dfrac{8}{12} \overset{?}{=} \dfrac{7}{12}$

$\dfrac{7}{12} \overset{\checkmark}{=} \dfrac{7}{12}$

2. $\dfrac{a}{4} - \dfrac{1}{5} + \dfrac{11}{20}$

At this point it is very important to distinguish between the example we have just completed, and the examples we did in the last section such as the one that follows.

EXAMPLE 2   *Combine.* $\dfrac{x}{4} - \dfrac{2}{3} + \dfrac{7}{12}$

Solution   Note that this is *not* an equation, and so we are not solving for x. We are going to use the LCD of 12, but *not* to eliminate the denominators. We cannot eliminate the denominators because we do not have an equation where we can multiply *both* sides. This example has only "one side." However, as we discussed in great detail in the last section, we do use the LCD to convert each fraction into an equivalent one with a denominator of 12.

$$\frac{x}{4} - \frac{2}{3} + \frac{7}{12} = \frac{x\,(3)}{4\,(3)} - \frac{2\,(4)}{3\,(4)} + \frac{7}{12}$$

$$= \frac{3x}{12} - \frac{8}{12} + \frac{7}{12}$$

$$= \frac{3x - 8 + 7}{12}$$

$$= \boxed{\frac{3x - 1}{12}} \qquad \textit{Since this was not an equation, we did not get a solution for } x. \qquad \blacksquare$$

Look over Examples 1 and 2 very carefully to make sure you see the difference between them.

---

3. $\dfrac{u - 4}{8} - u = \dfrac{5}{4}$

EXAMPLE 3   *Solve for a.* $\dfrac{a + 10}{6} + a = \dfrac{1}{2}$

Solution   Since the LCD is 6, we multiply both sides of the equation by 6, to "clear" the denominators.

$$\frac{a + 10}{6} + a = \frac{1}{2}$$

$$6\left(\frac{a + 10}{6} + a\right) = 6 \cdot \frac{1}{2}$$

$$\frac{6}{1} \cdot \left(\frac{a + 10}{6}\right) + 6 \cdot a = \frac{6}{1} \cdot \frac{1}{2}$$

$$\frac{\cancel{6}}{1} \cdot \left(\frac{a + 10}{\cancel{6}}\right) + 6a = \frac{\overset{3}{\cancel{6}}}{1} \cdot \frac{1}{\cancel{2}}$$

$$a + 10 + 6a = 3$$

$$7a + 10 = 3$$

$$7a = -7$$

$$\boxed{a = -1}$$

CHECK $a = -1$: $\dfrac{a + 10}{6} + a = \dfrac{1}{2}$

$$\dfrac{-1 + 10}{6} + (-1) \overset{?}{=} \dfrac{1}{2}$$

$$\dfrac{9}{6} - 1 \overset{?}{=} \dfrac{1}{2}$$

$$\dfrac{3}{2} - \dfrac{2}{2} \overset{?}{=} \dfrac{1}{2}$$

$$\dfrac{1}{2} \overset{\checkmark}{=} \dfrac{1}{2}$$ ∎

EXAMPLE 4 *Solve for a.* $\dfrac{1}{4}a + \dfrac{3a - 4}{10} = a - 4$

4. $\dfrac{5}{6}y + \dfrac{2y - 9}{15} = y - 1$

Solution The LCD for 4 and 10 is 20. We multiply both sides of the equation through by 20.

$$\dfrac{a}{4} + \dfrac{3a - 4}{10} = a - 4 \qquad \textit{Remember that } \dfrac{1}{4}a = \dfrac{1}{4} \cdot \dfrac{a}{1} = \dfrac{a}{4}.$$

$$20\left(\dfrac{a}{4} + \dfrac{3a - 4}{10}\right) = 20(a - 4)$$

$$\dfrac{20}{1} \cdot \dfrac{a}{4} + \dfrac{20}{1} \cdot \left(\dfrac{3a - 4}{10}\right) = 20(a - 4)$$

$$\dfrac{\overset{5}{\cancel{20}}}{1} \cdot \dfrac{a}{\cancel{4}} + \dfrac{\overset{2}{\cancel{20}}}{1} \cdot \left(\dfrac{3a - 4}{\cancel{10}}\right) = 20a - 80$$

$$5a + 2(3a - 4) = 20a - 80$$

$$5a + 6a - 8 = 20a - 80$$

$$11a - 8 = 20a - 80$$

$$\underline{-11a \qquad\quad -11a}$$

$$-8 = 9a - 80$$

$$\underline{+80 = \qquad\quad +80}$$

$$72 = 9a$$

$$\dfrac{72}{9} = \dfrac{9a}{9}$$

$$\boxed{8 = a}$$

CHECK $a = 8$: $\dfrac{1}{4}a + \dfrac{3a - 4}{10} = a - 4$

$$\dfrac{1}{4}(8) + \dfrac{3(8) - 4}{10} \overset{?}{=} 8 - 4$$

$$2 + \dfrac{24 - 4}{10} \overset{?}{=} 4$$

$$2 + \dfrac{20}{10} \overset{?}{=} 4$$

$$2 + 2 \overset{\checkmark}{=} 4$$ ∎

5. $\dfrac{6y + 7}{12} - \dfrac{3y + 3}{16} = \dfrac{3y + 11}{24}$

EXAMPLE 5 *Solve for y.* $\dfrac{8y}{9} - \dfrac{2y + 1}{12} = \dfrac{4y + 3}{18}$

Solution Here the LCD is not quite so obvious.

$$\left.\begin{array}{l} 9 = 3 \cdot 3 \\ 12 = 2 \cdot 2 \cdot 3 \\ 18 = 2 \cdot 3 \cdot 3 \end{array}\right\} \quad \textit{The LCD is } 2 \cdot 2 \cdot 3 \cdot 3 = 36.$$

$$36\left(\dfrac{8y}{9} - \dfrac{2y + 1}{12}\right) = 36\left(\dfrac{4y + 3}{18}\right)$$

$$\dfrac{36}{1} \cdot \dfrac{8y}{9} - \dfrac{36}{1} \cdot \left(\dfrac{2y + 1}{12}\right) = \dfrac{36}{1} \cdot \dfrac{4y + 3}{18}$$

$$\dfrac{\overset{4}{\cancel{36}}}{1} \cdot \dfrac{8y}{\cancel{9}} - \dfrac{\overset{3}{\cancel{36}}}{1} \cdot \left(\dfrac{2y + 1}{\cancel{12}}\right) = \dfrac{\overset{2}{\cancel{36}}}{1} \cdot \dfrac{4y + 3}{\cancel{18}}$$

$$4(8y) - 3(2y + 1) = 2(4y + 3) \qquad \textit{Watch the signs.}$$

$$32y - 6y - 3 = 8y + 6$$

$$26y - 3 = 8y + 6$$

$$\underline{-8y \qquad\quad -8y}$$

$$18y - 3 = 6$$

$$\underline{+3 \qquad +3}$$

$$18y = 9$$

$$\boxed{y = \dfrac{1}{2}}$$

CHECK $y = \frac{1}{2}$: $\dfrac{8y}{9} - \dfrac{2y + 1}{12} = \dfrac{4y + 3}{18}$

$$\dfrac{8\left(\dfrac{1}{2}\right)}{9} - \dfrac{2\left(\dfrac{1}{2}\right) + 1}{12} \overset{?}{=} \dfrac{4\left(\dfrac{1}{2}\right) + 3}{18}$$

$$\dfrac{4}{9} - \dfrac{1 + 1}{12} \overset{?}{=} \dfrac{2 + 3}{18}$$

$$\dfrac{4}{9} - \dfrac{2}{12} \overset{?}{=} \dfrac{5}{18}$$

$$\dfrac{4}{9} - \dfrac{1}{6} \overset{?}{=} \dfrac{5}{18}$$

$$\dfrac{8}{18} - \dfrac{3}{18} \overset{?}{=} \dfrac{5}{18}$$

$$\dfrac{5}{18} \overset{\checkmark}{=} \dfrac{5}{18}$$

■

As we saw in Chapter 3, our procedure for solving inequalities is basically the same as for solving equations, except that if we multiply or divide the inequality by a *negative* number we must *reverse* the inequality.

6. $\dfrac{4}{3} - \dfrac{a + 3}{6} \geq \dfrac{1}{2}$

EXAMPLE 6 *Solve for x.* $\dfrac{3}{5} - \dfrac{x - 3}{15} \leq \dfrac{1}{3}$

Solution The LCD is 15.

$$\frac{3}{5} - \frac{x-3}{15} \le \frac{1}{3}$$

$$15\left(\frac{3}{5} - \frac{x-3}{15}\right) \le 15 \cdot \frac{1}{3}$$

$$\frac{15}{1} \cdot \frac{3}{5} - \frac{15}{1} \cdot \left(\frac{x-3}{15}\right) \le \frac{15}{1} \cdot \frac{1}{3}$$

$$\frac{\overset{3}{\cancel{15}}}{1} \cdot \frac{3}{\cancel{5}} - \frac{\cancel{15}}{1} \cdot \left(\frac{x-3}{\cancel{15}}\right) \le \frac{\overset{5}{\cancel{15}}}{1} \cdot \frac{1}{\cancel{3}}$$

Be careful. The minus sign in front of the $x - 3$ belongs to both terms. It is incorrect to write $-x - 3$.

$$9 - (x - 3) \le \quad 5$$

The parentheses are necessary.

$$9 - x + 3 \le \quad 5$$

$$12 - x \le \quad 5$$

$$\underline{-12 \qquad \le -12}$$

$$-x \le -7$$

In order to get x alone, we can divide both sides by -1. Dividing by a negative number reverses the inequality.

$$\frac{-x}{-1} \ge \frac{-7}{-1}$$

$$\boxed{x \ge 7}$$

Recall that we cannot check every value of $x \ge 7$, so we check one value of x which is greater than or equal to 7 to verify that it is a solution and we check one value of $x < 7$ to verify that it is not.

CHECK $x = 7$: This should satisfy the inequality.

$$\frac{3}{5} - \frac{x-3}{15} \le \frac{1}{3}$$

$$\frac{3}{5} - \frac{7-3}{15} \overset{?}{\le} \frac{1}{3}$$

$$\frac{3}{5} - \frac{4}{15} \overset{?}{\le} \frac{1}{3}$$

$$\frac{9}{15} - \frac{4}{15} \overset{?}{\le} \frac{5}{15}$$

$$\frac{5}{15} \overset{\checkmark}{\le} \frac{5}{15}$$

CHECK $x = 6$: This should not satisfy the equation.

$$\frac{3}{5} - \frac{x-3}{15} \le \frac{1}{3}$$

$$\frac{3}{5} - \frac{6-3}{15} \overset{?}{\le} \frac{1}{3}$$

$$\frac{3}{5} - \frac{3}{15} \overset{?}{\le} \frac{1}{3}$$

$$\frac{9}{15} - \frac{3}{15} \overset{?}{\le} \frac{5}{15}$$

$$\frac{6}{15} \not\le \frac{5}{15}$$ ∎

EXAMPLE 7 *Solve for t.* $.23t + .7(t - 20) = 172$

7. $.6(x - 2) + .15x = 7.8$

Solution (If you would like to review decimal arithmetic see Section 0.5.)

While it is not necessary to rewrite this equation in fractional form, it will help us see that a problem involving decimals is handled in the same way as one involving fractions.

$$.23t + .7(t - 20) = 172$$

rewritten in fractional form becomes

$$\frac{23}{100}t + \frac{7}{10}(t - 20) = 172$$

We can now see that the LCD is 100. In other words, if we look at the original equation in decimal form, wanting to "clear" the decimals is equivalent to "clearing" the fractions. If we want to move the decimal point 1 place to the right we multiply through by 10; to move the decimal point 2 places to the right we multiply by 100; to move the decimal point 3 places to the right we multiply by 1,000, etc. In this example we want to move the decimal point 2 places to the right (so that .23 will become 23), and therefore we multiply both sides of the equation by 100.

$$100[.23t + .7(t - 20)] = 100(172)$$
$$100(.23t) + 100(.7)(t - 20) = 100(172)$$
$$23t + 70(t - 20) = 17,200$$
$$23t + 70t - 1,400 = 17,200$$
$$93t - 1,400 = 17,200$$
$$\underline{\quad +1,400 \quad +1,400}$$
$$93t = 18,600$$
$$\frac{93t}{93} = \frac{18,600}{93}$$
$$\boxed{t = 200}$$

CHECK $t = 200$:
$$.23t + .7(t - 20) = 172$$
$$.23(200) + .7(200 - 20) \stackrel{?}{=} 172$$
$$46 + .7(180) \stackrel{?}{=} 172$$
$$46 + 126 \stackrel{?}{=} 172$$
$$172 \stackrel{\checkmark}{=} 172$$
∎

Answers to Learning Checks in Section 4.4

1. $a = 3$ **2.** $\dfrac{5a + 7}{20}$ **3.** $u = -2$ **4.** $y = 12$ **5.** $y = \dfrac{1}{3}$
6. $a \leq 2$ **7.** $x = 12$

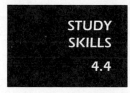

STUDY SKILLS 4.4

Taking an Algebra Exam

Dealing with Panic

In the first two chapters of this text we have given you advice on how to learn algebra. In the last chapter we discussed how to prepare for an algebra exam. If you followed this advice and put the proper amount of time to good use you should feel fairly confident and less anxious about the exam. But you may still find during the course of the exam that you are suddenly stuck or you "draw a blank." This may lead you to panic and say irrational things like "I'm stuck. . . . I can't do this problem. . . . I can't do any of these problems. . . . I'm going to fail this test." Your heart may start to beat faster and your breath may quicken. You are entering a panic cycle.

These statements are irrational. Getting stuck on a few problems does not mean that you cannot do any algebra. These statements only serve to interfere with your concentrating on the exam itself. How can you think about solving a problem while you are telling yourself that you cannot? The increased heart and breath rate are part of this cycle.

What we would like to do is to break this cycle. What we recommend that you do is first put aside the exam and silently say to yourself **STOP**! Then try to relax, clear your mind, and encourage yourself by saying to yourself such things as "This is only one (or a few) problems, not the whole test" or "I've done problems like this before, so I'll get the solution soon." (Haven't you ever talked to yourself this way before?)

Now take some slow deep breaths and search for some problems that you know how to solve and start with those. Build your concentration and confidence up slowly with more problems. When you are through with the problems you can complete go back to the ones you were stuck on. If you have the time, take a few minutes and rest your head on your desk, and then try again. But make sure you have checked the problems you have completed.

NOTE TO THE STUDENT

Use the space on this page to write down any questions you have or points you want to review with your instructor.

Exercises 4.4

Solve each of the following equations.

1. $\dfrac{x}{3} - 2 = \dfrac{2}{3}$

2. $\dfrac{x}{5} - 3 = \dfrac{4}{5}$

3. $\dfrac{3a}{4} + 2 = \dfrac{5}{4}$

4. $\dfrac{5a}{2} + 9 = \dfrac{3}{2}$

5. $\dfrac{u}{2} - \dfrac{u}{4} = 2$

6. $\dfrac{u}{3} - \dfrac{u}{9} = 4$

7. $\dfrac{y}{3} + \dfrac{y}{5} < \dfrac{8}{5}$

8. $\dfrac{y}{2} + \dfrac{y}{7} > \dfrac{9}{7}$

9. $x - \dfrac{2}{3}x = \dfrac{4}{3}$

10. $\dfrac{3}{5}x - 2x = \dfrac{14}{5}$

11. $\dfrac{a}{4} - \dfrac{a}{3} \geq \dfrac{5}{2}$

12. $\dfrac{a}{6} - \dfrac{a}{5} \leq \dfrac{4}{3}$

13. $.7x + .4x = 5.5$

14. $.6x + .8x = 8.4$

ANSWERS

1. _____

2. _____

3. _____

4. _____

5. _____

6. _____

7. _____

8. _____

9. _____

10. _____

11. _____

12. _____

13. _____

14. _____

15. _____

16. _____

17. _____

18. _____

19. _____

20. _____

21. _____

22. _____

23. _____

24. _____

25. _____

26. _____

15. $.3x - .25x = 2$

16. $.9x - .36x = 2.7$

17. $.8m + .05m = .34$

18. $.65m + 1.5m = .43$

19. $\dfrac{w+3}{4} = \dfrac{w+4}{3}$

20. $\dfrac{w-2}{6} = \dfrac{w-1}{9}$

21. $\dfrac{w+3}{4} + 1 = \dfrac{w+4}{3}$

22. $\dfrac{w-2}{6} - 2 = \dfrac{w-1}{9}$

23. $\dfrac{x+1}{2} + x = 11$

24. $\dfrac{x+3}{4} + x = 12$

25. $\dfrac{y}{6} - \dfrac{y-2}{4} > 1$

26. $\dfrac{y}{5} - \dfrac{y-1}{2} > 2$

27. $3 - \dfrac{a+1}{4} = \dfrac{a+4}{2}$

28. $4 - \dfrac{a+3}{10} = \dfrac{a+11}{5}$

27. _____

29. $\dfrac{2y-3}{2} - \dfrac{y-5}{3} = \dfrac{1}{6}$

30. $\dfrac{y+2}{14} - \dfrac{4y+1}{7} = 1$

28. _____

29. _____

31. $\dfrac{x+2}{3} - \dfrac{2x+3}{4} = \dfrac{x+4}{8}$

32. $\dfrac{5x-1}{2} - \dfrac{x-2}{5} = \dfrac{8x+11}{6}$

30. _____

31. _____

33. $\dfrac{t}{2} + \dfrac{t-1}{3} + \dfrac{t-6}{4} = t - 2$

34. $\dfrac{t}{6} + \dfrac{t+3}{5} + \dfrac{t-2}{10} = t - 6$

32. _____

33. _____

35. $.5(x+2) - .3(x-4) = 3$

36. $.6(x-4) - .4(x-5) = 4.6$

34. _____

35. _____

37. $3(y+2) + \dfrac{y+3}{5} = \dfrac{9y+8}{2}$

38. $5(y-3) + \dfrac{2-y}{3} = \dfrac{7y+1}{4}$

36. _____

37. _____

38. _____

39. $z + \dfrac{z+5}{3} - \dfrac{z-2}{6} = \dfrac{z+4}{4} + 1$

40. $5z - \dfrac{3-z}{2} + \dfrac{z+4}{5} = 8 - \dfrac{z+8}{3}$

39. _____

40. _____

41. $3 \le \dfrac{x}{3} - \dfrac{x+1}{2} \le 6$

42. $\dfrac{2}{5} < \dfrac{2x+3}{5} - \dfrac{3x+1}{7} < \dfrac{6}{7}$

41. _____

In the exercises that follow, if the exercise is an equation, solve it; if not, perform the indicated operations and express your answer as a single fraction.

42. _____

43. $\dfrac{x}{3} + \dfrac{x}{2} + \dfrac{x}{5}$

44. $x + \dfrac{x}{3} - \dfrac{x}{4}$

43. _____

44. _____

45. $\dfrac{x}{3} + \dfrac{x}{2} + \dfrac{x}{5} = 62$

46. $x + \dfrac{x}{3} - \dfrac{x}{4} = 26$

45. _____

46. _____

47. $\dfrac{x+5}{2} - \dfrac{x-1}{4} = 2$

48. $\dfrac{2x-1}{5} - \dfrac{x-7}{3} = 2$

47. _____

49. $\dfrac{x+5}{2} - \dfrac{x-1}{4}$

50. $\dfrac{2x-1}{5} - \dfrac{x-7}{3}$

48. _____

49. _____

50. _____

Ratio and Proportion

4.5

One of the most common and useful applications of the techniques we have developed so far is in dealing with the ideas of ratio and proportion.

A *ratio* is simply a fraction. If we say that the ratio of boys to girls in a certain class is 2 to 3, that means that there are 2 boys for every 3 girls. The ratio 2 to 3 is written as the fraction $\frac{2}{3}$ (sometimes it is also written 2 : 3). However, the ratio of girls to boys would be written $\frac{3}{2}$, since there are 3 girls for every 2 boys.

EXAMPLE 1 If a certain car dealer sold 350 domestic cars and 210 imported cars, find each of the following:

(a) The ratio of domestic cars sold to imported cars sold

(b) The ratio of imported cars sold to domestic cars sold

(c) The ratio of domestic cars sold to the total number of cars sold

Solution

(a) $\dfrac{\text{Number of domestic cars sold}}{\text{Number of imported cars sold}} = \dfrac{350}{210} = \dfrac{5 \cdot 70}{3 \cdot 70} = \boxed{\dfrac{5}{3}}$

(b) $\dfrac{\text{Number of imported cars sold}}{\text{Number of domestic cars sold}} = \dfrac{210}{350} = \dfrac{3 \cdot 70}{5 \cdot 70} = \boxed{\dfrac{3}{5}}$

(c) $\dfrac{\text{Number of domestic cars sold}}{\text{Total number of cars sold}} = \dfrac{350}{350 + 210} = \dfrac{350}{560} = \dfrac{5 \cdot 70}{8 \cdot 70} = \boxed{\dfrac{5}{8}}$ ∎

A *proportion* is an equation between two ratios. Frequently we can interpret given information in terms of a proportion, and the resulting equation is often easy to solve.

EXAMPLE 2 If the male to female ratio in a certain factory is 8 to 3 and there are 45 women in the factory, how many men are there?

Solution If we let x = number of men in the factory, then the given information of the ratio of male to female being 8 to 3 allows us to write the following proportion:

$$\frac{\text{Number of males}}{\text{Number of females}} = \frac{8}{3}$$

$$\frac{x}{45} = \frac{8}{3} \qquad \textit{Since we want to solve for x, we multiply both sides of the equation by 45.}$$

$$45 \cdot \frac{x}{45} = 45 \cdot \frac{8}{3}$$

$$\frac{\cancel{45}}{1} \cdot \frac{x}{\cancel{45}} = \frac{\overset{15}{\cancel{45}}}{1} \cdot \frac{8}{\cancel{3}}$$

$$\boxed{x = 120} \qquad \text{There are 120 males in the factory.} \quad ∎$$

✔ **LEARNING CHECKS**

1. At a small college 375 students live on campus while 875 live at home. Find:
   (a) The ratio of students who live on campus to those who live at home
   (b) The ratio of students who live at home to those who live on campus
   (c) The ratio of the students who live on campus to the total number of students

2. The ratio of children to adults in a certain housing development is 2 to 9. If there are 126 adults, how many children are there?

3. A poll of employees at a certain factory showed that 90 of the people asked were satisfied with their union representation. If the ratio of those who were not satisfied to those who were was 5 to 6, how many people were surveyed all together?

EXAMPLE 3   A recent survey found that of the total number of people surveyed, 192 preferred brand X. If the ratio of those who did not prefer brand X to those who did was 3 to 4, how many people were surveyed all together?

Solution   Let n = number of people who did not prefer brand X. Then our proportion is

$$\frac{\text{Number of people who did not prefer brand X}}{\text{Number of people who did prefer brand X}} = \frac{3}{4}$$

$$\frac{n}{192} = \frac{3}{4} \qquad \textit{To solve for n we multiply through by 192.}$$

$$\frac{\cancel{192}}{1} \cdot \frac{n}{\cancel{192}} = \frac{\overset{48}{\cancel{192}}}{1} \cdot \frac{3}{\cancel{4}}$$

$$n = 144$$

We are not finished yet. The example asks for the *total* number of people surveyed. Thus, the answer is

$$144 + 192 = \boxed{336}$$

An alternative solution is:

Let n = total number of people surveyed. Then

$$\frac{n}{192} = \frac{7}{4} \qquad \textit{Because out of 7 people surveyed, 4 preferred brand X}$$

$$\frac{\cancel{192}}{1} \cdot \frac{n}{\cancel{192}} = \frac{\overset{48}{\cancel{192}}}{1} \cdot \frac{7}{\cancel{4}}$$

$$\boxed{n = 336}$$ ∎

4. If there are 3.28 feet in 1 meter, how tall, in meters, is someone who is 6 feet tall?

EXAMPLE 4   If there are 2.54 centimeters to 1 inch, how wide, in inches, is a table which is 72 centimeters wide?

Solution   Let x = number of inches in 72 cm.

It is usually easier to solve a fractional equation if the variable is in the numerator. We therefore translate the given information into the following proportion:

$$\frac{\text{Length of the table in inches}}{72 \text{ cm.}} = \frac{1 \text{ inch}}{2.54 \text{ cm.}} \qquad \textit{Notice how the units in a proportion must agree.}$$

$$\frac{x}{72} = \frac{1}{2.54}$$

$$\frac{\cancel{72}}{1} \cdot \frac{x}{\cancel{72}} = \frac{72}{1} \cdot \frac{1}{2.54}$$

$$x = \frac{72}{2.54}$$

$$\boxed{x = 28.3 \text{ inches}}$$ ∎

5. If there are 3.785 liters in 1 gallon, how many gallons are there in 10 liters?

EXAMPLE 5   If there is .946 liter in 1 quart, how many ounces are there in a 2-liter bottle?

Solution   Let n = number of ounces in a 2-liter bottle. Our proportion is

$$\frac{\text{Number of ounces in 2 liters}}{2 \text{ liters}} = \frac{\text{Number of ounces in 1 quart}}{.946 \text{ liter}}$$

$$\frac{n}{2} = \frac{32}{.946}$$

$$\frac{\cancel{2}}{1} \cdot \frac{n}{\cancel{2}} = \frac{2}{1}\left(\frac{32}{.946}\right)$$

We must use the same units throughout the proportion. Since we are working in ounces, we must convert 1 quart into 32 ounces.

$$n = \frac{64}{.946}$$

$$\boxed{n = 67.7 \text{ oz.}}$$ ∎

In Chapter 8 we will examine other problems involving ratio and proportion.

STUDY SKILLS

4.5

Taking an Algebra Exam

A Few Other Comments About Exams

Do not forget to check over all your work as we have suggested on numerous occasions. Reread all directions and make sure that you have answered all the questions as directed.

If you are required to show your work (such as for partial credit), make sure that your work is neat. Do not forget to put your final answers where directed or at least indicate your answers clearly by putting a box or a circle around your answer. For multiple-choice tests be sure you have filled in the correct space.

One other bit of advice: Some students are unnerved when they see others finishing the exam early. They begin to believe that there may be something wrong with themselves because they are still working on the exam. They should not be concerned for there are some students who can do the work quickly and others who leave the exam early because they give up, not because the exam was too easy for them.

In any case, do not be in a hurry to leave the exam. If you are given 1 hour for the exam then take the entire hour. If you have followed the suggestions in this chapter such as checking your work, etc., and you still have time left over, relax for a few minutes and then go back and check over your work again.

✔ *Answers to Learning Checks in Section 4.5*

1. (a) $\frac{3}{7}$ (b) $\frac{7}{3}$ (c) $\frac{3}{10}$ **2.** 28   **3.** 165   **4.** 1.83 meters

5. 2.64 gallons

NOTE TO THE STUDENT

Use the space on this page to write down any questions you have or points you want to review with your instructor.

Exercises 4.5

In Exercises 1–10, write each of the phrases as a ratio.

1. 7 red to 5 black

2. 9 short to 2 long

3. 5 black to 7 red

4. 2 long to 9 short

5. 11 with to 5 without

6. 5 with to 11 without

7. x to $3x$

8. $3x$ to x

9. a to the sum of b and c

10. the sum of b and c to a

In Exercises 11–18, solve each proportion.

11. $\dfrac{x}{5} = \dfrac{12}{3}$

12. $\dfrac{x}{6} = \dfrac{10}{2}$

13. $\dfrac{a}{6} = \dfrac{5}{3}$

14. $\dfrac{a}{8} = \dfrac{9}{2}$

15. $\dfrac{y}{15} = \dfrac{20}{6}$

16. $\dfrac{y}{10} = \dfrac{35}{14}$

17. $\dfrac{y}{9} = \dfrac{4}{3}$

18. $\dfrac{y}{18} = \dfrac{10}{9}$

Solve each of the following exercises by first setting up a proportion. Round off your answers to the nearest hundredth.

19. A jar contains marbles in the ratio of 7 red to 5 black. If there are 210 black marbles, how many red marbles are there?

20. Repeat Exercise 19 with the ratio of red to black reversed.

21. On a certain test a math teacher found that the ratio of grades 90 or above to those below 90 was $\frac{3}{8}$. If 24 students got below 90, how many got 90 or above?

22. Repeat Exercise 21 if the ratio of grades 90 or above to those below 90 was 5 to 12.

1. _____
2. _____
3. _____
4. _____
5. _____
6. _____
7. _____
8. _____
9. _____
10. _____
11. _____
12. _____
13. _____
14. _____
15. _____
16. _____
17. _____
18. _____
19. _____
20. _____
21. _____
22. _____

ANSWERS

23. _____

24. _____

25. _____

26. _____

27. _____

28. _____

29. _____

30. _____

31. _____

32. _____

33. _____

34. _____

35. _____

36. _____

37. _____

38. _____

23. If the ratio of the length of a rectangle to its width is $\frac{9}{4}$, and the length is 18 cm., what is the width of the rectangle?

24. If the ratio of the width of a rectangle to its length is $\frac{3}{7}$, and the length is 35 mm., find the width of the rectangle.

25. If the sides of a rectangle are as shown in the accompanying diagram, what is the ratio of the shorter side to the longer side?

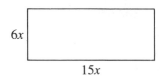

26. In Exercise 25, what is the ratio of the shorter side to the perimeter?

27. In a scale drawing, actual sizes are all diminished in the same proportion. If a 12-meter wall is represented by a 5-cm. length, what length would represent a 20-meter wall?

28. Repeat Exercise 27 if the 12-meter wall is represented by a 7-cm. length.

29. If there are 2.2 lb. in 1 kilogram, how many kilograms are there in 10 lb.?

30. If there are 1.06 quarts in 1 liter, how many liters are there in 2 gallons?

31. If there is .92 meter in 1 yard, how many yards are there in 100 meters?

32. There is approximately .625 mile in 1 kilometer. If a speedometer reads 55 mph, how would it read in kilometers per hour?

CALCULATOR EXERCISES

Solve each of the following proportions. Round off your answers to the nearest hundredth.

33. $\dfrac{x}{.43} = \dfrac{.26}{.9}$

34. $\dfrac{a}{2.61} = \dfrac{3.82}{7.41}$

35. $\dfrac{61.95}{t} = \dfrac{47.02}{8.8}$

36. $\dfrac{218.36}{14.61} = \dfrac{97.03}{w}$

37. $\dfrac{y + .3}{.7} = \dfrac{y - .2}{.9}$

38. $\dfrac{u - 2.6}{4.5} = \dfrac{u + 7.8}{6.6}$

Verbal Problems

4.6

Now that we have developed the ability to handle a wider range of first-degree equations in one variable, we can apply our knowledge to a greater variety of verbal problems. It is worthwhile repeating that no attempt is being made to make you an expert in any particular type of verbal problem. As we do more problems, we will see that while problems may seem at first to be very different, their solutions often show a similar structure.

As in our previous work with verbal problems, not only do we need to know how to translate certain phrases and sometimes certain mathematical formulas algebraically, but we also need to be able to apply our common sense to the problem as well.

Before we proceed with some examples, let's restate our suggested outline for solving verbal problems.

OUTLINE OF STRATEGY FOR SOLVING VERBAL PROBLEMS

1. Read the problem carefully, as many times as is necessary to understand what the problem is saying and what it is asking.

2. Use diagrams whenever you think it will make the given information clearer.

3. Ask whether there is some underlying relationship or formula you need to know. If not, then the words of the problem themselves give the required relationship.

4. Clearly identify the unknown quantity (or quantities) in the problem, and label it (them) using one variable.

   Step 4 is very important, and not always easy.

5. By using the underlying formula or relationship in the problem, write an equation involving the unknown quantity (or quantities).

   Step 5 is the *crucial step*.

6. Solve the equation.

7. Make sure you have answered the question that was asked.

8. Check the answer(s) in the original words of the problem.

EXAMPLE 1   If 3 more than three-fifths of a number is 1 less than that number, find the number.

Solution   Let x = the number. Next we translate the phrases in the problem.

$$\text{"three-fifths of the number"} = \frac{3}{5} \cdot x = \frac{3x}{5}$$

$$\text{"3 more than three-fifths of the number"} = \frac{3x}{5} + 3$$

$$\text{"1 less than the number"} = x - 1 \qquad \textbf{\textit{Remember:}} \quad \textit{not } 1 - x$$

✔ **LEARNING CHECKS**

1. If 8 less than four-thirds of a number is 2 more than the number, what is the number?

The word "is" is translated as "equal to," so that the equation is

$$\frac{3x}{5} + 3 = x - 1 \qquad \textit{We multiply through by 5 to clear the fraction.}$$

$$5\left(\frac{3x}{5} + 3\right) = 5(x - 1)$$

$$\frac{\cancel{5}}{1} \cdot \frac{3x}{\cancel{5}} + 5 \cdot 3 = 5x - 5$$

$$3x + 15 = 5x - 5$$

$$20 = 2x$$

$$\boxed{10 = x}$$

CHECK: $\dfrac{3}{5}$ of the number $= \dfrac{3}{5}(10) = 6.$

3 more than $\dfrac{3}{5}$ of the number $= 6 + 3 = 9.$

One less than the number $= 10 - 1 = 9.$ The answer checks. ∎

2. In a rectangle the length is 3 more than three-fourths the width. If the perimeter is 34 cm., find the dimensions of the rectangle.

EXAMPLE 2   In a certain triangle the medium side is 2 less than twice the shortest side, while the longest side is $2\frac{1}{2}$ times the shortest side. If the perimeter of the triangle is 2 more than 5 times the shortest side, find the lengths of the three sides of the triangle.

Solution   Since the sides of the triangle are all described in terms of the shortest side, let

$$x = \text{Length of the shortest side}$$
$$2x - 2 = \text{Length of the medium side}$$
$$\frac{5}{2}x = \text{Length of the longest side} \quad \textbf{\textit{Remember:}} \quad 2\tfrac{1}{2} = 2 + \tfrac{1}{2} = \tfrac{4}{2} + \tfrac{1}{2} = \tfrac{5}{2}$$

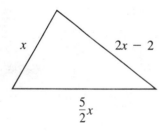

Figure 4.1 *Triangle for Example 2*

(See Figure 4.1.) The problem also tells us that the perimeter is 2 more than 5 times the shortest side, so that the perimeter is $5x + 2$. The fact that the perimeter is the length around the triangle gives us the equation

$$x + 2x - 2 + \frac{5}{2}x = 5x + 2 \qquad \textit{Combine like terms.}$$

$$3x - 2 + \frac{5}{2}x = 5x + 2 \qquad \textit{Multiply both sides by 2.}$$

$$2 \cdot 3x - 2 \cdot 2 + \cancel{2} \cdot \frac{5x}{\cancel{2}} = 2(5x + 2)$$

$$6x - 4 + 5x = 10x + 4$$

$$11x - 4 = 10x + 4$$

$$x = 8$$

Therefore, the shortest side $= \boxed{8}$

The medium side $= 2x - 2 = 2(8) - 2 = \boxed{14}$

The longest side $= \dfrac{5}{2}x = \dfrac{5}{2}(8) = \boxed{20}$

The check is left to the student.   ∎

EXAMPLE 3   Jane sold 60 tickets to a concert and collected a total of $319.50. If regular tickets cost $6.25 each while senior citizen tickets cost $4.75 each, how many of each type of ticket did she sell?

Solution

> Let n = Number of regular tickets sold.
>
> Then $60 - n$ = Number of senior citizen tickets sold.   (Why?)

(We could also have let

$$n = \text{Number of senior citizen tickets sold.}$$
$$60 - n = \text{Number of regular tickets sold.}$$

The value for n will not come out the same but the answer to the problem will.)

Since n tickets were sold at $6.25 each, Jane collected n times $6.25 for the regular tickets. By the same reasoning she collected $4.75 times $(60 - n)$ for the senior citizen tickets. We are told that Jane collected a total of $319.50, and so the equation is

$$6.25n \quad + \ 4.75(60 - n) = \quad 319.50$$

Value of regular tickets sold	Value of senior citizen tickets sold	Total value of tickets sold

$$6.25n + 4.75(60 - n) = 319.50 \qquad \textit{Multiply through by } 100.$$
$$625n + 475(60 - n) = 31{,}950$$
$$625n + 28{,}500 - 475n = 31{,}950$$
$$150n + 28{,}500 = 31{,}950$$
$$150n = 3{,}450$$
$$n = \frac{3{,}450}{150} = 23$$

Jane sold $\boxed{23 \text{ regular tickets}}$ and $60 - 23 = \boxed{37 \text{ senior citizen tickets}}$

CHECK:

> 23 tickets + 37 tickets = 60 tickets $\checkmark$
>
> 23 tickets @ $6.25 = 23(6.25) = \$143.75$
>
> 37 tickets @ $4.75 = 37(4.75) = \dfrac{\$175.75}{\$319.50} \checkmark$ ∎

EXAMPLE 4   A total of $6,000 was split into two investments. Part was invested in a bank account paying 7% interest per year and the remainder was invested in stocks paying 11% interest per year. If the total yearly interest from the two investments is $492, how much was invested in the bank account?

Solution   In order to solve this problem we must know how to compute simple interest and how to write a percentage as a decimal (see Section 0.6).

To compute simple interest we use the formula

$$I = Prt$$

where I = interest; P = principal (the amount invested); r = rate of interest, which is the percentage written as a decimal; and t = time (the number of time

---

3. Tim bought 32 boxes of paper goods for a total of $246. Some were boxes of paper plates costing $6 each and the rest were boxes of paper cups costing $9 each. How many boxes of each did he buy?

4. A certain amount of money is invested at 8% and twice that amount is invested at 10%. If the total annual income from the two investments is $980, how much is invested at each rate?

periods for which the interest is being computed). For example, if $800 is invested at 6% per year for 2 years, then

$$P = 800, \ r = .06 \ (6\% \text{ written as a decimal}), \text{ and } t = 2.$$

The formula would yield

$$I = (800)(.06)(2) = \$96$$

In this example

Let x = Amount invested at 7%   (not the interest, but the actual amount invested)

Then $6{,}000 - x$ = Amount invested at 11%   (because there was $6,000 invested all together).

Our equation relates how much was earned from each investment to the total amount earned.

$$\underbrace{.07x}_{\substack{\text{Interest from the} \\ \text{7\% investment}}} + \underbrace{.11(6{,}000 - x)}_{\substack{\text{Interest from the} \\ \text{11\% investment}}} = \underbrace{492}_{\text{Total interest}}$$

$$
\begin{aligned}
.07x + .11(6{,}000 - x) &= 492 \\
7x + 11(6{,}000 - x) &= 49{,}200 \\
7x + 66{,}000 - 11x &= 49{,}200 \\
-4x + 66{,}000 &= 49{,}200 \\
\underline{-66{,}000 \quad -66{,}000} & \\
-4x &= -16{,}800 \\
x &= \frac{-16{,}800}{-4}
\end{aligned}
$$

Multiply through by 100 to clear the decimals.

$$\boxed{x = \$4{,}200}$$ *This is the amount invested in the bank at 7%.*

CHECK: The amount invested at 11% is $6{,}000 - x = 6{,}000 - 4{,}200 = 1{,}800$.
$$4{,}200 + 1{,}800 \overset{\checkmark}{=} 6{,}000$$
$$.07(4{,}200) + .11(1{,}800) = 294 + 198 \overset{\checkmark}{=} 492 \qquad \blacksquare$$

5. How much of each of a 40% chlorine solution and a 65% chlorine solution must be mixed together to produce 60 liters of a 50% chlorine solution?

EXAMPLE 5   How many milliliters (ml.) of a 30% alcohol solution must be mixed with 50 ml. of a 70% alcohol solution to produce a 55% alcohol solution?

Solution   It may help to visualize the problem as shown in Figure 4.2.

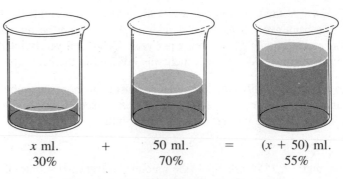

x ml.   +   50 ml.   = $(x + 50)$ ml.
30%   70%   55%

Figure 4.2 *Alcohol solutions for Example 5*

If you ask how much *alcohol* is in the second container in Figure 4.2, the answer is neither 50 ml. nor 70%; 50 ml. is the amount of *solution*, not the amount of actual alcohol, and 70% is a *percentage*, not an amount. To figure out how much actual alcohol is in the second container you compute

$$70\% \text{ of } 50 \text{ ml.} = .70(50) = 35 \text{ ml. of alcohol}$$

Using the same idea, we can write the following equation:

$$.30x \quad + \quad .70(50) \quad = \quad .55(x + 50)$$

| *Amount of alcohol* | *Amount of alcohol* | *Total amount of* |
| *in first container* | *in second container* | *alcohol in the solution* |

$$.30x + .70(50) = .55(x + 50) \qquad \textit{Multiply through by } 100.$$
$$30x + 70(50) = 55(x + 50)$$
$$30x + 3{,}500 = 55x + 2{,}750$$
$$750 = 25x$$
$$\boxed{30 \text{ ml.} = x}$$

The check is left to the student.   ∎

If we take a moment to look back at Examples 3, 4, and 5 (and, as you might also recognize, the coin problems from the last chapter), we can see that while these problems may have seemed to have nothing in common, once we analyzed them the resulting equations exhibited a very similar structure. The types of problems which result in such equations are often called ***value problems*** or ***mixture problems***. Looking for similarities in different situations is often more illuminating than looking for differences.

EXAMPLE 6   At 10 A.M. Gary leaves home on a trip driving at 48 mph. If Bill leaves from Gary's home 15 minutes later and drives the same route at 52 mph, how long will it take Bill to catch up with Gary?

Solution   Drawing a little diagram will help us see the relationship between the distances Gary and Bill travel (see Figure 4.3). As we can see, when Bill catches up to Gary they have travelled the same distance.

$$\text{Distance Gary travelled} = \text{Distance Bill travelled}$$
$$d_G = d_B$$

But we know that $d = rt$, so we compute each distance individually.

$$(\text{Rate for Gary}) \cdot (\text{Time for Gary}) = (\text{Rate for Bill}) \cdot (\text{Time for Bill})$$

Let t = Number of hours Bill drives the distance d_B.

Then $t + \dfrac{1}{4}$ = Number of hours Gary drives the distance d_B.

(*Note:*   Since Gary left 15 minutes before Bill, Gary drives 15 minutes *more* than Bill, and 15 minutes is $\frac{1}{4}$ of an hour. Because the rate is given in miles per *hour*, we convert the time of 15 minutes into one-quarter of an hour. We must use the same units throughout a problem.)

6. Arlene leaves home at 11:00 A.M. travelling at 50 kph. If Marge leaves Arlene's house 20 minutes later travelling at 60 kph in the opposite direction, at what time will they be 255 kilometers apart?

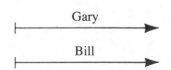

Figure 4.3   *Diagram for Example* 6

Thus, the equation is

$$48\left(t + \frac{1}{4}\right) = 52t$$

This equation is the algebraic translation of $r_{Gary} \cdot t_{Gary} = r_{Bill} \cdot t_{Bill}$

$$48t + 12 = 52t$$

$$12 = 4t$$

$$\boxed{3 = t}$$

It will take Bill 3 hours to catch up with Gary.

CHECK: In 3 hours Bill travels $3(52) = 156$ miles.

In $3\frac{1}{4}$ hours Gary travels $\left(3 + \frac{1}{4}\right)48 = 144 + 12 = 156$ miles. $\checkmark$ ■

7. An auto mechanic can install 9 batteries per hour while his apprentice can install only 4 per hour. If the apprentice starts installing batteries at 7:00 A.M. and is joined by the mechanic 40 minutes later, at what time will they have installed 20 batteries?

EXAMPLE 7   Harry can process 12 forms per hour while Susan can process 16 forms per hour. If Harry starts processing forms at 1:00 P.M. and Susan starts at 2:20 P.M., at what time will Susan have processed as many forms as Harry has?

Solution   The basic idea in the problem is that

$$\text{\# of forms processed} = \left(\begin{array}{c}\text{Rate at which they}\\\text{are processed}\end{array}\right) \cdot (\text{Time spent processing})$$

For example, if you process forms at the rate of 8 per hour for 3 hours you could process $8 \cdot 3 = 24$ forms.

The problem is asking at what time will they both have processed the same number of forms. That is, at what time will

$$\text{\# of forms Harry processed} = \text{\# of forms Susan processed}$$

In other words, at what time will

$$\left(\begin{array}{c}\text{Harry's}\\\text{processing rate}\end{array}\right) \cdot \left(\begin{array}{c}\text{Harry's}\\\text{processing time}\end{array}\right)$$

$$= \left(\begin{array}{c}\text{Susan's}\\\text{processing rate}\end{array}\right) \cdot \left(\begin{array}{c}\text{Susan's}\\\text{processing time}\end{array}\right)$$

Let $t = \text{\# of hours Susan works until she catches up to Harry.}$

Then $t + \frac{4}{3} = \text{\# of hours Harry works (because Harry works 1 hr. 20 min.}$

more than Susan; one hour and 20 min. $= 1\frac{1}{3}$ hr. $= \frac{4}{3}$ hr.)

Therefore, the equation is

$$12\left(t + \frac{4}{3}\right) = 16t$$

$$12t + 12 \cdot \frac{4}{3} = 16t$$

$$12t + 16 = 16t$$

$$16 = 4t$$

$$4 = t$$

It takes 4 hours for Susan to catch up to Harry. Since she started at 2:20, she will have processed as many forms as Harry at 6:20 P.M. . (Do not forget to answer the original question.)

The check is left to the student.   ■

Looking back at Examples 6 and 7, we can again see basic similarities in the structure of our solutions to these problems. Problems of this type are often called *rate–time* problems.

One final reminder: Do not just read a problem and give up. Make an honest effort at a solution. Time spent in an unsuccessful attempt at a solution is not wasted. With continued effort you will find that each attempt will bring you closer to a complete solution.

✔ *Answers to Learning Checks in Section 4.6*

1. 30   **2.** Width = 8 cm.; length = 9 cm.

3. 18 boxes of cups and 14 boxes of plates

4. $3,500 at 8% and $7,000 at 10%

5. 36 liters of 40% solution and 24 liters of 65% solution

6. 1:30 P.M.   **7.** 9:00 A.M.

NOTE TO THE STUDENT

Use the space on this page to write down any questions you have or points you want to review with your instructor.

Exercises 4.6

In each of the following exercises set up an equation or inequality, and solve the problem. Be sure to indicate clearly what quantity your variable represents.

1. If 5 more than two-thirds of a number is 9, find the number.

2. If 3 less than five-sevenths of a number is 12, find the number.

3. If 2 less than three-fourths of a number is 7 less than one-eighth of the number, find the number.

4. If 2 less than three-fourths of a number is less than one-eighth of the number, how large can the number be?

5. If the width of a rectangle is $\frac{1}{2}$ of its length and the perimeter is 36 meters, find the dimensions of the rectangle.

6. If the width of a rectangle is 1 more than $\frac{2}{3}$ of its length and the perimeter is 32 cm., what are the dimensions of the rectangle?

7. The medium side of a triangle is $\frac{3}{4}$ of the longest side, and the shortest side is $\frac{1}{2}$ of the medium side. If the perimeter of the triangle is 17 inches, find the lengths of the sides of the triangle.

8. The medium side of a triangle is 6 more than $\frac{1}{2}$ the shortest side, and the longest side is 4 times the shortest side. If the perimeter is 17 inches, find the lengths of the sides of the triangle.

9. An amusement park sells regular admission tickets as well as combination tickets which cover admission and a number of rides. The price of a regular admission ticket is $3 while a combination ticket costs $7. If a total of $1,990 was collected on the sale of 350 tickets, how many of each type were sold?

10. A plumber and her assistant work on a certain installation job. First the assistant does some preparatory work, and then the plumber completes the job alone. The plumber gets paid $18 per hour and the assistant gets paid $12 per hour. If the installation job took a total of 7 hours to complete, and the total labor cost for the plumber and assistant was $99, how many hours did each work on the job?

1. _____

2. _____

3. _____

4. _____

5. _____

6. _____

7. _____

8. _____

9. _____

10. _____

11. A collection of coins consisting of dimes and quarters has a value of $2.55. If there are 3 more than twice as many dimes as quarters, how many of each type of coin are there?

12. A collection of 40 coins consisting of nickels and dimes has a value of $2.65. How many of each type of coin are there?

11. _____

13. A certain machine can sort screws at the rate of 175 per minute. A newer, faster machine can do the sorting at the rate of 250 per minute. If the older machine begins sorting a batch of 13,675 screws at 10:00 A.M. and then 15 minutes later the newer machine joins in the sorting process, at what time will the sorting be completed?

12. _____

14. At 8:00 A.M. a line of 650 cars is waiting for passage through a toll barrier. At 8:00 A.M. a toll machine begins admitting cars at the rate of 15 per minute. At 8:10 A.M. a toll-taker comes on duty in the next booth and begins admitting cars at the rate of 10 per minute, and they continue working together until all 650 cars have been admitted. At what time will the last car pass through the toll barrier?

13. _____

15. A certain sum of money is invested at 8%, and $4,000 more than that amount is invested at 11%. If the annual interest from the two investments is $1,390, how much was invested at 11%?

14. _____

16. A certain sum was invested at 10% and $2,000 less than that amount was invested at 8%. If the annual interest from the two investments was $920, how much was invested at each rate?

15. _____

17. A total of $800 was split into two investments. Part paid 9% and the remainder paid 6%. If the annual interest from the two investments was $67.50, how much was invested at each rate?

16. _____

18. A total of $7,500 was invested as follows: a certain amount at 7%, twice that amount at 10%, and the remainder at 12%. If the annual interest from the three investments was $738, how much was invested at each rate?

17. _____

18. _____

19. A total of $6,000 is to be split into two investments, part at 8% and the remainder at 12%, in such a way that the annual interest is 9% of the amount invested. How should the $6,000 be split?

20. Repeat Exercise 19 if the total to be invested is $10,000, and the yearly interest is to be 11.8% of the amount invested.

21. How many milliliters of a 30% hydrochloric acid solution must be mixed with 30 milliliters of a 50% hydrochloric acid solution to produce a 45% solution of hydrochloric acid?

22. How many ounces of a 40% alcohol solution should be mixed with 60 ounces of a 70% alcohol solution to produce a 60% alcohol solution?

23. How much of each of a 25% salt solution and a 55% salt solution must be mixed together to produce 90 liters of a 50% salt solution?

24. How many ounces of each of a 20% and a 30% iodine solution need to be mixed together to produce 40 ounces of a 26% iodine solution?

25. How much pure antifreeze should be added to a radiator which contains 10 gallons of a 30% antifreeze solution to produce a 50% antifreeze solution? [*Hint:*   Pure antifreeze is 100% antifreeze.]

26. How much water should be added to a radiator which contains 10 gallons of an 80% antifreeze solution to dilute it to a 50% antifreeze solution? [*Hint:*   Pure water is 0% antifreeze.]

ANSWERS

19. _____

20. _____

21. _____

22. _____

23. _____

24. _____

25. _____

26. _____

27. How many pounds of candy selling at $3.75/lb. should be mixed with 35 pounds of candy selling at $5/lb. to produce a mixture which should sell at $4.25/lb.?

28. How many pounds of coffee beans selling at $1.65/lb. should be mixed with how many pounds of coffee beans selling for $2.25/lb. to produce a 24-lb. mixture which should sell for $2/lb.?

29. John and Susan leave their homes at 8:00 A.M., going toward each other along the same route which is 9 miles long. John walks at 4 mph while Susan jogs at 8 mph. At what time will they meet?

30. Two trains leave two cities which are 400 kilometers apart. They both leave at 11:00 A.M. travelling toward each other on parallel tracks. If one train travels at 70 kph and the other travels at 90 kph, at what time will they pass each other?

31. Repeat Exercise 29 if John leaves at 7:45 and Susan leaves at 8:00.

32. Repeat Exercise 30 if the slower train leaves at 9:40 A.M. and the faster train leaves at 11:00.

33. If David walks at the rate of 5 mph and jogs at the rate of 9 mph and it takes 2 hours to cover a distance of 16 miles, how much time was spent jogging?

34. A person drives from town A to town B at the rate of 50 mph and then flies back at the rate of 160 mph. If the total travelling time is 21 hours, how far is it from town A to town B?

27. _____

28. _____

29. _____

30. _____

31. _____

32. _____

33. _____

34. _____

CHAPTER 4 SUMMARY

After having completed this chapter you should be able to:

1. Use the Fundamental Principle of Fractions to reduce fractions (Section 4.1).

   For example:

   Reduce to lowest terms. $\dfrac{6x^2y^5}{8x^3y^2}$

   Solution: $\dfrac{6x^2y^5}{8x^3y^2} = \dfrac{2 \cdot 3 \cdot x^2 \cdot y^2 \cdot y^3}{2 \cdot 4 \cdot x^2 \cdot x \cdot y^2} = \boxed{\dfrac{3y^3}{4x}}$

2. Multiply and divide algebraic fractions (Section 4.2).

   For example:

   (a) $\dfrac{2a^2b}{9x} \cdot \dfrac{3x^2}{4b} = \dfrac{2a^2b}{9x} \cdot \dfrac{\overset{x}{3x^2}}{\underset{2}{4b}}$

   $= \boxed{\dfrac{a^2x}{6}}$

   (b) $\dfrac{8yz}{5x^3} \div (10xyz) = \dfrac{8yz}{5x^3} \cdot \dfrac{1}{10xyz}$

   $= \dfrac{\overset{4}{8yz}}{5x^3} \cdot \dfrac{1}{\underset{5}{10xyz}}$

   $= \boxed{\dfrac{4}{25x^4}}$

3. Find the LCD of several rational expressions (Section 4.3).

   For example:

   Find the LCD of the fractions. $\dfrac{2}{3xy^2}, \dfrac{1}{6x^2y}, \dfrac{3}{8x^3}$

   $\left.\begin{array}{l} 3xy^2 = 3xyy \\ 6x^2y = 2 \cdot 3xxy \\ 8x^3 = 2 \cdot 2 \cdot 2xxx \end{array}\right\}$ *The distinct factors are 2, 3, x, and y.*

   Following our outline we get that the LCD $= 24x^3y^2$.

4. Combine rational expressions, using the Fundamental Principle to build fractions where necessary (Section 4.3).

   For example:

   (a) $\dfrac{6x + 5}{2x^2} + \dfrac{x - 5}{2x^2} = \dfrac{6x + 5 + x - 5}{2x^2}$

   $= \dfrac{7x}{2x^2}$ *Reduce.*

   $= \boxed{\dfrac{7}{2x}}$

   (b) $\dfrac{2}{3xy^2} - \dfrac{1}{6x^2y} + \dfrac{3}{8x^3}$ *In (3) above we saw that the LCD for these fractions is $24x^3y^2$.*

   $= \dfrac{2\,(8x^2)}{3xy^2\,(8x^2)} - \dfrac{1\,(4xy)}{6x^2y\,(4xy)} + \dfrac{3\,(3y^2)}{8x^3\,(3y^2)}$

   $= \boxed{\dfrac{16x^2 - 4xy + 9y^2}{24x^3y^2}}$

5. Solve a first-degree equation or inequality in one variable with rational coefficients (Section 4.4).

For example: Solve for t.

$$\frac{t+1}{2} - \frac{t-2}{3} = 2$$ *Multiply through by the LCD, which is 6, to clear the fractions.*

$$\frac{\overset{3}{\cancel{6}}}{1} \cdot \frac{t+1}{\cancel{2}} - \frac{\overset{2}{\cancel{6}}}{1} \cdot \frac{t-2}{\cancel{3}} = 6 \cdot 2$$

$$3(t+1) - 2(t-2) = 12$$

$$3t + 3 - 2t + 4 = 12$$

$$t + 7 = 12$$

$$\boxed{t = 5}$$

6. Write and solve proportions (Section 4.5).

For example:

The ratio of those employees of a certain firm who have college degrees to those who do not is 6 to 5. If there are 180 employees who do not have college degrees, how many do?

Let n = # of employees who have a college degree.

$$\frac{n}{180} = \frac{6}{5}$$ *Multiply through by 180.*

$$n = 180 \cdot \frac{6}{5}$$

$$\boxed{n = 216}$$

7. Translate and solve a wide variety of verbal problems which give rise to first-degree equations and inequalities in one variable (Section 4.6).

For example:

The length of a rectangle is 4 more than $\frac{2}{3}$ the width. If the perimeter of the rectangle is 28 meters, find the dimensions of the rectangle.

Solution: The underlying idea is that the perimeter of a rectangle is twice the width plus twice the length:

$$2L + 2W = P$$

This problem says that the length is 4 more than $\frac{2}{3}$ the width.

If we let W = width,

then $\frac{2}{3} \cdot W + 4 = \frac{2W}{3} + 4$ = length.

If we like, we can express the same information in a diagram, as shown here.

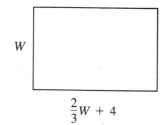

$\frac{2}{3}W + 4$

From the fact that the perimeter is equal to 28 we can write the equation

$$2\left(\frac{2W}{3} + 4\right) + 2W = 28$$

$$\frac{4W}{3} + 8 + 2W = 28$$ *Multiply both sides of the equation by 3 to clear the fraction.*

$$\frac{\cancel{3}}{1} \cdot \frac{4W}{\cancel{3}} + 3 \cdot 8 + 3 \cdot 2W = 3 \cdot 28$$

$$4W + 24 + 6W = 84$$

$$10W + 24 = 84$$

$$10W = 60$$

$$W = 6$$

Since the width is 6, the length is $\frac{2}{3}(6) + 4 = 4 + 4 = 8$ meters.

Thus, the solution is $\boxed{\text{width} = 6 \text{ meters; length} = 8 \text{ meters}}$

CHECK: The perimeter of the rectangle is $2(6) + 2(8) = 12 + 16 \overset{\checkmark}{=} 28$.

ANSWERS TO QUESTIONS FOR THOUGHT

1. The value of a fraction is unchanged when its numerator and denominator are both either multiplied or divided by the same nonzero amount.

2. We would conclude that $2 = 4$. It was incorrect to write $\dfrac{4 + \cancel{2}}{1 + \cancel{2}} = \dfrac{4}{1}$. The 2's cannot be reduced, since 2 is a common term, not a common factor.

3. (a) The factor of x in the reduced fraction should be in its numerator.

   (b) When both factors in the numerator are crossed out, a factor of 1 remains.

   (c) As in part (b), a factor of 1 remains in the numerator, so that the reduced fraction is $\dfrac{1}{3x^2y}$.

   (d) Crossing out x's amounts to crossing out terms rather than factors.

4. The fractions in (a), (c), (d), and (e) are equivalent. So are the fractions in (b), (f), and (g).

5. To divide by a fraction, multiply by its reciprocal. This rule works because division is defined to be the inverse of multiplication.

6. (a) The fraction $\dfrac{2y}{3x}$ must be inverted, leading to

$$\frac{3x}{2y} \cdot \frac{3x}{2y} = \frac{9x^2}{4y^2}$$

   (b) The factor of 5 is $\frac{5}{1}$, not $\frac{5}{5}$. Thus,

$$5 \cdot \frac{3x}{2} = \frac{5}{1} \cdot \frac{3x}{2} = \frac{5 \cdot 3x}{1 \cdot 2} = \frac{15x}{2}$$

7. The least common denominator (LCD) is the "smallest" expression that is exactly divisible by each of the denominators in a problem. We need the LCD in order to add or subtract two or more fractions. (Actually, a common denominator is really all that is needed, but the LCD is the most efficient one to use.)

8. Each of the original denominators is a factor of the LCD, so this is a common denominator. Since each factor of the LCD was chosen the maximum number of times that it appears as a factor in any one of the denominators, there are no extra factors. Thus, this must be the smallest common denominator possible.

9. (a) We must subtract the entire expression $(5 - x)$ from $x + 3$ in the numerator. This gives

$$\frac{x + 3 - (5 - x)}{x} = \frac{x + 3 - 5 + x}{x} = \frac{2x - 2}{x}$$

(b) We cannot cancel when performing addition.

(c) When building fractions, we must multiply both numerator and denominator by the same quantity. So,

$$\frac{5x}{2y} = \frac{5x(3x)}{2y(3x)} = \frac{15x^2}{6xy} \quad \text{and} \quad \frac{7y}{6x} = \frac{7y(y)}{6x(y)} = \frac{7y^2}{6xy}$$

Then

$$\frac{5x}{2y} + \frac{7y}{6x} = \frac{15x^2}{6xy} + \frac{7y^2}{6xy} = \frac{15x^2 + 7y^2}{6xy}$$

(d) The cancellation step undoes the building step and brings the problem back to its original form. The final step is incorrect, since we cannot add two fractions with unlike denominators.

(e) The cancellation is not allowed, since there are no common factors to be crossed out.

10. Reducing these fractions would reverse the building process and bring us back to the original problem.

CHAPTER 4 REVIEW EXERCISES

In Exercises 1–8, reduce each fraction to lowest terms.

1. $\dfrac{-18}{42}$

2. $-\dfrac{14}{35}$

3. $\dfrac{15x^6}{6x^2}$

4. $\dfrac{8a^3}{28a^9}$

5. $\dfrac{-10x^3y^5}{4xy^{10}}$

6. $\dfrac{24x^8y}{-15x^4y^5}$

7. $\dfrac{3t - 7t - t}{-2t^2 - 3t^2}$

8. $\dfrac{2w^2z + 4w^2z}{8wz^2 - 2wz^2}$

In Exercises 9–26, perform the indicated operations; express your final answer reduced to lowest terms.

9. $\dfrac{a}{4} \cdot \dfrac{a}{4}$

10. $\dfrac{a}{4} + \dfrac{a}{4}$

11. $\dfrac{7a}{6} - \dfrac{5a}{6}$

12. $\dfrac{7a}{6} \div \dfrac{5a}{6}$

13. $\dfrac{4x - 3}{6x} - \dfrac{x - 1}{6x}$

14. $\dfrac{7x - 2}{5x} - \dfrac{2x - 3}{5x}$

15. $\dfrac{2y^2 - 3y}{4} - \dfrac{y^2 - 3y}{4} + \dfrac{y^2}{4}$

16. $\left(\dfrac{2y^2 - 3y}{4} - \dfrac{y^2 - 3y}{4}\right) \div \dfrac{y^2}{4}$

17. $\left(\dfrac{x^2}{4} \cdot \dfrac{6}{xy^2}\right) \div (2xy)$

18. $\dfrac{x^2}{4} \div \left(\dfrac{6}{xy^2} \cdot (2xy)\right)$

ANSWERS

1. _____

2. _____

3. _____

4. _____

5. _____

6. _____

7. _____

8. _____

9. _____

10. _____

11. _____

12. _____

13. _____

14. _____

15. _____

16. _____

17. _____

18. _____

ANSWERS

19. _____

20. _____

21. _____

22. _____

23. _____

24. _____

25. _____

26. _____

27. _____

28. _____

29. _____

30. _____

31. _____

32. _____

19. $\dfrac{a}{2} \cdot \dfrac{a}{4}$

20. $\dfrac{a}{2} + \dfrac{a}{4}$

21. $\dfrac{x^2}{2} - \dfrac{x^2}{6} + \dfrac{x^2}{3}$

22. $\dfrac{x^2}{2} \cdot \dfrac{x^2}{6} \div \dfrac{x^2}{3}$

23. $\dfrac{4}{x^2} + \dfrac{3}{2x}$

24. $\dfrac{5}{3y^2} + \dfrac{3}{2y}$

25. $\dfrac{3}{4a^2b} - \dfrac{5}{6ab} + \dfrac{7}{8b^3}$

26. $\dfrac{3}{8rt^2} + \dfrac{7}{12rt} - \dfrac{5}{3t^3}$

In Exercises 27–36, solve the equation or inequality.

27. $\dfrac{x}{6} - \dfrac{1}{4} = \dfrac{7}{12}$

28. $\dfrac{x}{8} - \dfrac{5}{6} = \dfrac{1}{24}$

29. $\dfrac{t+1}{2} + \dfrac{t+2}{3} < \dfrac{t+7}{6}$

30. $\dfrac{2a+5}{4} + \dfrac{4a+1}{6} = 2$

31. $\dfrac{y+3}{5} - \dfrac{y-2}{3} = 1$

32. $\dfrac{z+4}{15} - \dfrac{z-4}{5} > \dfrac{z-8}{3}$

33. $\dfrac{x}{3} = \dfrac{x+1}{6}$ 34. $\dfrac{x}{3} = \dfrac{x}{6} + 1$

33. _____

35. $2x + .2(x + 6) = 10$ 36. $\dfrac{x}{2} + .3x = 16$

34. _____

In Exercises 37–41, solve each problem by first writing an appropriate equation or inequality.

37. If there are 28.4 grams in 1 ounce, how many ounces are there in 1 kilogram (1,000 grams)?

35. _____

38. If 1 less than three-fourths of a number is 4 less than the number, find the number.

36. _____

39. A total of $7,000 is split into three investments. A certain amount is invested at 6%, twice that amount at 7%, and the remainder at 8%. If the annual interest from the three investments is to be at least $500, what is the most that can be invested at 6%?

37. _____

38. _____

40. A collection of 30 coins consists of nickels, dimes, and quarters. There are twice as many dimes as nickels, and the rest are quarters. If the value of the collection is $3.50, how many of each type are there?

39. _____

41. After driving along at a certain rate of speed for 5 hours, Bill realizes that he could have covered the same distance in 3 hours if he had driven at 20 mph faster. What is his present speed?

40. _____

41. _____

ANSWERS

1. a. _____

b. _____

c. _____

d. _____

2. a. _____

b. _____

c. _____

d. _____

e. _____

f. _____

g. _____

h. _____

i. _____

j. _____

k. _____

CHAPTER 4 PRACTICE TEST

1. Reduce each of the following to lowest terms:

(a) $\dfrac{-10}{24}$

(b) $\dfrac{x^{10}}{x^2}$

(c) $\dfrac{-6a^6}{-3a^3}$

(d) $\dfrac{25r^2t^3}{-15r^4t}$

2. Perform the indicated operations and express your answer in lowest terms.

(a) $\dfrac{2y}{3x} \cdot \dfrac{9x^2}{4y^2}$

(b) $\dfrac{2y}{3x^2} \div \dfrac{8y^2}{9x^4}$

(c) $\dfrac{-3ab^2}{4a^2} \cdot \dfrac{2a^3}{3b} \cdot \dfrac{9a^2b}{-6a^2b^2}$

(d) $\dfrac{4xy^3}{9x^2} \div 18xy$

(e) $\dfrac{a}{5} + \dfrac{a}{5}$

(f) $\dfrac{a}{5} \cdot \dfrac{a}{5}$

(g) $\dfrac{9}{4x} + \dfrac{3}{4x}$

(h) $\dfrac{a}{3} + \dfrac{a}{8}$

(i) $\dfrac{5}{4x} + \dfrac{1}{6x}$

(j) $\dfrac{3}{10x^2y} + \dfrac{7}{3x}$

(k) $\dfrac{x^2 - 3x - 2}{8x} - \dfrac{6 - 3x + x^2}{8x}$

3. Solve each of the following equations or inequalities:

(a) $\dfrac{x}{3} + \dfrac{x}{5} = 8$

(b) $\dfrac{x-5}{2} + \dfrac{x}{5} \geq 8$

3. a. _____

(c) $\dfrac{a+3}{5} - \dfrac{a-2}{4} = 1$

(d) $.03t + .5t = 10.6$

b. _____

Solve each of the following problems algebraically.

4. If there are 1.61 kilometers in a mile, how many miles are there in 50 kilometers?

c. _____

d. _____

5. A number when added to $\frac{2}{3}$ of itself is 5 less than twice the number. Find the number.

4. _____

6. Advance sale tickets to a show cost $7.50 each, while tickets at the door cost $9 each. If $3,330 was collected on the sale of 400 tickets, how many tickets were sold at the door?

5. _____

6. _____

7. A total of $7,000 is split into two investments, one paying 8% interest and the other paying 13%. If the annual interest from the two investments is $750, how much is invested at each rate?

7. _____

8. How many ounces of a 20% sulfuric acid solution must be added to 24 ounces of a 65% sulfuric acid solution to produce a 50% solution?

8. _____

9. _____

9. Two people leave a town travelling in opposite directions. The first person leaves at 11:00 A.M. driving at 48 kph, while the second person leaves at 3:00 P.M. that afternoon and drives at 55 kph. At what time will they be 604 kilometers apart?

Exponents and Polynomials

✔ **LEARNING CHECKS**

1. $m^8 \cdot m^4$

In this chapter we will be reviewing much of the material we covered in Chapter 2. We will begin by reviewing our work with exponents, and use our knowledge as a basis on which to extend our understanding of exponents beyond the natural numbers.

As we move on to the material on simplifying algebraic expressions, we will introduce some new terminology as well as extend the scope of examples that we can handle.

Exponent Rules

5.1

Ever since we introduced exponent notation in Section 2.1, we have been continuously using exponents and the first rule for exponents.

Keep in mind that our definition of a^n is for n a positive integer. If an exponent tells us how many times the base appears as a factor in the product, then the exponent must be a counting number (positive integer).

For example, if we have to compute

$$(x^3)(x^4)$$

we immediately get an answer of x^7 often without realizing that we are using the first rule for exponents. After all, the first rule for exponents is just a formal statement of our ability to count factors.

Let's begin our discussion here by reviewing and restating the first rule for exponents.

EXAMPLE 1   *Simplify.* $a^6 \cdot a^2$

Solution

$$
\begin{aligned}
a^6 \cdot a^2 &= aaaaaa \cdot aa \\
&= a^{6+2} \\
&= \boxed{a^8}
\end{aligned}
$$
∎

EXPONENT RULE 1

$$a^m \cdot a^n = a^{m+n}$$

It is proven as follows:

$$a^m \cdot a^n = \underbrace{(aaa \cdots \cdot a)}_{m\ factors} \cdot \underbrace{(aaa \cdots \cdot a)}_{n\ factors} = \underbrace{(aaa \cdots \cdot a)}_{m\ +\ n\ factors}$$

Of course, as we have seen before, this rule extends to more than two powers of the same base:

$$a^m \cdot a^n \cdot a^p \cdots \cdot a^q = a^{m+n+p+\cdots+q}$$

Let's look at several more examples that will enable us to develop several other exponent rules.

EXAMPLE 2 *Simplify.* $(a^2)^6$

2. $(m^8)^4$

Solution

$$(a^2)^6 = a^2 \cdot a^2 \cdot a^2 \cdot a^2 \cdot a^2 \cdot a^2 \quad \textit{Use Exponent Rule 1.}$$
$$= a^{2+2+2+2+2+2} \qquad \textit{Write the repeated addition as multiplication.}$$
$$= a^{6 \cdot 2}$$
$$= \boxed{a^{12}} \qquad\qquad\qquad\qquad\qquad\qquad \blacksquare$$

We can generalize this to the rule stated in the box.

EXPONENT RULE 2

$$(a^m)^n = a^{n \cdot m}$$

Rule 2 is often called the ***power of a power rule***.

Sometimes rules 1 and 2 are confused (When do I add the exponents and when do I multiply them?). Keep these differences in mind:

Exponent Rule 1 says that when we multiply two powers of the same base we keep the base and add the exponents.

Exponent Rule 2 says that when we raise a power to a power we keep the base and multiply the exponents.

Even more basic, however, is the fact that if you are in doubt as to which rule to use, write down the expression and simply count factors. After all, that is where the rules come from in the first place.

EXAMPLE 3 *Simplify.* $\dfrac{a^6}{a^2}$

3. $\dfrac{m^8}{m^4}$

Solution

$$\frac{a^6}{a^2} = \frac{aaaaaa}{aa} \qquad \textit{Reduce.}$$
$$= \frac{\not a \not a aaaa}{\not a \not a}$$
$$= a^{6-2}$$
$$= \boxed{a^4} \qquad\qquad\qquad \blacksquare$$

Basically all we have done is cancel as many factors as we have in the denominator with the same number of factors in the numerator.

We can generalize this, as indicated in the box.

EXPONENT RULE 3

$$\frac{a^m}{a^n} = a^{m-n}, \qquad m > n \quad (a \neq 0)$$

In order to use rule 2 we must have $m > n$; otherwise, when we subtract the exponents, we will get a zero or negative exponent, which we have not yet defined. This restriction is a nuisance, and we will see how to remove it in the next section. However, this does not prevent us from reducing an expression like $\dfrac{a^2}{a^6}$ as we usually do:

$$\frac{a^2}{a^6} = \frac{aa}{aaaaaa} = \frac{\cancel{a}\cancel{a}}{\cancel{a}\cancel{a}aaaa} = \boxed{\frac{1}{a^4}}$$

Of course, as always, we restrict all variables from making any denominator equal to 0. Thus, in rule 3 we must have $a \neq 0$.

EXAMPLE 4 *Remove the parentheses.* $(ab)^5$

Solution

$$\begin{aligned}
(ab)^5 &= (ab)(ab)(ab)(ab)(ab) \qquad &\text{*We reorder and regroup using the Commutative}\\
&= (aaaaa)(bbbbb) &\text{and Associative Laws.}\\
&= \boxed{a^5b^5}
\end{aligned}$$

■

We can generalize this to

EXPONENT RULE 4

$$(ab)^n = a^n b^n$$

In words, rule 4 says that if a *product* is raised to a power, then each *factor* gets raised to that power.

EXAMPLE 5 *Remove the parentheses.* $\left(\dfrac{a}{b}\right)^5$

Solution

$$\begin{aligned}
\left(\frac{a}{b}\right)^5 &= \left(\frac{a}{b}\right) \cdot \left(\frac{a}{b}\right) \cdot \left(\frac{a}{b}\right) \cdot \left(\frac{a}{b}\right) \cdot \left(\frac{a}{b}\right) \qquad &\text{*Multiply the fractions.}\\
&= \boxed{\frac{a^5}{b^5}}
\end{aligned}$$

■

We can generalize this to

EXPONENT RULE 5

$$\left(\frac{a}{b}\right)^n = \frac{a^n}{b^n} \qquad b \neq 0$$

---

4. $(mn)^7$

5. $\left(\dfrac{m}{n}\right)^7$

In words, this says that if a *quotient* is raised to a power, then the numerator and the denominator each get raised to that power.

Let's summarize all five exponent rules so they will be easier to refer to.

EXPONENT RULES

For *m* and *n* positive integers:

Exponent Rule 1 $\quad a^m \cdot a^n = a^{m+n}$

Exponent Rule 2 $\quad (a^m)^n = a^{n \cdot m}$

Exponent Rule 3 $\quad \left(\dfrac{a^m}{a^n}\right) = a^{m-n} \quad m > n \quad (a \neq 0)$

Exponent Rule 4 $\quad (ab)^n = a^n b^n$

Exponent Rule 5 $\quad \left(\dfrac{a}{b}\right)^n = \dfrac{a^n}{b^n} \quad b \neq 0$

In the next few examples we will see how these rules work together to make it easier for us to simplify expressions involving exponents. Keep in mind that we will be showing *all* the steps in the solutions. As you get the hang of it, you can easily leave out or combine some of the steps.

EXAMPLE 6 *Simplify as completely as possible.*

(a) $(x^3 y^5)^4$ **(b)** $\left(\dfrac{s^3}{t^7}\right)^5$

6. (a) $(u^2 v^3)^5$

(b) $\left(\dfrac{r^2}{s^5}\right)^4$

Solution

(a) We begin by recognizing that basically we have a *product* raised to a power, and therefore our first step in the solution is to apply Exponent Rule 4.

$(x^3 y^5)^4$

> Rule 4 says that the outside exponent gets applied to each factor inside. We are viewing x^3 and y^5 as two separate factors.

$= (x^3)^4 (y^5)^4$ *Now use rule 2 on both factors.*

$= x^{4 \cdot 3} y^{4 \cdot 5}$

$= \boxed{x^{12} y^{20}}$

(b) This time we begin with rule 5 because basically we have a *quotient* raised to a power.

$\left(\dfrac{s^3}{t^7}\right)^5$

> According to rule 5 the outside exponent 5 applies to numerator and denominator.

$= \dfrac{(s^3)^5}{(t^7)^5}$ *Now use rule 2 in both numerator and denominator.*

$= \dfrac{s^{5 \cdot 3}}{t^{5 \cdot 7}}$

$= \boxed{\dfrac{s^{15}}{t^{35}}}$ ∎

7. $\dfrac{(y^3)^5 y^6}{(xy)^7}$

EXAMPLE 7 *Simplify as completely as possible.* $\dfrac{(x^2)^4 x^6}{(xy)^5}$

Solution   We begin by removing the parentheses in the numerator and the denominator.

$\dfrac{(x^2)^4 x^6}{(xy)^5}$ *Use rule 2 in the numerator and rule 4 in the denominator.*

$= \dfrac{x^{4\cdot2} x^6}{x^5 y^5}$

$= \dfrac{x^8 x^6}{x^5 y^5}$ *Use rule 1 in the numerator.*

$= \dfrac{x^{8+6}}{x^5 y^5}$

$= \dfrac{x^{14}}{x^5 y^5}$ *Use rule 3 on the x's.*

$= \dfrac{x^{14-5}}{y^5}$

$= \boxed{\dfrac{x^9}{y^5}}$ ∎

Clearly a number of steps here are usually mental steps. You should write down as many steps as *you* think are necessary.

Problems involving exponents often lend themselves to more than one method of solution, as the next example illustrates.

---

8. $\left(\dfrac{3a^4 b^2}{a^2 b}\right)^3$

EXAMPLE 8 *Simplify as completely as possible.* $\left(\dfrac{2x^3 y^5}{xy^2}\right)^4$

Solution   We offer two possible approaches.
Solution 1   offers the "inside-out" approach.

$\left(\dfrac{2x^3 y^5}{xy^2}\right)^4$ *We work inside the parentheses first. Since this is a quotient we apply rule 3.*

$= (2x^{3-1} y^{5-2})^4$

$= (2x^2 y^3)^4$ *Now we use rule 4 on the product.*

$= 2^4 (x^2)^4 (y^3)^4$ *We evaluate 2^4 and use rule 2.*

$= \boxed{16x^8 y^{12}}$

Solution 2   offers the "outside-in" approach.

$\left(\dfrac{2x^3 y^5}{xy^2}\right)^4$ *We view this expression as a quotient raised to a power, and so we use rule 5 to bring the outside exponent in on both numerator and denominator.*

$= \dfrac{(2x^3 y^5)^4}{(xy^2)^4}$ *Now we use rule 4 in the numerator and denominator.*

$= \dfrac{2^4 (x^3)^4 (y^5)^4}{x^4 (y^2)^4}$ *Now we use rule 2 on each power of a power.*

$= \dfrac{16x^{12} y^{20}}{x^4 y^8}$ *Now we use rule 3 on the x's and the y's.*

$= 16\left(\dfrac{x^{12}}{x^4}\right)\left(\dfrac{y^{20}}{y^8}\right)$

$$= 16x^{12-4}y^{20-8}$$

$$= \boxed{16x^8y^{12}} \qquad \blacksquare$$

EXAMPLE 9 *Simplify as completely as possible.* $\dfrac{(-3x^2y^3)^5}{6(xy)^3}$

9. $\dfrac{4(a^5b^2)^4}{(-2a^2b^3)^3}$

Solution   We begin by using rule 4 in both the numerator and the denominator.

$$\frac{(-3x^2y^3)^5}{6(xy)^3} = \frac{(-3)^5(x^2)^5(y^3)^5}{6x^3y^3} \qquad$$ *Be careful: the 6 in the denominator is outside the parentheses and therefore does not get raised to the third power. Also, do not try to reduce the -3 with the 6 since -3 is inside the parentheses and 6 is not. We evaluate $(-3)^5$ and use rule 2.*

$$= \frac{-243x^{10}y^{15}}{6x^3y^3} \qquad \text{Now we use rule 3.}$$

$$= \frac{-243}{6}x^{10-3}y^{15-3}$$

$$= \boxed{\frac{-81}{2}x^7y^{12}} \text{ which is the same as } \boxed{\frac{-81x^7y^{12}}{2}} \qquad \blacksquare$$

Knowing what the rules are does not make you good at using them. The only way to be sure you can apply the rules correctly is by doing the exercises.

✔ *Answers to Learning Checks in Section 5.1*

1. m^{12} **2.** m^{32} **3.** m^4 **4.** m^7n^7 **5.** $\dfrac{m^7}{n^7}$

6. (a) $u^{10}v^{15}$ (b) $\dfrac{r^8}{s^{20}}$ **7.** $\dfrac{y^{14}}{x^7}$ **8.** $27a^6b^3$ **9.** $-\dfrac{a^{14}}{2b}$

NOTE TO THE STUDENT

Use this space on this page to write down any questions you have or points you want to review with your instructor.

Exercises 5.1

Simplify each of the following as completely as possible.

1. x^3x^2

2. y^4y^3

3. $(x^3)^2$

4. $(y^4)^3$

5. x^3xx^5

6. a^3a^7a

7. 10^410^5

8. 7^37^8

9. 2^33^4

10. 3^25^3

11. $\dfrac{y^3y^5}{y^2y^4}$

12. $\dfrac{z^2z^7}{z^4z^6}$

13. $\dfrac{9u^9v^8}{3u^3v^4}$

14. $\dfrac{16r^{16}v^4}{8r^8v^8}$

15. $\dfrac{(a^3)^5}{(a^4)^2}$

16. $\dfrac{(x^4)^3}{(x^3)^2}$

17. $(-x^2)^4$

18. $-(x^2)^4$

19. $(x^2y)^2$

20. $(2xy^2)^3$

21. $(x^2y^3)^5$

22. $(a^2b^4)^3$

23. $(2r^3s^5)^4$

24. $2(r^3s^5)^4$

1. _____
2. _____
3. _____
4. _____
5. _____
6. _____
7. _____
8. _____
9. _____
10. _____
11. _____
12. _____
13. _____
14. _____
15. _____
16. _____
17. _____
18. _____
19. _____
20. _____
21. _____
22. _____
23. _____
24. _____

ANSWERS

25. _____

26. _____

27. _____

28. _____

29. _____

30. _____

31. _____

32. _____

33. _____

34. _____

35. _____

36. _____

37. _____

38. _____

39. _____

40. _____

41. _____

42. _____

25. $(-x^3y)^3$

26. $(-x^3y)^4$

27. $\left(\dfrac{x^3}{y^2}\right)^4$

28. $\left(\dfrac{u^5}{v^2}\right)^3$

29. $(2x^3)^4(3x^2)^2$

30. $(3a^2)^3(5a^3)^2$

31. $\dfrac{(3x^5y^4)^2}{9(x^3y)^3}$

32. $\dfrac{16(ab^2)^5}{(2ab^4)^2}$

33. $\left(\dfrac{2x^3y^4}{xy^6}\right)^5$

34. $\left(\dfrac{3a^5b^3}{a^3b^6}\right)^4$

35. $\left(\dfrac{-3a^2b^3}{2c}\right)^3$

36. $\left(\dfrac{4u^5v^2}{-5w^2}\right)^3$

37. $\dfrac{-3^2}{(-3)^2}$

38. $\dfrac{(-5)^2}{-5^2}$

39. $\dfrac{-x^2}{(-x)^2}$

40. $\dfrac{(-a)^4}{-a^4}$

41. $\dfrac{-2^4 + 3^2}{(-4 + 3)^2}$

42. $\dfrac{-3^2 + 4^2}{-2^2 - 3}$

QUESTIONS FOR THOUGHT

1. State in words the difference between the first and second rules for exponents.

2. Given Exponent Rule 3, what would you get as an answer to the following examples:

    (a) $\dfrac{x^8}{x^8}$ (b) $\dfrac{x^4}{x^7}$

3. State what is *wrong* with each of the following:

    (a) $(x^2)^3 \overset{?}{=} x^5$

    (b) $x^4 \cdot x^3 \overset{?}{=} x^{12}$

    (c) $\dfrac{x^6}{x^2} \overset{?}{=} x^3$

    (d) $x^2 + x^3 \overset{?}{=} x^5$

Zero and Negative Exponents

5.2

Thus far everything we have said about exponents has depended on the fact that our exponents are positive integers. We know, for example, that 5^4 means "take the number 5 as a factor 4 times."

All well and good, but what should 5^0 mean? It certainly cannot mean take the number 5 as a factor zero times—that would give us a blank page. It should be clear that whatever meaning we decide to give to 5^0, the exponent will no longer just be counting the factors for us.

In the last section we discussed the five basic exponent rules. These rules were a direct consequence of our ability to count factors. There (hopefully) was no mystery as to where those rules came from.

Since at this point we have no real motivation as to how we want to define a zero or a negative exponent, we will do an about-face. We will let the rules tell us how to define these new exponents in a reasonable way.

Let's be more specific. Consider the expression

$$\frac{5^3}{5^3}$$

We know from arithmetic that $\frac{5^3}{5^3} = 1$. On the other hand, if we want to be able to apply Exponent Rule 3, we must have

$$\frac{5^3}{5^3} = 5^{3-3} = 5^0$$

Therefore, in order to be consistent we must have $5^0 = 1$.

Similarly,

$$\frac{7^6}{7^6} = 1 \quad \text{and we want} \quad \frac{7^6}{7^6} = 7^{6-6} = 7^0$$

Again, to be consistent, we must have $7^0 = 1$.

Thus, it should not be surprising that we make the following definition.

DEFINITION $a^0 = 1$ for all $a \neq 0$. *Note:* 0^0 is undefined.

In words, this definition says that any nonzero quantity raised to the zero power is always equal to 1.

It is important to realize that a definition is neither right nor wrong—it just is. The proper question to ask about a definition is: "Is it useful?" The answer in this case is yes for many reasons, one of which is the following: Suppose someone decides to define $a^0 = 14$ because 14 happens to be his or her favorite number. This is perfectly legal, *but* causes the following difficulty:

Suppose we consider an expression involving exponents such as $a^0 a^4$. Using Exponent Rule 1 we get

$$a^0 a^4 = a^{0+4} = a^4$$

while using the "other definition" we get

$$a^0 a^4 \stackrel{?}{=} 14 a^4$$

which is not the same answer. The rule and this other definition are not consistent.

Something has to give. Either we would have to give up all the exponent rules, or we have to give up this other definition. Clearly, we do not want to give up the rules.

However, with our definition we do not face this problem. Using our definition, we get

$$a^0 a^4 = 1 \cdot a^4 = a^4$$

which is the same answer obtained using rule 1. It can be shown that our definition is consistent with all five of our exponent rules.

The same type of reasoning will lead us to a definition of negative exponents. We know that

$$\frac{5^3}{5^7} = \frac{5 \cdot 5 \cdot 5}{5 \cdot 5 \cdot 5 \cdot 5 \cdot 5 \cdot 5 \cdot 5} = \frac{\cancel{5} \cdot \cancel{5} \cdot \cancel{5}}{\cancel{5} \cdot \cancel{5} \cdot \cancel{5} \cdot 5 \cdot 5 \cdot 5 \cdot 5} = \frac{1}{5^4}$$

On the other hand, using Exponent Rule 3 on this same expression, we get

$$\frac{5^3}{5^7} = 5^{3-7} = 5^{-4}$$

Thus, for the rules and our definition to be consistent we want

$$5^{-4} = \frac{1}{5^4}$$

Similarly, we get

$$\frac{8^2}{8^5} = \frac{8 \cdot 8}{8 \cdot 8 \cdot 8 \cdot 8 \cdot 8} = \frac{\cancel{8} \cdot \cancel{8}}{\cancel{8} \cdot \cancel{8} \cdot 8 \cdot 8 \cdot 8} = \frac{1}{8^3}$$

while using Exponent Rule 3 gives us

$$\frac{8^2}{8^5} = 8^{2-5} = 8^{-3}$$

Based on this analysis we make the following definition.

DEFINITION For any integer n, $a^{-n} = \dfrac{1}{a^n}$, $a \neq 0$.

In words, the definition says that a^{-n} is the *reciprocal* of a^n, where a is any nonzero quantity. Remember that the reciprocal of x is $\dfrac{1}{x}$.

In all the examples that follow, we will assume that our variables are not equal to 0 so that we can apply the definitions.

✔ **LEARNING CHECKS**

1. (a) 5^0

(b) $5x^0$

(c) $(5x)^0$

(d) x^{-5}

(e) $3x^{-5}$

(f) 5^{-2}

(g) -5^{-2}

(h) $(-5)^{-2}$

EXAMPLE 1 Write each of the following expressions without zero or negative exponents.

(a) 8^0 **(b)** $4x^0$ **(c)** $(4x)^0$ **(d)** x^{-3}

(e) $4x^{-3}$ **(f)** 3^{-4} **(g)** -2^{-4} **(h)** $(-2)^{-4}$

Solution

(a) $8^0 = \boxed{1}$ *By the definition of a zero exponent*

(b) $4x^0 = 4 \cdot 1 = \boxed{4}$ *The zero exponent applies only to the x.*

(c) $(4x)^0 = \boxed{1}$ *The zero exponent applies to the entire $4x$.*

(d) $x^{-3} = \boxed{\dfrac{1}{x^3}}$ *By the definition of a negative exponent*

(e) $4x^{-3} = 4 \cdot \dfrac{1}{x^3} = \boxed{\dfrac{4}{x^3}}$ *The -3 exponent applies only to the x.*

(f) $3^{-4} = \dfrac{1}{3^4} = \boxed{\dfrac{1}{81}}$ *By the definition of a negative exponent*

(g) $-2^{-4} = -\dfrac{1}{2^4} = \boxed{-\dfrac{1}{16}}$ *The -4 exponent applies only to the 2.*

(h) $(-2)^{-4} = \dfrac{1}{(-2)^4} = \boxed{\dfrac{1}{16}}$ *By the definition of a negative exponent* ■

Note the difference between parts **(g)** and **(h)** of Example 1. Several comments are in order.

First, whenever we are simplifying an expression involving negative exponents, we expect the *final* answer to be expressed with positive exponents only. Second, as we stated earlier, all the exponent rules are totally consistent with our definitions of zero and negative exponents. In particular, Exponent Rule 3 can be used regardless of whether the exponent in the numerator is larger than the exponent in the denominator.

We are free to use the exponent rules whenever it is appropriate to do so. In reality, the hardest part of much of algebra is learning what is appropriate for a particular situation.

Let's summarize all the exponent rules again, this time removing the restriction in rule 3, and including our definitions of zero and negative exponents.

EXPONENT RULES

For m and n any integers:

     Definition $a^0 = 1$ for all $a \neq 0$

     Definition $a^{-n} = \dfrac{1}{a^n}$ for all $a \neq 0$

     Exponent Rule 1 $a^m \cdot a^n = a^{m+n}$

     Exponent Rule 2 $(a^m)^n = a^{n \cdot m}$

     Exponent Rule 3 $\left(\dfrac{a^m}{a^n}\right) = a^{m-n},$ $a \neq 0$

     Exponent Rule 4 $(ab)^n = a^n b^n$

     Exponent Rule 5 $\left(\dfrac{a}{b}\right)^n = \dfrac{a^n}{b^n},$ $b \neq 0$

2. $\dfrac{1}{u^{-4}}$

EXAMPLE 2   *Simplify.* $\dfrac{1}{a^{-3}}$

Solution

$$\dfrac{1}{a^{-3}}$$ *Using the definition of a^{-n} on a^{-3} we get*

$$= \dfrac{1}{\dfrac{1}{a^3}}$$ *Using the rule for dividing fractions, we invert the divisor and multiply.*

$$= 1 \cdot \dfrac{a^3}{1}$$

$$= \boxed{a^3}$$

We could also have gotten this answer just by thinking about what the example is saying. We have already said that a^{-n} is the reciprocal of a^n. Of course, the reverse is also true because we know that being a reciprocal is a two-way street. Just as $\frac{2}{3}$ is the reciprocal of $\frac{3}{2}$, so too $\frac{3}{2}$ is the reciprocal of $\frac{2}{3}$.

Thus, a^n and a^{-n} are reciprocals *of each other:* $\dfrac{1}{a^{-n}} = a^n$

Example 2 is asking for $\dfrac{1}{a^{-3}}$, which means the reciprocal of a^{-3} and therefore the answer must be a^3. ∎

---

3. $\dfrac{u^2}{u^8}$

EXAMPLE 3   *Simplify.* $\dfrac{x^3}{x^9}$

Solution

$$\dfrac{x^3}{x^9}$$ *Using rule 3 we get*

$$= x^{3-9}$$

$$= x^{-6}$$ *Now use the definition of a negative exponent.*

$$= \boxed{\dfrac{1}{x^6}}$$ ∎

If you are thinking to yourself "couldn't Example 3 have been done by simply reducing the fraction, as we did in the past?" you are perfectly correct. We used rule 3 just to illustrate it, not because it necessarily offered any advantages. However, if such an example involves negative exponents then the use of rule 3 (or the other exponent rules, if they apply) is advisable.

Another important thing to keep in mind is not to confuse a negative exponent with a negative answer. That is, do not confuse the sign of the exponent with the sign of the base. For example:

$$2^{-1} \neq -2$$

$$2^{-1} = \dfrac{1}{2}$$ *The negative exponent does not make anything negative, but rather it tells us to take the reciprocal of the base.*

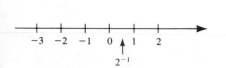
2^{-1}

EXAMPLE 4   Simplify as completely as possible.

(a) $x^{-3}x^{-9}$

(b) $\dfrac{x^{-3}}{x^{-9}}$

(c) $\dfrac{x^{-9}}{x^{-3}}$

(d) $(x^{-3})^{-9}$

4. (a) $a^{-2}a^{-6}$

(b) $\dfrac{a^{-2}}{a^{-6}}$

(c) $\dfrac{a^{-6}}{a^{-2}}$

(d) $(a^{-6})^{-2}$

Solution

Many students fall into a "one-over syndrome." That is, as soon as they see a negative exponent they use the definition

$$a^{-n} = \frac{1}{a^n}$$

While this is not wrong, it can often make the example very messy. It is generally better to apply any appropriate exponent rules *before* using the definition.

(a) $x^{-3}x^{-9}$ *Using rule 1, we get*

$= x^{-3+(-9)}$

$= x^{-12}$ *Since we want our final answer with positive exponents only, **now** we use the definition of a^{-n}.*

$= \boxed{\dfrac{1}{x^{12}}}$

(b) $\dfrac{x^{-3}}{x^{-9}}$ *Using rule 3, which says take the top exponent and subtract the bottom one, we get*

$= x^{-3-(-9)}$ *Remember we are subtracting a negative number.*

$= x^{-3+9}$

$= \boxed{x^6}$

(c) $\dfrac{x^{-9}}{x^{-3}}$ *Using rule 3, we get*

$= x^{-9-(-3)}$

$= x^{-6}$

$= \boxed{\dfrac{1}{x^6}}$

(d) $(x^{-3})^{-9}$ *Using rule 2, we get*

$= x^{(-9)(-3)}$

$= \boxed{x^{27}}$

Note that in parts **(b)** and **(d)** of this example we never had to use the definition of a negative exponent at all. Also, look at parts **(b)** and **(c)**. Do you see why the answers should come out to be reciprocals?   ■

Let's look at a few more examples similar to those of the last section, but now involving negative exponents as well. Wherever possible we will take the same approach as we did when the examples involved only positive exponents.

5. $(u^3v^{-1})^{-2}$

EXAMPLE 5 *Simplify.* $(x^{-3}y^2)^{-4}$

Solution

$$(x^{-3}y^2)^{-4}$$ *Using rule 4, we get*

$$= (x^{-3})^{-4}(y^2)^{-4}$$ *Using rule 2, we get*

$$= x^{12}y^{-8}$$ *Using the definition of a^{-n}, we get*

$$= x^{12} \cdot \frac{1}{y^8}$$

$$= \boxed{\frac{x^{12}}{y^8}}$$ ∎

6. $\dfrac{(3a^{-4}b^2)^{-2}}{(a^2b^3)^{-1}}$

EXAMPLE 6 *Simplify.* $\dfrac{(2x^2y^{-3})^{-1}}{(x^{-3}y)^{-2}}$

Solution

$$\frac{(2x^2y^{-3})^{-1}}{(x^{-3}y)^{-2}}$$ *Use rule 4 in numerator and denominator to get*

$$= \frac{2^{-1}(x^2)^{-1}(y^{-3})^{-1}}{(x^{-3})^{-2}y^{-2}}$$ *Using rule 2 in numerator and denominator, we get*

$$= \frac{2^{-1}x^{-2}y^3}{x^6y^{-2}}$$ *Using rule 3, we get*

$$= 2^{-1}x^{-2-6}y^{3-(-2)}$$

$$= 2^{-1}x^{-8}y^5$$ *Using the definition of a^{-n}, we get*

$$= \frac{1}{2} \cdot \frac{1}{x^8} \cdot \frac{y^5}{1}$$

$$= \boxed{\frac{y^5}{2x^8}}$$ ∎

7. $\left(\dfrac{m^2n^{-5}}{m^{-1}n^2}\right)^{-3}$

EXAMPLE 7 *Simplify.* $\left(\dfrac{x^{-6}y^3}{x^5y^{-1}}\right)^{-2}$

Solution Working from the inside out, we have

$$\left(\frac{x^{-6}y^3}{x^5y^{-1}}\right)^{-2}$$ *Use rule 3 inside the parentheses to get*

$$= (x^{-6-5}y^{3-(-1)})^{-2}$$

$$= (x^{-11}y^4)^{-2}$$ *Using rule 4, we get*

$$= (x^{-11})^{-2}(y^4)^{-2}$$ *Using rule 2, we get*

$$= x^{22}y^{-8}$$ *Using the definition of a^{-n}, we get*

$$= \boxed{\frac{x^{22}}{y^8}}$$ ∎

8. (a) $(a^{-1}b^{-1})^{-3}$

  (b) $a^{-3} + b^{-3}$

  (c) $(a + b)^{-3}$

EXAMPLE 8 Express using positive exponents only.

(a) $(x^{-2}y^{-2})^{-1}$ **(b)** $x^{-2} + y^{-2}$ **(c)** $(x^2 + y^2)^{-1}$

Solution Be careful! It is just as important to know when a particular exponent rule does *not* apply as to know when it does.

(a) $(x^{-2}y^{-2})^{-1}$ *Using rule 4, we get*

$$= (x^{-2})^{-1}(y^{-2})^{-1}$$

$$= \boxed{x^2y^2}$$

(b) $x^{-2} + y^{-2} = \boxed{\dfrac{1}{x^2} + \dfrac{1}{y^2}}$ *Obtained by using the definition of* a^{-n}

(c) Rule 4 *does not apply* here because of the addition inside the parentheses. Rule 4 applies only when there is a *product* in the parentheses.

$$(x^2 + y^2)^{-1} = \boxed{\dfrac{1}{x^2 + y^2}}$$ *Obtained by using the definition of* a^{-n}

Also note that the answers to parts **(b)** and **(c)** are not the same.   ∎

Answers to Learning Checks in Section 5.2

1. (a) 1   **(b)** 5   **(c)** 1   **(d)** $\dfrac{1}{x^5}$ **(e)** $\dfrac{3}{x^5}$ **(f)** $\dfrac{1}{25}$ **(g)** $-\dfrac{1}{25}$ **(h)** $\dfrac{1}{25}$

2. u^4 **3.** $\dfrac{1}{u^6}$ **4. (a)** $\dfrac{1}{a^8}$ **(b)** a^4 **(c)** $\dfrac{1}{a^4}$ **(d)** a^{12}

5. $\dfrac{v^2}{u^6}$ **6.** $\dfrac{a^{10}}{9b}$ **7.** $\dfrac{n^{21}}{m^9}$ **8. (a)** a^3b^3 **(b)** $\dfrac{1}{a^3} + \dfrac{1}{b^3}$ **(c)** $\dfrac{1}{(a+b)^3}$

NOTE TO THE STUDENT

Use the space on this page to write down any questions you have or points you want to review with your instructor.

Exercises 5.2

Exercises 1–4 consist of a number of related parts. Simplify each as completely as possible.

1. (a) $-3(2)$ **(b)** x^2x^{-3} **(c)** $(x^2)^{-3}$ **(d)** 2^{-3}

2. (a) $-3(-2)$ **(b)** $x^{-2}x^{-3}$ **(c)** $(x^{-2})^{-3}$ **(d)** $(-2)^{-3}$

Simplify each of the following expressions as completely as possible. Final answers should be expressed with positive exponents only. (Assume that all variables represent positive quantities.)

3. 8^0 **4.** $(-27)^0$ **5.** $5 \cdot 4^0$ **6.** $7 \cdot 7^0$

7. xy^0 **8.** $(xy)^0$ **9.** 5^{-2} **10.** a^{-4}

11. $\dfrac{1}{5^{-2}}$ **12.** $\dfrac{1}{a^{-4}}$ **13.** $x^{-4}x^4$ **14.** x^6x^{-6}

15. $x^{-4}x^{-6}$ **16.** $(x^{-4})^{-6}$ **17.** $a^2a^{-4}aa^{-7}$ **18.** $z^3z^2z^{-4}z^{-7}$

19. $10^6 10^{-5} 10^{-4}$ **20.** $10^{-2}10^{-4}10^9$ **21.** $(xy)^4$ **22.** $(xy)^{-5}$

23. $2a^{-3}$ **24.** $(2a)^{-3}$ **25.** $-3y^{-2}$ **26.** $(-3y)^{-2}$

27. xy^{-1} **28.** $(xy)^{-1}$ **29.** $(a^{-4}b^3)^{-2}$ **30.** $(a^5b^{-2})^{-3}$

31. $(3x^{-2}y^3z^{-4})^2$ **32.** $(5x^2y^{-3}z^4)^{-2}$ **33.** $4(x^{-1}y)^{-3}$ **34.** $(4x^{-4}y)^3$

35. $\dfrac{x^5}{x^2}$ **36.** $\dfrac{x^{-5}}{x^{-2}}$

1. a. _____
   b. _____
   c. _____
   d. _____
2. a. _____
   b. _____
   c. _____
   d. _____
3. _____
4. _____
5. _____
6. _____
7. _____
8. _____
9. _____
10. _____
11. _____
12. _____
13. _____
14. _____
15. _____
16. _____
17. _____
18. _____
19. _____
20. _____
21. _____
22. _____
23. _____
24. _____
25. _____
26. _____
27. _____
28. _____
29. _____
30. _____
31. _____
32. _____
33. _____
34. _____
35. _____
36. _____

ANSWERS

37. _____

38. _____

39. _____

40. _____

41. _____

42. _____

43. _____

44. _____

45. _____

46. _____

47. _____

48. _____

49. _____

50. _____

51. _____

52. _____

53. _____

54. _____

37. $\dfrac{-3a^{-3}}{9a^9}$

38. $\dfrac{-4a^{-4}}{2a^2}$

39. $x^{-2} + y^{-1}$

40. $x^{-1} + y^{-2}$

41. $\dfrac{x^4 x^{-10}}{x^{-2} x^{-5}}$

42. $\dfrac{a^{-3}a}{a^6 a^{-3}}$

43. $\dfrac{x^4 y^{-10}}{x^{-2} y^{-5}}$

44. $\dfrac{a^{-3}b}{a^6 b^{-3}}$

45. $\dfrac{12(10^{-3})}{4(10^{-7})}$

46. $\dfrac{18(10^{-4})}{3(10^2)}$

47. $\left(\dfrac{a^{-2}}{a^3}\right)^{-3}$

48. $\left(\dfrac{y^{-3}}{y^{-5}}\right)^{-2}$

49. $\dfrac{(x^2 y^{-1})^{-1}}{(x^3 y^{-2})^2}$

50. $\dfrac{(ay^{-2})^{-3}}{(a^4 y^{-3})^2}$

51. $\left(\dfrac{2m^{-2}n^{-3}}{m^{-6}n^{-1}}\right)^{-2}$

52. $\left(\dfrac{3r^{-4}s^{-1}}{r^{-8}s^{-3}}\right)^{-3}$

53. $\left(\dfrac{x^{-1}y^{-2}}{3x^{-2}y^{-3}}\right)^{-1}$

54. $\left(\dfrac{a^{-2}b^{-4}}{5a^{-4}b^{-8}}\right)^{-1}$

QUESTIONS FOR THOUGHT

4. Discuss the difference between $\dfrac{x^6}{x^4}$ and $\dfrac{x^6}{x^{-4}}$.

5. Discuss the difference between 3^{-1} and -3.

Scientific Notation

5.3

There are many occasions (especially in science-related fields) when we may come across either very large or very small numbers. For example, we may read statements such as:

The mass of the earth is approximately
5,980,000,000,000,000,000,000,000,000 kilograms

The wavelength of yellow-green light is approximately .0000006 centimeter.

Writing numbers like this in their decimal form can be quite messy and is very prone to errors. Using our knowledge of integer exponents, we can describe an alternative form for writing such numbers which makes them easier to work with. This other form is called *scientific notation*.

Scientific notation is a concise way of expressing very large or very small numbers. Before we describe scientific notation, let's recall some basic facts from arithmetic using the language of exponents.

Multiplying a number by 10 (that is, 10^1) moves the decimal point 1 place to the *right*.

Multiplying a number by 100 (that is, 10^2) moves the decimal point 2 places to the *right*.

Multiplying a number by 1,000 (that is, 10^3) moves the decimal point 3 places to the *right*.

On the other hand,

Dividing a number by 10 (or multiplying by $\frac{1}{10}$, which is the same thing as multiplying by 10^{-1}) moves the decimal point 1 place to the *left*.

Dividing a number by 100 (or multiplying by $\frac{1}{100}$, which is the same as multiplying by 10^{-2}) moves the decimal point 2 places to the *left*.

Dividing a number by 1,000 (or multiplying by $\frac{1}{1,000}$, which is the same thing as multiplying by 10^{-3}) moves the decimal point 3 places to the *left*.

To be more specific, suppose we have a number, say 5.49; let's see what happens to the number as we multiply it by these various powers of 10.

Original number	Multiplied by	Becomes
5.49	$1 = 10^0$	5.49
5.49	$10 = 10^1$	54.9
5.49	$100 = 10^2$	549.
5.49	$1,000 = 10^3$	5,490.
5.49	$\frac{1}{10} = 10^{-1}$	.549
5.49	$\frac{1}{100} = 10^{-2}$	.0549
5.49	$\frac{1}{1,000} = 10^{-3}$	.00549

Standard notation is the decimal notation we normally use. Numbers such as

143.4,   7.956,   and   .00538

are written in standard notation.

DEFINITION   We say that a number is written in *scientific notation* if the number is of the form

$$a \times 10^n \quad \text{where } 1 \le a < 10, \quad \text{and } n \text{ is an integer.}$$

This is just about the only place in algebra where the symbol "×" is used to indicate multiplication. The reason it is used here is because the "·" can too easily be misread as a decimal point.

Verify for yourself that the two numbers on each line of the following display are equal.

Number in standard notation	*Number in scientific notation*
48,500	4.85×10^4
3,756	3.756×10^3
980	9.8×10^2
72	$7.2 \times 10^1 = 7.2 \times 10$
6.5	$6.5 \times 10^0 = 6.5 \times 1 = 6.5$
.432	4.32×10^{-1}
.0999	9.99×10^{-2}
.005	5×10^{-3}
.00012	1.2×10^{-4}

Note: A number is in scientific notation if its decimal point is immediately to the right of its first nonzero digit. Thus, the number 6.5 is the same in both columns because any real number between 1 and 10 is already in scientific notation. The 10^0 is usually not written but it is understood.

Here are some numbers not in scientific notation (and why):

$.095 \times 10^4$ because .095 is not between 1 and 10.   (It should be written 9.5×10^2.)

62×10^{-3} because 62 is not between 1 and 10.   (It should be written 6.2×10^{-2}.)

Scientific notation is also used by many calculators and computers. When a number is either too large or too small to be displayed on a calculator, the number might appear on the display as

$$6.4203 \text{ E } 9 \quad \text{or} \quad 2.38 \text{ E} - 15$$

These expressions are actually shorthand forms of scientific notation (the E stands for exponent).

$6.4203 \text{ E } 9$ means 6.4203×10^9.

$2.38 \text{ E} -15$ means 2.38×10^{-15}.

✔ **LEARNING CHECKS**

1. Convert 5.89×10^{-3} and 5.89×10^3 to standard notation.

EXAMPLE 1   Convert 2.91×10^5 and 2.91×10^{-5} into standard notation.

Solution   Converting from scientific notation into standard notation is quite straightforward because, as we have already pointed out, multiplying by 10 moves the decimal point one place to the right, while dividing by 10 moves the decimal point one place to the left.

$$2.91 \times 10^5 = 2.91 \times 10 \times 10 \times 10 \times 10 \times 10 = 291,000$$

*Note that the decimal point has moved 5 places to the **right**.*

$$2.91 \times 10^{-5} = 2.91 \times \frac{1}{10} \times \frac{1}{10} \times \frac{1}{10} \times \frac{1}{10} \times \frac{1}{10} = .0000291$$

*Note that the decimal point has moved 5 places to the **left**.*

As you can see, we may bypass the middle step by observing that the *number* of places the decimal point is moved is given by the exponent of 10; the *direction* in which it is moved is determined by the sign of the exponent.

In 2.91×10^5 the exponent is $+5$ and the decimal point moves 5 places to the right.

In 2.91×10^{-5} the exponent is -5 and the decimal point moves 5 places to the left.   ■

EXAMPLE 2   Convert 2,830 into scientific notation.

Solution   Converting from standard notation to scientific notation requires a little more thought. The decimal point in 2,830 is understood to be to the right of the 0. In order to be in scientific notation the decimal point must be to the immediate right of the first nonzero digit, which in this case means between the 2 and the 8. But moving the decimal point there will, of course, change the value of the number. We compensate for this change by putting in the "correcting" power of 10.

$$2,830 \quad = \quad 2.83 \quad \times \quad 10^3$$

We want to move the decimal point 3 places to the left.   *We put the decimal point where we want it.*   *This power of 10 compensates by putting the decimal point back where it really belongs, 3 places to the right.*

The net effect is that we have not changed the value of the number; we have changed only its appearance—which was the whole idea.

Thus, our answer is $\boxed{2.83 \times 10^3}$ ■

EXAMPLE 3   Convert .072 into scientific notation.

Solution   We want the decimal point to be between the 7 and the 2.

$$.072 \quad = \quad 7.2 \quad \times \quad 10^{-2}$$

We want to move the decimal point 2 places to the right.   *We put the decimal point where we want it.*   *This power of 10 compensates by putting the decimal point back where it really belongs, 2 places to the left.*

Thus, our answer is $\boxed{7.2 \times 10^{-2}}$ ■

Scientific notation often allows us to carry out complicated arithmetic problems more easily.

2. Convert 47,200 into scientific notation.

3. Convert .00631 into scientific notation.

4. Compute using scientific notation. $\dfrac{(.08)(9,000)}{(600)(.0005)}$

EXAMPLE 4 *Compute the following using scientific notation.*

$$\frac{(360)(.004)}{(.0002)(600,000)}$$

Solution First we convert each number into scientific notation, and then we group the numbers so as to make the computation easiest.

$$\frac{(360)(.004)}{(.0002)(600,000)} = \frac{(3.6 \times 10^2)(4 \times 10^{-3})}{(2 \times 10^{-4})(6 \times 10^5)}$$

$$= \frac{(3.6)(4)}{(2)(6)} \times \frac{10^2 10^{-3}}{10^{-4} 10^5}$$

Reduce the first fraction; use rule 1 on the second.
$10^2 10^{-3} = 10^{2+(-3)} = 10^{-1}$
$10^{-4} 10^5 = 10^{-4+5} = 10^1$

$$= \frac{\overset{1.2}{\cancel{(3.6)}}\overset{2}{\cancel{(4)}}}{\underset{3}{\cancel{(2)}\cancel{(6)}}} \times \frac{10^{-1}}{10^1}$$

Use rule 3 on the 10's.

$$= 1.2 \times 10^{-1-1}$$

$$= \boxed{1.2 \times 10^{-2}}$$

You might have noticed that this computation would have been simpler if we had written 360 as

$$36 \times 10 \quad \text{rather than} \quad 3.6 \times 10^2$$

Strictly speaking, 36×10 is not scientific notation because 36 is not a number between 1 and 10. Nevertheless, you should feel free to use a "modified" form of scientific notation in your computations. ∎

---

5. If 6.02×10^{23} atoms of iron have a mass of approximately 56 grams, what is the mass of 1 atom of iron?

EXAMPLE 5 The mass of a hydrogen atom is approximately 1.67×10^{-24} gram. Based on this figure, compute the number of atoms in 1 gram of hydrogen. Express your answer in scientific notation.

Solution We want to know how many times does 1.67×10^{-24} gram go into 1 gram. Thus, this example is actually a division problem.

$$1 \div (1.67 \times 10^{-24}) = \frac{1}{1.67 \times 10^{-24}}$$

$$= \frac{1}{1.67} \times \frac{1}{10^{-24}}$$

$$= \frac{1}{1.67} \times 10^{24}$$

We compute 1 divided by 1.67, and round off to 3 places.

$$= .599 \times 10^{24}$$

This answer is not in scientific notation.

$$= \boxed{5.99 \times 10^{23}}$$

∎

✔ **Answers to Learning Checks in Section 5.3**

1. .00589 and 5,890   **2.** 4.72×10^4

3. 6.31×10^{-3} **4.** 2.4×10^3 **5.** 9.3×10^{-23} gram

Exercises 5.3

In Exercises 1–28, convert each number into scientific notation.

1. 4,530

2. 1,250

3. .0453

4. .0125

5. .00007

6. .0004

7. 7,000,000

8. 400,000

9. 85,370

10. 12,340

11. .0085370

12. .0001234

13. 90

14. 70

15. 9

16. 7

17. .9

18. .7

19. .09

20. .07

21. .00000003

22. .0000002

23. 28

24. 37

25. 47.5

26. 52.4

27. 9,727.3

28. 111.12

In Exercises 29–42, convert each number into standard notation.

29. 2.8×10^4

30. 5×10^3

31. 2.8×10^{-4}

32. 5×10^{-3}

33. 4.29×10^7

34. 1.76×10^{-5}

35. 4.29×10^{-7}

36. 1.76×10^5

37. 3.52×10^{-3}

38. 6.81×10^{-2}

39. 3.5286×10^5

40. $.0527 \times 10^4$

41. $.026 \times 10^{-3}$

42. $78,951 \times 10^{-5}$

1. _____
2. _____
3. _____
4. _____
5. _____
6. _____
7. _____
8. _____
9. _____
10. _____
11. _____
12. _____
13. _____
14. _____
15. _____
16. _____
17. _____
18. _____
19. _____
20. _____
21. _____
22. _____
23. _____
24. _____
25. _____
26. _____
27. _____
28. _____
29. _____
30. _____
31. _____
32. _____
33. _____
34. _____
35. _____
36. _____
37. _____
38. _____
39. _____
40. _____
41. _____
42. _____

ANSWERS

In Exercises 43–50, *do your computation using scientific notation.*

43. $(.004)(250)$

44. $(600)(.0015)$

45. $\dfrac{.003}{6{,}000}$

46. $\dfrac{80}{.0002}$

47. $\dfrac{(480)(.008)}{(.24)(4{,}000)}$

48. $\dfrac{(.0075)(6{,}400)}{(.032)(250)}$

49. $\dfrac{(.0036)(.005)}{(.01)(.06)}$

50. $\dfrac{(2{,}400)(1{,}500)}{(90{,}000)(4{,}000)}$

43. _____

44. _____

45. _____

46. _____

47. _____

48. _____

49. _____

50. _____

51. _____

52. _____

53. _____

54. _____

55. _____

56. _____

57. _____

58. _____

59. _____

60. _____

In the following exercises, do your computations using scientific notation.

51. The first paragraph of this section states the approximate mass of the earth. Write this number in scientific notation.

52. The first paragraph of this section states the approximate wavelength of yellow-green light. Write this number in scientific notation.

53. If one atom of iron has a mass of 9.3×10^{-23} gram, what is the mass of 80,000 atoms?

54. If the mass of one atom of iron is 9.3×10^{-23} gram, how many atoms are there in 1 gram?

55. Atomic measurements or other very small distances are often measured in units called *angstroms*. One angstrom, which is written 1 Å, is equal to .00000001 cm. Write this number in scientific notation.

56. One angstrom is equal to what part of a meter? A kilometer?

57. If 1 ton is equal to 888.9 kilograms, what is the weight of Earth in tons? (Use the result of Exercise 51.)

58. Light travels at a speed of approximately 186,000 miles per second. How far does light travel in 1 year? (This *distance* is called 1 light-year.)

59. The Mt. Palomar Observatory has a 200-inch mirror telescope that can see a star 5 billion light-years away. Use the result of Exercise 58 to compute this distance in miles.

60. If light travels 5.87×10^{12} miles in 1 year, how long will it take light to reach us from a star that is 3×10^{22} miles away?

QUESTION FOR THOUGHT

6. What can you say about the sign of the exponent of 10 of a number written in scientific notation, if the number is bigger than 1? If the number is smaller than 1? Why?

Introduction to Polynomials

5.4

Thus far much of our attention in this book has been focused on expressions made up of single terms. Our fractional expressions contained mostly single terms in the numerator and in the denominator. We have not yet talked about multiplying expressions involving more than one term. Now we are going to broaden our scope by looking at expressions which involve more than one term.

Terminology

On several occasions we have pointed out the important distinction between *terms* and *factors*. Let's begin by giving a formal definition of a particular type of term.

> **DEFINITION** A *monomial* is an algebraic expression which is either a constant or a product of constants and variables.

The following are all examples of monomials:

$$4x^3y^2 \qquad -5t^6 \qquad \frac{2}{3}a^2$$

Note that a monomial is simply a special kind of *term*. Adding and subtracting monomials gives us more complex expressions.

> **DEFINITION** A *polynomial* is the sum of one or more monomials.

For example, all the following are polynomials:

$3x^2 - 5xy$ is often called a *binomial* because it has two terms.

$-5a^3 + 2ab + b^2$ is often called a *trinomial* because it has three terms.

$2y$ is a polynomial with just one term (a monomial). It is usually just called a *term*.

The following are just two examples of *nonpolynomial* expressions:

$\dfrac{3}{x + 1}$ because polynomials do not allow variables in denominators.

3^x because polynomials do not allow variables as exponents.

Besides classifying polynomials by the number of terms (monomial, binomial, trinomial), we can also classify them by what is known as their *degree*.

> **DEFINITION** The *degree* of a nonzero *monomial* is the sum of the exponents of the variables.
> The degree of a nonzero constant is 0.
> The degree of the number 0 is undefined.

For example:

The degree of the monomial $-2a^4$ is 4.   (*There is only one exponent of a variable.*)

The degree of the monomial $7x^3y^2$ is 5.   (*The sum of the exponents is $2 + 3$.*)

The degree of the constant monomial 6 is 0 (because we can write $6 = 6x^0$) and the same is true for any constant.

The degree of the number 0 is undefined.

The reason we must specify "nonzero" is the definition and the reason the degree of the number 0 is undefined is that we can write

$$0 = 0x^8 \quad \text{or} \quad 0 = 0x^{17} \quad \text{or} \quad 0 = 0x^{35}$$

and so we cannot possibly tell what the degree of the number 0 is. Consequently, the degree of 0 is undefined.

DEFINITION   The *degree of a polynomial* is the highest degree of any monomial in it.

For example, the polynomial $5x^4 - 7x^2y + 2x$ has three terms whose individual degrees are

$$\underbrace{5x^4}_{\textit{Degree is } 4} \qquad \underbrace{-7x^2y}_{\substack{\textit{Degree is } 3 \\ 2 + 1 = 3}} \qquad \underbrace{+2x}_{\textit{Degree is } 1}$$

Thus, the overall degree of the polynomial is 4, which is the highest degree of any term in it.

The idea of degree provides us with a useful way of categorizing expressions and equations. In fact, we have already used the word *degree* in Chapter 3 when we learned about solving first-*degree* equations.

Most often we are interested in the degree of a polynomial involving only one variable, which is simply the highest exponent of the variable which appears. The expression

$$3x^5 - 4x^3 + 8x^2 + 7$$

is a fifth-degree polynomial, because the highest exponent of the variable x is 5. When a polynomial is written in this way, with the degrees of the terms in descending order, starting with the highest-degree term and ending with the lowest-degree term, the polynomial is said to be in **standard form** for polynomials.

As we have discussed previously, every term has a coefficient. In the polynomial mentioned above, $3x^5 - 4x^3 + 8x^2 + 7$, there are four terms and their coefficients are 3, -4, 8, and 7, respectively. Note that 7, because it is a constant term, is both a term and a coefficient.

Sometimes we have occasion to talk about "missing terms." For example, in the polynomial we have been talking about

$$3x^5 - 4x^3 + 8x^2 + 7$$

the fourth-degree and first-degree terms do not appear.

If we want to think of, say, a fifth-degree polynomial as always containing all the terms from degree 5 down to the constant, then we can say that the fourth-degree and first-degree terms here have a coefficient of 0. In other words, we think of

$$3x^5 - 4x^3 + 8x^2 + 7 \quad \text{as} \quad 3x^5 + 0x^4 - 4x^3 + 8x^2 + 0x + 7$$

When we write the polynomial with the "missing terms" appearing with 0 coefficients, the polynomial is said to be in **complete standard form.**

EXAMPLE 1 Given the polynomial $4x^3 - 5x + 9$:

(a) What is the degree of each term?

(b) What is the degree of the polynomial?

(c) Write the polynomial in complete standard form. What are its coefficients?

Solution

(a) The polynomial $4x^3 - 5x + 9$ consists of three terms.
The degree of the first term, $4x^3$, is 3.
The degree of the second term, $-5x$, is 1.
The degree of the third term, 9, is 0.

(b) According to the definition of the degree of a polynomial, we take the highest degree of any term in it. Thus, the degree of $4x^3 - 5x + 9$ is 3.

(c) We write the polynomial in complete standard form, which means that all the terms from the highest degree down to the constant are accounted for.

$$4x^3 - 5x + 9 = 4x^3 + 0x^2 - 5x + 9$$

Now we see that:

The coefficient of $4x^3$ is 4.

The coefficient of $0x^2$ is 0.

The coefficient of $-5x$ is -5.

9 is both a term and a coefficient.   *Think of 9 as $9 \cdot 1$.*   ∎

EXAMPLE 2 *Add the polynomials.* $(5x^2 - 4x + 3) + (3x^2 - 2x - 7)$

Solution The use of our new terminology does not alter the fact that we have done many examples of this type before (see Section 2.4). Since the parentheses are nonessential, we remove them and regroup the terms.

$$(5x^2 - 4x + 3) + (3x^2 - 2x - 7) = 5x^2 + 3x^2 - 4x - 2x + 3 - 7$$

Combine like terms.

$$= \boxed{8x^2 - 6x - 4}$$ ∎

EXAMPLE 3 *Subtract the polynomials.* $(3x^2y - 5xy^2) - (x^2 - xy^2 - x^2y)$

Solution Recall that when we have a minus sign in front of a parenthesis, we think of it as a -1 multiplying the parenthesis and so we "distribute the minus sign."

$$(3x^2y - 5xy^2) - (x^2 - xy^2 - x^2y) = (3x^2y - 5xy^2) - 1(x^2 - xy^2 - x^2y)$$

The first parentheses are nonessential.

$$= 3x^2y - 5xy^2 - x^2 + xy^2 + x^2y$$

Combine like terms.

$$= \boxed{4x^2y - 4xy^2 - x^2}$$ ∎

Sometimes examples such as these are stated verbally.

✔ **LEARNING CHECKS**

1. Given the polynomial $5x^4 - 3x^2 + 7x - 5$:

   (a) What is the degree of each term?

   (b) What is the degree of the polynomial?

   (c) Write the polynomial in complete standard form. What are the coefficients of each term?

2. $(6a^3 - 5a + 7) + (a^2 - 4a - 8)$

3. $(6a^2b^3 - a^4) - (a^4 + a^2b^3 - a^4b)$

4. Subtract $t^2 - t + 3$ from $2t^2 - 3t - 6$.

EXAMPLE 4   Subtract $5a^3 - a + 3$ from $a^3 - a^2 - a - 4$.

Solution   As usual, when we deal with subtraction we have to be careful. The example tells us to subtract the first polynomial, $5a^3 - a + 3$, *from* the second polynomial, $a^3 - a^2 - a - 4$. Algebraically we must write the second polynomial first, as well as put parentheses around the first polynomial.

$$\underbrace{(a^3 - a^2 - a - 4)} - \underbrace{(5a^3 - a + 3)} \quad \textit{Distribute the minus sign.}$$

Subtract this
from

$$= a^3 - a^2 - a - 4 - 5a^3 + a - 3 \quad \textit{Combine like terms.}$$

$$= \boxed{-4a^3 - a^2 - 7} \quad \blacksquare$$

5. Subtract the sum of $u^2 - 3u$ and $u + 8$ from $u^2 + 6u - 10$.

EXAMPLE 5   Subtract $x^2 - 3x$ from the sum of $3x^2 - 5x$ and $2x + 3$.

Solution   This example gets translated as

$$(3x^2 - 5x) + (2x + 3) - (x^2 - 3x)$$

The first two sets of parentheses are optional; the third is essential.

$$(3x^2 - 5x) + (2x + 3) - (x^2 - 3x)$$

Distribute the minus sign. Watch out for the sign where the arrow points.

$$= 3x^2 - 5x + 2x + 3 - x^2 + 3x$$

$$= \boxed{2x^2 + 3} \quad \blacksquare$$

It is possible to add and subtract polynomials in a vertical format as well. Since it does not seem to offer any significant advantages, we will illustrate the vertical format in the next example, but we will use primarily the horizontal format we have been using thus far. However, if you happen to prefer the vertical format, feel free to use it.

6. (a) *Add.* $5a^2 + 6a - 1, a^3 - a,$ $a^3 + a^2 - 2a + 3$

(b) Subtract $7 - a^2$ from $a^2 - 10$.

EXAMPLE 6

(a) *Add.* $3x^3 - 5x + 2,$ $x^2 - 3x,$ $x^3 - x^2 - 2x + 7$

(b) Subtract $2x^2 - x + 3$ from $x^2 - 2$.

Solution

(a) Using the vertical format, we line up the polynomials with like terms directly above each other, then we add up the like terms in each column.

$$
\begin{array}{r}
3x^3 \quad\quad - 5x + 2 \\
x^2 - 3x \quad\quad \\
x^3 - x^2 - 2x + 7 \\
\hline
4x^3 \quad\quad - 10x + 9
\end{array}
$$

Thus, our answer is $\boxed{4x^3 - 10x + 9}$.

(b) Subtraction requires a bit more care, since *all* the terms in the bottom polynomial must be subtracted from the top one.

$$
\begin{array}{r}
x^2 \quad\ - 2 \\
-\ \underline{2x^2 - x + 3}
\end{array}
$$

We must now change *all* the signs of the polynomial being subtracted and add.

$$
\begin{array}{r}
x^2 \quad\ - 2 \\
\underline{-2x^2 + x - 3} \\
-x^2 + x - 5
\end{array}
$$

Thus, our answer is $\boxed{-x^2 + x - 5}$. ∎

In Section 2.2 we discussed evaluating algebraic expressions for certain replacement values of the variables. The process is exactly the same for polynomials.

EXAMPLE 7 Evaluate the polynomial $a^3 + 3a^2 - 5a + 10$ for $a = -3$.

7. Evaluate $3t^2 - 4t + 1$ for $t = -2$.

Solution We replace each occurrence of the variable, a, with -3.

$$
\begin{aligned}
a^3 + 3a^2 - 5a + 10 &= (\ \ )^3 + 3(\ \ )^2 - 5(\ \ ) + 10 \\
&= (-3)^3 + 3(-3)^2 - 5(-3) + 10 \qquad \textit{Evaluate powers} \\
&= -27 + 3(9) + 15 + 10 \qquad\qquad \textit{first, then multiply.} \\
&= -27 + 27 + 15 + 10 \\
&= \boxed{25}
\end{aligned}
$$

∎

✔ *Answers to Learning Checks in Section 5.4*

1. (a) 4, 2, 1, 0   **(b)** 4

  (c) $5x^4 + 0x^3 - 3x^2 + 7x - 5$. The coefficients are 5, 0, -3, 7, -5.

2. $6a^3 + a^2 - 9a - 1$ **3.** $5a^2b^3 - 2a^4 + a^4b$ **4.** $t^2 - 2t - 9$ **5.** $8u - 18$

6. (a) $2a^3 + 6a^2 + 3a + 2$ **(b)** $2a^2 - 17$ **7.** 21

NOTE TO THE STUDENT

Use the space on this page to write down any questions you have or points you want to review with your instructor.

Exercises 5.4

In Exercises 1–14, *answer the following questions:*

(a) How many terms are there?
(b) What is the degree of each term?
(c) What is the degree of the polynomial?

1. $3x^5$

2. $-4x^7$

3. $3x + 4$

4. $5y - 7$

5. $x^2 + y^3$

6. $2w^3 - v^5$

7. x^2y^3

8. $2w^3v^5$

9. 8

10. -12

11. $2x^3 - 5x^2 + x$

12. $5a^4 - 3a^3 + 26a$

13. $2x^3 + y^5$

14. $3m^2n^5$

In Exercises 15–20, *answer the following questions:*

(a) What is the degree of each term?
(b) What is the degree of the polynomial?
(c) Using *complete standard form*, what is the coefficient of each term?

15. $x^2 - 5x + 6$

16. $y^2 + 3y - 7$

17. $x^2 + 4$

18. $3y^2 - 8$

19. $x^3 - 1$

20. $1 - x^5$

In Exercises 21–36, *perform the indicated operations and simplify.*

21. $(a^3 + 7) + (5 - 2a^3)$

22. $(5w^4 - w^3 - w) + (2w^3 - 3w^2 - 5w)$

23. $(3u^2 - 2u + 7) - (u^3 - u^2 + 7u)$

24. $(5w^4 - w^3 - w) - (2w^3 - 3w^2 - 5w)$

25. $(4t^3 - t) + (t^2 + t) - (t^3 - t^2)$

26. $(s^4 - 6s^3) - (s^3 - s^2 + s) - (2s^3 - s^4)$

1. _____
2. _____
3. _____
4. _____
5. _____
6. _____
7. _____
8. _____
9. _____
10. _____
11. _____
12. _____
13. _____
14. _____
15. _____
16. _____
17. _____
18. _____
19. _____
20. _____
21. _____
22. _____
23. _____
24. _____
25. _____
26. _____

ANSWERS

27. _____

28. _____

29. _____

30. _____

31. _____

32. _____

33. _____

34. _____

35. _____

36. _____

37. _____

38. _____

39. _____

40. _____

41. _____

42. _____

27. $(r^3s^2 - r^2s^3 - 2r^2s^2) + (5r^2s^3 - r^2s^2 - r^3s^2)$

28. $(x^2y + 3xy - x^2y^2) - (x^2y - 5x^2y^2 - xy^2)$

29. $4(w^3 - w^2 + 7) + 5(w^3 - w - 2)$

30. $8(t^4 - t^3 - 2t) - 2(t^3 - t^4 - 8t)$

31. *Add.* $x^2 + 3x - 7$, $5x - x^2$, $3x^2 - x - 2$

32. *Add.* $x^3 + x^2 - 5x + 9$, $3x^3 - 6x^2 - x - 4$, $2x^3 - 5x^2 - 6x - 2$

33. Subtract $x^2 - 7x + 3$ from $2x^2 - 3x + 5$.

34. Subtract $2w^3 + w^2 + 6$ from $w^3 - w^2 + 2w$.

35. Subtract $a^3 - a^2 - b + b^2$ from the sum of $a^3 - b^2$ and $a^2b + 2b^2$.

36. Subtract the sum of $x^2 - 3x + 2$ and $4 - 3x$ from $9 - x^2$.

In Exercises 37–42, evaluate each polynomial for the given values.

37. $x^2 - x + 3$ for $x = -5$

38. $2a^2 - 5a + 7$ for $a = -2$

39. $y^4 + y^3 + y^2 + y + 1$ for $y = -3$

40. $2s^3 - 4s^2 + 6s - 8$ for $s = 3$

41. $-3x^2y + 5xy^2$ for $x = 2, y = -1$

42. $5w^2v^3 - wv^2 - 8v$ for $w = 2, v = -2$

QUESTIONS FOR THOUGHT

7. Explain the difference between a *factor* and a *term*.

8. Is 3 a factor of the expression $6x + 8$? Explain why or why not.

9. Is 2 a factor of the expression $6x + 8$? Explain why or why not.

Multiplying Polynomials

5.5

Back in Chapter 2 we learned how to multiply monomials, and how to multiply a polynomial by a monomial. Let's begin by reviewing these procedures in Example 1.

EXAMPLE 1 *Multiply.* **(a)** $5x(2x^3 + 3x^2)$ **(b)** $5x(2x^3)(3x^2)$

Solution
(a) This example calls for the use of the Distributive Law.

$$5x(2x^3 + 3x^2) = 5x \cdot 2x^3 + 5x \cdot 3x^2$$
$$= \boxed{10x^4 + 15x^3}$$

(b) This is a product of monomials. We use the Commutative and Associative Laws to rearrange and regroup the factors, and then simplify using Exponent Rule 1.

$$5x(2x^3)(3x^2) = (5 \cdot 2 \cdot 3)(x \cdot x^3 \cdot x^2)$$
$$= \boxed{30x^6}$$

Note that the Distributive Law does *not* apply here. ∎

Multiplying a polynomial by another polynomial requires the repeated use of the Distributive Law. However, by analyzing the process, we can develop a mechanical procedure for multiplying any two polynomials. For example, in order to multiply

$$(x + 4)(x^2 + 3x - 7)$$

let's think of this product as $(a + b) \cdot c$. In other words, consider the expression $x^2 + 3x - 7$ as one factor, and distribute it over $x + 4$:

$$\underbrace{(x + 4)}_{\substack{is\ like \\ (a + b)}}\underbrace{(x^2 + 3x - 7)}_{\substack{is\ like \\ c}}$$

In the same way that we can distribute the c in $(a + b)c$ to get $a \cdot c + b \cdot c$, we distribute $(x^2 + 3x - 7)$:

$$(x + 4)(x^2 + 3x - 7) = x(x^2 + 3x - 7) + 4(x^2 + 3x - 7)$$

Now we use the Distributive Law again on each set of parentheses.

$$= x^3 + 3x^2 - 7x + 4x^2 + 12x - 28$$

Combine like terms.

$$= \boxed{x^3 + 7x^2 + 5x - 28}$$

If we carefully examine our first application of the Distributive Law we see that it causes each term in the first polynomial to multiply each term in the second polynomial. That is, each term in $x^2 + 3x - 7$ is multiplied by x *and* is multiplied by 4.

This same procedure works regardless of the number of terms in each polynomial.

LEARNING CHECKS

1. **(a)** $2a(3a^3 - 5a)$
   (b) $2a(3a^3)(-5a)$

> **RULE FOR MULTIPLYING POLYNOMIALS**
>
> In order to multiply two polynomials, multiply each term in the first polynomial by each term in the second polynomial.

2. $(3x - 2)(4x^2 - 5x + 6)$

EXAMPLE 2　*Multiply and simplify.*　$(2x - 5)(3x^2 + x - 4)$

Solution　We follow the rule for multiplying polynomials and multiply each term in the first polynomial by each term in the second.

$$(2x - 5)(3x^2 + x - 4) = 2x(3x^2 + x - 4) - 5(3x^2 + x - 4)$$

Distribute the $2x$ and the -5.

$$= 6x^3 + 2x^2 - 8x - 15x^2 - 5x + 20$$

Combine like terms.

$$= \boxed{6x^3 - 13x^2 - 13x + 20} \qquad \blacksquare$$

It is also possible to do polynomial multiplication in a vertical format, as follows:

$$
\begin{array}{r}
3x^2 + x - 4 \\
2x - 5 \\
\hline
-15x^2 - 5x + 20 \quad \leftarrow \textit{Obtained by multiplying top row by } -5 \\
6x^3 + 2x^2 - 8x \quad \leftarrow \textit{Obtained by multiplying top row by } 2x \\
\hline
6x^3 - 13x^2 - 13x + 20 \quad \leftarrow \textit{Obtained by adding like terms which} \\
\textit{are lined up in columns}
\end{array}
$$

While the vertical format does sometimes offer advantages, at other times it may not. If the polynomials involve several variables, it may not be clear how to order the polynomials; or the vertical format may involve many columns. Additionally, in the next chapter when we talk about factoring polynomials, we will work exclusively with the horizontal format. Consequently, if you prefer the vertical format feel free to use it, but we will stick to the horizontal format.

3. $(t^2 - t + 1)(t^2 + t - 1)$

EXAMPLE 3　*Multiply and simplify.*　$(2x^2 + xy - xy^2)(x^2y - xy + y^2)$

Solution　Following our outline, we multiply each term in the first polynomial by each term in the second.

$$(2x^2 + xy - xy^2)(x^2y - xy + y^2)$$
$$= \underbrace{2x^2(x^2y) - 2x^2(xy) + 2x^2(y^2)}_{\textit{Multiplying by } 2x^2} + \underbrace{xy(x^2y) - xy(xy) + xy(y^2)}_{\textit{Multiplying by } xy}$$
$$\underbrace{-xy^2(x^2y) + xy^2(xy) - xy^2(y^2)}_{\textit{Multiplying by } -xy^2}$$
$$= 2x^4y - 2x^3y + 2x^2y^2 + x^3y^2 - x^2y^2 + xy^3 - x^3y^3 + x^2y^3 - xy^4$$

The only like terms are $2x^2y^2$ and $-x^2y^2$.

$$= \boxed{2x^4y - 2x^3y + x^2y^2 + x^3y^2 + xy^3 - x^3y^3 + x^2y^3 - xy^4} \qquad \blacksquare$$

Even though Example 3 was messy because of the "bookkeeping" involved, keep in mind that we are always doing *one* multiplication at a time.

EXAMPLE 4 *Multiply and simplify.* $(x + 7)(x + 4)$

4. $(a + 8)(a + 5)$

Solution While this example is much less complex that the previous one, we proceed in exactly the same way. We multiply each term in the first binomial by each term in the second.

$$(x + 7)(x + 4) = x \cdot x + x \cdot 4 + 7 \cdot x + 7 \cdot 4 \qquad \textit{Combine like terms.}$$
$$= \boxed{x^2 + 11x + 28} \qquad \blacksquare$$

This type of example, in which we multiply two similar binomials, comes up very frequently in algebra—so frequently, in fact, that we need to be able to multiply them rapidly, accurately, and often mentally. In the next chapter, when we discuss factoring, this ability will be particularly important.

In light of this, let's analyze the last example carefully.

$$(x + 7)(x + 4) = \underbrace{x^2}_{\substack{\textit{Product of first} \\ \textit{terms in each} \\ \textit{set of parentheses}}} + \underbrace{4x + 7x}_{\substack{\textit{"Cross" terms}}} + \underbrace{28}_{\substack{\textit{Product of second} \\ \textit{or "last" terms} \\ \textit{in each set of parentheses}}}$$

The key thing to notice here is that in this type of situation the cross terms are alike and can be combined. Sometimes cross terms are called the "*outer*" and "*inner*" terms, and the entire multiplication process of *two binomials* is called the **FOIL** method (see Figure 5.1).

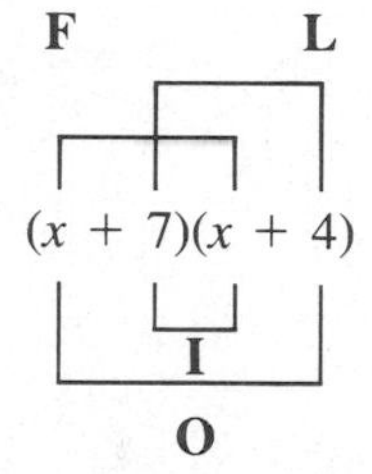

FOIL   stands for   First   Outer   Inner   Last

Giving the multiplication process a name does not tell us how to carry it out. It is important to keep in mind that we still have to do the multiplication—each term in the first binomial times each term in the second—whether you call it the FOIL method or not. The name is simply a device to remember the method and help us to be systematic in carrying it out.

It is also very important to remember that the name FOIL applies only to the product of two *binomials*.

Since we will often be multiplying binomials, we would like to shorten the FOIL method by doing part of the multiplication mentally. Let's illustrate what is meant by doing the multiplication mentally.

Figure 5.1 *FOIL method*

EXAMPLE 5 *Multiply and simplify.*

(a) $(x + 5)(x - 3)$ **(b)** $(3y - 4)(y - 6)$
(c) $(3a - 4)(2a + 7)$ **(d)** $(x + 2y)(x - 6y)$

5. (a) $(a + 6)(a - 2)$
   (b) $(4a - 5)(a - 1)$
   (c) $(6a - 1)(3a + 4)$
   (d) $(a + 3b)(a - 5b)$

Solution Basically, doing the problem mentally means that we carry the "cross" terms in our head, and write down only the result of combining the like terms.

(a) $(x + 5)(x - 3) = x^2 \boxed{- 3x + 5x} - 15$ *The shaded portion is the step we do mentally.*

$$= \boxed{x^2 + 2x - 15}$$

6. $(t + 4)^3$

(b) $(3y - 4)(y - 6) = 3y^2 - 18y - 4y + 24$

$$= \boxed{3y^2 - 22y + 24}$$

(c) $(3a - 4)(2a + 7) = 6a^2 + 21a - 8a - 28$

$$= \boxed{6a^2 + 13a - 28}$$

(d) $(x + 2y)(x - 6y) = x^2 - 6xy + 2xy - 12y^2$

$$= \boxed{x^2 - 4xy - 12y^2}$$ ∎

If we carry the cross terms mentally, multiplying binomials requires writing only the final answer.

EXAMPLE 6 *Multiply and simplify.* $(x - 3)^3$

Solution

$$(x - 3)^3 = (x - 3)(x - 3)(x - 3)$$

We can choose to begin by multiplying either the first two factors of $(x - 3)$ or the second two factors of $(x - 3)$. We will start with the second two.

$(x - 3)^3 = (x - 3)(x^2 - 6x + 9)$ *Obtained by multiplying* $(x - 3)(x - 3)$ *mentally*

$$= \underbrace{x^3 - 6x^2 + 9x}_{\substack{\text{The result of} \\ x(x^2 - 6x + 9)}} - \underbrace{3x^2 + 18x - 27}_{\substack{\text{The result of} \\ -3(x^2 - 6x + 9)}}$$ *Now we combine like terms.*

$$= \boxed{x^3 - 9x^2 + 27x - 27}$$

Notice that after the first step in the solution, since we were not multiplying binomials any longer, we made no attempt to multiply mentally. ∎

7. $(3a - 2)(a + 3) - (a + 4)(a - 3)$

EXAMPLE 7 *Multiply and simplify.* $(2x + 3)(x - 4) - (x + 5)(x - 2)$

Solution Watch out for the minus sign between the two products! It applies to the entire result of multiplying $(x + 5)(x - 2)$.

$(2x + 3)(x - 4) - (x + 5)(x - 2)$

$$= 2x^2 - 8x + 3x - 12 - (x^2 - 2x + 5x - 10)$$

↑ **This** *minus sign forces us to put in the parentheses.*

$$= 2x^2 - 5x - 12 - (x^2 + 3x - 10)$$

Remove parentheses by distributing the minus sign.

$$= 2x^2 - 5x - 12 - x^2 - 3x + 10$$

Combine like terms.

$$= \boxed{x^2 - 8x - 2}$$ ∎

We began this chapter discussing the various exponent rules, and we pointed out at the time the importance of knowing when it is appropriate to use the rules and when not to. The next example serves to highlight this point.

EXAMPLE 8   Perform the indicated operations and simplify.

(a) $(x^3y)^2$ (b) $(x^3 + y)^2$

8. (a) $(ab^4)^2$

(b) $(a + b^4)^2$

Solution   Note that we take entirely different approaches to the two parts because they are entirely different examples.

(a) Since this is *not* a binomial raised to a power, but rather a single term raised to a power, we can apply Exponent Rule 2 and square each *factor*.

$$(x^3y)^2 = (x^3)^2y^2 \qquad \textit{By Exponent Rule 4}$$
$$= \boxed{x^6y^2} \qquad \textit{By Exponent Rule 2}$$

(b) Since this is a binomial being raised to a power, we must use our method for multiplying polynomials (call it FOIL, if you like). Exponent Rule 4 *does not* apply.

$$(x^3 + y)^2 = (x^3 + y)(x^3 + y)$$
$$= x^6 + x^3y + x^3y + y^2 \qquad \textit{Combine like terms.}$$
$$= \boxed{x^6 + 2x^3y + y^2}$$

✔ *Answers to Learning Checks in Section 5.5*

1. (a) $6a^4 - 10a^2$ (b) $-30a^5$ **2.** $12x^3 - 23x^2 + 28x - 12$

3. $t^4 - t^2 + 2t - 1$ **4.** $a^2 + 13a + 40$

5. (a) $a^2 + 4a - 12$ (b) $4a^2 - 9a + 5$ (c) $18a^2 + 21a - 4$

(d) $a^2 - 2ab - 15b^2$

6. $t^3 + 12t^2 + 48t + 64$ **7.** $2a^2 + 6a + 6$

8. (a) a^2b^8 (b) $a^2 + 2ab^4 + b^8$

NOTE TO THE STUDENT

Use the space on this page to write down any questions you have or points you want to review with your instructor.

Exercises 5.5

Multiply and simplify each of the following. Whenever possible, do the multiplication of two binomials mentally.

1. $3x(5x^3)(4x^2)$

2. $2y^3(5y^2)(3y)$

3. $3x(5x^3 + 4x^2)$

4. $2y^3(5y^2 + 3y)$

5. $4xy(3yz)(-5xz)$

6. $8ab(2ac - 3bc)$

7. $4xy(3yz - 5xz)$

8. $8ab(2ac)(-3bc)$

9. $3x^2(x + 3y) + 4xy(x - 3y)$

10. $5rs(r - 2s) + r^2(3s - 4rs)$

11. $5xy^2(xy - y) - 2y(x^2y^2 - xy^2)$

12. $7r^2s(r^2 - s^2) - 2rs(r^2s - rs^2)$

13. $(x + 2)(x^2 - x + 3)$

14. $(m + 3)(m^2 - 2m + 5)$

15. $(y - 5)(y^2 + 2y - 6)$

16. $(n - 4)(n^2 + 7n + 1)$

1. _____

2. _____

3. _____

4. _____

5. _____

6. _____

7. _____

8. _____

9. _____

10. _____

11. _____

12. _____

13. _____

14. _____

15. _____

16. _____

ANSWERS

17. _____

18. _____

19. _____

20. _____

21. _____

22. _____

23. _____

24. _____

25. _____

26. _____

27. _____

28. _____

29. _____

30. _____

31. _____

32. _____

17. $(3x - 2)(x^2 + 3x - 5)$

18. $(4a - 3)(a^2 - 7a - 3)$

19. $(5z + 2)(3z^2 + 2z + 8)$

20. $(2c - 1)(6c^2 - 3c - 1)$

21. $(x + y)(x^2 - xy + y^2)$

22. $(m - n)(m^2 + mn + n^2)$

23. $(x^2 + x + 1)(x^2 + x - 1)$

24. $(u^2 + 2u + 1)(u^2 - 2u + 1)$

25. $(x + 5)(x + 3)$

26. $(y + 4)(y + 6)$

27. $(x - 5)(x - 3)$

28. $(y - 4)(y - 6)$

29. $(x - 5)(x + 3)$

30. $(y + 4)(y - 6)$

31. $(x + 5)(x - 3)$

32. $(y - 4)(y + 6)$

33. $(a + 8b)(a - 5b)$

34. $(x - 12y)(x + 3y)$

35. $(3x - 4)(4x - 1)$

36. $(2y - 5)(3y + 4)$

37. $(x^2 + 3)(x^2 + 2)$

38. $(t^2 - 5)(t^2 - 4)$

39. $(x + 7)(x + 7)$

40. $(x - 8)(x - 8)$

41. $(x + 7)(x - 7)$

42. $(x - 8)(x + 8)$

43. $(x - 4)^2$

44. $(a + 6)^2$

45. $(x + 2)^3$

46. $(x - 4)^3$

ANSWERS

33. _____

34. _____

35. _____

36. _____

37. _____

38. _____

39. _____

40. _____

41. _____

42. _____

43. _____

44. _____

45. _____

46. _____

ANSWERS

47. _____

48. _____

49. _____

50. _____

51. _____

52. _____

53. _____

54. _____

55. _____

56. _____

57. _____

58. _____

59. _____

60. _____

61. _____

62. _____

47. $2x^2(x + 4)(x - 8)$

48. $4z(3z - 5)(2z + 7)$

49. $3x(5x - 6)(3x - 2)$

50. $z^2(2z - 3)(3z - 4)$

51. $(x + 4)(x - 3) + (x - 6)(x - 2)$

52. $(y - 3)(y + 6) + (y + 2)(y - 9)$

53. $(a - 5)(a - 4) - (a - 3)(a - 2)$

54. $(y - 6)(y - 1) - (y - 3)(y - 5)$

55. $(x - 6)^2 - (x + 6)^2$

56. $(2x - 3)^2 - (2x + 3)^2$

CALCULATOR EXERCISES

Perform the indicated operations and simplify as completely as possible. Round off your answers to the nearest thousandth.

57. $.8x(2.4x^2 - 3x + 6.1)$

58. $3.1xy(5.4x^2y + 6.4xy - 9.2y^2)$

59. $(.3x + .8)(.2x - .5)$

60. $(2.3a - .8)(7.5a - 14.6)$

61. $(.01x - 2.5)(.7x - 12.6)$

62. $(28.02t - 3.81w)(12.36t + 32.6w)$

QUESTION FOR THOUGHT

10. Given the examples $(xy)^2$ and $(x + y)^2$ explain:

(a) How are the two examples similar?

(b) How are they different?

(c) How should each one be multiplied out?

CHAPTER 5 SUMMARY

After having completed this chapter you should be able to:

1. Apply the definition of zero and negative exponents (Section 5.2).

   For example:

   Evaluate.

   (a) $4^0 = 1$

   (b) $4^{-3} = \dfrac{1}{4^3} = \boxed{\dfrac{1}{64}}$

2. Apply the various exponent rules to simplify expressions involving integer exponents (Sections 5.1, 5.2).

   For example:

$$\frac{(4x^{-2}y^3)^2}{2(x^3y^{-5})^{-2}} \qquad \textit{First use Exponent Rule 4}$$

$$= \frac{4^2(x^{-2})^2(y^3)^2}{2(x^3)^{-2}(y^{-5})^{-2}} \qquad \textit{Next use Exponent Rule 2.}$$

$$= \frac{16x^{-4}y^6}{2x^{-6}y^{10}} \qquad \textit{Now use Exponent Rule 3.}$$

$$= \frac{16}{2}x^{-4-(-6)}y^{6-10}$$

$$= 8x^2y^{-4} \qquad \textit{Use the definition of } a^{-n}.$$

$$= \boxed{\frac{8x^2}{y^4}}$$

3. Write and use *scientific notation* (Section 5.3).

   For example:

   (a) $28{,}340 = 2.834 \times 10^4$

   (b) $.02834 = 2.834 \times 10^{-2}$

   (c) *Compute.*
$$\frac{(.00008)(2{,}500)}{.005} = \frac{(8 \times 10^{-5})(2.5 \times 10^3)}{5 \times 10^{-3}}$$

$$= \frac{8(2.5)}{5} \times \frac{10^{-5}10^3}{10^{-3}}$$

$$= \frac{20}{5} \times \frac{10^{-2}}{10^{-3}}$$

$$= 4 \times 10^{-2-(-3)}$$

$$= 4 \times 10$$

$$= \boxed{40}$$

4. Write a polynomial in *complete standard form,* identify all its coefficients, and find its degree (Section 5.4).

   For example:

   The polynomial $2x^4 - 3x^2 + x^3 - 4$ is written in complete standard form as

$$2x^4 + x^3 - 3x^2 + 0x - 4$$

The coefficients are 2, 1, -3, 0, and -4, respectively.

The degree of the polynomial is 4.

5. Add and subtract polynomials (Section 5.4).

   For example:

   (a) $(2x^3 + 3xy - y^2) + (x^3y - xy + 5y^2) = \boxed{2x^3 + x^3y + 2xy + 4y^2}$

   (b) Subtract $x^2 - 4x$ from $3x - 5x$

   $$3x^2 - 5x - (x^2 - 4x) = 3x^2 - 5x - x^2 + 4x$$
   $$= \boxed{2x^2 - x}$$

6. Multiply polynomials in general, and binomials mentally (Section 5.5).

   For example:

   (a) $(3x - 4)(2x^2 - 5x - 3)$

   Each term in the first set of parentheses multiplies each term in the second set of parentheses.

   $$(3x - 4)(2x^2 - 5x - 3) = 6x^3 - 15x^2 - 9x - 8x^2 + 20x + 12$$
   $$= \boxed{6x^3 - 23x^2 + 11x + 12}$$

   (b) *Multiply mentally.* $(2x - 3)(x + 5)$

   $$(2x - 3)(x + 5) = 2x \cdot x \;\boxed{+\, 2x \cdot 5 - 3 \cdot x}\; - 3 \cdot 5$$
   $$= 2x^2 \;\boxed{+\, 10x - 3x}\; - 15$$
   $$= \boxed{2x^2 + 7x - 15}$$

ANSWERS TO QUESTIONS FOR THOUGHT

1. When we multiply two powers of the same base, we keep the base and add the exponents. When we raise a power to a power, we keep the base and multiply the exponents.

2. **(a)** $\dfrac{x^8}{x^8} = x^{8-8} = x^0$ **(b)** $\dfrac{x^4}{x^7} = x^{4-7} = x^{-3}$

3. **(a)** According to Exponent Rule 2, we must multiply the exponents, not add them.

   (b) According to Exponent Rule 1, we must add the exponents, not multiply them.

   (c) According to Exponent Rule 3, we must subtract the exponents, not divide them.

   (d) Since x^2 and x^3 are unlike terms, we cannot combine them when they are added.

4. $\dfrac{x^6}{x^4}$ requires us to divide x^6 by x^4, whereas

   $$\frac{x^6}{x^{-4}} = \frac{x^6}{\dfrac{1}{x^4}} = x^6 \cdot \frac{x^4}{1} \text{ asks us to multiply } x^6 \text{ by } x^4.$$

5. When -1 appears in the exponent, it tells us to take the reciprocal of the base. Thus, $3^{-1} = \frac{1}{3}$. When the minus sign appears in front of the 3, it tells us to take the opposite of 3. Put another way, 3^{-1} is the multiplicative inverse of 3, while -3 is the additive inverse of 3.

6. If the number is bigger than 1, the exponent cannot be negative; if the number is smaller than 1, the exponent must be negative.

7. In a sum, the expressions to be added are called terms; in a product, the expressions to be multiplied are called factors.

8. 3 is not a factor of the expression $6x + 8$ because 3 does not exactly divide 8.

9. 2 is a factor of the expression $6x + 8$ because 2 does exactly divide both $6x$ and 8. Here, we can write $6x + 8 = 2(3x + 4)$.

10. (a) Both examples require us to find the square of an expression involving x and y.

   (b) The first example asks us to square a product; the second asks us to square a sum.

   (c) $(xy)^2 = x^2 y^2$ by Exponent Rule 4.
   $(x + y)^2 = x^2 + 2xy + y^2$ by the FOIL method.

NOTE TO THE STUDENT

Use the space on this page to write down any questions you have or points you want to review with your instructor.

CHAPTER 5 REVIEW EXERCISES

In Exercises 1–12, simplify the given expression as completely as possible. Express final answers with positive exponents only.

1. 3^{-4}

2. $4^0 + 8 \cdot 4^0 + 4^{-1} + 12 \cdot 4^{-2}$

3. $(3^{-1} + 2^{-2})^2$

4. $(3^{-1} + 2^{-2})^{-1}$

5. $\dfrac{(xy^2)^3}{(x^2y)^4}$

6. $\dfrac{(x^3)^2(y^2)^4}{(x^3y)^2}$

7. $\dfrac{(3x^3y^2)^4}{9(x^2y^4)^3}$

8. $x^{-2}x^{-3}$

9. $(x^{-2})^{-3}$

10. $\dfrac{x^{-2}y^{-5}}{x^{-4}y^{-3}}$

11. $\left(\dfrac{2x^{-2}x^3}{x^{-3}}\right)^{-2}$

12. $\dfrac{(x^2y^{-3})^{-3}}{(x^{-1}y^{-2})^{-4}}$

In Exercises 13–16, write the given number in scientific notation.

13. 58,700,000

14. .00587

15. .000002

16. 7,000

In Exercises 17–20, write the given number in standard notation.

17. 2.56×10^{-3}

18. 8.79×10^5

19. 5.773×10^8

20. 7.447×10^{-8}

In Exercises 21–24, compute using scientific notation.

21. $(.008)(250,000)$

22. $(3,600)(.0005)$

23. $\dfrac{.001}{.000025}$

24. $\dfrac{(28,500)(.004)}{.0002}$

ANSWERS

1. _____

2. _____

3. _____

4. _____

5. _____

6. _____

7. _____

8. _____

9. _____

10. _____

11. _____

12. _____

13. _____

14. _____

15. _____

16. _____

17. _____

18. _____

19. _____

20. _____

21. _____

22. _____

23. _____

24. _____

In Exercises 25–34, answer the following questions:

(a) How many terms are there?

(b) What is the degree of each term?

(c) What is the degree of the polynomial?

25. $x^2 + 3x - 7$ **26.** $t^3 + t^2 - 3t + 9$

27. $3x^3y - 5y^2 + 6xy$ **28.** $-5x^5 + 3x^2y^4 - 6x^2 + 2y$

29. $8x - 5$ **30.** $3 - 4t$

31. 9 **32.** 0

33. $(3x^5)(2x^3)$ **34.** $3x^5 + 2x^3$

In Exercises 35–38, write the given polynomial in complete standard form.

35. $2x^3 - 7x^2 + 4$ **36.** $3t^5 - t^2 - 10$

37. $y^2 + y^5 - 2y - 1$ **38.** $1 - x^4$

In Exercises 39–74, perform the indicated operations and simplify as completely as possible.

39. $(3x^2 - 5x + 7) + (5x - x^2 - 5)$

40. $(5y^4 - y^2 + 9y) + (2y^2 - y^4 - y)$

41. $(3x^2 - 5x + 7) - (5x - x^2 - 5)$

42. $(5y^4 - y^2 + 9y) - (2y^2 - y^4 - y)$

43. $2(x^2y - xy^2 - 5x^2y^2) + 3(xy^2 - x^2y + x^2y^2)$

44. $4(m^2 - 3m^2n) + 6(m^2n - 2m^2)$

45. $2(x^2y - xy^2) - 5x^2y^2 - 3(xy^2 - x^2y + x^2y^2)$

46. $3(r^2s - rs^2) - r^2s^2 - 4(rs^2 - r^2s^2)$

47. $2a^2(a - 3b) + 4a(a^2 + ab) - 2(a^3 - a^2b)$

48. $8mn(m - mn) - n^2(n - m) - (m^2n - mn^2)$

49. Subtract $x^2 - 4x$ from $x^2 + 4x$.

48. _____

49. _____

50. Subtract $3a^2 - b^2$ from $8a^2 - 6b^2$.

50. _____

51. Subtract $3x - 5$ from the sum of $x^2 + 4x - 3$ and $2x^2 - x - 2$.

51. _____

52. _____

52. Subtract the sum of $2a^3 + a + 5$ and $4a - a^2$ from $a^2 - 4$.

53. _____

53. $(x + 4)(x - 7)$ **54.** $(a - 5)(a - 4)$

54. _____

55. _____

55. $(2x - 3)(4x - 5)$ **56.** $(5x - 4)(6x - 1)$

56. _____

57. _____

57. $(3a - 4b)(2a + 5b)$ **58.** $(4x - 3y)(7x - 2y)$

58. _____

59. _____

59. $(x + 2)(x - 3)(x + 1)$ **60.** $(x - 3)(x - 4)(x - 2)$

60. _____

61. $(x + 6)^2$ **62.** $(3x - 2y)^2$

61. _____

62. _____

63. $(x - 5)^3$ **64.** $(2x - 1)^3$

63. _____

64. _____

65. $3x^2(x - 4)(x + 2)$ **66.** $2y(3y + 1)(y - 5)$

65. _____

66. _____

67. $(x - 5)(x + 5)$

68. $(3x + 2y)(3x - 2y)$

67. _____

68. _____

69. $(x + 2)(x^2 - 3x + 4)$

70. $(x - 3y)(x^2 + xy - 4y^2)$

69. _____

70. _____

71. $(x^2 + 2x - 1)(x^2 + 2x + 1)$

72. $(y^2 - 3y - 4)(y^2 - 3y + 4)$

71. _____

72. _____

73. $(2x - 3)(x + 4) - (x - 2)(x - 1)$

74. $(x - 3)^2 - (x - 2)^2$

73. _____

74. _____

CHAPTER 5 PRACTICE TEST

1. *Evaluate.* $5^0 + 2^{-2} + 4^{-1}$

In Exercises 2–5, simplify as completely as possible. Express final answers with positive exponents only.

2. $\dfrac{(x^4)^2(xy)^3}{x^3y^5}$

3. $x^{-4}x^{-5}$

4. $(x^{-4})^{-5}$

5. $\dfrac{(2x^{-3}y^4)^4}{4(x^{-2}y^{-1})^3}$

6. Given the polynomial $5x^4 - x^3 + 2x + 7$:
   (a) How many terms are there?
   (b) What is the coefficient of the third-degree term? The second-degree term?
   (c) What is the degree of the polynomial?

In Exercises 7–13, perform the indicated operations and simplify as completely as possible.

7. $-3x^2y(4x^2y)(-2x^3)$

8. $-3x^2y(4x^2y - 2x^3)$

9. $2x(x^2 - y) - 3(x - xy) - (2x^3 - 3x)$

10. $(3x - 2)(4x^2 - 5x + 6)$

11. $3x^2(2x - y) - xy(x + y)$

12. Subtract $x^2 - 4x - 5$ from $2x^2 - 3x - 4$.

13. $(a - 1)^2 - (a + 1)^2$

14. Write in scientific notation.
    (a) .00316
    (b) 31,600

15. *Compute using scientific notation.* $\dfrac{(.24)(5{,}000)}{.006}$

1. _____

2. _____

3. _____

4. _____

5. _____

6. a. _____

   b. _____

   c. _____

7. _____

8. _____

9. _____

10. _____

11. _____

12. _____

13. _____

14. a. _____

    b. _____

15. _____

NOTE TO THE STUDENT

Use the space on this page to write down any questions you have or points you want to review with your instructor.

Factoring

During the course of our discussion of multiplying polynomials, we placed particular emphasis on the ability to multiply *binomials* quickly and accurately. In this chapter we turn our attention to reversing this process. That is, given a polynomial, we want to be able to factor it into a product of monomials and/or binomials.

There are many reasons we are interested in being able to factor polynomials. Recall, for example, that the Fundamental Principle of Fractions allows us to reduce common *factors*, but not common terms. Therefore, if we want to reduce fractions involving polynomial expressions, we need to be able to put them into factored form. In addition, we shall see in Chapter 10 that factoring will offer one possible method for solving second-degree equations.

Before proceeding to the material on factoring we will first take a more detailed look at multiplying binomials.

Special Products

6.1

In Section 5.5 we learned a mechanical procedure for multiplying two polynomials. When this procedure was applied to multiplying out two binomials we often called it the FOIL method.

Since the structure of products of binomials plays an extremely important role in the discussion to follow, we analyze this structure in detail in the next few examples.

LEARNING CHECKS

1. (a) $(x - 7)(x + 4)$
   (b) $(x + r)(x + s)$

EXAMPLE 1   *Multiply and simplify.*

(a) $(x + 5)(x - 3)$ (b) $(x + m)(x + n)$

Solution   Our rule for multiplying polynomials says that each term in the first polynomial multiplies each term in the second.

(a) $(x + 5)(x - 3) = \underbrace{x \cdot x}_{F} \quad \underbrace{-3 \cdot x}_{O} \quad \underbrace{+5 \cdot x}_{I} \quad \underbrace{-15}_{L}$

$$= x^2 - 3x + 5x - 15$$
$$= x^2 + (-3 + 5)x - 15$$
$$= \boxed{x^2 + 2x - 15}$$

(b) $(x + m)(x + n) = x \cdot x + n \cdot x + m \cdot x + m \cdot n$

We factor out the common factor of x from the cross terms.

$$= \boxed{x^2 + (n + m)x + mn}$$

Note that the coefficient of x is the *sum* of the second terms in each binomial $(-3 + 5$ or $n + m)$, while the last term is the *product* of the second terms in each binomial $(-3(5)$ or $m \cdot n)$.   ∎

Similarly,

$$(x + 2)(x + 4) = x^2 + (2 + 4)x + (2)(4) = x^2 + 6x + 8$$
$$(x - 5)(x + 3) = x^2 + (-5 + 3)x + (-5)(+3) = x^2 - 2x - 15$$
$$(x - 7)(x - 3) = x^2 + (-7 - 3)x + (-7)(-3) = x^2 - 10x + 21$$

EXAMPLE 2 *Multiply and simplify.*

(a) $(x - 5)^2$ (b) $(x + p)^2$

Solution

(a) $(x - 5)^2 = (x - 5)(x - 5)$

$\qquad\qquad = x^2 - 5x - 5x + 25$

$\qquad\qquad = \boxed{x^2 - 10x + 25}$

(b) $(x + p)^2 = (x + p)(x + p)$

$\qquad\qquad = x^2 + px + px + p^2 \qquad px + px = 2px$

$\qquad\qquad = \boxed{x^2 + 2px + p^2}$

The expressions $x^2 - 10x + 25$ and $x^2 + 2px + p^2$ are called **perfect square trinomials** (or usually just **perfect squares** for short) because they are the result of squaring a binomial, just as the number 36 is called a perfect square because it is the result of squaring 6. ∎

Part **(b)** of Example 2 tells us that

$$(x + p)^2 = x^2 + 2px + p^2$$

If we like, we can think of this as a "formula" for squaring a binomial. In words, it says that the square of a binomial is the square of the first term, plus twice the product of the first and second terms, plus the square of the second term.

That is, we can square binomials such as $(x + 4)^2$ and $(x - 9)^2$ by modeling them after part **(b)**:

$(x + p)^2 = x^2 + 2px + p^2 \qquad (x + p)^2 = x^2 + 2px + p^2$

$(x + 4)^2 = x^2 + 2 \cdot 4x + 4^2 \qquad (x - 9)^2 = x^2 + 2(-9)x + (-9)^2$

$\qquad = \boxed{x^2 + 8x + 16} \qquad\qquad = \boxed{x^2 - 18x + 81}$

On the other hand, we can equally well just multiply out $(x + 4)^2$ or $(x - 9)^2$ by using FOIL the way we usually do. However, regardless of which method we use, we want to be able to write out the final product in one step, *without* writing down the middle terms.

The reason this particular product is called "special" is that it comes up frequently enough to make it worth recognizing. In addition, perfect squares will play an important role in Chapter 10 where we discuss solving second-degree equations.

EXAMPLE 3 *Multiply and simplify.*

(a) $(3x^2y)^2$ (b) $(3x^2 + y)^2$

Solution It is very important that you recognize the difference between parts (a) and (b). Part (a) is *not* the square of a binomial. It is a *product* raised to a power and so can be handled by using the exponent rules.

(a) $(3x^2y)^2 = 3^2(x^2)^2y^2$ *Obtained by using Exponent Rule 4*

$\qquad = \boxed{9x^4y^4}$

Part **(b)**, on the other hand, *is* the square of a *sum*, and so we cannot use Exponent Rule 4. (The exponent rules pertain only to products and quotients,

2. (a) $(x + 6)^2$

(b) $(x - n)^2$

3. (a) $(5xy^3)^2$

(b) $(5x + y^3)^2$

not sums and differences.) We must multiply out using FOIL, or by following our perfect square form.

(b) $(3x^2 + y)^2 = (3x^2 + y)(3x^2 + y)$
$$= 9x^4 + 3x^2y + 3x^2y + y^2 \quad \text{or} \quad 9x^4 + 2(3x^2)(y) + y^2$$
$$= \boxed{9x^4 + 6x^2y + y^2} \qquad \blacksquare$$

Be careful! Confusing expressions of types **(a)** and **(b)** in Example 3, and therefore improperly applying the exponent rules, is a *very common* error. Always keep in mind that

$$(x + y)^2 \neq x^2 + y^2$$
but rather
$$(x + y)^2 = x^2 + 2xy + y^2$$

4. (a) $(r - 7)(r + 7)$
  (b) $(r + s)(r - s)$

EXAMPLE 4   *Multiply and simplify.*

(a) $(x + 5)(x - 5)$ **(b)** $(x + a)(x - a)$

Solution

(a) $(x + 5)(x - 5) = x^2 - 5x + 5x - 25$
$$= \boxed{x^2 - 25}$$
Note that the cross terms drop out.

(b) $(x + a)(x - a) = x^2 - ax + ax - a^2$
$$= \boxed{x^2 - a^2} \qquad \blacksquare$$

An expression such as $x^2 - 25$ or $x^2 - a^2$ is called the **difference of two squares** (for obvious reasons). This type of expression is given a special name because even though it is the product of two similar binomials (which usually results in a trinomial), we get only two terms due to the cross terms dropping out.

A note of caution is in order here. It is very easy to confuse expressions such as

$$(x - y)^2 \quad \text{with} \quad x^2 - y^2$$

They are *not* the same:

$$(x - y)^2 = (x - y)(x - y) = x^2 - 2xy + y^2 \qquad \textit{This is a perfect square.}$$
$$\downarrow \qquad \downarrow$$
Same signs

which is *not* the same as

$$x^2 - y^2 = (x + y)(x - y) \qquad \textit{This is the difference of two squares.}$$
$$\downarrow \qquad \downarrow$$
Opposite signs

5. (a) $(3x - 5)(2x - 3)$
  (b) $(my - n)(ry + s)$

EXAMPLE 5   *Multiply and simplify.*

(a) $(2x - 3)(5x + 4)$ **(b)** $(ax + b)(cx + d)$

Solution

(a) $(2x - 3)(5x + 4) = (2x)(5x) + 4(2x) - 3(5x) - 3 \cdot 4$
$$= 10x^2 + 8x - 15x - 12$$
$$= \boxed{10x^2 - 7x - 12}$$

(b) $(ax + b)(cx + d) = (ax)(cx) + d(ax) + b(cx) + bd$
$$= acx^2 + adx + bcx + bd \qquad \textit{Factor out x from the}$$
$$= \boxed{acx^2 + (ad + bc)x + bd} \qquad \textit{cross terms.} \qquad \blacksquare$$

Perhaps part **(b)** of Example 5 should not be called a special product at all, but rather should be called a *general product*. Nevertheless, we include it because it illustrates how the coefficient of the middle term can be the result of the interaction of the coefficients in the binomials. We will have much more to say about this situation in Section 6.4.

These "special binomial products" are summarized in the box for ease of reference.

SPECIAL BINOMIAL PRODUCTS

$$(x + m)(x + n) = x^2 + (n + m)x + mn$$
For example: $(x + 6)(x + 5) = x^2 + 11x + 30$

Perfect square $\qquad (x + p)^2 = x^2 + 2px + p^2$
For example: $\qquad (x + 7)^2 = x^2 + 14x + 49$

Difference of two squares $\qquad (x + r)(x - r) = x^2 - r^2$
For example: $\qquad (x + 7)(x - 7) = x^2 - 49$

$$(ax + b)(cx + d) = acx^2 + (ad + bc)x + bd$$
For example: $(3x + 4)(2x + 5) = 6x^2 + 23x + 20$

While memorizing these forms is certainly *not* necessary, understanding and recognizing them will be very helpful in the work ahead. The best way to develop this understanding and recognition is by working out lots of exercises.

✔ *Answers to Learning Checks in Section 6.1*

1. (a) $x^2 - 3x - 28$ **(b)** $x^2 + (r + s)x + rs$

2. (a) $x^2 + 12x + 36$ **(b)** $x^2 - 2nx + n^2$

3. (a) $25x^2y^6$ **(b)** $25x^2 + 10xy^3 + y^6$ **4. (a)** $r^2 - 49$ **(b)** $r^2 - s^2$

5. (a) $6x^2 - 19x + 15$ **(b)** $mry^2 + (ms - nr)y - ns$

NOTE TO THE STUDENT

Use the space on this page to write down any questions you have or points you want to review with your instructor.

Exercises 6.1

Multiply out each of the following. As you work out the problems, identify those exercises which are either a perfect square or the difference of two squares.

1. $(x + 4)(x + 3)$ **2.** $(x + 10)(x + 2)$

3. $(x - 4)(x - 3)$ **4.** $(x - 10)(x - 2)$

5. $(x + 4)(x - 3)$ **6.** $(x + 10)(x - 2)$

7. $(x - 4)(x + 3)$ **8.** $(x - 10)(x + 2)$

9. $(x + 6)(x + 2)$ **10.** $(x + 5)(x + 4)$

11. $(x - 6)(x - 2)$ **12.** $(x - 5)(x - 4)$

13. $(x + 6)(x - 2)$ **14.** $(x + 5)(x - 4)$

15. $(x - 6)(x + 2)$ **16.** $(x - 5)(x + 4)$

17. $(a + 8)(a + 8)$ **18.** $(t - 6)(t - 6)$

19. $(a - 8)(a - 8)$ **20.** $(t + 6)(t + 6)$

21. $(a + 8)(a - 8)$ **22.** $(t + 6)(t - 6)$

23. $(c - 4)^2$ **24.** $(z + 9)^2$

25. $(c + 4)^2$ **26.** $(z - 9)^2$

27. $(c + 4)(c - 4)$ **28.** $(z - 9)(z + 9)$

1. _____
2. _____
3. _____
4. _____
5. _____
6. _____
7. _____
8. _____
9. _____
10. _____
11. _____
12. _____
13. _____
14. _____
15. _____
16. _____
17. _____
18. _____
19. _____
20. _____
21. _____
22. _____
23. _____
24. _____
25. _____
26. _____
27. _____
28. _____

ANSWERS

29. _____

30. _____

31. _____

32. _____

33. _____

34. _____

35. _____

36. _____

37. _____

38. _____

39. _____

40. _____

41. _____

42. _____

43. _____

44. _____

45. _____

46. _____

47. _____

48. _____

49. _____

50. _____

51. _____

52. _____

29. $(3x + 4)(x + 7)$

30. $(2y - 5)(y - 3)$

31. $(3x + 7)(x + 4)$

32. $(2y - 3)(y - 5)$

33. $(3x + 4)(x - 7)$

34. $(2y + 5)(y - 3)$

35. $(3x - 4)(x + 7)$

36. $(2y + 3)(y - 5)$

37. $(3x + 4)(5x + 7)$

38. $(2y - 5)(4y - 3)$

39. $(3x + 7)(5x + 4)$

40. $(2y - 3)(4y - 5)$

41. $(3x + 4)(5x - 7)$

42. $(2y - 5)(4y + 3)$

43. $(3x - 4)(5x + 7)$

44. $(2y + 5)(4y - 3)$

45. $(2a + 5)^2$

46. $(3y - 4)^2$

47. $(2a + 5)(2a - 5)$

48. $(3y + 4)(3y - 4)$

49. $(3xy)^2$

50. $(3x + y)^2$

51. $(x^3 + y^2)^2$

52. $(x^3y^2)^2$

QUESTIONS FOR THOUGHT

1. Multiply out $(x + 6)(x + 4)$ and $(x - 6)(x - 4)$. What is the effect of switching both + signs to − signs?

2. The two examples $(x + 6)(x - 4)$ and $(x - 6)(x + 4)$ also have their signs "switched." What about the middle terms of the resulting trinomials?

3. Look back through this exercise set and identify those *pairs* of exercises which have "switched signs," such as $(x + 6)(x - 4)$ and $(x - 6)(x + 4)$. Are the middle terms always the opposite sign? Will this always be the case? Why?

Common Factors

6.2

When we first discussed the Distributive Law back in Chapter 2, we mentioned and illustrated the fact that it can be used in two ways. We can use it to multiply out, in which case we remove parentheses, or we can use it to factor, in which case we create parentheses.

Even though we know that it makes no difference whether we read an equality statement from left to right or from right to left, when we talk about multiplying out we write the Distributive Law as

$$a(b + c) = ab + ac$$

while when we talk about factoring we write it

$$ab + ac = a(b + c)$$

Factoring an expression changes it from a *sum* into a *product,* and as we have already pointed out, there are numerous situations in which having a product is helpful.

Throughout our discussion, whenever we talk about factoring an expression, we always mean using integers only. In other words, if we list the factors of 5 we would *not* list $\frac{1}{2}$ times 10.

The most basic type of factoring, which we have already discussed briefly in Section 2.3, involves the direct application of the Distributive Law. For example, if we are interested in factoring the expression $12x + 30$, then we write (and think)

$$12x + 30 = \boxed{6 \cdot 2x + 6 \cdot 5}$$

$$= \boxed{6(2x + 5)}$$

We have used the Distributive Law to "take out" the common factor of 6.

This procedure is usually called *taking out the common factor.* We can, of course, check our answer immediately by multiplying out $6(2x + 5)$ and verifying that we get the original expression $12x + 30$.

Before we proceed any further we need to lay down some ground rules. Whenever we are asked to factor an expression, the intention is to factor it as completely as possible. This means that the expression (or expressions) remaining inside parentheses have *no* common factors remaining (other than 1 or -1, of course).

Thus, if we had factored

$$12x + 30 = 2(6x + 15)$$

we have a factorization, but it is *incomplete,* because both $6x$ and 15 still have a common factor of 3.

EXAMPLE 1 *Factor as completely as possible.*

(a) $8x^3 + 20x - 28$ **(b)** $6x^2 - 12x$

Solution In general, it is probably easiest to begin by first determining the greatest common numerical factor, then the greatest common x factor, then the greatest common y factor, etc. Then we put all the common factors together to get the overall greatest common factor (GCF for short), of the entire polynomial.

 **LEARNING CHECKS**

1. (a) $9a^4 - 12a^3 - 18$

   (b) $10a^3 + 15a^2$

(a) In $8x^3 + 20x - 28$, the GCF of 8, 20, and -28 is 4. Since there is no common x factor (the -28 does not have an x factor in it), the GCF for the entire polynomial is 4.

$$8x^3 + 20x - 28 = \boxed{4 \cdot 2x^3 + 4 \cdot 5x - 4 \cdot 7}$$

Take out the common factor of 4.

$$= \boxed{4(2x^3 + 5x - 7)}$$

(b) Following the same outline, we see that the GCF for 6 and 12 is 6. Since x^2 is two factors of x, and x is one factor of x, the GCF of x^2 and x is x (they have *one* factor of x in common). Thus, the GCF of $6x^2 - 12x$ is $6x$.

$$6x^2 - 12x = \boxed{6x \cdot x - 6x \cdot 2} \quad \textit{Take out the common factor of } 6x.$$

$$= \boxed{6x(x - 2)}$$

As always, we check our factorization by multiplying out. However, remember that this check does not guarantee that we have the *complete* factorization. ∎

Basically, as Example 1 illustrates, factoring out the greatest common factor is a two-step process. First we determine the GCF, and second, we determine what factors remain in each term of the polynomial after the GCF is taken outside the parentheses.

2. $18mn^2 + 24m^3n - 12m^2n^2$

EXAMPLE 2 *Factor as completely as possible.* $12a^3b - 8a^2c^2 + 6ab^3$

Solution

The GCF of 12, -8, and 6 is 2.

The GCF of a^3, a^2, and a is a.

There is no common factor for b and c because the second term has no b factors, and the first and third terms have no c factors.

Thus, the GCF of the entire polynomial is $2a$.

$$12a^3b - 8a^2c^2 + 6ab^3 = \boxed{2a \cdot 6a^2b - 2a \cdot 4ac^2 + 2a \cdot 3b^3}$$

$$= \boxed{2a(6a^2b - 4ac^2 + 3b^3)} \quad ∎$$

3. $16a^3b^2 + 8ab^2$

EXAMPLE 3 *Factor as completely as possible.* $15x^3y^4 - 5x^2y^3$

Solution

The GCF of 15 and 5 is 5.

The GCF of x^3 and x^2 is x^2.

The GCF of y^4 and y^3 is y^3.

Therefore, the GCF of the entire polynomial is $5x^2y^3$.

$$15x^3y^4 - 5x^2y^3 = \boxed{5x^2y^3 \cdot 3xy - 5x^2y^3 \cdot 1}$$

Do not forget the understood factor of 1.

$$= \boxed{5x^2y^3(3xy - 1)}$$

It is a very common error to neglect putting the 1 into the parentheses, so be careful. (If you check your factorization by multiplying out, then you cannot possibly leave out the 1.) ∎

Factoring by Grouping

When we factor out a common factor it is not necessary that it be a *monomial*.

EXAMPLE 4 *Factor as completely as possible.* $a(x + 3) - b(x + 3)$

4. $r(t + 5) + s(t + 5)$

Solution The entire factor of $x + 3$ is a common factor to both terms.

$$a(x + 3) - b(x + 3) = a\ (x + 3)\ - b\ (x + 3)$$

Factor out the common factor of $x + 3$.

$$= (x + 3)\ (a - b)$$

$$= \boxed{(x + 3)(a - b)}$$ ∎

Sometimes we must group the terms in order to see a common factor.

EXAMPLE 5 *Factor as completely as possible.* $x^2 + 4x + xy + 4y$

5. $a^2 + 5a + ab + 5b$

Solution It is not readily apparent how to factor this entire expression, as there is no common factor. However, sometimes when we group the terms we do get common factors.

If we group the first two terms and the last two terms, each group will have a common factor. Even more, the resulting expression will also have a common factor.

$$x^2 + 4x + xy + 4y = (x^2 + 4x) + (xy + 4y)$$

We factor out a common factor of x from the first group, and a common factor of y from the second group.

$$= x(x + 4) + y(x + 4)$$

Now there is a common factor of $x + 4$.

$$= x\ (x + 4)\ + y\ (x + 4)$$

Factor out the $(x + 4)$.

$$= \boxed{(x + 4)(x + y)}$$ ∎

EXAMPLE 6 *Factor as completely as possible.* $a^2 - ab - 3a + 3b$

6. $y^2 - 2yz - 3y + 6z$

Solution We will again begin by splitting the four terms into two groups.

$$a^2 - ab - 3a + 3b = (a^2 - ab) + (-3a + 3b)$$

If we factor out a common factor of a from the first group, and a common factor of 3 from the second group, we will not readily see that we again have a common factor. In other words we would get

$$= a(a - b) + 3(-a + b)$$

> *We do not see any further common factors. Instead, let's factor out -3 from the second group. We get*

$$= a(a - b) - 3(a - b)$$

$$\uparrow$$

> *Watch for **this** sign!*
> *Now we see a common factor of $a - b$.*

$$= \boxed{(a - b)(a - 3)}$$ ∎

✔ ***Answers to Learning Checks in Section 6.2***

1. (a) $3(3a^4 - 4a^3 - 6)$ (b) $5a^2(2a + 3)$ **2.** $6mn(3n + 4m^2 - 2mn)$

3. $8ab^2(2a^2 + 1)$ **4.** $(t + 5)(r + s)$ **5.** $(a + 5)(a + b)$

6. $(y - 2z)(y - 3)$

Exercises 6.2

Factor each of the following as completely as possible. If the expression is not factorable, say so. Try factoring by grouping where it might help.

1. $5x + 20$

2. $4x + 28$

3. $8a - 12$

4. $6y - 15$

5. $3a + 6b - 8c$

6. $9m - 12n + 8p$

7. $x^2 + 3x$

8. $y^2 + 6y$

9. $a^2 + a$

10. $t^2 - t$

11. $x^2 - 5x + xy$

12. $a^2 - 3ab - 5a$

13. $3c^6 - 6c^3$

14. $5y^5 - 10y^2$

15. $x^2y - xy^2$

16. $a^3b + ab^3$

17. $6x^2 + 3x$

18. $4u^3 - 8u^4$

19. $8x^3y^2 - 25z^4$

20. $9mn^3 - 16p^5$

21. $12c^3d^5 + 4c^2d^3$

22. $5x^3y - 15x^4y^2$

23. $x^2y^3 - y^2z^4 + x^3z^2$

24. $6a^2b + 10a^3c^2 - 9b^2c^3$

1. _____
2. _____
3. _____
4. _____
5. _____
6. _____
7. _____
8. _____
9. _____
10. _____
11. _____
12. _____
13. _____
14. _____
15. _____
16. _____
17. _____
18. _____
19. _____
20. _____
21. _____
22. _____
23. _____
24. _____

ANSWERS

25. _____

26. _____

27. _____

28. _____

29. _____

30. _____

31. _____

32. _____

33. _____

34. _____

35. _____

36. _____

37. _____

38. _____

39. _____

40. _____

25. $2x^2yz^3 + 8xyz^2 - 10x^2y^2z^2$

26. $6m^3n^2p^4 - 15m^2n^3p^2 + 12mn^2p^3$

27. $6u^3v^2 + 18u^3v^3 - 12u^3v^5$

28. $9w^2z^3 - 3wz + 6wz^4$

29. $x(x - 5) + 4(x - 5)$

30. $a(a + 7) + 3(a + 7)$

31. $y(y + 6) - 3(y + 6)$

32. $z(z - 3) - 5(z - 3)$

33. $x^2 + 8x + xy + 8y$

34. $a^2 - 6a + ab - 6b$

35. $m^2 + mn + 9m + 9n$

36. $w^2 - rw + 10w - 10r$

37. $x^2 - xy - 4x + 4y$

38. $y^2 + wy - 7y - 7w$

39. $3x^2y + 6xy - 5x - 10$

40. $8a^2 - 4ab - 6a + 3b$

QUESTION FOR THOUGHT

4. Which of the following is in completely factored form?

    (a) $x^2 + 5x + 6$, $x(x + 5) + 6$, $(x + 2)(x + 3)$

    (b) $x^3y^2 + x^2y^3$, $xy(x^2y + xy^2)$, $x^2y^2(x + y)$

Factoring Trinomials

6.3

Factoring a polynomial by taking out a common factor, as we did in the last section, is a fairly straightforward mechanical process.

Now let's turn our attention to factoring trinomials such as

$$x^2 + 5x + 6$$

The first thing we notice is that there is no common factor. If there were a common factor, we would certainly factor that out first.

Were it not for our experience in multiplying out binomials in Sections 5.5 and 6.1, we might simply say that $x^2 + 5x + 6$ cannot be factored. However, we have seen many examples of two binomials multiplying out to give answers of the form $x^2 + 5x + 6$. It is therefore reasonable to ask:

Can we construct two binomials so that their product

$$(? \quad ?)(? \quad ?) = x^2 + 5x + 6?$$

With a little bit of trial and error, we might very quickly arrive at

$$(x + 3)(x + 2) = x^2 + 5x + 6$$

as our answer. (*Check it!*) However, we want to analyze this example very carefully so that we can develop a systematic approach to factoring trinomials.

One of our "special products" in Section 6.1 was the multiplication of two simple binomials.

$$(x + m)(x + n) = x^2 + nx + mx + mn \qquad \text{\textit{Recall that nx and mx are called}}$$
$$= x^2 + (n + m)x + mn \qquad \text{\textit{the "cross terms."}}$$

Note that the coefficient of x (the first-degree term) is the *sum* of m and n, while the last term is the *product* of m and n.

Let's analyze the factorization of $x^2 + 5x + 6$. To factor $x^2 + 5x + 6$ into the product of two binomials, we know that in order to get x^2 as the first term, the binomials must look like

$$(x \quad)(x \quad)$$

Now let's focus on the signs that will go into each set of parentheses, without regard to the number which will go into each.

The $+$ sign in front of the 6 tells us that the two signs in the parentheses must be the *same*, either both $+$ signs or both $-$ signs. Do you see why? Since the $+6$ is the *product* of m and n, m and n must have the same signs (for if their signs were opposite, their product would be negative).

Now the $+$ sign in front of the 5 tells us how the two cross terms add up. Since we already know that the signs are the same, and the $+5$ tells us that m and n must add up to $+5$, then both numbers *must* be positive. (If they were both negative, they would add up to a negative.)

To summarize, looking at $x^2 + 5x + 6$, we see

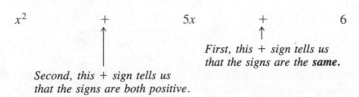

$$x^2 \qquad + \qquad 5x \qquad + \qquad 6$$

First, this $+$ sign tells us that the signs are the ***same.***

Second, this $+$ sign tells us that the signs are both positive.

Thus far our analysis has told us that

$$x^2 + 5x + 6 = (x \quad)(x \quad)$$
$$= (x + \quad)(x + \quad)$$

Now we are ready to find the numbers to be inserted into the parentheses. We are looking for two numbers which *multiply* to 6. The possible pairs of factors are 6 and 1, or 3 and 2. Does one of these pairs *also* add up to 5? Yes. 3 and 2.

Thus, the final result for the factorization is

$$x^2 + 5x + 6 = \boxed{(x + 3)(x + 2)}$$

Of course, $(x + 2)(x + 3)$ is equally correct. The *order* of the factors is irrelevant.

Let's try some examples.

✔ **LEARNING CHECKS**

1. $t^2 + 6t - 16$

EXAMPLE 1   *Factor as completely as possible.* $x^2 - 5x - 6$

Solution   As always, we first look for any common factors, but there are none. Therefore, we try to build two binomials which multiply out to $x^2 - 5x + 6$. We know that we must have

$$x^2 - 5x - 6 = (x \quad)(x \quad)$$

What do the signs tell us? The minus sign in front of the 6 tells us that the signs in the parentheses must be *opposite;* one must be + and one must be −, because −6 is the product of the two numbers. Thus, we already know that

$$x^2 - 5x - 6 = (x + \quad)(x - \quad)$$

Notice that where we put the + and − signs does not matter, because in $(x \quad)(x \quad)$ the two parentheses are identical.

Since the signs of the two numbers we put into the parentheses are opposite, when we add the two numbers we are actually getting a *difference* from the cross terms. The minus sign in front of the 5 tells us that the result of this difference must be negative. Therefore, we know that the cross term with the larger absolute value must get the minus sign. (Remember that when we add numbers with opposite signs, we keep the sign of the number with the larger absolute value.) As in the last example, since we want the product of the two numbers to be 6 (by putting in the + and − signs we have already taken care of the fact that we want the 6 to be negative), the possible pairs of factors are 6 and 1, or 3 and 2.

Some students say to themselves that they want 5 to be the coefficient of *x,* and so immediately choose 3 and 2 as the factors. This is incorrect, because it neglects the fact that the 5 must result from a *difference,* and so 3 and 2 do not work. However, 6 and 1 do work. Where we put the 6 and where we put the 1 *does* matter since the two parentheses are not identical (one has a + sign in it, the other a − sign in it). We already know that we want the cross term with the larger absolute value to be negative. Therefore, we have

$$x^2 - 5x - 6 = (x \quad)(x \quad)$$
$$= (x + \quad)(x - \quad)$$
$$= \boxed{(x + 1)(x - 6)}$$

Again, the order of the factors does not matter as long as you have the same two factors, $x + 1$ and $x - 6$.   ∎

If we take the time to analyze the signs, we know whether we are looking for a sum or a difference from the cross terms. This usually makes it easier to find the correct factors (if there are any).

It is possible to look at this last example and simply say we are looking for two numbers whose product is -6 and whose sum is -5, and therefore the numbers are -6 and $+1$. This works well for examples where the leading coefficient (that is, the coefficient of x^2) is 1, but not otherwise.

We have taken the time to analyze the signs in order to lay the groundwork for the more complicated factoring work ahead.

EXAMPLE 2 *Factor each of the following as completely as possible.*

(a) $x^2 - 7x + 6$ **(b)** $x^2 + 5x - 6$ **(c)** $x^2 - x - 6$

2. (a) $t^2 - 8t + 15$
   (b) $t^2 + 10t - 24$
   (c) $t^2 + 4t - 12$

Solution Again, we always begin by looking for any common factors. There are none in any of the three examples.

(a) In $x^2 - 7x + 6$, the $+6$ tells us that the signs are the same; the -7 tells us that they must both be negative.

$$x^2 - 7x + 6 = (x \quad)(x \quad)$$
$$= (x - \quad)(x - \quad)$$
$$= \boxed{(x - 6)(x - 1)}$$

*We need factors of 6 which **add** to 7 (because the signs are the same). 6 and 1 work.*

(b) In $x^2 + 5x - 6$, the -6 tells us that the signs are opposite; the $+5$ tells us that we want the cross term with larger absolute value to be positive.

$$x^2 + 5x - 6 = (x \quad)(x \quad)$$
$$= (x + \quad)(x - \quad)$$
$$= \boxed{(x + 6)(x - 1)}$$

*We need factors of 6 whose **difference** is 5 (because the signs are opposite). 6 and 1 work, with the 6 getting the + sign.*

(c) In $x^2 - x - 6$, the -6 again tells us that the signs will be opposite, and the -1 coefficient of x tells us that we want the cross term with larger absolute value to be negative.

$$x^2 - x - 6 = (x \quad)(x \quad)$$
$$= (x + \quad)(x - \quad)$$
$$= \boxed{(x + 2)(x - 3)}$$

*We need factors of 6 whose **difference** is 1 (because the signs are opposite). 3 and 2 work, with the 3 getting the − sign.* ∎

Remember always to check your factorization by multiplying out.

EXAMPLE 3 *Factor as completely as possible.* $x^2 + 4x + 5$

3. $a^2 - 6a + 7$

Solution First of all, there are no common factors. Next, we try to build two binomials.

$$x^2 + 4x + 5 = (x \quad)(x \quad)$$
$$= (x + \quad)(x + \quad) \quad \text{\textit{Do you see why there must be two + signs?}}$$

The only possible pair of factors for 5 is 5 and 1; therefore, the *only* possible factorization is

$$(x + 5)(x + 1)$$

But this does not work. Check it! Consequently, we say that $x^2 + 4x + 5$ cannot be factored. ∎

4. $28 + t^2 - 11t$

EXAMPLE 4 *Factor as completely as possible.* $2x + x^2 - 15$

Solution It is generally easier to factor a trinomial when it is in standard form. You will recall that standard form basically means that the polynomial is written from the highest power of the variable to the lowest. Therefore, we first reorder the terms.

$$
\begin{aligned}
2x + x^2 - 15 &= x^2 + 2x - 15 & &\textit{There are no common factors.} \\
&= (x \quad)(x \quad) & &\textit{The product is } -15 \textit{ so the signs are } + \\
& & &\textit{and } -. \\
&= (x + \quad)(x - \quad) & &\textit{We need numbers whose product is 15} \\
& & &\textit{and whose } \textbf{difference} \textit{ is 2.} \\
&= \boxed{(x + 5)(x - 3)}
\end{aligned}
$$ ∎

5. $4y^2 + 24y + 20$

EXAMPLE 5 *Factor as completely as possible.* $6a^2 - 18a + 12$

Solution If you immediately attempt to build the two binomials, this example quickly becomes much more complicated than is necessary. In addition, you will probably not get a complete factorization. Our first step should always be to look for any common factors. There is a common factor of 6.

$$
\begin{aligned}
6a^2 - 18a + 12 &= 6(a^2 - 3a + 2) & &\textit{Now we try to factor further.} \\
&= 6(a \quad)(a \quad) \\
&= 6(a - \quad)(a - \quad) & &\textit{Do you see why both signs must be} \\
& & &\textit{negative? We need two numbers} \\
& & &\textit{whose product is 2 and whose } \textbf{sum} \\
& & &\textit{is 3. They are 2 and 1.} \\
&= \boxed{6(a - 2)(a - 1)} & &\textit{Do not forget the 6.}
\end{aligned}
$$ ∎

6. $z^2 - 5z$

EXAMPLE 6 *Factor as completely as possible.* $x^2 + 6x$

Solution Resist the temptation to immediately write down $(x \quad)(x \quad)$. We do not need to construct two binomials in this case.

> *Remember:* Always look for common factors *first*.

$$
x^2 + 6x = \boxed{x(x + 6)}
$$ ∎

7. $a^2 - 6ab + 9b^2$

EXAMPLE 7 *Factor as completely as possible.* $x^2 + 8xy + 16y^2$

Solution We can apply the same basic approach here. However, instead of having an x term and a numerical term in each binomial, we will have an x term and a y term.

Since there is no common factor we proceed to try to build our two binomials.

$$
\begin{aligned}
x^2 + 8xy + 16y^2 &= (x \quad)(x \quad) \\
&= (x + \quad)(x + \quad)
\end{aligned}
$$

We need two numbers whose product is 16 and whose sum is 8. They are 4 and 4.

$$
= \boxed{(x + 4y)(x + 4y)} \quad \text{or} \quad \boxed{(x + 4y)^2}
$$

Notice that we did not need to recognize that $x^2 + 8xy + 16y^2$ is a perfect square in order to factor it. ∎

EXAMPLE 8 *Factor as completely as possible.* $x^2 - 16$

8. $u^2 - 81$

Solution This is sort of a special case in that $x^2 - 16$ is not a trinomial—the middle term is missing. Actually, we can think of the middle term as having a coefficient of 0.

$$x^2 - 16 = x^2 + 0x - 16$$

Thus, we are looking for two numbers whose product is -16 and whose sum is 0.

$x^2 - 16 = (x \qquad)(x \qquad)$ *The signs are opposite because the product must*
$\qquad = (x + \quad)(x - \quad)$ *be -16.*
$\qquad = \boxed{(x + 4)(x - 4)}$ *$+4$ and -4 work.*

You may recall that we called this type of expression the *difference of two squares*. It is usually easy to recognize because of its appearance. ■

One final comment: It is possible to factor trinomials by listing *all* the possible pairs of binomial factors, and then checking to see if any of them work. Clearly, if this method is chosen, being able to multiply out the binomials mentally is extremely helpful.

In the next section we will look at more complex factoring examples.

✔ *Answers to Learning Checks in Section 6.3*

1. $(t + 8)(t - 2)$

2. (a) $(t - 5)(t - 3)$ **(b)** $(t + 12)(t - 2)$ **(c)** $(t + 6)(t - 2)$

3. Not factorable   **4.** $(t - 7)(t - 4)$ **5.** $4(y + 5)(y + 1)$

6. $z(z - 5)$ **7.** $(a - 3b)(a - 3b)$ **8.** $(u + 9)(u - 9)$

NOTE TO THE STUDENT

Use the space on this page to write down any questions you have or points you want to review with your instructor.

Exercises 6.3

Factor each of the following expressions as completely as possible. If an expression is not factorable, say so.

1. $x^2 + 3x$

2. $x^2 + 4x$

3. $x^2 + 3x + 2$

4. $x^2 + 4x + 3$

5. $x^2 - 3x + 2$

6. $x^2 - 4x + 3$

7. $x^2 + 3x - 2$

8. $x^2 - 4x - 3$

9. $x^2 + x - 2$

10. $x^2 - 2x - 3$

11. $x^2 - x - 2$

12. $x^2 + 2x - 3$

13. $a^2 + 8a + 12$

14. $a^2 + 7a + 12$

15. $a^2 - a - 12$

16. $a^2 + 4a - 12$

17. $a^2 - a + 12$

18. $a^2 - 4a - 12$

19. $a^2 - 12a$

20. $a^2 + 12a$

21. $a - 12 + a^2$

22. $12 - 8a + a^2$

23. $x^2 - 3xy + 2y^2$

24. $x^2 + 6xy - 7y^2$

25. $a^2 + 10a + 24$

26. $a^2 - 10a - 24$

27. $y^2 + 12y + 36$

28. $t^2 - 12t + 36$

29. $y^2 - 36$

30. $t^2 + 36$

1. _____
2. _____
3. _____
4. _____
5. _____
6. _____
7. _____
8. _____
9. _____
10. _____
11. _____
12. _____
13. _____
14. _____
15. _____
16. _____
17. _____
18. _____
19. _____
20. _____
21. _____
22. _____
23. _____
24. _____
25. _____
26. _____
27. _____
28. _____
29. _____
30. _____

ANSWERS

31. _____

32. _____

33. _____

34. _____

35. _____

36. _____

37. _____

38. _____

39. _____

40. _____

41. _____

42. _____

43. _____

44. _____

45. _____

46. _____

47. _____

48. _____

49. _____

50. _____

51. _____

52. _____

53. _____

54. _____

31. $x^2 - 7x - 18$

32. $m^2 + 6m - 18$

33. $r^2 - 3rs - 10s^2$

34. $r^2 + 9rs - 10s^2$

35. $c^2 - 6c + 5$

36. $c^2 - 13c + 12$

37. $4x^2 + 8x + 4$

38. $6x^2 - 30x + 36$

39. $x^2 - 30 + x$

40. $x^2 - 30 + 7x$

41. $2x^2 - 50$

42. $3x^2 - 27$

43. $x^2 - x - 20$

44. $x^2 - 8x - 20$

45. $x^2 - x + 20$

46. $x^2 - 8x + 20$

47. $y^2 + 11y + 28$

48. $y^2 - 13y - 48$

49. $2y^2 + 2y - 84$

50. $3y^2 - 6y - 72$

51. $49 - d^2$

52. $t^2 - 1$

53. $49 + d^2$

54. $t^2 + 1$

QUESTIONS FOR THOUGHT

5. Find *all* integers k so that $x^2 + kx + 10$ can be factored.

6. Find *all* integers b so that $x^2 + bx - 10$ can be factored.

7. Can you find all integers c so that $x^2 + 5x + c$ can be factored? Why or why not?

More Factoring

In order to expand and adapt our factoring skills to more complex situations, let's begin by reviewing a multiplication example.

EXAMPLE 1 *Multiply and simplify.* $(3x - 4)(2x + 5)$

Solution

$$(3x - 4)(2x + 5) = 6x^2 + 15x - 8x - 20$$
$$= \boxed{6x^2 + 7x - 20}$$

Notice that because the coefficients of x in the original parentheses were not 1, the coefficient 7, of x, in our final answer is no longer just the sum of the numbers -4 and $+5$. Rather, the 7 results from the interaction of the factors of 6 (3 and 2) with the factors of 20 (4 and 5). ■

The interaction seen in Example 1 is what makes examples where the leading coefficient (that is, the coefficient of x^2) is not 1 more difficult to factor. The following examples illustrate how we can handle these more complicated situations.

EXAMPLE 2 *Factor as completely as possible.* $2x^2 - 7x + 3$

Solution As usual, the first step is to look for any common factors. In this case there are none.

Because the coefficient of x^2 is 2, and 2 has only one possible pair of factors, 2 and 1, we can begin to analyze this example in the same way that we analyzed examples in the last section.

$$2x^2 - 7x + 3 = (2x \quad)(x \quad) \quad \textit{Next we analyze the signs.}$$
$$= (2x - \quad)(x - \quad)$$

Even though there is only one possible pair of factors for 3, 3 and 1, where we put them does make a difference. This is because of how they will interact with the factors of 2. If we like, we can simply list all the possible binomials, and check to see if any one of them works. The possible pairs of binomials are:

$$(2x - 3)(x - 1) = 2x^2 - 5x + 3$$
$$(2x - 1)(x - 3) = 2x^2 - 7x + 3$$

Thus, the answer is

$$2x^2 - 7x + 3 = \boxed{(2x - 1)(x - 3)}$$ ■

EXAMPLE 3 *Factor as completely as possible.* $3a^2 + 20a + 12$

Solution There is no common factor.

Analyzing the possible factors of 3 and the signs, we see that if we are to find two binomial factors we must have

$$3a^2 + 20a + 12 = (3a \quad)(a \quad)$$
$$= (3a + \quad)(a + \quad)$$

We are looking for the factors of 3 and 12 to interact and give us a *sum* of 20. The possible factors of 12 are $12 \cdot 1$, $6 \cdot 2$, and $4 \cdot 3$. Listing the possible binomial pairs, we obtain:

$$(3a + 12)(a + 1) = 3a^2 + 15a + 12$$
$$(3a + 1)(a + 12) = 3a^2 + 37a + 12$$
$$(3a + 6)(a + 2) = 3a^2 + 12a + 12$$
$$(3a + 2)(a + 6) = 3a^2 + 20a + 12$$

Thus, the correct factorization is

$$3a^2 + 20a + 12 = \boxed{(3a + 2)(a + 6)}$$ ■

4. $8t^2 - 2t - 15$

EXAMPLE 4   *Factor as completely as possible.* $10x^2 + 7x - 12$

Solution   Again, there are no common factors.

Since *both* 10 and 12 contain several pairs of factors, this example has numerous possibilities to consider right at the beginning. However, our analysis of the signs does tell us that the signs are opposite, so we know we are looking for a *difference*.

We list the possible binomial pairs using $5x$ and $2x$:

$$(5x + 1)(2x - 12) = 10x^2 - 58x - 12$$
$$(5x - 1)(2x + 12) = 10x^2 + 58x - 12^{\ddagger}$$

$$(5x + 12)(2x - 1) = 10x^2 + 19x - 12$$
$$(5x - 12)(2x + 1) = 10x^2 - 19x - 12$$

$$(5x + 2)(2x - 6) = 10x^2 - 26x - 12$$
$$(5x - 2)(2x + 6) = 10x^2 + 26x - 12^{\ddagger}$$

$$(5x + 6)(2x - 2) = 10x^2 + 2x - 12$$
$$(5x - 6)(2x + 2) = 10x^2 - 2x - 12^{\ddagger}$$

$$(5x + 3)(2x - 4) = 10x^2 - 14x - 12$$
$$(5x - 3)(2x + 4) = 10x^2 + 14x - 12^{\ddagger}$$

$$(5x + 4)(2x - 3) = 10x^2 - 7x - 12$$
$$(5x - 4)(2x + 3) = 10x^2 + 7x - 12$$

Thus, the complete factorization is

$$10x^2 + 7x - 12 = \boxed{(5x - 4)(2x + 3)}$$

In order to make the factorization process as short as possible, several comments are in order here.

First, while the positions of the + and − signs do make a difference, we recognize that a difference which yields +7 is also useful. We note that switching the + and − signs in the binomials simply switches the sign of the middle term. If we can get a middle term of $-7x$ from $(5x + 4)(2x - 3)$, then we can get a middle term of $+7x$ by simply switching the signs to $(5x - 4)(2x + 3)$. Therefore, we do not have to write each binomial pair *twice* with the signs interchanged.

Second, notice that even though we wanted to end up with a $+7x$ from our difference, it was not the larger number (4) which got the + sign, but rather the larger cross term (the 3 times $5x$) which got the + sign.

Third, if you are saying to yourself "Isn't this just a trial-and-error process?", you are absolutely right. We are simply trying to make the procedure as systematic as we can.

Fourth, as you practice more, and get better at factoring, you will be able to make an educated guess as to which pairs of factors are most likely to work, and start your list with them. Thus, in this example pairing up 12 with any of the other factors gives a fairly large product as compared to the 7 we are looking for, and so makes it less likely to work. We could have started our list of possible binomial pairs by using $3 \cdot 4$ rather than $12 \cdot 1$ to make a product of 12.

Fifth, the fact that there was no common factor to begin with eliminates some of the lines from the table. Do you see why $(10x - 2)(x + 6)$ could not possibly work *without* computing the middle term, which is $58x$?

Answer: $10x - 2$ has a common factor of 2, but the original trinomial had no common factor! If $10x - 2$ were one of the factors of $10x^2 + 7x - 12$, then 2 would also have to be a common factor of $10x^2 + 7x - 12$. Thus, since $10x^2 + 7x - 12$ has no common factors, $10x - 2$ cannot possibly be a factor.

In fact, if you look at our list of possible binomial factors all the lines marked with a "‡" could have been eliminated from consideration for exactly the same reason. Using these ideas we can substantially narrow down the trial-and-error process. In this example our list could have had two possible pairs instead of twelve! ∎

EXAMPLE 5 *Factor as completely as possible.* $12x^2 + 24x - 36$

5. $8t^2 - 24t - 32$

Solution After the last example, you might sigh looking at all the possibilities here. In fact, if you start looking for two binomials immediately, there are thirty possible binomial pairs!

Actually, this example is quite simple if we remember to begin by looking for any common factors. We first take out the common factor of 12.

$$12x^2 + 24x - 36 = 12(x^2 + 2x - 3)$$
$$= 12(x \quad)(x \quad)$$
$$= 12(x + \quad)(x - \quad)$$
$$= \boxed{12(x + 3)(x - 1)}$$

We need two numbers whose product is 3 and whose difference is 2.

∎

REMEMBER

Always look for common factors first.

EXAMPLE 6 *Factor as completely as possible.* $7x - x^2 - 10$

6. $20 + a - a^2$

Solution Since there is no common factor, we can proceed directly to try to build two binomial factors.

As we mentioned in the last section, it is best to rewrite the polynomial in standard form as

$$-x^2 + 7x - 10$$

Since we are more familiar factoring trinomials whose leading coefficient is $+1$ rather than -1, we can first factor out a -1 so that $-x^2$ becomes $+x^2$. That is,

$$-x^2 + 7x - 10 = -(x^2 - 7x + 10)$$

Factoring out -1 has the same effect as multiplying by -1; it changes the sign of each term inside the parentheses.

Now we can proceed as before. The entire solution appears as follows:

$$\begin{aligned} 7x - x^2 - 10 &= -x^2 + 7x - 10 \\ &= -(x^2 - 7x + 10) \\ &= -(x \quad)(x \quad) \\ &= -(x - \quad)(x - \quad) \\ &= \boxed{-(x - 5)(x - 2)} \end{aligned}$$

Do not forget the minus sign. ∎

Having done all these examples, let's pause and outline our approach to factoring.

OUTLINE FOR FACTORING POLYNOMIALS

1. Take out common factors, if any.
2. Make sure that the polynomial is in standard form, preferably with the leading coefficient positive. Factor out -1, if necessary.
3. Factor the remaining polynomial into two binomials, if possible.
4. Check your factorization by multiplying out the factors.

---

7. $4u^3 + 8u^2 - 60u$

EXAMPLE 7 *Factor as completely as possible.* $\quad 5a^3 - 45a^2 - 50a$

Solution Following our outline, we first take out the common factor of $5a$.

$$\begin{aligned} 5a^3 - 45a^2 - 50a &= 5a(a^2 - 9a - 10) \\ &= 5a(a \quad)(a \quad) \\ &= 5a(a + \quad)(a - \quad) \\ &= \boxed{5a(a + 1)(a - 10)} \end{aligned}$$

∎

8. $2a^3b^2 - 50ab^4$

EXAMPLE 8 *Factor as completely as possible.* $\quad 3x^3y - 27xy^3$

Solution The greatest common factor is $3xy$.

$$\begin{aligned} 3x^3y - 27xy^3 &= 3xy(x^2 - 9y^2) \\ &= \boxed{3xy(x + 3y)(x - 3y)} \end{aligned}$$

Within the parentheses we recognize the difference of two squares.

∎

Being able to factor trinomials (particularly when the leading coefficient is 1 or a prime number) is a very useful skill. We will begin to see how factoring can be used in the next section.

However, what is true for most algebraic skills pertains even more so to factoring. Acquiring the ability to factor accurately and rapidly requires lots of practice. Just because you understand *how* an expression was factored does not necessarily mean that you could factor it yourself. There is absolutely no substitute for doing many exercises.

   One final comment: At this point it is easy to get "factorization of the brain" and think that *everything* has to be factored. This is not so. Factoring is just a tool which we will soon learn to use. However, if you are simplifying an expression such $3x + 2x + 10$ and get $5x + 10$, this is a completely acceptable answer. You do not have to factor it to $5(x + 2)$, unless you have a reason to.

✔ *Answers to Learning Checks in Section 6.4*

1. $15x^2 + 11x - 12$ **2.** $(3t + 2)(t + 1)$ **3.** $(2y - 3)(y - 4)$

4. $(4t + 5)(2t - 3)$ **5.** $8(t - 4)(t + 1)$ **6.** $-(a - 5)(a + 4)$

7. $4u(u - 3)(u + 5)$ **8.** $2ab^2(a + 5b)(a - 5b)$

NOTE TO THE STUDENT

Use the space on this page to write down any questions you have or points you want to review with your instructor.

Exercises 6.4

Factor each of the following as completely as possible. If the polynomial is not factorable, say so.

1. $x^2 + 3x$ **2.** $3x^2 - 6x$ **3.** $x^2 + 3x + 2$

4. $3x^2 - 6x + 3$ **5.** $x^2 + 3x - 2$ **6.** $t^2 - 4t + 5$

7. $3x^2 + 8x + 4$ **8.** $3x^2 - 7x + 4$ **9.** $2x^2 + 11x + 12$

10. $3x^2 + 13x + 12$ **11.** $2x^2 + 10x + 12$ **12.** $3x^2 + 12x + 12$

13. $5x^2 - 27x + 10$ **14.** $7x^2 - 50x + 7$ **15.** $5x^2 - 15x + 10$

16. $7x^2 - 56x + 49$ **17.** $2y^2 - y - 6$ **18.** $3y^2 - y - 10$

19. $5a^2 + 9a - 18$ **20.** $5a^2 + 2a - 16$ **21.** $2t^2 + 7t + 6$

22. $3t^2 + 5t - 12$ **23.** $2t^2 + 6t + 6$ **24.** $3t^2 + 6t - 12$

25. $3w^2 - 6w - 30$ **26.** $5z^2 - 15z - 60$ **27.** $3x^2 - 4x + 2$

28. $5x^2 - 10x + 4$ **29.** $3x^2 - 14xy + 15y^2$ **30.** $3x^2 + 13xy + 12y^2$

1. _____
2. _____
3. _____
4. _____
5. _____
6. _____
7. _____
8. _____
9. _____
10. _____
11. _____
12. _____
13. _____
14. _____
15. _____
16. _____
17. _____
18. _____
19. _____
20. _____
21. _____
22. _____
23. _____
24. _____
25. _____
26. _____
27. _____
28. _____
29. _____
30. _____

ANSWERS

31. _____
32. _____
33. _____
34. _____
35. _____
36. _____
37. _____
38. _____
39. _____
40. _____
41. _____
42. _____
43. _____
44. _____
45. _____
46. _____
47. _____
48. _____
49. _____
50. _____
51. _____
52. _____
53. _____
54. _____
55. _____
56. _____
57. _____
58. _____

31. $6a^2 + 17a + 10$ **32.** $6a^2 + 19a + 10$ **33.** $6a^2 + 17a - 10$

34. $6a^2 - 7a - 10$ **35.** $6a^2 - 18a - 24$ **36.** $6a^2 - 24a - 30$

37. $x^2 - 36y^2$ **38.** $x^2 - 16y^2$ **39.** $4x^2 - 36y^2$

40. $4x^2 - 16y^2$ **41.** $x^3 + 5x^2 - 24x$ **42.** $x^3 + 2x^2 - 24x$

43. $x^2 + 5x^2 - 24x$ **44.** $x^3 + 2x^3 - 24x$ **45.** $4x^4 - 24x^3 + 32x^2$

46. $6x^5 + 18x^4 - 60x^3$ **47.** $6x^2y - 8xy^2 + 12xy$ **48.** $10a^2b^3 + 15ab^2 - 20a^3b^2$

49. $3x^2 - 7x - 48$ **50.** $3x^2 - 70x - 48$ **51.** $8x^2 - 32x$

52. $8x^2 - 32$ **53.** $2x - x^2 + 15$ **54.** $48 + 8a - a^2$

55. $84xy - 16x^2y - 4x^3y$ **56.** $27x^2y^2 + 6x^3y^2 - x^4y^2$ **57.** $-x^2 + 25$

58. $-w^2 - 16$

QUESTION FOR THOUGHT

8. As was mentioned in this section, if a polynomial *does not* have a common factor then we can eliminate any binomial factors which *do* have a common factor. For example, the trinomial $6x^2 - 5x - 4$ has no common factors. Why does this imply that $(3x - 2)(2x + 2)$ should not even be considered as a possible factorization? Of the twelve possible binomial pairs we might list, how many can be eliminated from consideration in this way?

Dividing Polynomials

6.5

In Chapter 5 we discussed various procedures with polynomials. We talked about how to add, subtract, and multiply polynomials. We postponed considering division until now so that we could see how division and factoring often go hand in hand.

When we divide a polynomial by a *monomial* the process is quite straightforward, and is basically just an application of the Distributive Law. That is,

$$\frac{a + b}{c} = \frac{1}{c}(a + b) = \frac{1}{c} \cdot a + \frac{1}{c} \cdot b = \frac{a}{c} + \frac{b}{c}$$

Alternatively, we can say that this is just the reverse of combining fractions with the same denominators.

In any case, the result is the same. Each term in the numerator is divided by the denominator.

EXAMPLE 1   *Divide.* $\dfrac{4x^5 + 8x^3 - 6x}{2x}$

Solution   We offer two methods of solution.

Method 1.   Just as we can combine several fractions which have the same denominator into a single fraction, we can also break up a single fraction with several terms in the numerator into separate fractions.

$$\frac{4x^5 + 8x^3 - 6x}{2x} = \frac{4x^5}{2x} + \frac{8x^3}{2x} - \frac{6x}{2x} \qquad \textit{Now we reduce each fraction.}$$

$$= \frac{\overset{2\,x^4}{\cancel{4x^5}}}{\cancel{2x}} + \frac{\overset{4\,x^2}{\cancel{8x^3}}}{\cancel{2x}} - \frac{\overset{3}{\cancel{6x}}}{\cancel{2x}}$$

$$= \boxed{2x^4 + 4x^2 - 3}$$

If you use this method, **be careful.** Be sure to divide *each* term in the numerator by the denominator.

Method 2.   Division can sometimes be accomplished by reducing the original fraction. However, we know that according to the Fundamental Principle of Fractions, we are allowed to cancel common factors only. Thus, we have our first application of factoring. We begin by factoring the numerator.

$$\frac{4x^5 + 8x^3 - 6x}{2x} = \frac{2x(2x^4 + 4x^2 - 3)}{2x}$$

$$= \frac{\cancel{2x}(2x^4 + 4x^2 - 3)}{\cancel{2x}}$$

$$= \boxed{2x^4 + 4x^2 - 3} \qquad \blacksquare$$

Let's try another example.

2. $\dfrac{18a^2b + 12a^3b^2 - 10a^2b^2}{4a^2b}$

EXAMPLE 2 *Divide.* $\dfrac{15x^2y^3 - 5xy^2 + 10x^3y}{5x^2y^2}$

Solution

Method 1.   We break up the fraction.

$$\frac{15x^2y^3 - 5xy^2 + 10x^3y}{5x^2y^2} = \frac{15x^2y^3}{5x^2y^2} - \frac{5xy^2}{5x^2y^2} + \frac{10x^3y}{5x^2y^2}$$

$$= \frac{\overset{3}{\cancel{15}}\,x^2\,\overset{y}{\cancel{y^3}}}{\cancel{5}\,\cancel{x^2}\,\cancel{y^2}} - \frac{\cancel{5}\,\cancel{x}\,\cancel{y^2}}{\cancel{5}\,\cancel{x^2}\,\cancel{y^2}}\underset{x}{} + \frac{\overset{2x}{\cancel{10}}\,\cancel{x^3}\,\cancel{y}}{\cancel{5}\,\cancel{x^2}\,\cancel{y^2}}\underset{y}{}$$

$$= \boxed{3y - \frac{1}{x} + \frac{2x}{y}}$$

Method 2.   We first factor the numerator. The GCF is $5xy$.

$$\frac{15x^2y^3 - 5xy^2 + 10x^3y}{5x^2y^2} = \frac{5xy(3xy^2 - y + 2x^2)}{5x^2y^2}$$

$$= \frac{\cancel{5}\,\cancel{x}\,\cancel{y}(3xy^2 - y + 2x^2)}{\underset{x\;y}{\cancel{5}\,\cancel{x^2}\,\cancel{y^2}}}$$

$$= \boxed{\frac{3xy^2 - y + 2x^2}{xy}}\quad\blacksquare$$

Although the answers obtained by the two methods look different, they are in fact equal. In Example 2, if we combine the fractions in the answer obtained by method 1 into a single fraction, we will get the answer obtained by method 2. Try it! Both forms of the answer are acceptable, although for most purposes the form of the answer obtained by method 2 is preferred.

While method 2 can *sometimes* be used when dividing a polynomial by a *polynomial*, we will postpone considering such an approach until our second meeting with fractions in Chapter 7. Instead, we will outline a mechanical procedure for dividing a polynomial by a polynomial. This procedure basically follows the same lines as long division for numbers does.

For example, if we are to divide $\frac{389}{12}$ as a long division problem we would write:

Which is really just shorthand for:

$$
\begin{array}{r}
32 \\
12\overline{)389} \\
36 \\
\hline
29 \\
24 \\
\hline
5
\end{array}
\qquad\qquad
\begin{array}{r}
30\;+\;2 \\
10 + 2\,\overline{)300 + 80 + 9} \\
300 + 60 \\
\hline
20 + 9 \\
20 + 4 \\
\hline
5
\end{array}
$$

            ← *We divide* 10 *into* 300.

            ← *We divide* 10 *into* 20.

Note that we focus our attention on dividing by 10, but then we must multiply the number in the quotient by 10 *and* by 2.

We will follow basically the same outline for polynomials. If we wish to divide

$$\frac{x^3 + 5x^2 + 3x - 6}{x + 2}$$

we write $\dfrac{x^3 + 5x^2 + 3x - 6}{x + 2} = x + 2\overline{)x^3 + 5x^2 + 3x - 6}$

$x + 2\overline{)x^3 + 5x^2 + 3x - 6}$ We begin by asking: How many times does x go into x^3? **Answer:** x^2

$\begin{array}{r} x^2 \qquad\qquad\qquad\quad \\ x + 2\overline{)x^3 + 5x^2 + 3x - 6} \end{array}$ Now we multiply $x + 2$ by x^2.

$\begin{array}{r} x^2 \qquad\qquad\qquad\quad \\ x + 2\overline{)x^3 + 5x^2 + 3x - 6} \\ \underline{-(x^3 + 2x^2)\qquad\qquad} \\ 3x^2 + 3x \qquad\quad \end{array}$ **Subtract**, and bring down the next term, $3x$.

Up to here is one complete step in the division process. We repeat this process by asking: How many times does x goes into $3x^2$?   **Answer:** $3x$

$\begin{array}{r} x^2 + 3x \qquad\qquad\quad \\ x + 2\overline{)x^3 + 5x^2 + 3x - 6} \\ \underline{-(x^3 + 2x^2)\qquad\qquad} \\ 3x^2 + 3x \qquad\quad \\ \underline{-(3x^2 + 6x)\qquad} \\ -3x - 6 \end{array}$ We multiplied $3x$ times $x + 2$. We **subtract** and bring down the next term.

Next we ask: How many times does x go into $-3x$? **Answer:** -3

$\begin{array}{r} x^2 + 3x \quad - 3 \qquad \\ x + 2\overline{)x^3 + 5x^2 + 3x - 6} \\ \underline{-(x^3 + 2x^2)\qquad\qquad} \\ 3x^2 + 3x \qquad\quad \\ \underline{-(3x^2 + 6x)\qquad} \\ -3x - 6 \\ \underline{-(-3x - 6)} \\ 0 \end{array}$ **Subtract** from the line above.

We can, of course, check this answer by multiplying $(x + 2)(x^2 + 3x - 3)$.

$$(x + 2)(x^2 + 3x - 3) = x^3 + 3x^2 - 3x + 2x^2 + 6x - 6$$
$$= x^3 + 5x^2 + 3x - 6$$

Note that the division process ended when the degree of the divisor was greater than the degree of the new dividend.

Before we begin the long division process, we require that the **dividend** (the polynomial we divide into) be in **complete standard form.** Complete standard form (see Section 5.4) basically means that all the terms from highest degree down to lowest are accounted for. We illustrate this in the next example.

EXAMPLE 3   *Divide.* $\dfrac{4x^3 - 7x - 3}{2x + 3}$

3. $\dfrac{6x^3 - 13x^2 + 4}{3x - 2}$

Solution   We will write out the example in long division format with the dividend (numerator) in complete standard form. Note that we fill in the "missing" x^2 term with coefficient 0.

$$
\require{enclose}
\begin{array}{r}
2x^2 - 3x\ + 1 \\
2x + 3\enclose{longdiv}{4x^3 + 0x^2 - 7x - 3} \\
-(4x^3 + 6x^2) \quad\quad\quad\quad \\
\hline
-6x^2 - 7x \quad\quad \\
-(-6x^2 - 9x) \quad \\
\hline
2x - 3 \\
-(2x + 3) \\
\hline
-6
\end{array}
$$

Subtract.

Let's check this example. As with arithmetic division we must check that Dividend = (Divisor)(Quotient) + Remainder.

$$
\begin{aligned}
4x^3 - 7x - 3 &\overset{?}{=} (2x + 3)(2x^2 - 3x + 1) - 6 \\
&\overset{?}{=} 4x^3 - 6x^2 + 2x + 6x^2 - 9x + 3 - 6 \\
&\overset{\checkmark}{=} 4x^3 - 7x - 3
\end{aligned}
$$

■

EXAMPLE 4 *Divide.* $\dfrac{2x^4 + x^2 + 3}{x - 1}$

Solution Again, we write out the example in long division format with the numerator in complete standard form.

$$
\begin{array}{r}
2x^3 + 2x^2 + 3x\ + 3 \\
x - 1\enclose{longdiv}{2x^4 + 0x^3 + \ x^2 + 0x + 3} \\
-(2x^4 - 2x^3) \quad\quad\quad\quad\quad\quad \\
\hline
2x^3 + \ x^2 \quad\quad\quad\quad \\
-(2x^3 - 2x^2) \quad\quad\quad\quad \\
\hline
3x^2 + 0x \quad\quad \\
-(3x^2 - 3x) \quad\quad \\
\hline
3x\ + 3 \\
-(3x\ - 3) \\
\hline
6
\end{array}
$$

← *Remainder, R*

*Subtract means change the signs of the line you are subtracting, and **add**.*

We can leave the answer in this form, or we can write

$$
\frac{2x^4 + x^2 + 3}{x - 1} = 2x^3 + 2x^2 + 3x + 3 + \frac{6}{x - 1}
$$

We leave the check as an exercise for the student.   ■

4. $\dfrac{x^4 - 3x + 4}{x + 2}$

✔ *Answers to Learning Checks in Section 6.5*

1. $4t^3 - 2t^2 - 6$ **2.** $\dfrac{9}{2} + 3ab - \dfrac{5b}{2}$ **3.** $2x^2 - 3x - 2$

4. $x^3 - 2x^2 + 4x - 11, R = 26$

Exercises 6.5

Divide each of the following. Use the long division process wherever it is appropriate.

1. $\dfrac{3x + 12}{6}$

2. $\dfrac{5a + 20}{15}$

3. $\dfrac{t^2 - 6t}{6t}$

4. $\dfrac{u^2 - 8u}{4u}$

5. $\dfrac{3x^2y - 9xy^2}{3xy}$

6. $\dfrac{6m^3n^2 - 12mn^3}{2mn}$

7. $\dfrac{3x^2y - 9xy^2}{6x^2y^2}$

8. $\dfrac{6m^3n^2 - 12mn^3}{8m^2n^3}$

9. $\dfrac{10a^2b^3c - 15ab^2c^2 - 20a^3b^2c^3}{5ab^2c}$

10. $\dfrac{12x^3y^2z^4 - 24x^2y^3z^2 + 18x^3y^4z^2}{12xyz}$

11. $\dfrac{x^2 - 3x + 2}{x + 2}$

12. $\dfrac{x^2 - 4x - 3}{x - 2}$

13. $\dfrac{t^2 - 3t - 10}{t - 5}$

14. $\dfrac{t^2 + 2t - 8}{t + 4}$

15. $\dfrac{w^2 + 4w - 21}{w + 3}$

16. $\dfrac{z^2 - 6z - 18}{z - 2}$

17. $\dfrac{2x^2 - 3x + 7}{x - 1}$

18. $\dfrac{3x^2 - 4x + 7}{x - 3}$

1. _____
2. _____
3. _____
4. _____
5. _____
6. _____
7. _____
8. _____
9. _____
10. _____
11. _____
12. _____
13. _____
14. _____
15. _____
16. _____
17. _____
18. _____

19. $\dfrac{y^3 + y^2 + y - 14}{y - 2}$

20. $\dfrac{y^3 + 3y^2 - 4y - 12}{y + 2}$

19. _____

20. _____

21. $\dfrac{a^2 + 2a^3 - 3a + 2}{a + 1}$

22. $\dfrac{3a^2 + a^3 + 5 - 6a}{a - 1}$

21. _____

22. _____

23. $\dfrac{x^3 - x^2 + 36}{x + 3}$

24. $\dfrac{x^3 - 5x + 4}{x - 1}$

23. _____

24. _____

25. $\dfrac{x^4 - 16}{x - 2}$

26. $\dfrac{x^4 - 1}{x - 1}$

25. _____

26. _____

27. $\dfrac{3x^3 + 14x^2 + 2x - 4}{3x + 2}$

28. $\dfrac{2x^3 - 3x^2 - x + 12}{2x + 3}$

27. _____

28. _____

29. $\dfrac{4t^3 - 33t + 24}{2t - 5}$

30. $\dfrac{20t^3 + 33t^2 - 4}{5t + 2}$

29. _____

30. _____

After having completed this chapter you should be able to:

1. Recognize certain "special binomial products" (Section 6.1).

For example:

(a) $(x + 5)^2 = (x + 5)(x + 5) = x^2 + 10x + 25$ is a **perfect square**.

(b) $(a - 8)(a + 8) = a^2 - 64$ is the **difference of two squares**.

2. Factor polynomials (Sections 6.2, 6.3, 6.4).

For example:

(a) $8x^2y - 12xy^3 = \boxed{4xy(2x - 3y^2)}$

(b) $\begin{aligned} x^2 - 3x - 10 &= (x \quad\ )(x \quad\ ) \\ &= (x + \quad)(x - \quad) \\ &= \boxed{(x + 2)(x - 5)} \end{aligned}$

(c) $x^2 - 9 = \boxed{(x + 3)(x - 3)}$

(d) $\begin{aligned} 4x^2y + 24xy + 36y &= 4y(x^2 + 6x + 9) \\ &= 4y(x \quad\ )(x \quad\ ) \\ &= 4y(x + \quad)(x + \quad) \\ &= \boxed{4y(x + 3)(x + 3)} \end{aligned}$

3. Divide polynomials, using long division where necessary (Section 6.5).

For example:

(a) $\begin{aligned} \frac{12x^2 - 8x}{4x} &= \frac{12x^2}{4x} - \frac{8x}{4x} \\ &= \frac{\overset{3\ x}{\cancel{12x^2}}}{\cancel{4x}} - \frac{\overset{2}{\cancel{8x}}}{\cancel{4x}} \\ &= \boxed{3x - 2} \end{aligned}$

(b) $\dfrac{x^2 - 8x - 4}{x - 4} = $

$$
\begin{array}{r}
x \quad\ - 4 \\
x - 4 \overline{\smash{\big)}\, x^2 - 8x - 4} \\
-(x^2 - 4x) \\
\hline
-4x - 4 \\
-(-4x + 16) \\
\hline
-20 \quad \textit{Remainder}
\end{array}
$$

1. $(x + 6)(x + 4) = x^2 + 4x + 6x + 24 = x^2 + 10x + 24$

$(x - 6)(x - 4) = x^2 - 4x - 6x + 24 = x^2 - 10x + 24$

The effect of switching both $+$ signs to $-$ signs is to change the sign of the cross term from $+$ to $-$.

2. $(x + 6)(x - 4) = x^2 - 4x + 6x - 24 = x^2 + 2x - 24$

$(x - 6)(x + 4) = x^2 + 4x - 6x - 24 = x^2 - 2x - 24$

The middle terms of the resulting trinomials have opposite signs.

3. (5) and (7); (6) and (8); (13) and (15); (14) and (16); (33) and (35); (34) and (36); (41) and (43); (42) and (44). The middle terms will always have opposite signs, since

$$(x + a)(x - b) = x^2 + (a - b)x - ab$$

$$(x - a)(x + b) = x^2 + (-a + b)x - ab$$

and $(a - b)$ and $(-a + b)$ are always opposites of one another.

4. (a) $(x + 2)(x + 3)$

(b) $x^2y^2(x + y)$

5. The factors of 10 are 1 and 10 and 2 and 5. Since the last term is positive, the signs in the parentheses must be the same. Possibilities:

$$(x + 1)(x + 10) = x^2 + \mathbf{11}x + 10$$

$$(x - 1)(x - 10) = x^2 - \mathbf{11}x + 10$$

$$(x + 2)(x + 5) = x^2 + \mathbf{7}x + 10$$

$$(x - 2)(x - 5) = x^2 - \mathbf{7}x + 10$$

Therefore, the possible values of k are 11, -11, 7, and -7.

6. As in the previous question, the factors of 10 are 1 and 10 and 2 and 5. Since the last term is negative, the signs in the parentheses must be opposite. Possibilities:

$$(x + 1)(x - 10) = x^2 - \mathbf{9}x - 10$$

$$(x - 1)(x + 10) = x^2 + \mathbf{9}x - 10$$

$$(x + 2)(x - 5) = x^2 - \mathbf{3}x - 10$$

$$(x - 2)(x + 5) = x^2 + \mathbf{3}x - 10$$

Therefore, the possible values of b are -9, 9, -3, and 3.

7. There are infinitely many such integers. We can always find such a c by choosing two integers that differ by 5 and forming the product

$$(x + \text{larger integer})(x - \text{smaller integer})$$

For example, since 9 and 4 differ by 5, $(x + 9)(x - 4) = x^2 + 5x - 36$ will give the value $c = -36$. Clearly, we can find two integers that differ by 5 in infinitely many ways. (Interestingly, if we ask for all *positive* integers c with this property, there are only two answers: $c = 4$ and $c = 6$.)

8. The proposed factor $2x + 2$ has a common factor of 2, which would imply that the original trinomial $6x^2 - 5x - 4$ has a common factor of 2. This is not the case. We can eliminate eight possible factorizations of $6x^2 - 5x - 4$ in this way. In addition to $(3x - 2)(2x + 2)$, we can eliminate

$(3x - 1)(2x + 4)$ $(3x + 1)(2x - 4)$

$(3x + 2)(2x - 2)$ $(6x - 2)(x + 2)$

$(6x + 2)(x - 2)$ $(6x - 4)(x + 1)$

$(6x + 4)(x - 1)$

This leaves four possibilities: $(6x - 1)(x + 4)$, $(6x + 1)(x - 4)$, $(3x + 4)(2x - 1)$, and $(3x - 4)(2x + 1)$, the last of which is the correct one.

CHAPTER 6 REVIEW EXERCISES

In Exercises, 1–18, multiply and simplify.

1. $(x + 5)(x + 7)$ **2.** $(a + 6)(a + 3)$

3. $(x - 5)(x - 7)$ **4.** $(a - 6)(a - 3)$

5. $(x + 5)(x - 7)$ **6.** $(a - 6)(a + 3)$

7. $(x - 5)(x + 7)$ **8.** $(a + 6)(a - 3)$

9. $(x - 5)(x - 5)$ **10.** $(a + 6)(a + 6)$

11. $(x - 5)(x + 5)$ **12.** $(a + 6)(a - 6)$

13. $(x + 9y)(x - 9y)$ **14.** $(a - 7b)(a + 7b)$

15. $(2x + 3)(x - 7)$ **16.** $(3a - 4)(a + 6)$

17. $(5x - 2)(3x + 4)$ **18.** $(4a + 3b)(7a - 2b)$

ANSWERS

1. _____
2. _____
3. _____
4. _____
5. _____
6. _____
7. _____
8. _____
9. _____
10. _____
11. _____
12. _____
13. _____
14. _____
15. _____
16. _____
17. _____
18. _____

In Exercises 19–44, factor the given expressions as completely as possible.

19. $x^2 + 7x + 12$

20. $x^2 - 7x + 12$

21. $x^2 + 7x$

22. $x^2 + 12$

23. $x^2 - 13x + 12$

24. $x^2 - x - 12$

25. $x^2 - 6xy - 27y^2$

26. $r^2 - 8rt + 12t^2$

27. $x^2 - 64$

28. $16x^2 - 64$

29. $2x^2 + 9x + 10$

30. $2x^2 + 8x - 10$

31. $3x^2 - 6x - 24$

32. $3x^2 - 14x - 24$

33. $6a^2 + 36a + 48$

34. $6a^2 + 41a + 48$

35. $5x^3y - 80xy^3$

36. $6m^2n - 8mr^3 + 8n^2r$

37. $x^2 + 9x$

38. $x^2 + 9$

39. $25t^2 - 1$

40. $25t^2 - 100$

41. $30 - x^2 + x$

42. $8 + x - 3x^2$

39. _____

40. _____

43. $12x - 3x^2 - 9$

44. $20 + 16x - 4x^2$

41. _____

42. _____

43. _____

In Exercises 45–52, divide. Use long division where necessary.

45. $\dfrac{x^2y - xy^2}{xy}$

46. $\dfrac{6r^2t - 4rt^3 + 10r^2t^4}{2rt^2}$

44. _____

45. _____

47. $\dfrac{x^2 - 4x - 5}{x - 1}$

48. $\dfrac{y^3 - y^2 + y - 1}{y - 2}$

46. _____

47. _____

49. $\dfrac{2x^3 - 4x - 4}{x - 3}$

50. $\dfrac{6x^3 - 13x^2 + 11x - 10}{3x - 5}$

48. _____

49. _____

51. $\dfrac{x^3 + 8}{x + 2}$

52. $\dfrac{16x^4 - 64}{x - 2}$

50. _____

51. _____

52. _____

CHAPTER 6 PRACTICE TEST

1. _____

In Problems 1–11, factor as completely as possible. If not factorable, say so.

1. $6x^3 + 12x^2 - 15x$ **2.** $4x^2y - 8xy^2 - 2xy$

2. _____

3. _____

3. $x^2 + 9x + 8$ **4.** $x^2 - 9xy - 10y^2$

4. _____

5. $4x^2 - 20x$ **6.** $5x^3 - 45x$

5. _____

6. _____

7. $6x^2 + 24x + 18$ **8.** $2x^2 - 7x - 15$

7. _____

8. _____

9. $x^2 + 4x + 4$ **10.** $6x^2 + 5x - 6$

9. _____

11. $x^2y^2 - 9$ **12.** *Divide.* $\dfrac{12r^3t^2 - 18r^2t^2 + 20r^3t^4}{4r^3t^3}$

10. _____

11. _____

13. *Divide.* $\dfrac{2x^3 - 5x + 6}{x - 2}$

12. _____

13. _____

CUMULATIVE REVIEW
Chapters 4–6

In Exercises 1–4, reduce the given fractions to lowest terms.

1. $\dfrac{-24}{42}$

2. $\dfrac{9x^2}{15x^6}$

3. $\dfrac{36s^8t^9}{20s^9t^8}$

4. $\dfrac{3a - 8a - a}{6a^2 + 2a^2 + 4a^2}$

In Exercises 5–32, perform the indicated operations and simplify as completely as possible.

5. $\dfrac{6x}{25} \cdot \dfrac{10}{x}$

6. $\dfrac{6x}{25} \div \dfrac{10}{x}$

7. $\dfrac{6x}{25} + \dfrac{10}{x}$

8. $\dfrac{7}{2x} - \dfrac{3}{2x}$

9. $\dfrac{3t - 5}{6t^2} + \dfrac{9t + 5}{6t^2}$

10. $\dfrac{6u - 7}{10u^3} - \dfrac{4u - 7}{10u^3}$

11. $(x + 8)(x - 5)$

12. $(x + 8)(x^2 + x - 5)$

13. $(a + b + c)(a + b - c)$

14. $2z(z - 3)(z + 6)$

15. $\dfrac{12x^3y^2}{35z^2} \div \dfrac{20xy}{14z}$

16. $\dfrac{8uv^3}{9w^6} \cdot \dfrac{27w^2}{36v^6}$

17. $(5a - 3c)(4a + 3c)$

18. $2(t^2 - 3t) - t(t - 5)$

19. $\dfrac{5}{3x} - \dfrac{7}{2x}$

20. $\dfrac{9}{3y} + \dfrac{11}{6z}$

1. _____
2. _____
3. _____
4. _____
5. _____
6. _____
7. _____
8. _____
9. _____
10. _____
11. _____
12. _____
13. _____
14. _____
15. _____
16. _____
17. _____
18. _____
19. _____
20. _____

21. $\dfrac{5}{6x^2y} - \dfrac{9}{10xy^3}$

22. $\dfrac{5}{6x^2y} \cdot \dfrac{9}{10xy^3}$

23. $(x + 3)(x - 12) + (x + 6)^2$

24. $(x - 8)(x - 2) - (x - 4)^2$

25. $\left(8 \cdot \dfrac{4}{x}\right) \div \dfrac{16}{x^2}$

26. $\left(\dfrac{5}{a} - \dfrac{1}{a}\right) \cdot \dfrac{a^2}{12}$

27. $(a - 3)(2a + 3)(2a - 3)$

28. $(x + y)(x - 1)(x - y)(x + 1)$

29. $2 + \dfrac{3}{x} - \dfrac{1}{x^2}$

30. $\dfrac{5}{st} - \dfrac{1}{6s^2} + \dfrac{3}{8st^2}$

31. Find the sum of $x^2 - xy + 3y^2$, $5x^2 - 8y^2$, and $y^2 - 6x^2$.

32. Subtract $3a^3 - 4a + 7$ from $a^3 - a - 2$.

33. (a) What is the degree of the polynomial $5x^4 - 3x^2 + 6x - 1$?

　　(b) What is the coefficient of the second-degree term?

34. Write the polynomial $4 - x + 3x^3$ in complete standard form.

In Exercises 35–40, divide the given polynomials. Use long division where necessary.

35. $\dfrac{x^2 + 8x}{2x}$

36. $\dfrac{8r^2s - 12rs^3 - 4r^3s^2}{6r^2s^2}$

37. $\dfrac{y^2 - 3y + 4}{y - 2}$

38. $\dfrac{2a^3 - 5a^2 - 5a + 6}{a - 3}$

39. $\dfrac{18x^3 - 5x - 28}{3x - 4}$

40. $\dfrac{t^4 - t^2 - 6}{t + 2}$

In Exercises 41–52, *simplify the expression as completely as possible. Final answers should be expressed with positive exponents only.*

41. $5^0 + 2^{-3} + 2^{-4}$

42. $3^{-1} + 6 \cdot 3^{-2}$

43. $\dfrac{(x^2)^3}{x^2 x^3}$

44. $\dfrac{a^{-5}}{a^{-4}}$

45. $\dfrac{(2x^3)^4}{4(x^5)^3}$

46. $\dfrac{(x^{-2}y^3)^{-4}}{(xy^{-2})^{-5}}$

47. $\dfrac{(3a^{-3}t^2)^{-3}}{(a^{-1}t^{-2})^2}$

48. $\left(\dfrac{5x^{-2}y^3}{x^{-1}y^{-2}}\right)^{-3}$

49. *Write in scientific notation.*   .000439

50. *Write in scientific notation.*   578,000

51. $\dfrac{(4 \times 10^{-3})(5 \times 10^4)}{2 \times 10^{-3}}$

52. *Evaluate.* $\dfrac{(.0006)(4,000)}{(.024)(50,000)}$

In Exercises 53–60, *solve the given equation.*

53. $\dfrac{x}{3} - \dfrac{x}{4} = \dfrac{x - 4}{6}$

54. $\dfrac{t + 3}{8} + \dfrac{t - 2}{6} = \dfrac{t - 7}{12}$

55. $\dfrac{a}{5} - \dfrac{a}{6} = \dfrac{a}{30}$

56. $z - \dfrac{z}{4} = \dfrac{3z}{8}$

57. $\dfrac{7 - 2y}{4} - \dfrac{5 - 4y}{6} = \dfrac{8y + 5}{9}$

58. $\dfrac{x}{5} - \dfrac{x}{3} = \dfrac{1}{2}$

59. $.8x - .07(x - 5) = 58.75$

60. $\dfrac{2}{3}(x + 1) - \dfrac{1}{2}(x - 7) = 7$

ANSWERS

41. _____

42. _____

43. _____

44. _____

45. _____

46. _____

47. _____

48. _____

49. _____

50. _____

51. _____

52. _____

53. _____

54. _____

55. _____

56. _____

57. _____

58. _____

59. _____

60. _____

ANSWERS

61. _____

62. _____

63. _____

64. _____

65. _____

66. _____

67. _____

68. _____

69. _____

70. _____

71. _____

72. _____

73. _____

74. _____

75. _____

76. _____

77. _____

78. _____

79. _____

80. _____

81. _____

82. _____

83. _____

84. _____

85. _____

In Exercises 61–80, factor the polynomial as completely as possible.

61. $x^2 + 6x + 5$ **62.** $x^2 + 6x$ **63.** $x^2 - 5x + 6$

64. $x^2 - 5x - 6$ **65.** $6x^3y - 12xy^2 - 9x^2y$ **66.** $10m^3n^5 - 5m^2n^3$

67. $u^2 - 49$ **68.** $4a^2 - 24a + 36$ **69.** $2r^2 + r - 15$

70. $t^4 - 36t^2$ **71.** $5t^2 + 10t + 15$ **72.** $x^3y - xy^3$

73. $6x^2 - 17xy + 12y^2$ **74.** $24 + 10x - x^2$ **75.** $x^2 + 16x$

76. $x^2 + 16$ **77.** $x^2 + ax + xy + ay$ **78.** $a^2 - 3a + az - 3z$

79. $x^2 - 4x - ax + 4a$ **80.** $x^8 - y^8$

81. A total of $2,850 was collected on the sale of 360 tickets to an art show. If some of the tickets cost $6.25 each and the rest cost $8.75 each, how many of each type were sold?

82. If there are 454 grams in 1 pound, how many grams are there in 15 ounces? [*Note:* There are 16 ounces in 1 pound.]

83. During an election the ratio of the number of people voting for party A to those voting for party B was 8 to 5. If party B received 15,700 votes, how many votes did party A receive?

84. Irma is going to make three investments paying 8%, 9%, and 10%. She will invest twice as much at 9% as at 8%, and $1,000 more at 10% than at 9%. If the yearly interest from the three investments is $3,090, how much is invested all together?

85. How long would it take someone driving at 80 kph to overtake someone with a 15-minute head start driving at 65 kph?

CUMULATIVE PRACTICE TEST
Chapters 4–6

In Problems 1–16, perform the indicated operations and simplify as completely as possible. Final answers should be reduced to lowest terms, and be expressed with positive exponents only.

1. $(x - 2y)(x^2 - 3xy - y^2)$

2. $\dfrac{4x^{-8}y^6}{6x^{-4}y^{-2}}$

3. $\dfrac{12s^2t^3}{5d^2} \cdot \dfrac{15d^5}{9st^4}$

4. $2a(3a - 5) + (a - 6)(a - 4)$

5. $(x - 5)^2 - (x + 5)^2$

6. $\dfrac{11a}{9x} - \dfrac{a}{9x} + \dfrac{5a}{9x}$

7. $\dfrac{3x - 7}{2y} - \dfrac{9x - 7}{2y}$

8. $\dfrac{(2x^3)^{-4}}{(x^{-3})^3}$

9. $\dfrac{5}{6ab^2} + \dfrac{4}{9b}$

10. $\dfrac{12s^2t^5 - 8s^3t^2}{6s^3t^3}$

11. Use long division to find the quotient and remainder. $\dfrac{4x^3 - 3x^2 + 5x - 20}{x - 2}$

12. Use long division to find the quotient and remainder. $\dfrac{x^4 - x^2 - 12}{x + 2}$

13. $\dfrac{16x^3y^5}{9z^4} \div (36xyz)$

14. Write in scientific notation.

(a) .000916

(b) 916,000

15. Compute using scientific notation. $\dfrac{(.008)(25,000)}{(6,000)(.00015)}$

ANSWERS

1. _____

2. _____

3. _____

4. _____

5. _____

6. _____

7. _____

8. _____

9. _____

10. _____

11. _____

12. _____

13. _____

14. a. _____

b. _____

15. _____

ANSWERS

16. _____

17. _____

18. _____

19. _____

20. _____

21. _____

22. _____

23. _____

24. _____

25. _____

26. _____

27. _____

28. _____

29. _____

30. _____

16. Subtract the sum of $x^3 - 3x$ and $x^2 - 5x$ from $x^3 - x^2 - x + 1$.

17. *Solve for a.* $\dfrac{a}{6} - \dfrac{a}{9} = 18$

18. *Solve for t.* $\dfrac{2t - 3}{4} - \dfrac{t - 2}{8} = t + 2$

In Exercises 19–26, factor the given polynomial as completely as possible.

19. $x^2 - 10x - 24$

20. $6a^2b^5 - 3ab^3$

21. $2t^2 + 5t - 12$

22. $6x^2 - 36x + 72$

23. $3x^3y - 12xy^3$

24. $a(a + 5) - 7(a + 5)$

25. $x^2 - 3x - xy + 3y$

26. $2u^4 - 32$

Solve each of the following problems algebraically. Be sure to clearly label what your variable represents.

27. The width of a rectangle is 5 more than one-third its length. If the perimenter is 34 meters, what are its dimensions?

28. Jamie bought 28 stamps for a total of $5.35. Some were 22¢ stamps and the rest were 13¢ stamps. How many of each were bought?

29. A computer consultant charges $40 per hour for her time, and $24 per hour for her assistant's time. On a certain project the assistant worked for 4 hours, after which time he was joined by the consultant and they completed the job together. If the total bill for the job was $480, how many hours did they work together?

30. Terry and Tom leave together from the same location walking in opposite directions. Terry leaves at 11:00 A.M. walking at 6 kph. Twenty minutes later Tom leaves walking at 8 kph. At what time will they be 9 kilometers apart?

More Rational Expressions

Reducing Rational Expressions

7.1

✔ LEARNING CHECKS

1. $\dfrac{6x^2 - 20x}{10x}$

In Chapter 4 we went through a rather detailed discussion of rational expressions. While we did not mention it explicitly at the time, our presentation was focused primarily on rational expressions whose numerators and denominators were monomials. To a great extent, that restriction automatically simplified much of our work.

In this chapter we are going to repeat much of what we did in Chapter 4, only this time we will be working with fractions whose numerators and/or denominators are usually polynomials of more than one term. For example, we will be working with fractions such as

$$\frac{3x^2}{x^2 + 3} \qquad \text{and} \qquad \frac{x + 2}{x^2 - x - 6}$$

As you will see very shortly, the factoring techniques we learned in Chapter 6 will play an extremely important role. As we stressed in Section 4.1, the basic idea underlying almost all our work with fractions is the Fundamental Principle of Fractions.

FUNDAMENTAL PRINCIPLE OF FRACTIONS

$$\frac{a}{b} = \frac{a \cdot k}{b \cdot k} \quad \text{where } b, k \neq 0$$

In words again, the Fundamental Principle says two things:

- Reading from left to right, it says that the value of a fraction is not changed when both numerator and denominator are *multiplied* by the same nonzero *factor*. We call this process **building the fraction.**
- Reading from right to left, it says that a common *factor* can be cancelled from both the numerator and the denominator without changing the value of the fraction. We call this process **reducing the fraction.**

The crucial idea is that we can build or reduce with common factors, but *not* with common terms.

EXAMPLE 1 *Reduce to lowest terms.* $\dfrac{3x^2 + 12x}{12x}$

Solution This type of example is the scene of one of the most common errors in algebra—that of improperly reducing a fraction. The following common "solutions" or variations thereof are *incorrect*!

Incorrect solution 1 $\dfrac{3x^2 + 12x}{12x} = \dfrac{3x^2 + \cancel{12x}}{\cancel{12x}} = 3x^2 \quad \text{or} \quad 3x^2 + 1$

Incorrect solution 2 $\dfrac{3x^2 + 12x}{12x} = \dfrac{\overset{x}{\cancel{3x^2}} + 12x}{\underset{4}{\cancel{12x}}} = \dfrac{x + 12x}{4} = \dfrac{13x}{4}$

These "solutions" are wrong because they both attempt to reduce a *term* with a *factor*. In this example, the $12x$ in the denominator is a factor, while the $12x$ in the numerator is a term.

The Fundamental Principle says that we can reduce *common factors only*. Whatever it is that we are trying to reduce must be a common *factor* of both the numerator *and* the denominator.

Now that we have seen two methods that do not work, what is the correct approach? Since the Fundamental Principle allows us to reduce only common factors, we should try to factor the numerator.

$$\frac{3x^2 + 12x}{12x} = \frac{3x(x + 4)}{12x} \qquad \textbf{\textit{Take out the common factor of }} 3x.$$

$$= \frac{\cancel{3}x(x + 4)}{\underset{4}{\cancel{12}x}}$$

$$= \boxed{\frac{x + 4}{4}}$$

Note that this is the final answer. The 4's cannot be reduced because the 4 in the numerator is not a factor. ∎

It is important to point out that the various procedures we learn are not isolated ideas but often merely variations on a theme. For instance, we have just completed an example on reducing fractions. We could also have looked at Example 1 and thought of it as an example on dividing polynomials. So we could have attempted a solution as was described in Section 6.5.

$$\frac{3x^2 + 12x}{12x} = \frac{3x^2}{12x} + \frac{12x}{12x} \qquad \textbf{\textit{We divided each term in the numerator by }} 12x.$$

$$= \frac{\overset{x}{\cancel{3}x^2}}{\underset{4}{\cancel{12}x}} + \frac{12x}{12x} \qquad \textbf{\textit{Reduce each fraction individually.}}$$

$$= \boxed{\frac{x}{4} + 1}$$

Is this answer of $\frac{x}{4} + 1$ the same as our first answer of $\frac{x + 4}{4}$? If we take our second answer and combine the two terms by finding a common denominator, we get

$$\frac{x}{4} + 1 = \frac{x}{4} + \frac{4}{4} = \frac{x + 4}{4}$$

and we see that our two answers are in fact the same. Recognizing that they are actually two forms of the same expression makes it even less likely that you will try to reduce $\frac{x + 4}{4}$ by cancelling the 4's.

EXAMPLE 2 *Reduce to lowest terms.* $\dfrac{2y^2 + 4y}{y^2 - 4}$

2. $\dfrac{5a^2 + 25a}{a^2 - 25}$

Solution As you can see, it almost seems as if these examples are designed to lead you astray (look how neatly the y^2 is lined up over the y^2, and the $4y$ over the 4). Are you tempted to cancel? *You must be strong and resist the temptation.*

> **REMEMBER**
>
> Reduce only common *factors*, not common terms.

We must have both the numerator and the denominator in factored form *before* we can reduce. You may recall our mentioning in Chapter 6 that there are a number of situations in which it is advantageous to have an expression in factored form. Working with fractions is clearly one such situation.

$$\frac{2y^2 + 4y}{y^2 - 4}$$

In the numerator we factor out the common factor of 2y. The denominator factors as the difference of two squares.

$$= \frac{2y(y + 2)}{(y + 2)(y - 2)}$$

We cancel the entire common factor of y + 2 from numerator and denominator.

$$= \boxed{\frac{2y}{y - 2}}$$

    As we have seen previously, the factoring process can sometimes involve more than one step.

3. $\dfrac{4y^2 + 32y + 64}{y^3 - y^2 - 20y}$

EXAMPLE 3   *Reduce to lowest terms.* $\dfrac{2x^2 - 12x + 18}{x^3 - x^2 - 6x}$

Solution   Whenever we try to factor, we always look for common factors first.

$$\frac{2x^2 - 12x + 18}{x^3 - x^2 - 6x} = \frac{2(x^2 - 6x + 9)}{x(x^2 - x - 6)}$$

Now we try to factor further.

$$= \frac{2(x - 3)(x - 3)}{x(x - 3)(x + 2)}$$

$$= \frac{2(x - 3)(x - 3)}{x(x - 3)(x + 2)}$$

$$= \boxed{\frac{2(x - 3)}{x(x + 2)}} \quad \text{or} \quad \boxed{\frac{2x - 6}{x^2 + 2x}}$$

Either answer is acceptable.

4. $\dfrac{z^2 - 8z - 20}{z^2 - z - 20}$

EXAMPLE 4   *Reduce to lowest terms.* $\dfrac{x^2 + 6x + 8}{x^2 - 3x - 4}$

Solution   We cannot reduce this fraction in its present form so we try to factor it first.

$$\frac{x^2 + 6x + 8}{x^2 - 3x - 4} = \frac{(x + 4)(x + 2)}{(x - 4)(x + 1)}$$

This *cannot* be reduced since there are no common factors. Therefore, the original fraction is already in lowest terms. Our answer is either

$$\boxed{\frac{x^2 + 6x + 8}{x^2 - 3x - 4}} \quad \text{or} \quad \boxed{\frac{(x + 4)(x + 2)}{(x - 4)(x - 1)}}$$

As usual, there may be some preliminary steps before we are ready to factor.

EXAMPLE 5 *Reduce to lowest terms.* $\dfrac{x^3 + 2x^2 + 3x^2}{x^2 + 3x - 5 + x}$

5. $\dfrac{2x^2 - 4x - x^2 - 12}{x^3 - 4x^2 - 2x^2}$

Solution Before plunging ahead we should take a moment to recognize that both the numerator and the denominator can be simplified by combining like terms.

$$\frac{x^3 + 2x^2 + 3x^2}{x^2 + 3x - 5 + x} = \frac{x^3 + 5x^2}{x^2 + 4x - 5} \qquad \textit{Now we factor numerator and denominator.}$$

$$= \frac{x^2(x + 5)}{(x + 5)(x - 1)}$$

$$= \frac{x^2\cancel{(x + 5)}}{\cancel{(x + 5)}(x - 1)}$$

$$= \boxed{\frac{x^2}{x - 1}} \qquad\qquad\qquad ■$$

EXAMPLE 6 *Reduce to lowest terms.* $\dfrac{x^3 y^2}{x^3 y - xy^3}$

6. $\dfrac{a^2 b^4}{a^3 b - 4ab^3}$

Solution We see that the numerator consists of factors but that the denominator does not. We begin by taking out the common factor of xy from the denominator.

$$\frac{x^3 y^2}{x^3 y - xy^3} = \frac{x^3 y^2}{xy(x^2 - y^2)} \qquad \textit{We can factor the denominator further.}$$

$$= \frac{x^3 y^2}{xy(x - y)(x + y)}$$

$$= \frac{\overset{x^2}{\cancel{x^3}}\overset{y}{\cancel{y^2}}}{\cancel{xy}(x - y)(x + y)}$$

$$= \boxed{\frac{x^2 y}{(x - y)(x + y)}} \quad \text{or} \quad \boxed{\frac{x^2 y}{x^2 - y^2}}$$

If you are looking at this solution and asking yourself what was the point of factoring $x^2 - y^2$, give yourself a pat on the back. In fact, if we recognize in the second step that there is no possibility of cancelling one of the *binomial* factors of $x^2 - y^2$ with a *monomial* factor in the numerator, then we can omit this step entirely. However—*be careful*. It is better to factor an expression and not use the factorization, than not to factor and consequently miss an opportunity to reduce it.   ■

One last comment: Keep in mind that back in Chapter 6, just factoring an expression was an entire problem in and of itself. We have now reached the point where factoring is just one part of a larger problem.

✔ *Answers to Learning Checks in Section 7.1*

1. $\dfrac{3x - 10}{5}$ **2.** $\dfrac{5a}{a - 5}$ **3.** $\dfrac{4(y + 4)}{y(y - 5)}$ **4.** $\dfrac{(z - 10)(z + 2)}{(z - 5)(z + 4)}$ **5.** $\dfrac{x + 2}{x^2}$

6. $\dfrac{ab^3}{a^2 - 4b^2}$

NOTE TO THE STUDENT

Use the space on this page to write down any questions you have or points you want to review with your instructor.

Exercises 7.1

Reduce each of the following as completely as possible.

1. $\dfrac{8x^3y^{10}}{10x^6y^5}$

2. $\dfrac{12a^8b^3}{9a^4b^9}$

3. $\dfrac{5x - 7x}{x^2 - 7x^2}$

4. $\dfrac{3m^3 - 8m^3}{6m^2 - 4m^2}$

5. $\dfrac{6x^2(x + 4)^5}{9x^3(x + 4)}$

6. $\dfrac{15y^4(y - 2)^2}{10y(y - 2)^6}$

7. $\dfrac{12a^2b + 6c^3}{8a^2b + 4c^3}$

8. $\dfrac{12a^2b(6c^3)}{8a^2b(4c^3)}$

9. $\dfrac{3x - 6}{5x - 10}$

10. $\dfrac{4x + 8}{7x + 14}$

11. $\dfrac{3x - 6}{6x - 12}$

12. $\dfrac{4x + 8}{12x - 24}$

ANSWERS

13. $\dfrac{3x - 6}{6x + 12}$

14. $\dfrac{4x + 8}{12x + 24}$

13. _____

14. _____

15. $\dfrac{5y}{10y + 20}$

16. $\dfrac{-3a}{6a - 18}$

15. _____

16. _____

17. $\dfrac{6x + 18}{x^2 - 9}$

18. $\dfrac{8x - 16}{x^2 - 4}$

17. _____

18. _____

19. $\dfrac{6x^2 + 18}{x^2 - 9}$

20. $\dfrac{8x^2 - 16}{x^2 - 4}$

19. _____

20. _____

21. $\dfrac{t^2 + 3t}{t^2 + 3t - 10}$

22. $\dfrac{m^2 - 2m}{m^2 - 2m - 8}$

21. _____

22. _____

23. $\dfrac{2x^2 + x + x}{4x^3 + 4x^2}$

24. $\dfrac{7x - x}{3x^2 + x^2}$

23. _____

24. _____

25. $\dfrac{y^2 - 5y - 6}{y^2 - 12y + 36}$

26. $\dfrac{y^2 - 8y + 16}{y^2 - 10y + 24}$

27. $\dfrac{s^2 - 2s - 15}{s^2 - 6s + 5}$

28. $\dfrac{z^2 + 10z + 24}{z^2 + 5z + 4}$

29. $\dfrac{x^2(x + 3)(x - 4)}{x^4 - 4x^3}$

30. $\dfrac{2y(y + 2)(y - 5)}{4y^3 + 8y^2}$

31. $\dfrac{3a^2 + a - 2}{a^2 - a - 2}$

32. $\dfrac{2c^2 - 5c + 3}{c^2 - 2c - 3}$

33. $\dfrac{4x^2 + 7x - 2}{x^2 + 4x + 4}$

34. $\dfrac{6x^2 - 7x - 5}{4x^2 - 1}$

35. $\dfrac{x^2 - x + 3x - 8}{x^2 + 4x}$

36. $\dfrac{a^2 - 2a - 10 - a}{a^2 - 2a + a^2 - 8a}$

37. $\dfrac{6x^2 - 12x - 18}{3x^2 - 9x - 30}$

38. $\dfrac{4x^2 + 12x - 16}{2x^2 + 8x}$

25. _____

26. _____

27. _____

28. _____

29. _____

30. _____

31. _____

32. _____

33. _____

34. _____

35. _____

36. _____

37. _____

38. _____

ANSWERS

39. $\dfrac{6x^2 - 5x^2 - 4}{x^2 - 6x + 8}$

40. $\dfrac{6x^2 - 23x - 4}{x^2 - 6x + 8}$

39. _____

41. $\dfrac{x^2 - 7x + 10}{x^2 - 7x + 12}$

42. $\dfrac{x^2 - 9x + 18}{x^2 - 9x + 14}$

40. _____

41. _____

43. $\dfrac{y^3 - y^2 - 2y}{6y^2 - 24}$

44. $\dfrac{z^3 - 2z^2 - 3z}{12z^2 + 36z}$

42. _____

45. $\dfrac{c^2 - 9c}{c^3 - 9c^2}$

46. $\dfrac{x^4 - 9x^2}{4x^2 - 8x - 12}$

43. _____

44. _____

QUESTION FOR THOUGHT

45. _____

1. A student was asked to reduce two fractions and proceeded as follows:

   (a) $\dfrac{x^2 + 5}{x} = \dfrac{\cancel{x^2}^{x} + 5}{\cancel{x}} = x + 5$

   (b) $\dfrac{x^2 + y^2}{x + y} = \dfrac{\cancel{x^2}^{x} + \cancel{y^2}^{y}}{\cancel{x} + \cancel{y}} = x + y$

   Discuss what the student did in each case and whether or not it was correct.

46. _____

Multiplying and Dividing Rational Expressions

7.2

In the last section we saw how to put our factoring techniques to use in reducing rational expressions. Here we will use these techniques to take another look at multiplying and dividing algebraic fractions.

We will continue to operate under the same ground rules we established earlier. That is, whenever we work with fractions, we expect our final answers to be reduced to lowest terms.

As always when discussing fractions, variables are allowed only those values for which the denominator is not equal to 0.

EXAMPLE 1 *Multiply and simplify.* $\dfrac{x^2}{6} \cdot \dfrac{10}{x^2 - 5x}$

Solution As we saw in Chapter 4, we much prefer to reduce *before* we multiply. This preference is even stronger now that the expressions which appear are more complex than before.

$$\dfrac{x^2}{6} \cdot \dfrac{10}{x^2 - 5x} \qquad \text{\textit{We begin by factoring the second denominator.}}$$

$$= \dfrac{x^2}{6} \cdot \dfrac{10}{x(x - 5)} \qquad \text{\textit{Reduce common factors.}}$$

$$= \dfrac{\overset{x}{\cancel{x^2}}}{\underset{3}{\cancel{6}}} \cdot \dfrac{\overset{5}{\cancel{10}}}{\cancel{x}(x - 5)}$$

$$= \boxed{\dfrac{5x}{3(x - 5)}} \quad \text{or} \quad \boxed{\dfrac{5x}{3x - 15}}$$

Note that even though we could have reduced the 6 with the 10 immediately, we prefer to have everything in factored form before we begin to cancel. ∎

EXAMPLE 2 *Multiply and simplify.* $\dfrac{4x}{x^2 - 4} \cdot \dfrac{x^2 - 5x + 6}{4x^2 - 12x}$

Solution We begin by factoring wherever possible so that we can reduce before we multiply out the polynomials.

$$\dfrac{4x}{x^2 - 4} \cdot \dfrac{x^2 - 5x + 6}{4x^2 - 12x} = \dfrac{4x}{(x + 2)(x - 2)} \cdot \dfrac{(x - 3)(x - 2)}{4x(x - 3)}$$

$$= \dfrac{\cancel{4x}}{(x + 2)\cancel{(x - 2)}} \cdot \dfrac{\cancel{(x - 3)}\cancel{(x - 2)}}{\cancel{4x}\cancel{(x - 3)}}$$

$$= \boxed{\dfrac{1}{x + 2}} \qquad \text{\textit{Do not forget the understood}} \atop \text{\textit{factor of 1 in the numerator.}} \quad ∎$$

EXAMPLE 3 *Multiply and simplify.* $5x \cdot \dfrac{x^2}{x^2 - 5x}$

✔ **LEARNING CHECKS**

1. $\dfrac{a^3}{9} \cdot \dfrac{12}{a^2 + 6a}$

2. $\dfrac{3x}{x^2 - 9} \cdot \dfrac{x^2 + 3x - 18}{3x^2 + 18x}$

3. $8x^2 \cdot \dfrac{x}{x^4 - 8x^2}$

Solution Since this example involves fractions, we write down the understood denominator of 1.

$$5x \cdot \frac{x^2}{x^2 - 5x} = \frac{5x}{1} \cdot \frac{x^2}{x^2 - 5x} \qquad \text{Factor the denominator.}$$

$$= \frac{5x}{1} \cdot \frac{x^2}{x(x - 5)} \qquad \text{Reduce.}$$

$$= \frac{5\cancel{x}}{1} \cdot \frac{x^2}{\cancel{x}(x - 5)}$$

$$= \boxed{\frac{5x^2}{x - 5}}$$

Note that we chose to cancel the factor of x in the denominator with the factor of x in the first numerator. We could just as well have reduced it with the x^2 in the second numerator. ∎

Now let's turn our attention to division.

4. $\dfrac{x^2 - 4y^2}{x^2 + 3xy + 2y^2} \div \dfrac{4x - 8y}{2x + 4y}$

──────────

EXAMPLE 4 *Divide and simplify.* $\dfrac{x^2 - y^2}{x^2 - 2xy + y^2} \div \dfrac{12x + 12y}{6x - 6y}$

Solution We follow the rule for dividing fractions which says "invert the divisor and multiply."

$$\frac{x^2 - y^2}{x^2 - 2xy + y^2} \div \frac{12x + 12y}{6x - 6y} = \frac{x^2 - y^2}{x^2 - 2xy + y^2} \cdot \frac{6x - 6y}{12x + 12y} \qquad \text{Next factor.}$$

$$= \frac{(x + y)(x - y)}{(x - y)(x - y)} \cdot \frac{6(x - y)}{12(x + y)} \qquad \text{Now reduce.}$$

$$= \frac{\cancel{(x + y)}\cancel{(x - y)}}{\cancel{(x - y)}\cancel{(x - y)}} \cdot \frac{\cancel{6}(x - y)}{\underset{2}{\cancel{12}}(x + y)}$$

$$= \boxed{\frac{1}{2}} \qquad\qquad\qquad ∎$$

5. $\dfrac{3a^2 - 2a - 8}{a^2 - 4a + 4} \div (15a + 20)$

──────────

EXAMPLE 5 *Divide and simplify.* $\dfrac{2x^2 + x - 15}{x^2 + 6x + 9} \div (8x - 20)$

Solution We think of $8x - 20$ as $\dfrac{8x - 20}{1}$, and follow the rule for division.

$$\frac{2x^2 + x - 15}{x^2 + 6x + 9} \div \frac{8x - 20}{1} = \frac{2x^2 + x - 15}{x^2 + 6x + 9} \cdot \frac{1}{8x - 20} \qquad \text{Next factor.}$$

$$= \frac{(2x - 5)(x + 3)}{(x + 3)(x + 3)} \cdot \frac{1}{4(2x - 5)} \qquad \text{Now reduce.}$$

$$= \frac{\cancel{(2x - 5)}\cancel{(x + 3)}}{(x + 3)\cancel{(x + 3)}} \cdot \frac{1}{4\cancel{(2x - 5)}}$$

$$= \boxed{\frac{1}{4(x + 3)}} \quad \text{or} \quad \boxed{\frac{1}{4x + 12}} \qquad ∎$$

As we have pointed out on numerous occasions, it is very important to read a problem carefully before plunging into it. See what the problem involves so that you can apply the *appropriate* procedures. Keep this in mind as you study the next two examples and their solutions.

EXAMPLE 6 *Perform the indicated operations and simplify.*

$$\left(\frac{8}{x} \cdot \frac{x}{2}\right) \div \frac{x+4}{16}$$

Solution We work inside the parentheses first—the operation within the parentheses is *multiplication*.

$$\left(\frac{8}{x} \cdot \frac{x}{2}\right) \div \frac{x+4}{16} = \left(\frac{\overset{4}{\cancel{8}}}{\cancel{x}} \cdot \frac{\cancel{x}}{\cancel{2}}\right) \div \frac{x+4}{16}$$

$$= 4 \div \frac{x+4}{16}$$

$$= \frac{4}{1} \cdot \frac{16}{x+4} \qquad \textit{We cannot reduce anything, so we multiply.}$$

$$= \boxed{\frac{64}{x+4}}$$

EXAMPLE 7 *Perform the indicated operations and simplify.*

$$\left(\frac{8}{x} - \frac{x}{2}\right) \div \frac{x+4}{16}$$

Solution Again we begin to work inside the parentheses—this time the operation within the parentheses is *subtraction*, which requires a common denominator. The common denominator within the parentheses is $2x$.

$$\left(\frac{8}{x} - \frac{x}{2}\right) \div \frac{x+4}{16} = \left(\frac{8(2)}{2x} - \frac{x(x)}{2x}\right) \div \frac{x+4}{16} \qquad \textit{Subtract the fractions.}$$

$$= \frac{16 - x^2}{2x} \div \frac{x+4}{16} \qquad \textit{Next use the rule for division.}$$

$$= \frac{16 - x^2}{2x} \cdot \frac{16}{x+4} \qquad \textit{Now factor.}$$

$$= \frac{(4+x)(4-x)}{2x} \cdot \frac{16}{x+4} \qquad \textit{Next reduce.}$$

$$= \frac{\cancel{(4+x)}(4-x)}{\cancel{2}x} \cdot \frac{\overset{8}{\cancel{16}}}{\cancel{x+4}} \qquad \begin{array}{l}\textit{Note that } x+4 \textit{ and} \\ 4+x \textit{ are the same.}\end{array}$$

$$= \boxed{\frac{8(4-x)}{x}} \quad \text{or} \quad \boxed{\frac{32-8x}{x}} \qquad \blacksquare$$

Even though Examples 6 and 7 look very similar, they are actually quite different. Again we emphasize that you must read problems carefully to understand what the problem is asking and what procedures are involved.

If you had difficulty with carrying out the subtraction in the first step of Example 7, it is strongly suggested that you review Section 4.3 before going on to the next section.

6. $\left(\dfrac{y}{12} \div \dfrac{3}{y}\right) \cdot \dfrac{6}{y+6}$

7. $\left(\dfrac{y}{12} - \dfrac{3}{y}\right) \cdot \dfrac{6}{y+6}$

Exercises 7.2

In each of the following exercises, perform the indicated operations and simplify as completely as possible.

1. $\dfrac{8x^2y^3}{9yz^3} \cdot \dfrac{12a^2z}{2ax^2}$

2. $\dfrac{10r^3s}{6mr} \cdot \dfrac{9m^4}{4mrs}$

3. $\dfrac{3st^2}{5p} \div \dfrac{15t^2}{s^2}$

4. $\dfrac{4s^3t}{7a} \div \dfrac{14s^2}{t}$

5. $\dfrac{12x^2y^3}{5z} \cdot (30xyz)$

6. $\dfrac{18x^3y^2}{7z^4} \div (14x^2y^2z^2)$

7. $(28a^2b^3z^4) \div \dfrac{4a}{7b}$

8. $(28a^2b^3z^4) \cdot \dfrac{4a}{7b}$

9. $\dfrac{x^2 + 4x}{x^2} \cdot \dfrac{x}{x^2 + 6x + 5}$

10. $\dfrac{y^2 - 4y}{y^4} \cdot \dfrac{y^3}{y^2 - 3y - 4}$

11. $\dfrac{x^2 + 4x}{x^2 + 4x + 4} \cdot \dfrac{x^2 - 4}{x^2 - 4x + 4}$

12. $\dfrac{a^2 - 4a}{a^2 - 4a - 12} \cdot \dfrac{a^2 + 5a + 6}{a^2 - a - 12}$

1. _____

2. _____

3. _____

4. _____

5. _____

6. _____

7. _____

8. _____

9. _____

10. _____

11. _____

12. _____

ANSWERS

13. $\dfrac{r^2 - 4r - 5}{2r - 10} \div \dfrac{r^2 - 3r + 2}{4r^2}$

14. $\dfrac{m^2 + 3m - 10}{5m^2} \div \dfrac{m^2 + 5m}{m^2 + 10m + 25}$

13. _____

14. _____

15. $\dfrac{m^2}{m^2 + 3m} \div \dfrac{3m^2}{m^2 + 6m}$

16. $\dfrac{3x^2}{x^2 + 3x} \div \dfrac{x^2 + 6x}{6x^2}$

15. _____

16. _____

17. $\dfrac{x^2 + 3x + 2}{x^2 + 2x} \cdot \dfrac{x}{x^2 + 2}$

18. $\dfrac{x^2 - 3x + 2}{x^2 - 2x} \cdot \dfrac{2x}{x^2 - 2x}$

17. _____

18. _____

19. $\dfrac{y^2 - 3y - 4}{y^2 - 2y - 8} \cdot \dfrac{y^2 + 4y + 4}{y^2 - 8y + 16}$

20. $\dfrac{y^2 - 3y - 4}{y^2 - 2y - 8} \div \dfrac{y^2 + 4y + 4}{y^2 - 8y + 16}$

19. _____

20. _____

21. $\dfrac{x^2 + 2x}{x^2 - x - 2} \cdot \dfrac{x - 2}{x}$

22. $\dfrac{x^2 - 4x - 5}{2x^2 - 10x} \div \dfrac{x + 1}{8x}$

21. _____

22. _____

23. $\dfrac{2x^2 + x - 15}{x^2 - 9} \cdot \dfrac{6x^2 + 7x + 1}{2x^2 - 3x - 5}$

24. $\dfrac{4z^2 + 12z + 5}{2z^2 + z - 1} \div \dfrac{4z^2 - 25}{2z^2 - 3z + 1}$

25. $\dfrac{4a}{a + 4} \cdot \dfrac{a + 5}{5a} \div \dfrac{a^2 + 6a + 5}{a^2 + 5a + 4}$

26. $\dfrac{3y}{y - 3} \div \dfrac{y^2 - 3y + 2}{y^2 - 4y + 3} \cdot \dfrac{y - 2}{2y}$

27. $\dfrac{x}{2} \div \dfrac{2}{x} \cdot \dfrac{x^2 - 16}{x^2 - 4x}$

28. $\dfrac{a}{8} \div \left(\dfrac{8}{a^2} \div \dfrac{a^2 - 64}{a^2 - 8a} \right)$

29. $\dfrac{t^2 + 2t}{t^2 + 2t + 1} \div \dfrac{2t^2 + 7t + 6}{t^2 + t}$

30. $\dfrac{u^2 + 6u + 9}{u^2 + 9} \cdot \dfrac{4u^2 + 36}{2u + 6}$

31. $\dfrac{2x^2 + 6x + 4x}{x^2 - 25} \cdot \dfrac{(x + 5)^2}{4x - 2x}$

32. $\dfrac{z^2 + 5z + z}{(z + 6)^2} \div \dfrac{z^2 + z}{z^2 + 7z + 6}$

ANSWERS

23. _____

24. _____

25. _____

26. _____

27. _____

28. _____

29. _____

30. _____

31. _____

32. _____

ANSWERS

33. $\dfrac{x^3y - xy^3}{8x^2y + 4xy^2} \div \dfrac{(x - y)^2}{2x^2 + 3xy + y^2}$

34. $\dfrac{r^3t - rt^2}{rt^2 - r^2t} \cdot \dfrac{t^2 - 2rt + r^2}{r^2 - t}$

33. _____

34. _____

35. $\left(\dfrac{x}{3} \cdot \dfrac{x}{4}\right) \cdot \dfrac{12}{x^2 - 3x}$

36. $\left(\dfrac{x}{3} - \dfrac{x}{4}\right) \cdot \dfrac{12}{x^2 - 3x}$

35. _____

36. _____

37. $\left(\dfrac{c}{2} + \dfrac{c}{5}\right) \div \dfrac{c^2 + 7c}{10}$

38. $\left(\dfrac{c}{2} \cdot \dfrac{c}{5}\right) \div \dfrac{c^2 + 7c}{10}$

37. _____

38. _____

39. $\left(\dfrac{w}{4} - \dfrac{9}{w}\right) \cdot \left(\dfrac{w + 10}{2w} - \dfrac{5}{w}\right)$

40. $\left(\dfrac{5t + 8}{4t} - \dfrac{t + 2}{t}\right) \div \left(\dfrac{t}{8} - \dfrac{2}{t}\right)$

39. _____

40. _____

Adding and Subtracting Rational Expressions

7.3

In this section we extend the procedures we developed in Section 4.3. You are strongly urged to review that section before continuing.

You will recall that addition and subtraction of fractions are quite straight-forward when the fractions have the same denominator. We have the general rule:

$$\frac{a}{d} + \frac{b}{d} - \frac{c}{d} = \frac{a + b - c}{d}$$

In other words, if the denominators are the same just combine the numerators and keep the common denominator.

It was mentioned only briefly then that even when the denominators are the same there may still be some things to watch out for. We elaborate on this idea a bit more now.

EXAMPLE 1   *Combine and simplify.* $\dfrac{2x - 3}{3x - 6} + \dfrac{x + 3}{3x - 6}$

Solution   Since the denominators are the same, we can proceed directly using the rule for addition of fractions.

$$\frac{2x - 3}{3x - 6} + \frac{x + 3}{3x - 6} = \frac{2x - 3 + x + 3}{3x - 6} \qquad \textit{Combine like terms.}$$

$$= \frac{3x}{3x - 6}$$
We want to reduce this fraction if we can. Therefore, we must first factor the denominator.

$$= \frac{3x}{3(x - 2)}$$

$$= \frac{\cancel{3}x}{\cancel{3}(x - 2)}$$

$$= \boxed{\frac{x}{x - 2}} \qquad\qquad ∎$$

EXAMPLE 2   *Combine and simplify.* $\dfrac{5x^2 - 12}{2x^2 + 8x} - \dfrac{2x^2 + 7}{2x^2 + 8x} + \dfrac{3 - 2x^2}{2x^2 + 8x}$

Solution   Since the denominators are the same there is a tendency to "copy" across the numerators of the example. This is incorrect because the minus sign in front of the second fraction must be applied to the *entire* numerator of the second fraction.

$$\frac{5x^2 - 12}{2x^2 + 8x} - \frac{2x^2 + 7}{2x^2 + 8x} + \frac{3 - 2x^2}{2x^2 + 8x} = \frac{5x^2 - 12 - (2x^2 + 7) + 3 - 2x^2}{2x^2 + 8x}$$

The parentheses around the $2x^2 + 7$ are inserted to make sure we distribute the minus sign.

$$= \frac{5x^2 - 12 - 2x^2 \overset{\swarrow}{-} 7 + 3 - 2x^2}{2x^2 + 8x}$$

The arrow indicates where a sign error often occurs if you do not insert the parentheses in the previous step. Now we combine like terms in the numerator.

✔ LEARNING CHECKS

1. $\dfrac{5x + 4}{2x + 8} + \dfrac{3x - 4}{2x + 8}$

2. $\dfrac{a^2 + 8}{3a^2 - 6a} + \dfrac{a - 5}{3a^2 - 6a} - \dfrac{2a + 5}{3a^2 - 6a}$

$$= \frac{x^2 - 16}{2x^2 + 8x}$$

Next we factor in the hope that we can reduce.

$$= \frac{(x + 4)(x - 4)}{2x(x + 4)}$$

$$= \frac{\cancel{(x + 4)}(x - 4)}{2x\cancel{(x + 4)}}$$

$$= \boxed{\frac{x - 4}{2x}}$$

Be careful: Watch out for the sign at the spot indicated by the arrow in this example. Whenever the numerator of a fraction contains more than one term and the fraction is preceded by a minus sign, the minus sign must be distributed to *each* term in the numerator. ∎

3. $\dfrac{5z^2}{20z} + \dfrac{9}{20z}$

EXAMPLE 3 *Combine and simplify.* $\dfrac{2x^2}{10x} + \dfrac{7}{10x}$

Solution Your first instinct might be to reduce the first fraction immediately:

$$\frac{2x^2}{10x} = \frac{\overset{x}{\cancel{2x^2}}}{\underset{5}{\cancel{10x}}} = \frac{x}{5}$$

This is not wrong, it is just inappropriate. Since the example is telling us to combine the fractions, having both denominators the same is of greater importance than having each fraction reduced to lowest terms. If we reduce first we lose the common denominator. Always take a moment to think about what the problem is asking, before deciding on the steps to follow. The solution to this example is

$$\frac{2x^2}{10x} + \frac{7}{10x} = \boxed{\frac{2x^2 + 7}{10x}}$$ ∎

When we are asked to combine fractions whose denominators are not the same, we first *build* the fractions so that they have a common denominator, and then we proceed as before.

Let's review the mechanical procedure we developed in Chapter 4 for finding the least common denominator (LCD).

TO FIND THE LCD

Step 1 Factor each denominator as completely as possible.

Step 2 The LCD consists of the product of each *distinct* factor taken the maximum number of times it appears in any one denominator.

Now that we are working with polynomial denominators we have to carry out this process very carefully.

EXAMPLE 4 *Combine and simplify.* $\dfrac{7}{3x} - \dfrac{2}{x+3}$

Solution Since the denominators are not the same, our first step is to build the fractions to have the same denominators.

As usual, since we prefer to use the LCD, we follow the outline given in the box. The first step in our outline is unnecessary in this example since the denominators are already in factored form.

In order to carry out the second step, we need to identify the distinct factors. Consequently, it is crucial to be able to distinguish terms from factors.

■
$$\dfrac{7}{3x} \qquad - \qquad \dfrac{2}{x+3}$$
$$\downarrow \qquad\qquad\qquad \downarrow$$

The factors are   *The only factor is*
3 and x.   *x + 3.*

Therefore, the distinct factors are 3, x, and $x + 3$.

In $3x$ $\begin{cases} 3 \text{ appears as a } factor \text{ once.} \\ x \text{ appears as a } factor \text{ once.} \\ x + 3 \text{ appears as a } factor \text{ zero times.} \end{cases}$

In $x + 3$ $\begin{cases} 3 \text{ appears as a } factor \text{ zero times.} \\ x \text{ appears as a } factor \text{ zero times.} \\ x + 3 \text{ appears as a } factor \text{ once.} \end{cases}$

Again we must emphasize that x and 3 are not *factors* of $x + 3$.

■ Following our outline, we take each of 3, x, and $x + 3$ as a factor once for the LCD.

The LCD is $3x(x + 3)$.

Next we use the Fundamental Principle to build each of the original fractions into an equivalent one with the LCD as the denominator. We do this by filling in the missing factors in both the numerator and the denominator.

We put the LCD in each denominator. Whatever factor we have inserted in the denominator, we must also insert in the numerator.

$$\dfrac{7}{3x} - \dfrac{2}{x+3}$$

$$= \dfrac{7\,(x+3)}{3x\,(x+3)} - \dfrac{2\,(3x)}{3x\,(x+3)}$$

$\underbrace{\qquad\qquad}$ $\underbrace{\qquad\qquad}$
Here the missing   *Here the missing*
factor was x + 3.   *factor was 3x.*

$$= \dfrac{7(x+3) - 2(3x)}{3x(x+3)}$$

In the numerator, multiply out and combine like terms.

$$= \dfrac{7x + 21 - 6x}{3x(x+3)}$$

$$= \boxed{\dfrac{x + 21}{3x(x+3)}}$$

■

4. $\dfrac{5}{8x} + \dfrac{2}{x+8}$

5. $\dfrac{4}{u^2 + 6u} + \dfrac{5}{6u^2}$

EXAMPLE 5 *Combine and simplify.* $\dfrac{2}{7x^2} + \dfrac{3}{x^2 - 7x}$

Solution As before, since the denominators are not the same we want to find the LCD. However, since the denominators are not both in factored form, our first step is to factor the second denominator.

$$\frac{2}{7x^2} + \frac{3}{x^2 - 7x} = \frac{2}{7x^2} + \frac{3}{x(x - 7)}$$

Now we are ready to determine the LCD.

$$\blacksquare \qquad \frac{2}{7x^2} \quad + \quad \frac{3}{x(x - 7)}$$
$$\downarrow \qquad\qquad \downarrow$$

The factors are   *The factors are*
7 and x.   *x and x − 7.*

Therefore, the distinct factors are 7, x, and $x - 7$.

In $7x^2$ $\begin{cases} 7 \text{ appears as a } \textit{factor} \text{ once.} \\ x \text{ appears as a } \textit{factor} \text{ twice.} \\ x - 7 \text{ appears as a } \textit{factor} \text{ zero times.} \end{cases}$

In $x(x - 7)$ $\begin{cases} 7 \text{ appears as a } \textit{factor} \text{ zero times.} \\ x \text{ appears as a } \textit{factor} \text{ once.} \\ x - 7 \text{ appears as a } \textit{factor} \text{ once.} \end{cases}$

▪ Following our outline, each distinct factor is taken the *maximum* number of times it appears. Therefore, we take 7 once, x twice, and $x - 7$ once.

The LCD is $7x^2(x - 7)$.

Go over this process of finding the LCD to make sure you follow the procedure. Next we build the fractions.

$$\frac{2}{7x^2} + \frac{3}{x^2 - 7x} = \frac{2}{7x^2} + \frac{3}{x(x - 7)}$$

Whatever factors we have inserted in the denominator, we must also insert in the numerator.

$$= \frac{2\,(x - 7)}{7x^2(x - 7)} + \frac{3\,(7x)}{7x^2(x - 7)}$$

Here the missing   *Here the missing*
factor was x − 7.   *factor was 7x.*

$$= \frac{2(x - 7) + 3(7x)}{7x^2(x - 7)} \qquad \textit{In the numerator, multiply out and combine like terms.}$$

$$= \frac{2x - 14 + 21x}{7x^2(x - 7)}$$

$$= \boxed{\frac{23x - 14}{7x^2(x - 7)}} \qquad\qquad\qquad \blacksquare$$

6. $\dfrac{5}{x^2 - x - 6} + \dfrac{4}{x^2 + 2x} - \dfrac{3}{x^2 - 3x}$

EXAMPLE 6 *Combine and simplify.* $\dfrac{7}{x^2 + x - 12} - \dfrac{4}{x^2 + 4x} + \dfrac{3}{x^2 - 3x}$

Solution To find the LCD we begin by factoring each denominator.

$$\frac{7}{x^2 + x - 12} - \frac{4}{x^2 + 4x} + \frac{3}{x^2 - 3x}$$

$$= \frac{7}{(x + 4)(x - 3)} - \frac{4}{x(x + 4)} + \frac{3}{x(x - 3)} \qquad \begin{array}{l} \textit{Following our outline,} \\ \textit{we find the LCD is} \\ x(x + 4)(x - 3). \end{array}$$

$$= \underbrace{\frac{7(x)}{x(x + 4)(x - 3)}}_{\substack{\textit{Here the missing} \\ \textit{factor was x.}}} - \underbrace{\frac{4(x - 3)}{x(x + 4)(x - 3)}}_{\substack{\textit{Here the missing} \\ \textit{factor was x − 3.}}} + \underbrace{\frac{3(x + 4)}{x(x + 4)(x - 3)}}_{\substack{\textit{Here the missing} \\ \textit{factor was x + 4.}}}$$

$$= \frac{7x - 4(x - 3) + 3(x + 4)}{x(x + 4)(x - 3)}$$

$$= \frac{7x - 4x + 12 + 3x + 12}{x(x + 4)(x - 3)} \qquad \begin{array}{l} \textit{Watch the signs!} \\ \textit{Next combine like terms.} \end{array}$$

$$= \frac{6x + 24}{x(x + 4)(x - 3)} \qquad \textit{Then factor the numerator.}$$

$$= \frac{6(x + 4)}{x(x + 4)(x - 3)} \qquad \textit{Next reduce.}$$

$$= \frac{6\cancel{(x + 4)}}{x\cancel{(x + 4)}(x - 3)}$$

$$= \boxed{\frac{6}{x(x - 3)}} \quad \text{or} \quad \boxed{\frac{6}{x^2 - 3x}}$$

It is worthwhile to go back and look at the LCD to see that it contains within it all of the original denominators.

As this example shows, it clearly does not pay to multiply out the LCD until we are sure that we cannot reduce. In fact, we can leave our answer in factored form. ∎

EXAMPLE 7 *Perform the indicated operations and simplify.*

$$\frac{x^2 - 3x}{x^2 - 6x + 9} \cdot \frac{3x - 9}{9x}$$

7. $\dfrac{z^2 + 10z + 25}{z^2 + 5z} \div \dfrac{z^2 - 25}{z}$

Solution We do begin by factoring, but there is no LCD necessary in this example, because it is a multiplicaton problem, *not* addition or subtraction. Remember to look at the example before you start working on it.

$$\frac{x^2 - 3x}{x^2 - 6x + 9} \cdot \frac{3x - 9}{9x} = \frac{x(x - 3)}{(x - 3)(x - 3)} \cdot \frac{3(x - 3)}{9x}$$

$$= \frac{\cancel{x}\cancel{(x - 3)}}{\cancel{(x - 3)}\cancel{(x - 3)}} \cdot \frac{\cancel{3}\cancel{(x - 3)}}{\underset{3}{\cancel{9}\cancel{x}}}$$

$$= \boxed{\frac{1}{3}} \qquad \blacksquare$$

Complex Fractions

Recall that a complex fraction is a fraction which contains fractions within it. We can apply the methods we have learned in this chapter to simplify complex fractions.

8. $\dfrac{1 + \dfrac{3}{x}}{1 - \dfrac{9}{x^2}}$

EXAMPLE 8   *Simplify as completely as possible.* $\dfrac{\dfrac{4}{y} + \dfrac{4}{x}}{\dfrac{1}{y^2} - \dfrac{1}{x^2}}$

Solution   Keep in mind that a complex fraction is just an alternate way of writing division of fractions. Thus, the complex fraction given in this example means

$$\left(\frac{4}{y} + \frac{4}{x}\right) \div \left(\frac{1}{y^2} - \frac{1}{x^2}\right)$$

We offer two possible solutions.

Solution 1   We basically do what the example tells us to do. Following the order of operations, we add the two fractions in the numerator, and subtract the two fractions in the denominator. We then obtain a straightforward division problem to which we apply the rule for dividing fractions. In order to add the fractions in the numerator we use an LCD of xy; in order to subtract the fractions in the denominator we use an LCD of x^2y^2.

$\dfrac{\dfrac{4}{y} + \dfrac{4}{x}}{\dfrac{1}{y^2} - \dfrac{1}{x^2}} = \dfrac{\dfrac{4(x)}{xy} + \dfrac{4(y)}{xy}}{\dfrac{x^2}{x^2y^2} - \dfrac{y^2}{x^2y^2}}$ *The LCD for the numerator is xy; for the denominator it is x^2y^2.*

$= \dfrac{\dfrac{4x + 4y}{xy}}{\dfrac{x^2 - y^2}{x^2y^2}}$ *Remember: This means* $\dfrac{4x + 4y}{xy} \div \dfrac{x^2 - y^2}{x^2y^2}$.

We use the rule for dividing fractions: "Invert and multiply."

$= \dfrac{4x + 4y}{xy} \cdot \dfrac{x^2y^2}{x^2 - y^2}$ *Factor and reduce.*

$= \dfrac{4(x + y)}{xy} \cdot \dfrac{x^2y^2}{(x + y)(x - y)}$

$= \dfrac{4\cancel{(x + y)}}{\cancel{xy}} \cdot \dfrac{\cancel{x^2y^2}^{\,x\,y}}{\cancel{(x + y)}(x - y)}$

$= \boxed{\dfrac{4xy}{x - y}}$

Solution 2   We can do this example a bit differently by applying the Fundamental Principle of Fractions to the entire complex fraction.

We will call the denominators of fractions *within* the complex fraction **minor denominators.** If we multiply the numerator and denominator of the complex fraction by the LCD of all the minor denominators, we will obtain a simple fraction. We can then proceed to reduce it (if possible).

$\dfrac{\dfrac{4}{y} + \dfrac{4}{x}}{\dfrac{1}{y^2} - \dfrac{1}{x^2}}$ *The LCD of **all** the minor denominators is x^2y^2.*

We multiply the numerator and denominator of the complex fraction by x^2y^2.

$$= \frac{\dfrac{x^2 y^2}{1}\left(\dfrac{4}{y} + \dfrac{4}{x}\right)}{\dfrac{x^2 y^2}{1}\left(\dfrac{1}{y^2} - \dfrac{1}{x^2}\right)}$$

Use the Distributive Law.

$$= \frac{\dfrac{x^2 y^2}{1} \cdot \dfrac{4}{y} + \dfrac{x^2 y^2}{1} \cdot \dfrac{4}{x}}{\dfrac{x^2 y^2}{1} \cdot \dfrac{1}{y^2} - \dfrac{x^2 y^2}{1} \cdot \dfrac{1}{x^2}}$$

Reduce.

$$= \frac{\dfrac{x^2 \overset{y}{\cancel{y^2}}}{1} \cdot \dfrac{4}{\cancel{y}} + \dfrac{\overset{x}{\cancel{x^2}} y^2}{1} \cdot \dfrac{4}{\cancel{x}}}{\dfrac{x^2 \cancel{y^2}}{1} \cdot \dfrac{1}{\cancel{y^2}} - \dfrac{\cancel{x^2} y^2}{1} \cdot \dfrac{1}{\cancel{x^2}}}$$

$$= \frac{4x^2 y + 4xy^2}{x^2 - y^2}$$

Factor and reduce.

$$= \frac{4xy\cancel{(x + y)}}{\cancel{(x + y)}(x - y)}$$

$$= \boxed{\dfrac{4xy}{x - y}}$$

The basic difference in the two approaches is that the method used in solution 1 works on the numerator and the denominator of the complex fraction separately, while the second method works on the entire complex fraction at once.

Be careful using the method of solution 2, which requires us to use the Fundamental Principle to multiply the numerator and denominator of the complex fraction by the *same* quantity (the LCD of all minor denominators).

You should choose the method you feel most comfortable with.   ■

✔ *Answers to Learning Checks in Section 7.3*

1. $\dfrac{4x}{x + 4}$ 2. $\dfrac{a + 1}{3a}$ 3. $\dfrac{5z^2 + 9}{20z}$ 4. $\dfrac{21x + 40}{8x(x + 8)}$

5. $\dfrac{29u + 30}{6u^2(u + 6)}$ 6. $\dfrac{6}{x(x + 2)}$ 7. $\dfrac{1}{z - 5}$ 8. $\dfrac{x}{x - 3}$

NOTE TO THE STUDENT

Use the space on this page to write down any questions you have or points you want to review with your instructor.

Exercises 7.3

In each of the following exercises, perform the indicated operations and simplify your answer as completely as possible.

1. $\dfrac{5x - 2}{4x + 8} + \dfrac{3x + 2}{4x + 8}$

2. $\dfrac{10x + 3}{5x - 10} + \dfrac{5x - 3}{5x - 10}$

3. $\dfrac{5x - 2}{4x + 8} - \dfrac{3x + 2}{4x + 8}$

4. $\dfrac{10x + 3}{5x - 10} - \dfrac{5x - 3}{5x - 10}$

5. $\dfrac{x + 7}{x + 2} + \dfrac{x + 3}{x + 2} + \dfrac{x - 2}{x + 2}$

6. $\dfrac{a - 6}{a + 3} + \dfrac{a + 9}{a + 3} + \dfrac{a + 6}{a + 3}$

7. $\dfrac{x + 7}{x + 2} - \dfrac{x + 3}{x + 2} + \dfrac{x - 2}{x + 2}$

8. $\dfrac{a - 6}{a + 3} + \dfrac{a + 9}{a + 3} - \dfrac{a + 6}{a + 3}$

9. $\dfrac{y^2}{4y} + \dfrac{2y}{4y}$

10. $\dfrac{2z}{10z^2} + \dfrac{5}{10z^2}$

11. $\dfrac{5}{2x} + \dfrac{4}{x + 2}$

12. $\dfrac{8}{3x} + \dfrac{6}{x + 3}$

13. $\dfrac{5}{2x} \cdot \dfrac{4}{x + 2}$

14. $\dfrac{8}{3x} \cdot \dfrac{6}{x + 3}$

1. _____

2. _____

3. _____

4. _____

5. _____

6. _____

7. _____

8. _____

9. _____

10. _____

11. _____

12. _____

13. _____

14. _____

15. $\dfrac{2}{x+2} + \dfrac{3}{x+3}$

16. $\dfrac{4}{x-4} + \dfrac{2}{x-2}$

17. $\dfrac{7}{a+7} - \dfrac{5}{a+5}$

18. $\dfrac{9}{a-3} - \dfrac{4}{a+4}$

19. $\dfrac{4}{3x^2} - \dfrac{2}{x^2+3x}$

20. $\dfrac{5}{6r^2} + \dfrac{1}{r^2-6r}$

21. $\dfrac{3}{4a^2} + \dfrac{1}{6a-18}$

22. $\dfrac{5}{6p} - \dfrac{4}{9p-18}$

23. $\dfrac{4}{x^2+4x} - \dfrac{2}{x^2-4x}$

24. $\dfrac{3}{x^2-6x} - \dfrac{2}{x^2+6x}$

25. $2 - \dfrac{x}{x-1}$

26. $3 - \dfrac{x}{x-2}$

27. $2 \cdot \dfrac{x}{x-1}$

28. $3 \cdot \dfrac{x}{x-2}$

29. $\dfrac{3}{x^2 - 16} - \dfrac{3}{2x^2 + 8x}$

30. $\dfrac{2}{x^2 - 9} - \dfrac{2}{3x^2 + 9x}$

31. $\dfrac{12}{x^2 + x - 2} - \dfrac{4}{x^2 - x}$

32. $\dfrac{4}{x^2 - 2x - 3} - \dfrac{2}{x^2 - 4x + 3}$

33. $\dfrac{x}{x^2 + 6x + 9} + \dfrac{1}{x^2 + 4x + 3}$

34. $\dfrac{y}{y^2 - 3y - 10} + \dfrac{3}{y^2 + 4y + 4}$

35. $\dfrac{3a + 6}{3a + 2} \cdot \dfrac{a + 1}{a + 2}$

36. $\dfrac{3a + 6}{3a + 2} + \dfrac{a + 1}{a + 2}$

37. $5 - \dfrac{1}{3x - 6} + \dfrac{3}{x^2 - 2x}$

38. $7 + \dfrac{3}{4x + 12} - \dfrac{2}{x^2 + 3x}$

39. $\dfrac{5}{x^2 + x - 6} - \dfrac{3}{x^2 + 3x} + \dfrac{2}{x^2 - 2x}$

40. $\dfrac{3}{4x^2 - 36} + \dfrac{7}{6x^2 - 18x}$

29. _____

30. _____

31. _____

32. _____

33. _____

34. _____

35. _____

36. _____

37. _____

38. _____

39. _____

40. _____

ANSWERS

41. _____

42. _____

43. _____

44. _____

45. _____

46. _____

47. _____

48. _____

41. $\left(\dfrac{x}{2} - \dfrac{2}{x}\right) \div \dfrac{x^2 - 2x}{x^2}$

42. $\left(\dfrac{2}{4y - 3} - \dfrac{1}{3y}\right) \cdot \dfrac{20y - 15}{4y^2 - 9}$

43. $\dfrac{\dfrac{1}{x} - 1}{\dfrac{2}{x}}$

44. $\dfrac{\dfrac{5}{a^2}}{2 + \dfrac{3}{a}}$

45. $\dfrac{1 - \dfrac{1}{a}}{\dfrac{1}{a} - \dfrac{1}{a^2}}$

46. $\dfrac{2 - \dfrac{3}{c}}{2 + \dfrac{3}{c}}$

47. $\dfrac{\dfrac{a}{2} - \dfrac{b}{4}}{\dfrac{4}{b^2} - \dfrac{1}{a^2}}$

48. $\dfrac{\dfrac{1}{3} - \dfrac{1}{x}}{\dfrac{2}{9x} - \dfrac{2}{3x^2}}$

QUESTION FOR THOUGHT

2. Multiply out $\left(x + \dfrac{2}{x}\right)\left(x - \dfrac{3}{x}\right)$ in the following two ways.

(a) First, multiply out the two binomials using FOIL.

(b) Second, combine the fractions within the parentheses, and then multiply the resulting fractions.

Which method did you find easier? If the instructions were to express your answer as a single fraction, which method would you choose in general?

Fractional Equations

7.4

Next we bring our technique to bear on solving equations involving fractional expressions. As we shall soon see, equations which have variables in their denominators will give us something additional to take into account.

In Section 4.4, we saw that we can use the LCD to clear the denominators from an equation and give us an equivalent equation which is much easier to solve. Let's review that process in our first example.

EXAMPLE 1   *Solve for t.* $\dfrac{3t - 5}{4} - \dfrac{t + 1}{6} = \dfrac{t - 2}{3}$

Solution   Since we want to solve for t, we want to produce (if possible) an equivalent equation without fractional expressions. Therefore, we multiply both sides of the equation by the LCD to "clear" the denominators.

$$\frac{3t - 5}{4} - \frac{t + 1}{6} = \frac{t - 2}{3}$$

The LCD is 12. We multiply both sides of the equation by 12.

$$12\left(\frac{3t - 5}{4} - \frac{t + 1}{6}\right) = 12\left(\frac{t - 2}{3}\right)$$

$$\frac{12}{1} \cdot \frac{3t - 5}{4} - \frac{12}{1} \cdot \frac{t + 1}{6} = \frac{12}{1} \cdot \frac{t - 2}{3}$$

$$\frac{\overset{3}{\cancel{12}}}{1} \cdot \frac{3t - 5}{\cancel{4}} - \frac{\overset{2}{\cancel{12}}}{1} \cdot \frac{t + 1}{\cancel{6}} = \frac{\overset{4}{\cancel{12}}}{1} \cdot \frac{t - 2}{\cancel{3}}$$

*We distribute the 12 to each term and reduce. If you want to shorten the amount of writing by doing these steps mentally, **watch out** for a sign error where the arrow points.*

$$3(3t - 5) - 2(t + 1) = 4(t - 2)$$

Remove parentheses.

$$9t - 15 - 2t - 2 = 4t - 8$$

Combine like terms.

$$7t - 17 = 4t - 8$$

Isolate t.

$$3t = 9$$

$$\boxed{t = 3}$$

CHECK $t = 3$: $\dfrac{3(3) - 5}{4} - \dfrac{(3) + 1}{6} \overset{?}{=} \dfrac{(3) - 2}{3}$

$$\frac{9 - 5}{4} - \frac{4}{6} \overset{?}{=} \frac{1}{3}$$

$$\frac{4}{4} - \frac{2}{3} \overset{?}{=} \frac{1}{3}$$

$$1 - \frac{2}{3} \overset{\checkmark}{=} \frac{1}{3}$$

We take basically the same approach when there are variables in the denominator.

EXAMPLE 2   *Solve for x.* $\dfrac{2}{3} + \dfrac{1}{x - 4} = \dfrac{7}{3x - 12}$

LEARNING CHECKS

1. $\dfrac{2x - 7}{3} - \dfrac{x + 1}{8} = \dfrac{x - 4}{4}$

2. $\dfrac{4}{5} + \dfrac{2}{x - 3} = \dfrac{14}{5x - 15}$

Solution In order to find the LCD, we first factor each denominator.

$$\frac{2}{3} + \frac{1}{x-4} = \frac{7}{3x-12}$$

$$\frac{2}{3} + \frac{1}{x-4} = \frac{7}{3(x-4)}$$

We multiply both sides of the equation by the LCD, which is $3(x-4)$, to clear the denominators.

$$3(x-4)\left(\frac{2}{3} + \frac{1}{x-4}\right) = 3(x-4) \cdot \frac{7}{3(x-4)}$$

Each term gets multiplied by $3(x-4)$.

$$\frac{3(x-4)}{1} \cdot \frac{2}{3} + \frac{3(x-4)}{1} \cdot \frac{1}{x-4} = \frac{3(x-4)}{1} \cdot \frac{7}{3(x-4)}$$

$$\frac{\cancel{3}(x-4)}{1} \cdot \frac{2}{\cancel{3}} + \frac{3(\cancel{x-4})}{1} \cdot \frac{1}{\cancel{x-4}} = \frac{\cancel{3}(\cancel{x-4})}{1} \cdot \frac{7}{\cancel{3}(\cancel{x-4})}$$

$$2(x-4) + 3 = 7$$

$$2x - 8 + 3 = 7$$

$$2x - 5 = 7$$

$$2x = 12$$

$$\boxed{x = 6}$$

CHECK $x = 6$: $\quad \dfrac{2}{3} + \dfrac{1}{(6)-4} \overset{?}{=} \dfrac{7}{3(6)-12}$

$$\frac{2}{3} + \frac{1}{2} \overset{?}{=} \frac{7}{18-12}$$

$$\frac{4}{6} + \frac{3}{6} \overset{\checkmark}{=} \frac{7}{6}$$

∎

3. $\dfrac{6}{a+2} - \dfrac{2}{a} + \dfrac{3}{4a}$

EXAMPLE 3 *Combine and simplify.* $\quad \dfrac{2}{x} - \dfrac{3}{x+1} + \dfrac{5}{4x}$

Solution We must repeat the warning we gave in Section 4.4 about not confusing equations with expressions. This example is *not* an equation, and therefore we cannot clear the denominators. Notice that even the instructions are telling us that this is not an equation (it does not say solve). We will use the LCD in this example, but to *build* the fractions, not clear the denominators. We proceed as we did in the last section.

$$\frac{2}{x} - \frac{3}{x+1} + \frac{5}{4x} \qquad \text{The LCD is } 4x(x+1).$$

$$= \frac{2 \cdot 4(x+1)}{4x(x+1)} - \frac{3(4x)}{4x(x+1)} + \frac{5(x+1)}{4x(x+1)}$$

$$= \frac{8(x+1) - 12x + 5(x+1)}{4x(x+1)}$$

$$= \frac{8x + 8 - 12x + 5x + 5}{4x(x+1)}$$

$$= \boxed{\frac{x+13}{4x(x+1)}}$$

∎

EXAMPLE 4   *Solve for x.* $\dfrac{6}{x-3} + 5 = \dfrac{x+3}{x-3}$

4. $\dfrac{8}{x-2} + 3 = \dfrac{x+6}{x-2}$

Solution   We proceed as before to clear the denominators by multiplying both sides of the equation by the LCD, which is $x-3$.

$$\frac{6}{x-3} + 5 = \frac{x+3}{x-3}$$

$$(x-3)\left(\frac{6}{x-3} + 5\right) = (x-3)\left(\frac{x+3}{x-3}\right)$$

$$\frac{x-3}{1} \cdot \frac{6}{x-3} + \frac{x-3}{1} \cdot 5 = \frac{x-3}{1} \cdot \frac{x+3}{x-3}$$

$$\frac{\cancel{x-3}}{1} \cdot \frac{6}{\cancel{x-3}} + \frac{x-3}{1} \cdot \frac{5}{1} = \frac{\cancel{x-3}}{1} \cdot \frac{x+3}{\cancel{x-3}}$$

$$6 + 5(x-3) = x+3$$

$$6 + 5x - 15 = x+3$$

$$5x - 9 = x+3$$

$$4x = 12$$

$$x = 3 \qquad \textit{Are you wondering why there is no box?}$$

CHECK $x = 3$: $\dfrac{6}{3-3} + 5 \stackrel{?}{=} \dfrac{6}{3-3}$

$$\frac{6}{0} + 5 \stackrel{?}{=} \frac{6}{0}$$

$\frac{6}{0}$ is undefined. We are never allowed to divide by 0. Therefore, $x = 3$ is *not* a solution.

What happened? Our logic tells us that *if* there is a solution to the original equation, then it must be $x = 3$. Since we see that $x = 3$ is not a solution, that must mean that the original equation has *no* solutions. How is it possible for us to solve an equation properly (we really did not make any mistakes) and yet get an answer that does not work?

The difficulty lies in the first step of the solution, where we multiplied both sides of the equation by $x - 3$. Back in Section 3.2 we saw that if we want to be sure that we obtain an *equivalent* equation, then we cannot multiply both sides of the equation by 0. If x is equal to 3, then $x - 3 = 0$. Therefore, when we multiplied by $x - 3$ we were actually multiplying by 0, and we got an equation which was no longer equivalent to the original equation.

Thus, when we multiply an equation by a variable quantity which might be equal to 0, we must be sure to check our answers in the *original* equation. ***This check is not optional***—it is a necessary step in the solution. We are not checking for errors, but rather to see if we have obtained a valid solution.

An alternative way of saying this is that since $x - 3$ is a denominator of the original equation, and since we are never allowed to divide by 0, $x = 3$ was disqualified from consideration as a solution *at the outset*.   ∎

EXAMPLE 5   *Solve for y.* $\dfrac{2}{y-1} - \dfrac{4}{3y} = \dfrac{1}{y^2 - y}$

5. $\dfrac{2}{u+1} - \dfrac{3}{4u} = \dfrac{1}{u^2+u}$

Solution   We will use the LCD to clear denominators. In order to find the LCD we make sure each denominator is in factored form.

$$\frac{2}{y-1} - \frac{4}{3y} = \frac{1}{y^2 - y}$$

$$\frac{2}{y-1} - \frac{4}{3y} = \frac{1}{y(y-1)} \qquad \begin{array}{l} LCD = 3y(y-1) \\ \textit{We multiply each term} \\ \textit{by the LCD.} \end{array}$$

$$\frac{3y(y-1)}{1} \cdot \frac{2}{y-1} - \frac{3y(y-1)}{1} \cdot \frac{4}{3y} = \frac{3y(y-1)}{1} \cdot \frac{1}{y(y-1)}$$

$$\frac{3y\cancel{(y-1)}}{1} \cdot \frac{2}{\cancel{y-1}} - \frac{\cancel{3y}(y-1)}{1} \cdot \frac{4}{\cancel{3y}} = \frac{3\cancel{y}\cancel{(y-1)}}{1} \cdot \frac{1}{\cancel{y}\cancel{(y-1)}}$$

$$6y - 4(y-1) = 3$$

$$6y - 4y + 4 = 3$$

$$2y + 4 = 3$$

$$2y = -1$$

$$\boxed{y = -\frac{1}{2}}$$

CHECK $y = -\frac{1}{2}$:

$$\frac{2}{-\frac{1}{2} - 1} - \frac{4}{3\left(-\frac{1}{2}\right)} \overset{?}{=} \frac{1}{\left(-\frac{1}{2}\right)^2 - \left(-\frac{1}{2}\right)}$$

$$\frac{2}{-\frac{3}{2}} - \frac{4}{-\frac{3}{2}} \overset{?}{=} \frac{1}{\frac{1}{4} + \frac{1}{2}}$$

$$\frac{2}{1} \cdot \frac{-2}{3} - \frac{4}{1} \cdot \frac{-2}{3} \overset{?}{=} \frac{1}{\frac{3}{4}}$$

$$\frac{-4}{3} + \frac{8}{3} \overset{?}{=} 1 \cdot \frac{4}{3}$$

$$\frac{4}{3} \overset{\checkmark}{=} \frac{4}{3}$$ ∎

A final comment about the word "check" is in order. A check can serve two purposes. We can check an equation in order to see whether we have made any algebraic or arithmetic errors. As far as this purpose is concerned, you may think the word "check" was misapplied in Example 5, because the check was more difficult than the original problem. If the check had turned out wrong you might have believed it more likely that your error occurred in the check rather than in the process of solving the equation. However, since the fact remains that the only way to be sure we have a correct answer is to check it, we must weigh the confidence we have in our check against the confidence we have in our solution.

The second purpose a check can serve is to ensure that we have obtained a valid answer to our original problem. Thus, in Example 5 we must *at least* verify that $y = -\frac{1}{2}$ is a valid solution in that it does not make any of the denominators in the original equation equal to 0. Of course, just because we verify that $y = -\frac{1}{2}$ is a valid answer, does *not* mean that it is a correct answer as well. The only way to verify that an answer actually works is by substituting it into the original equation.

✔ *Answers to Learning Checks in Section 7.4*

1. $x = 5$ **2.** $x = 4$

3. $\dfrac{19a - 10}{4a(a + 2)}$ **4.** No solutions

5. $u = \dfrac{7}{5}$

Exercises 7.4

Solve each of the following equations and check.

1. $\dfrac{x}{5} + \dfrac{x-1}{4} + \dfrac{x-3}{2} = 3$

2. $\dfrac{x}{6} + \dfrac{x-3}{3} + \dfrac{x-2}{4} = 3$

3. $\dfrac{a+5}{2} = \dfrac{a+2}{5}$

4. $\dfrac{a-3}{7} = \dfrac{a-7}{3}$

5. $\dfrac{x+2}{3} + 1 = \dfrac{x+3}{2} - 1$

6. $\dfrac{y-5}{5} - 3 = \dfrac{y+7}{6} - 4$

7. $\dfrac{2r+1}{3} - \dfrac{r+1}{5} = \dfrac{r+8}{6}$

8. $\dfrac{5a-8}{4} - \dfrac{a+4}{8} = \dfrac{a}{2}$

9. $\dfrac{3}{x} - \dfrac{2}{3} = \dfrac{2}{x}$

10. $\dfrac{5}{y} - \dfrac{3}{5} = \dfrac{4}{y}$

1. _____

2. _____

3. _____

4. _____

5. _____

6. _____

7. _____

8. _____

9. _____

10. _____

ANSWERS

11. _____

12. _____

13. _____

14. _____

15. _____

16. _____

17. _____

18. _____

19. _____

20. _____

11. $\dfrac{4}{t-2} + \dfrac{3}{t} = \dfrac{1}{2t}$

12. $\dfrac{2}{a+3} + \dfrac{1}{a} = \dfrac{4}{3a}$

13. $\dfrac{4}{x-1} - \dfrac{5}{8} = \dfrac{3}{2x-2}$

14. $\dfrac{10}{y+3} - \dfrac{3}{5} = \dfrac{10y+1}{3y+9}$

15. $\dfrac{8}{x-2} + 3 = \dfrac{x+6}{x-2}$

16. $4 + \dfrac{5}{a+3} = \dfrac{a+8}{a+3}$

17. $\dfrac{m+3}{6} - \dfrac{m+4}{8} = m$

18. $\dfrac{5}{4t} - \dfrac{2}{t} - \dfrac{1}{2} = -\dfrac{1}{8}$

19. $\dfrac{5}{x^2-x} - \dfrac{1}{2x-2} = \dfrac{1}{x}$

20. $\dfrac{11}{x^2-9} - \dfrac{7}{2x+6} = \dfrac{2}{x+3}$

21. $\dfrac{x}{x-5} + \dfrac{3}{2} = \dfrac{5}{x-5}$

22. $\dfrac{c-5}{10} - \dfrac{c-2}{4} = c$

21. _____

23. $\dfrac{a}{2} + \dfrac{2}{a-2} = \dfrac{a-4}{2}$

24. $\dfrac{y}{3} - \dfrac{3}{y} = \dfrac{y-3}{3}$

22. _____

23. _____

25. $\dfrac{5}{x^2 - 2x - 3} = \dfrac{4}{x^2 - 3x - 4}$

26. $\dfrac{6}{r^2 - 9} = \dfrac{3}{r^2 - 3r}$

24. _____

25. _____

27. $\dfrac{9}{x^2 - 3x + 2} - \dfrac{2}{x-1} = \dfrac{1}{x-2}$

28. $\dfrac{5}{y^2 - 3y - 10} + \dfrac{3}{y-5} = \dfrac{-1}{4y+8}$

26. _____

27. _____

28. _____

ANSWERS

29. _____

30. _____

31. _____

32. _____

33. _____

34. _____

35. _____

36. _____

In each of the following, if the exercise is an equation, solve it and check. Otherwise, perform the indicated operations and simplify.

29. $\dfrac{x}{2} + \dfrac{x}{3} + \dfrac{x}{4}$

30. $\dfrac{x}{2} + \dfrac{x}{3} = \dfrac{x}{4}$

31. $\dfrac{x}{2} \cdot \dfrac{x}{3} \cdot \dfrac{x}{4}$

32. $\dfrac{x}{2} - \dfrac{x}{3} - \dfrac{4}{x}$

33. $\dfrac{x+1}{2} - \dfrac{3}{x} = \dfrac{x}{2}$

34. $\dfrac{x+1}{2} - \dfrac{2}{x} - \dfrac{x}{2}$

35. $\dfrac{x+1}{2} \cdot \dfrac{3}{x} = \dfrac{2}{x}$

36. $\dfrac{x+1}{2} \cdot \dfrac{3}{x} \div \dfrac{2}{x}$

Literal Equations

7.5

Thus far when we have been asked to solve an equation which has a unique solution, our answer has been a number. Solving such an equation means finding its *numerical* solution—that is, the number which satisfies the equation. (Keep in mind that we have restricted our attention to first-degree equations in one variable.) For example, to solve the equation $3x - 5 = x + 7$, we apply the procedures we have learned and obtain the solution $x = 6$.

However, if an equation has more than one variable, then solving the equation takes on an entirely different meaning. An equation which contains more than one variable (letter) is often called a ***literal equation.*** The reason is that, as we shall soon see, when we solve such an equation for one of its variables, we will get a literal solution rather than a numerical one.

If a literal equation is of the form where one of the variables is totally isolated on one side of the equation, then we say that the equation is solved *explicitly* for that variable. For example, the equation

$$z = 3x - 4$$

is solved explicitly for z, while the *same* equation written in the form

$$z - 3x + 4 = 0$$

is not solved explicitly for either variable.

While we cannot be asked to solve a literal equation in order to get a numerical solution, we can be asked to solve it explicitly for a particular variable.

EXAMPLE 1 Solve the equation $3x + 5y = 15$
(a) explicitly for x (b) explicitly for y

Solution

(a) To solve the equation explicitly for x means we want to isolate x on one side of the equation.

$$3x + 5y = 15$$
$$\underline{\quad -5y \qquad -5y}$$
$$3x \qquad = 15 - 5y$$
$$\frac{3x}{3} = \frac{15 - 5y}{3}$$
$$\boxed{x = \frac{15 - 5y}{3}}$$

(b) To solve explicitly for y means we want to isolate y on one side of the equation.

$$3x + 5y = 15$$
$$\underline{-3x \qquad\qquad -3x}$$
$$5y = 15 - 3x$$
$$\frac{5y}{5} = \frac{15 - 3x}{5}$$
$$\boxed{y = \frac{15 - 3x}{5}}$$ ∎

✔ LEARNING CHECKS

1. Solve $2s + 7t = 14$
(a) explicitly for s
(b) explicitly for t

2. Solve for u.

$2u + 5v - 3 = v - 3u$

EXAMPLE 2   Solve the equation $r + 4s = s - 4r + 2$ for r.

Solution   When asked to solve for r, it is understood that we want to solve explicitly for r. Just as we did with numerical equations, we isolate the variable we are solving for. In this case we focus our attention on isolating r.

$$
\begin{array}{rl}
r + 4s = & s - 4r + 2 \\
- 4s = & -4s \\
\hline
r = & -3s - 4r + 2 \\
+4r & +4r \\
\hline
5r = & -3s + 2 \\
\dfrac{5r}{5} = & \dfrac{-3s + 2}{5} \\
\boxed{r = \dfrac{-3s + 2}{5}}
\end{array}
$$

Do not stop here! r must appear on one side only.

If we had decided to isolate r on the right-hand side of the equation, we could have proceeded as follows:

$$
\begin{array}{rl}
r + 4s = & s - 4r + 2 \\
- s - 2 = -s & - 2 \\
\hline
r + 3s - 2 = & -4r \\
-r & - r \\
\hline
3s - 2 = & -5r \\
\dfrac{3s - 2}{5} = & \dfrac{-5r}{-5} \\
\boxed{-\dfrac{3s - 2}{5}} = & r
\end{array}
$$

Subtract s and 2 from both sides.

Which side you decide to isolate the variable on is not important. What is important is that you recognize that the two answers are equivalent.

Our first answer is

$$
\frac{-3s + 2}{5} = \frac{-(3s - 2)}{5} = -\frac{3s - 2}{5}
$$

which is our second answer. Keep this in mind when you check your answers with those in the answer key. ∎

Sometimes literal equations which have some "real-life" interpretation are called **formulas**. For instance, the following formula for *simple* interest

$$
A = P(1 + rt)
$$

allows us to compute the amount, A, of money in an account if the original amount, P (P stands for principal), is invested at a rate of $r\%$ per year (r is written as a decimal) for t years.

For example, if \$1,000 is invested at 8% for 3 years, then

$$
P = 1,000 \qquad r = .08 \qquad t = 3
$$

and the amount of money, A, in the account after 3 years is

$$A = 1,000[1 + .08(3)]$$
$$A = 1,000(1.24)$$
$$A = \$1,240$$

This formula is fine if we are given P, r, t, and want to compute A. But what if we are given A, P, t, and we want to compute r? In that case we would much prefer to have a formula which is solved explicitly for r rather than A.

EXAMPLE 3   Solve $A = P(1 + rt)$ for r.

Solution   To isolate r, we can begin by first multiplying out the parentheses.

$$A = P(1 + rt)$$
$$A = P + Prt \qquad \textit{Isolate the r term.}$$
$$A - P = Prt \qquad \textit{Divide both sides by what is multiplying r.}$$
$$\frac{A - P}{Pt} = \frac{Prt}{Pt}$$

$$\boxed{\frac{A - P}{Pt} = r}$$ ∎

Naturally, this same procedure can be applied to literal inequalities as well.

EXAMPLE 4   *Solve for a.* $2a - 5b < 6a + 8c$

Solution

$$2a - 5b < 6a + 8c$$
$$\underline{-6a \qquad\qquad -6a}$$
$$-4a - 5b < 8c$$
$$\underline{\quad +5b \qquad +5b}$$
$$-4a < 8c + 5b$$
$$\frac{-4a}{-4} > \frac{8c + 5b}{-4}$$
$$\uparrow$$

*Remember that when we divide both sides of an inequality by a negative number, we must **reverse** the inequality symbol.*

$$\boxed{a > -\frac{8c + 5b}{4}}$$ ∎

Sometimes additional steps are necessary in order to isolate a particular variable.

EXAMPLE 5   *Solve for x.* $ax + b = cx + d$

Solution   Let's compare solving this equation for x with solving the equation $6x + 5 = 4x + 7$ for x. Notice that we will follow virtually the same procedure in both cases. The steps in the two solutions correspond to each other. First we collect all the x terms on one side of the equation and all other terms on the opposite side of the equation.

3. Solve $A = P(1 + rt)$ for t.

4. Solve for m. $3m - 4n \geq 8m + 6p$

5. Solve for y. $xy - z = 5y + w$

$$6x + 5 = 4x + 7$$
$$\underline{-4x \qquad\qquad -4x}$$
$$6x - 4x + 5 = \qquad 7$$
$$\underline{\quad -5 \qquad -5}$$
$$6x - 4x = 7 - 5$$

The next step we usually do mentally. Let's show it, without doing the arithmetic.

$$(6 - 4)x = 7 - 5$$

$$\frac{(6 - 4)x}{6 - 4} = \frac{7 - 5}{6 - 4}$$

$$\boxed{x = 1}$$

$$ax + b = cx + d$$
$$\underline{-cx \qquad\qquad -cx}$$
$$ax - cx + b = \qquad d$$
$$\underline{\qquad\quad -b \qquad -b}$$
$$ax - cx = d - b$$

Now what? We want to isolate x. Let's try factoring out x.

$$x(a - c) = d - b$$

$$\frac{x(a - c)}{a - c} = \frac{d - b}{a - c}$$

$$\boxed{x = \frac{d - b}{a - c}}$$

Since we are not allowed to divide by 0, this answer assumes that $a \neq c$. ■

✔ *Answers to Learning Checks in Section 7.5*

1. (a) $s = \dfrac{14 - 7t}{2}$ **(b)** $t = \dfrac{14 - 2s}{7}$ **2.** $u = \dfrac{-4v + 3}{5}$ **3.** $t = \dfrac{A - P}{Pr}$

4. $m \leq -\dfrac{4n + 6p}{5}$ **5.** $y = \dfrac{w + z}{x - 5}$

Exercises 7.5

Solve each of the following equations or inequalities explicitly for the indicated variable.

1. $4x - 3y = 12$ for x **2.** $4x - 3y = 12$ for y

3. $3x + 6y = 18$ for y **4.** $3x + 6y = 18$ for x

5. $a - 3b = 2a - b + 5$ for a **6.** $a - 3b = 2a - b + 5$ for b

7. $3(m + 2p) = 4(p - m)$ for m **8.** $3(m + 2p) > 4(p - m)$ for p

9. $5x - 7z + 12 = 3y - 7x + 2z$ for z **10.** $5x - 7z + 12 = 3y - 7x + 2z$ for y

11. $\dfrac{x}{2} + \dfrac{y}{3} = \dfrac{x}{3} + \dfrac{y}{4} - \dfrac{1}{6}$ for x **12.** $\dfrac{x + y}{3} = \dfrac{x}{3} + \dfrac{y}{4} - \dfrac{x}{6} + 2$ for y

13. $ax + b = 2x + d$ for x **14.** $ma + r = 4a - s$ for a

15. $(x + a)(y + b) = c$ for x **16.** $(2x + a)(3x - b) = c$ for a

ANSWERS

1. _____
2. _____
3. _____
4. _____
5. _____
6. _____
7. _____
8. _____
9. _____
10. _____
11. _____
12. _____
13. _____
14. _____
15. _____
16. _____

Each of the following is a formula either from mathematics, or the physical or social sciences. Solve each of the formulas for the indicated variable.

17. $y = mx + b$ for x

18. $y = mx + b$ for m

19. $A = \frac{1}{2}h(b_1 + b_2)$ for h

20. $A = \frac{1}{2}h(b_1 + b_2)$ for b_2

21. $A = P(1 + rt)$ for P

22. $A = P(1 + rt)$ for t

23. $C = \frac{5}{9}(F - 32)$ for F

24. $F = \frac{9}{5}C + 32$ for C

25. $S = S_0 + v_0 t + \frac{1}{2}gt^2$ for v_0

26. $S = S_0 + v_0 t + \frac{1}{2}gt^2$ for g

27. $\frac{x - \mu}{s} < 2$ for x (Assume $s > 0$)

28. $\frac{x - \mu}{s} < 2$ for μ (Assume $s > 0$)

29. $\frac{1}{f} = \frac{1}{f_1} + \frac{1}{f_2}$ for f

30. $\frac{1}{f} = \frac{1}{f_1} + \frac{1}{f_2}$ for f_1

Verbal Problems

7.6

We close this chapter with yet another look at verbal problems. We will look at a variety of examples; some relate to problems we have already studied, but extend the ideas a bit further.

You may find it helpful to review the material in Sections 3.6 and/or 4.6 before continuing here.

EXAMPLE 1   A rectangle has a perimeter of 52 cm. The ratio of its length to its width is 9 to 4. Find its dimensions.

Solution   The fact that the ratio of the length to the width is 9 to 4 means that we can label the rectangle as shown in Figure 7.1. Notice that

$$\frac{9x}{4x} = \frac{9\cancel{x}}{4\cancel{x}} = \frac{9}{4}$$

Since the perimeter is given as 52 cm., our equation is

$$2L + 2W = 52$$

$$\boxed{2(9x) + 2(4x) = 52}$$

$$18x + 8x = 52$$

$$26x = 52$$

$$x = 2$$

Be careful! $x = 2$ is not the answer to the problem. Looking at the figure, we see that

$$\boxed{\text{length} = 18 \text{ cm}} \quad \text{and} \quad \boxed{\text{width} = 8 \text{ cm}}$$

CHECK: $\dfrac{18 \text{ cm}}{8 \text{ cm}} = \dfrac{18}{8} = \dfrac{9}{4}$ so the ratio is correct.

$2(18) + 2(8) = 36 + 16 = 52$ cm. and so the perimeter is also correct. ∎

EXAMPLE 2   A metal bar weighing 160 kg. is divided into two parts, so that the ratio of the weight of the lighter part to that of the heavier part is 3 to 5. Find the weight of each part.

Solution   This example can be done in exactly the same way as the previous one was. (Try it.) However, we offer an alternative solution.

If we let x = # of kilograms in the lighter part of the bar,

then $160 - x$ = # of kilograms in the heavier part of the bar

(because Heavier part = Total − Lighter part)

The information given in the problem is that

$$\frac{\text{Lighter part}}{\text{Heavier part}} = \frac{3}{5}$$

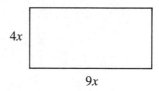

✔ LEARNING CHECKS

1. A rectangle has a perimeter of 100 centimeters. If the ratio of the width to the length is 3 to 7, find its dimensions.

Figure 7.1   *Rectangle for Example 1*

2. A 60-pound bag of beans is divided into two parts so that the ratio of the heavier part to the lighter part is 3 to 2. Find the weight of each part.

Therefore, our equation is

$$\boxed{\frac{x}{160 - x} = \frac{3}{5}}$$

The LCD is 5(160 − x).
We multiply through by the LCD to clear the denominators.

$$\frac{5(160 - x)}{1} \cdot \frac{x}{160 - x} = \frac{3}{5} \cdot \frac{5(160 - x)}{1}$$

$$5x = 3(160 - x)$$
$$5x = 480 - 3x$$
$$8x = 480$$
$$x = 60$$

Therefore,

the lighter part = $\boxed{60 \text{ kg.}}$ and the heavier part = $\boxed{100 \text{ kg.}}$

CHECK: $\dfrac{60}{100} = \dfrac{3}{5}$ so the ratio checks, and clearly the weights add up to 160 kg.

There was no reason we had to let x represent the weight of the lighter part. It is important to realize that we could have let x represent the number of kilograms in the heavier part. In that case we would have gotten a different answer for x (x would have come out to be 100 kg. instead of 60 kg.), *but* our final answer to the verbal problem would have been the same. ∎

3. What number must be added to the numerator and denominator of $\frac{5}{9}$ to obtain a fraction whose value is $\frac{5}{6}$?

EXAMPLE 3 Susie Slugger has compiled a batting average of .300 by getting 90 hits in 300 at bats ($\frac{90}{300} = \frac{3}{10} = .300$). How many consecutive hits must Susie get in order to raise her batting average to .400? (The answer may surprise you.)

Solution A hitter's batting average is computed as $\dfrac{\text{\# of hits}}{\text{\# of at bats}}$.

Let x = # of consecutive hits Susie needs to raise her average to

$$.400 = \frac{400}{1,000} = \frac{2}{5}$$

Then Susie will have $300 + x$ at bats, so her average will be

$$\frac{\text{\# of hits}}{\text{\# of at bats}} = \frac{90 + x}{300 + x}$$

and since we want her average to be $.400 = \frac{400}{1,000} = \frac{2}{5}$, our equation is

$$\boxed{\frac{90 + x}{300 + x} = \frac{2}{5}}$$

The LCD is 5(300 + x).

$$\frac{5(300 + x)}{1} \cdot \frac{90 + x}{300 + x} = \frac{5(300 + x)}{1} \cdot \frac{2}{5}$$

$$5(90 + x) = 2(300 + x)$$
$$450 + 5x = 600 + 2x$$
$$3x = 150$$
$$\boxed{x = 50}$$

Susie needs 50 hits in a row to raise her average to .400.

CHECK: If Susie should perform the incredibly remarkable feat of getting 50 hits in a row, she would have $90 + 50 = 140$ hits, in $300 + 50 = 350$ at bats, which would give her a batting average of $\frac{140}{350} = \frac{14}{35} = \frac{2}{5} = .400$. ∎

EXAMPLE 4 One clerk can perform a job in 5 hours, while a second clerk can do the same job in 7 hours. How long would it take them to do the job if they work together?

Solution We must make two assumptions here. The first is that the clerks work in a totally cooperative fashion. The second is that they are working at a constant rate. These may not be completely realistic assumptions, but we make them nonetheless.)

Our basic approach is to analyze what *part* of the job each person does.

Let $x = $ # of hours it takes for them to do the job together.

If it takes the first clerk 5 hours to do the job, then he or she does

$\frac{1}{5}$ of the job in 1 hour.

$\frac{2}{5}$ of the job in 2 hours.

$\frac{3}{5}$ of the job in 3 hours.

$\frac{x}{5}$ of the job in x hours.

Similarly, if it takes the second clerk 7 hours to do the same job, then he or she does

$\frac{1}{7}$ of the job in 1 hour.

$\frac{2}{7}$ of the job in 2 hours.

$\frac{3}{7}$ of the job in 3 hours.

$\frac{x}{7}$ of the job in x hours.

If we add the amounts of work done by each clerk we must get 1. That is, together they have 1 complete job.

Thus, our equation is

$$\boxed{\frac{x}{5} + \frac{x}{7} = 1} \qquad \text{The LCD is 35.}$$

$$\frac{\overset{7}{\cancel{35}}}{1} \cdot \frac{x}{\cancel{5}} + \frac{\overset{5}{\cancel{35}}}{1} \cdot \frac{x}{\cancel{7}} = 35 \cdot 1$$

$$7x + 5x = 35$$

$$12x = 35$$

$$x = \frac{35}{12} = 2\frac{11}{12} \text{ hours} = \boxed{2 \text{ hours and 55 minutes}}$$

CHECK: The check is left to the student. ∎

4. Sam can build a bookcase in 6 hours while Jim can build the same bookcase in 9 hours. How long will it take them to build the bookcase working together?

5. A salesperson can drive 280 km. at a certain rate of speed. He calculates that if he increased his speed by 30 kph he could cover 400 km. in the same amount of time. Find the faster rate of speed.

EXAMPLE 5   Robert and Jan were both competing in long-distance races. Robert's race was 150 kilometers long, while Jan's race was 180 kilometers long. They both completed their races in the same amount of time. If Jan's pace was 2 kilometers per hour faster than Robert's, how fast was Robert's pace?

Solution   We will use the relationship

$$d = r \cdot t$$

Let r = Rate for Robert.

Then $r + 2$ = Rate for Jan.

Since their *times* are the same, we have

$$t_{\text{Robert}} = t_{\text{Jan}}$$

We will use the formula $d = rt$ in the form $\dfrac{d}{r} = t$.

$$\frac{d_{\text{Robert}}}{r_{\text{Robert}}} = \frac{d_{\text{Jan}}}{r_{\text{Jan}}}$$

$$\frac{150}{r} = \frac{180}{r + 2} \qquad \text{\textit{The LCD is} } r(r + 2).$$

$$\frac{r(r + 2)}{1} \cdot \frac{150}{r} = \frac{180}{r+2} \cdot \frac{r(r + 2)}{1}$$

$$150(r + 2) = 180r$$

$$150r + 300 = 180r$$

$$300 = 30r$$

$$10 = r$$

Thus, Robert's rate was $\boxed{10 \text{ kph}}$

CHECK: If Robert's rate was 10 kph he completed his 150-kilometer race in $\frac{150}{10} = 15$ hours. Jan's rate was 2 kph faster, so Jan's rate was 12 kph. Therefore, Jan completed the 180-kilometer race in $\frac{180}{12} = 15$ hours as well.   ∎

Sometimes a verbal problem consists of nothing more than reading a problem and substituting values.

6. What is the initial velocity of a ball if it is thrown from a height of 2,000 feet and reaches a height of 1,200 feet after 8 seconds?

EXAMPLE 6   A basic formula from physics states that the height, s, of a free-falling object (neglecting air resistance) is given by the formula

$$s = s_0 + v_0 t + \frac{1}{2}gt^2$$

where

s_0 = Initial height

v_0 = Initial velocity

t = Time in seconds

g = Acceleration due to gravity = $-32\dfrac{\text{ft}}{\text{sec}^2}$

What is the height of a ball dropped from a height of 2,000 feet after 10 seconds?

Solution If we strip away all the extra information in the problem, what the problem is asking us is:

Find s when $s_0 = 2{,}000$, $t = 10$, and $g = -32$.

But what about v_0? We are told that the ball is being dropped, which implies that there is *no* initial velocity. In other words, $v_0 = 0$.

We substitute all these values into the given formula and we get:

$$s = s_0 + v_0 t + \frac{1}{2}gt^2$$

$$s = 2{,}000 + 0(10) + \frac{1}{2}(-32)(10)^2$$

$$s = 2{,}000 - 1{,}600$$

$$\boxed{s = 400 \text{ feet}}$$

The ball is 400 feet above the ground after 10 seconds. ■

✔ **Answers to Learning Checks in Section 7.6**

1. 15 cm. by 35 cm.   **2.** 36 lb. and 24 lb.   **3.** 15   **4.** $3\frac{3}{5}$ hours

5. 100 kph   **6.** $v_0 = 28$ ft/sec

NOTE TO THE STUDENT

Use the space on this page to write down any questions you have or points you want to review with your instructor.

Exercises 7.6

Solve each of the following problems algebraically.

1. The ratio of the width of a rectangle to its length is 3 to 7, and its perimeter is 100 meters. What are its dimensions?

    1. _____

2. The length of a rectangle is 3 more than its width. The ratio of the length of the rectangle to the *perimeter* is 1 to 3. Find the dimensions of the rectangle.

    2. _____

3. Of 200 people surveyed, the proportion of those who preferred brand X to those who did not was 13 to 12. How many people preferred brand X?

    3. _____

4. In a recent survey 300 more people preferred brand X than the famous national brand. If the ratio of those who did not prefer brand X to those who did was 5 to 7, how many people were surveyed all together?

    4. _____

5. Suppose Joe Slugger's batting average is .300 as the result of 120 hits in 400 at bats. How many hits must he get in his next 50 at bats to raise his average to .400? Look at your answer. What does it imply?

    5. _____

6. Joe Slugger has 90 hits in 300 at bats resulting in a batting average of .300. How many hits must he get in his next 100 at bats in order to raise his average to .400? (You might guess an answer first; the actual answer may surprise you.)

    6. _____

7. Bill can mow a lawn in 3 hours, while Sandy can do the same job in 2 hours. How long will it take them to mow the lawn together?

7. _____

8. Judy can paint her house in 8 days, while Jane could do the same job in 10 days. How long will it take them to paint the house working together?

8. _____

9. An electrician works twice as fast as his apprentice. Together they can complete a rewiring job in 6 hours. How long would it take each of them working alone?

9. _____

10. A physics professor can perform an experiment three times as fast as her graduate assistant. Together they can perform the experiment in 3 hours. How long would it take each of them working alone?

10. _____

11. A plane travels 100 kph faster than a train. The plane covers 500 kilometers in the same time that the train covers 300 kilometers. Find the speed of each.

11. _____

12. A swimmer can swim freestyle 2 meters per second faster than she can swim breaststroke. If she covers 96 meters swimming freestyle in the same time she covers 64 meters doing the breaststroke, how fast does she swim each stroke?

12. _____

13. A man can row at the rate of 4 mph in still water. He can row 8 miles upstream (against the current) in the same time that he can row 24 miles downstream (with the current). What is the speed of the current?

13. _____

14. A family drives to its vacation home at the rate of 90 kph, and returns home at the rate of 80 kph. If the trip returning takes 2 hours longer than the trip going, how far is the trip to the vacation home?

14. _____

15. Ronnie walks over to a friend's house at the rate of 6 kph, and jogs home at the rate of 14 kph. If the total time, walking and jogging, is 3 hours, how far is it to the friend's house?

15. _____

16. A plane's air speed (speed in still air) is 500 kph. The plane covers 1,120 kilometers with a tailwind in the same time it covers 880 kilometers with a headwind (against the wind). What is the speed of the wind?

16. _____

17. One number is three times another, and the sum of their reciprocals is $\frac{5}{3}$. Find the numbers.

17. _____

18. One number is twice another, and the sum of their reciprocals is 2. Find the numbers.

18. _____

19. If the same number is added to the numerator and the denominator of $\frac{3}{5}$, the resulting fraction has the value $\frac{5}{6}$. Find the number.

19. _____

20. What number must be added to both the numerator and the denominator of $\frac{7}{9}$ to obtain a fraction whose value is $\frac{2}{3}$?

20. _____

21. The denominator of a fraction is 2 more than its numerator, and the reciprocal of the fraction is equal to itself. Find the fraction.

21. _____

22. One number is 5 times another and the difference of their reciprocals is $\frac{2}{5}$. Find the numbers.

22. _____

23. The surface area, S, of a closed right circular cylinder is given by the formula

$$S = 2\pi rh + 2\pi r^2$$

where r is the radius and h is the height. Find S, if $r = 6$ cm. and $h = 15$ cm. (You may leave your answer in terms of π.)

23. _____

24. Using the same formula as in Exercise 23, find h, if $S = 351.68$ sq. cm., $r = 4$ cm., and $\pi \approx 3.14$. ("$\approx$" means approximately equal.)

24. _____

CHAPTER 7 SUMMARY

After having completed this chapter, you should be able to:

1. Reduce a fraction by first factoring numerator and denominator where necessary, and then cancelling common factors (Section 7.1).

   For example:

   $$\frac{2x^2 - 6x}{x^2 + x - 12} = \frac{2x(x - 3)}{(x + 4)(x - 3)} = \frac{2x\cancel{(x - 3)}}{(x + 4)\cancel{(x - 3)}} = \boxed{\frac{2x}{x + 4}}$$

2. Multiply and divide fractions whose numerators and/or denominators are polynomials (Section 7.2).

   For example:

   $$\frac{3x^2}{9 - x^2} \div \frac{6x - 2x^2}{x^2 + 4x + 3} = \frac{3x^2}{(3 - x)(3 + x)} \cdot \frac{(x + 3)(x + 1)}{2x(3 - x)}$$

   We have factored where possible, changed division to multiplication, and inverted the divisor.

   $$= \frac{3\overset{x}{\cancel{x}^2}}{(3 - x)\cancel{(3 + x)}} \cdot \frac{\cancel{(x + 3)}(x + 1)}{2\cancel{x}(3 - x)}$$

   $$= \boxed{\frac{3x(x + 1)}{2(3 - x)^2}}$$

3. Find the LCD for polynomial denominators (Section 7.3).

   For example: If the expressions $3x^2$, $x^2 + 3x$, and $x^2 + 2x - 3$ appear as denominators, what is their LCD?

   We first put each polynomial in factored form.

   $$3x^2 = 3x^2$$
   $$x^2 + 3x = x(x + 3)$$
   $$x^2 + 2x - 3 = (x + 3)(x - 1)$$

   The distinct factors are 3, x, $x + 3$, and $x - 1$.

   According to our outline, the LCD is $\boxed{3x^2(x + 3)(x - 1)}$

4. Combine fractions with polynomial denominators (Section 7.3).

   For example:

   $$\frac{3}{x^2 + 2x} + \frac{5}{x^2 - 4} = \frac{3}{x(x + 2)} + \frac{5}{(x + 2)(x - 2)} \qquad LCD: x(x + 2)(x - 2)$$

   $$= \frac{3(x - 2)}{x(x - 2)(x + 2)} + \frac{\cdot 5(x)}{x(x + 2)(x - 2)}$$

   Here the missing factor was $x - 2$.   *Here the missing factor was x.*

   $$= \frac{3x - 6 + 5x}{x(x + 2)(x - 2)}$$

   $$= \boxed{\frac{8x - 6}{x(x + 2)(x - 2)}}$$

5. Solve and check a fractional equation involving polynomial denominators (Section 7.4).

For example: Solve for a and check. $\dfrac{5}{6} - \dfrac{3}{a+2} = \dfrac{8}{15}$

$$\text{LCD} = 30(a+2)$$

$$\frac{30(a+2)}{1} \cdot \frac{5}{6} - \frac{30(a+2)}{1} \cdot \frac{3}{a+2} = \frac{30(a+2)}{1} \cdot \frac{8}{15}$$

$$\frac{\overset{5}{\cancel{30}(a+2)}}{1} \cdot \frac{5}{\cancel{6}} - \frac{\cancel{30(a+2)}}{1} \cdot \frac{3}{\cancel{a+2}} = \frac{\overset{2}{\cancel{30}(a+2)}}{1} \cdot \frac{8}{\cancel{15}}$$

$$25(a+2) - 90 = 16(a+2)$$
$$25a + 50 - 90 = 16a + 32$$
$$25a - 40 = 16a + 32$$
$$9a = 72$$
$$\boxed{a = 8}$$

CHECK $a = 8$:

$$\frac{5}{6} - \frac{3}{(8)+2} \overset{?}{=} \frac{8}{1\text{:}}$$

$$\frac{5}{6} - \frac{3}{10} \overset{?}{=} \frac{8}{1\text{:}}$$

$$\frac{25}{30} - \frac{9}{30} \overset{?}{=} \frac{1\text{(}}{3\text{(}}$$

$$\frac{16}{30} \overset{\checkmark}{=} \frac{1\text{(}}{3\text{(}}$$

6. Solve a literal equation explicitly for a specified variable (Section 7.5).
For example: Solve for y. $3x - 4y = r - 2(y + x + 5)$

$$
\begin{array}{rl}
3x - 4y = & r - 2(y + x + 5) \\
3x - 4y = & r - 2y - 2x - 10 \qquad \text{\textit{We get the y's}} \\
\underline{ +4y \qquad\quad +4y } & \quad \text{\textit{together.}} \\
3x = & r + 2y - 2x - 10 \quad \text{\textit{We isolate the y term.}} \\
\underline{+2x - r + 10 \quad -r + 2x + 10} & \\
5x - r + 10 = & 2y \\
\end{array}
$$

$$\boxed{\dfrac{5x - r + 10}{2} = y}$$

7. Solve verbal problems which give rise to fractional equations (Section 7.6).

ANSWERS TO QUESTIONS FOR THOUGHT

1. The student was incorrect in each case. Neither of the fractions can be reduced because neither contains a common factor.

2. (a) $\left(x + \dfrac{2}{x}\right)\left(x - \dfrac{3}{x}\right) = x^2 - \dfrac{x}{1}\left(\dfrac{3}{x}\right) + \dfrac{x}{1}\left(\dfrac{2}{x}\right) - \left(\dfrac{2}{x}\right)\left(\dfrac{3}{x}\right)$

$$= x^2 - 3 + 2 - \dfrac{6}{x^2} = x^2 - 1 - \dfrac{6}{x^2}$$

(b) $x + \dfrac{2}{x} = \dfrac{x}{1} + \dfrac{2}{x} = \dfrac{x(x)}{1(x)} + \dfrac{2}{x} = \dfrac{x^2}{x} + \dfrac{2}{x} = \dfrac{x^2 + 2}{x}$

$x - \dfrac{3}{x} = \dfrac{x}{1} - \dfrac{3}{x} = \dfrac{x(x)}{1(x)} - \dfrac{3}{x} = \dfrac{x^2}{x} - \dfrac{3}{x} = \dfrac{x^2 - 3}{x}$

Then $\left(x + \dfrac{2}{x}\right)\left(x - \dfrac{3}{x}\right) = \dfrac{x^2 + 2}{x} \cdot \dfrac{x^2 - 3}{x} = \dfrac{(x^2 + 2)(x^2 - 3)}{x^2}$

If the instructions ask for the answer in the form of a single fraction, we get such an answer directly in part (b). In part (a), there would be further work to do.

CHAPTER 7 REVIEW EXERCISES

In Exercises 1–8, *reduce the fraction to lowest terms. If the fraction cannot be reduced, then say so.*

1. $\dfrac{x^2}{x^2 + 2}$

2. $\dfrac{2x^2 + 12x}{6x^2}$

3. $\dfrac{x^2 + 3x - 4}{x^2 - 16}$

4. $\dfrac{2x^2 + 7x - 15}{4x^2 - 9}$

5. $\dfrac{a^2 + 8a + 16}{a^2 + 6a + 8}$

6. $\dfrac{x^2y - xy^2}{x^2 + xy - 2y^2}$

7. $\dfrac{3z^2 - 12}{3z^2 + 9z + 6}$

8. $\dfrac{c^2 + 16}{3c^3 - 12c^2}$

In Exercises 9–18, *perform the indicated operations and simplify as completely as possible.*

9. $\dfrac{x}{x + 2} + \dfrac{x}{2}$

10. $\dfrac{a^2}{a^2 + 4a} \cdot \dfrac{a^2 - 16}{4a^2}$

11. $\dfrac{3}{2x + 4} + \dfrac{6}{x^2 + 2x}$

12. $\dfrac{3}{2x + 4} \div \dfrac{6}{x^2 + 2x}$

13. $\dfrac{x^2 - 5x - 6}{2x - 12} \div \dfrac{x^2 + 2x + 1}{8x^2}$

14. $\dfrac{1}{y} + \dfrac{2}{y + 2} - \dfrac{3}{y + 6}$

15. $\dfrac{5}{z^2 + z - 6} - \dfrac{3}{z^2 + 3z}$

16. $\dfrac{x^2 + xy - 2y^2}{x^2 + 4xy + 4y^2} \div (x^2 - y^2)$

17. $2 + \dfrac{3}{x + 2} - \dfrac{1}{x}$

18. $\left(\dfrac{x}{3} - \dfrac{3}{x}\right) \div \dfrac{2x^2 - 6x}{x^2}$

ANSWERS

1. _____
2. _____
3. _____
4. _____
5. _____
6. _____
7. _____
8. _____
9. _____
10. _____
11. _____
12. _____
13. _____
14. _____
15. _____
16. _____
17. _____
18. _____

In Exercises 19–26, solve the given equations.

19. $\dfrac{5}{x} - \dfrac{2}{3x} = \dfrac{13}{6}$

20. $\dfrac{3}{a+2} - \dfrac{5}{2a+4} = \dfrac{1}{2}$

21. $\dfrac{y+2}{y} + \dfrac{4}{y+2} = \dfrac{y}{y+2}$

22. $\dfrac{2}{x^2-4} - \dfrac{3}{x+2} = \dfrac{4}{x-2}$

23. $\dfrac{x+2}{x-3} + \dfrac{4}{3} = \dfrac{5}{x-3}$

24. $\dfrac{7}{6} - \dfrac{3}{a+2} = \dfrac{11}{12}$

25. $3x - 4y + 7 = 8x - 7y + 3$ for x

26. $\dfrac{1}{x} - \dfrac{2}{a} = \dfrac{3}{c}$ for a

Solve each of the following verbal problems algebraically.

27. A bag contains 80 more red marbles than black ones. If the ratio of red marbles to black ones is 7 to 5, how many marbles are there in the bag all together?

28. Leslie can row at the rate of 4 mph. If a trip 8 miles upstream takes twice as long as the trip downstream, what is the rate of the current?

29. Together John and Susan can complete a job of posting 1,500 notices in 4 hours. If Susan alone could do the job in 6 hours, how long would it take John to do the job alone?

30. What number must be added to the numerator and the denominator of $\frac{8}{13}$ so that the resulting fraction is equal to $\frac{4}{5}$?

19. _____

20. _____

21. _____

22. _____

23. _____

24. _____

25. _____

26. _____

27. _____

28. _____

29. _____

30. _____

1. *Reduce to lowest terms.* $\dfrac{4x^2}{x^2 - 4x}$

2. *Reduce to lowest terms.* $\dfrac{x^2 - 9y^2}{3x^2 - 9xy}$

In Exercises 3–7, perform the indicated operations and simplify as completely as possible.

3. $\dfrac{3}{x + 3} + \dfrac{2}{x + 2}$

4. $\dfrac{x^2 - 4}{x^2 - 4x + 4} \div \dfrac{4x + 8}{(x - 2)^2}$

5. $\dfrac{5}{2x} - \dfrac{10}{x^2 + 4x}$

6. $\dfrac{2x^2 - 8x}{x^2 - 16} \cdot \dfrac{x^2 + 8x + 16}{x^2}$

7. $\dfrac{2x - 5}{x^2 - 3x} - \dfrac{3x + 7}{x^2 - 3x} + \dfrac{6x - 3}{x^2 - 3x}$

1. _____

2. _____

3. _____

5. _____

6. _____

ANSWERS

8. *Solve for x.* $\dfrac{3}{x-2} - \dfrac{4}{x} = \dfrac{1}{2x}$

9. *Solve for t.* $at + b = \dfrac{3t}{2} + 7$

8. _____

10. *Solve for c.* $\dfrac{3c}{c-2} + 4 = \dfrac{c+4}{c-2}$

9. _____

Solve each of the following problems algebraically.

11. A person travels from town A to town B and back in a total of 14 hours. If the average speed going was 45 kph and the average speed returning was 60 kph, how far is it from town A to town B?

10. _____

12. The formula for the area, *A*, of a trapezoid is

$$A = \frac{1}{2}h(b_1 + b_2)$$

where h = height, and b_1, b_2 are the bases. If $A = 105$ sq. cm., $h = 3$ cm., and $b_1 = 10$ cm., find b_2.

11. _____

12. _____

Graphing and Systems of Linear Equations

In the last chapter we had occasion to briefly discuss equations containing more than *one* variable (***literal equations***). We saw that we cannot "solve" such an equation in the same way that we have learned to solve equations such as $5x - 4 = 3x + 2$, and obtain the numerical solution $x = 3$. We could graph this "solution set" as follows:

In this chapter we will focus our attention on a particular type of literal equation by examining first-degree equations in *two* variables. Our discussion will show that our understanding of what a numerical solution to an equation is will have to be modified. We will also see that in order to graph their solution sets, the number line will no longer be sufficient.

The Rectangular (Cartesian) Coordinate System

8.1

Let's begin by considering the following equation in *two* variables:

$$2x + y = 6$$

What does it mean to have a solution to this equation? A moment's thought will make us realize that a *single* solution to this equation consists of *two* numbers—a value for x together with a value for y. In other words, *a solution to this equation consists of a pair of numbers*.

Thus, in the equation $2x + y = 6$, we can easily "see" some pairs of numbers which work:

$$x = 1, \quad y = 4 \quad \text{works because} \quad 2(1) + 4 = 6.$$
$$x = 3, \quad y = 0 \quad \text{works because} \quad 2(3) + 0 = 6.$$
$$x = -1, \quad y = 8 \quad \text{works because} \quad 2(-1) + 8 = 6.$$

On the other hand, we can generate solutions to this equation by simply picking a value for either x or y and solving for the other variable. For example, we may choose $x = 2$, and then substitute $x = 2$ into the equation and solve for y:

$$2x + y = 6$$
$$2(2) + y = 6$$
$$4 + y = 6$$
$$\boxed{y = 2} \qquad \textit{Thus, we know that } x = 2, y = 2 \textit{ is a solution.}$$

If instead we choose $y = 6$, then we substitute $y = 6$ into the equation and solve for x:

$$2x + y = 6$$
$$2x + 6 = 6$$
$$2x = 0$$
$$\boxed{x = 0} \qquad \textit{Thus, we know that } x = 0, y = 6 \textit{ is a solution.}$$

The important thing to realize is that in this way we can generate an unlimited number of solutions to this equation. Even though this equation is not always true (for example, $x = 4$, $y = -1$ does not work), it does have *infinitely* many solutions.

One way of keeping track of the solutions is in a table. We could list the solutions mentioned above as follows:

x	y
1	4
3	0
-1	8
2	2
0	6

Since we cannot list an infinite number of solutions, we are going to develop an alternative way of listing all the solutions. We are going to draw a "picture" of the solution set. In order to do this we need to leave our equation for a little while, but we will return to it at the end of this section.

The Rectangular Coordinate System

The number line is often called a ***one-dimensional coordinate system.*** The word *coordinate* basically means location. *One-dimensional* means that we can move only right or left on the number line. We cannot move off the number line.

On the number line we determine the location of a point by using a *single* number. This number tells us the point's distance and direction from 0 on the number line. Thus, -6 is called the coordinate of a point which is 6 units to the left of 0.

In order to be able to describe locations on a flat surface, which is usually called a ***two-dimensional plane,*** we will need two coordinates. Using two coordinates to describe a location is a common practice with which we are all familiar. For example, to locate a seat in a theater, your ticket usually contains two "coordinates" such as A-4 or G-26. The first coordinate is the letter, which indicates the row. The row closest to the front is usually lettered A, and all remaining rows behind it are ordered alphabetically. The second coordinate is the number indicating your position in the row, usually starting from the left side of the theater (see Figure 8.1 on page 436). Note that we need *two* coordinates to determine an exact seat in the theater.

Naming points in the plane utilizes the same idea. Starting from some agreed-upon point of reference, we use two coordinates: one coordinate indicates right or left, the other indicates up or down.

We build our system in the following way. We begin with a number line,

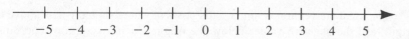

and then we draw another number line perpendicular to the first, so that their 0 points coincide (see Figure 8.2).

We call the horizontal number line the ***x-axis*** and the vertical number line the ***y-axis.*** Together they are called the ***coordinate axes.*** The point of intersection of the coordinate axes is called the ***origin.***

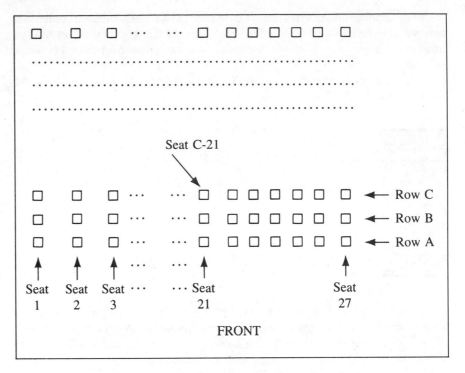

Figure 8.1   *Typical arrangement of seats in a theater*

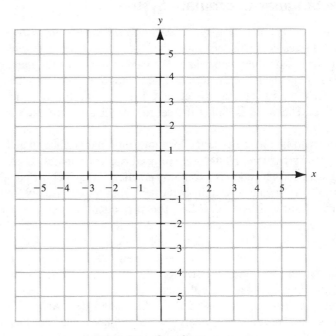

Figure 8.2   *Coordinate axes*

The usual convention is to mark the positive values on the *y*-axis above the origin, and the negative values below the origin. Such a system of coordinate axes is called a ***rectangular*** (or ***Cartesian,*** named after the famous French philosopher and mathematician René Descartes) coordinate system.

To describe the location of a point in the plane, we need *two* numbers, which are called an ***ordered pair.*** An ordered pair of numbers is a pair of numbers enclosed in parentheses and separated by a comma in the following way: (*x, y*).

We imagine ourselves situated at the origin. The first number, called the *x-coordinate*, tells us how far to move right or left: Positive means right, and negative means left. The second number, called the **y-coordinate**, tells us how far to move up or down: Positive means up, and negative means down.

For example, Figure 8.3 shows the location of the point (2, 3). To locate the point (2, 3) we start at the originl and move 2 units right and then 3 units up. Of course, we could also first move 3 units up and then 2 units right. What is important to remember is that the *x*-coordinate tells us "right–left," while the *y*-coordinate tells us "up–down."

Locating a point on a plane is called *plotting* the point.

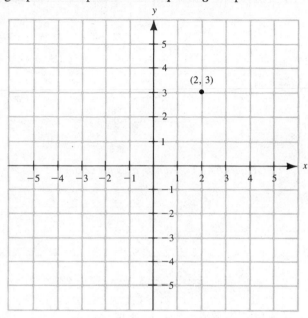

Figure 8.3 *Location of* (2, 3)

EXAMPLE 1   Plot each of the following points.

(a) (3, 4)   (b) (4, 3)   (c) (−3, 5)

(d) (−5, −4)   (e) (2, −2)   (f) (0, −5)

Solution   See Figure 8.4 on page 438. Note that points (3, 4) and (4, 3) are *not* the same. The order in an ordered pair is important.   ∎

EXAMPLE 2   Plot the following points.

(a) (−6, 0)   (b) (0, 4)   (c) (0, 0)

Solution

(a) For the point (−6, 0), the *x*-coordinate, −6, tells us to move 6 units to the left, and the *y*-coordinate, 0, tells us to move 0 units in the vertical direction. Thus, the point (−6, 0) is on the *x*-axis, 6 units to the left of the origin (see Figure 8.5). In fact, any time the *y*-coordinate of a point is 0, the point must lie on the *x*-axis.

(b) For the point (0, 4), the *x*-coordinate, 0, tells us to move 0 units in the horizontal direction, and the *y*-coordinate, 4, tells us to move up 4 units above the origin. Thus, the point (0, 4) is on the *y*-axis, 4 units above the origin. In fact, any time the *x*-coordinate of a point is 0, the point must lie on the *y*-axis.

✔ LEARNING CHECKS

1. Plot each point.
   (a) (1, 5)   (b) (5, 1)
   (c) (2, −4)   (d) (−2, 3)
   (e) (−4, 0)   (f) (−1, −3)

2. Plot each point.
   (a) (5, 0)
   (b) (0, − 3)
   (c) (−2, −2)

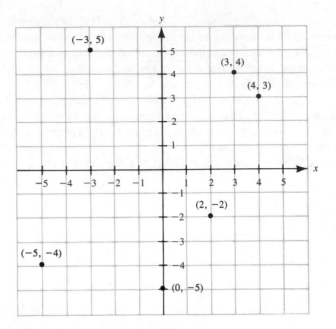

Figure 8.4 *Coordinate axes for Example 1*

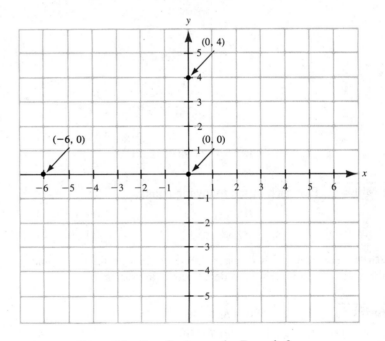

Figure 8.5 *Coordinate axes for Example 2*

(c) The point (0, 0) is the origin. ■

As Figure 8.6 illustrates, the coordinate axes divide the plane into four *quadrants,* which are numbered in a counterclockwise direction starting in the upper right quadrant.

3. Determine the quadrant each point falls in.

(a) (−3, 4)

(b) (−2, −5)

EXAMPLE 3 Indicate which quadrant contains each of the following points.

(a) (2, −5)

(b) (−3, 7)

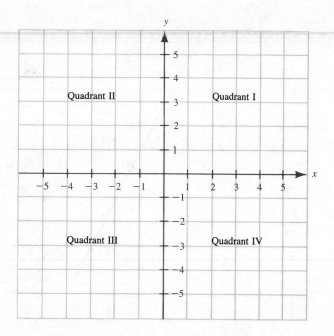

Figure 8.6 *Division of plane into quadrants*

Solution

(a) Since the point (2, −5) has a positive *x*-coordinate and a negative *y*-coordinate, we must move right and down, which puts us into Quadrant IV.

(b) Since the point (−3, 7) has a negative *x*-coordinate and a positive *y*-coordinate, we must move left and up, which puts us into Quadrant II.

Note that a point which lies on either the *x*-axis or the *y*-axis does *not* lie in any of the quadrants. ∎

Let us return now to our equation $2x + y = 6$. Using the notation of ordered pairs, we now have another way of recording those *x* and *y* values which *together* are a solution to the equation. Since we have agreed that in our ordered pair notation, the first coordinate stands for *x* and the second for *y*, instead of saying that $x = 1$, $y = 4$ is a solution to the equation, we can say that the ordered pair (1, 4) is a solution to the equation.

In our table of solutions to the equation $2x + y = 6$ we can add a column which records the same solutions in ordered pair notation:

x	*y*	(*x*, *y*)
1	4	(1, 4)
3	0	(3, 0)
−1	8	(−1, 8)
2	2	(2, 2)
0	6	(0, 6)

We plot these points in Figure 8.7 on page 440. This picture suggests very strongly that the points lie on a straight line and that we should "connect the

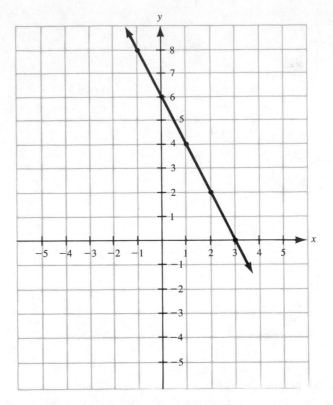

Figure 8.7 *Solutions to the equation* $2x + y = 6$

dots." In fact, this turns out to be true, which means that the straight line is a "picture" of the solution set.

We will continue this discussion of straight lines in the next section.

Answers to Learning Checks in Section 8.1

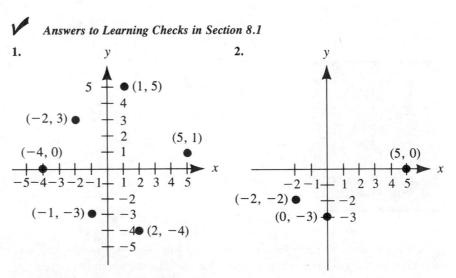

3. (a) Quadrant II   **(b)** Quadrant III

Exercises 8.1

In Exercises 1–20, *plot the given point.*

1. (2, 1)

2. (1, 2)

3. (−2, 1)

4. (1, −2)

5. (2, −1)

6. (−1, 2)

7. (−2, −1)

8. (−1, −2)

9. (4, 0)

10. (0, 4)

11. (−4, 0)

12. (0, −4)

13. (0, 3)

14. (6, 0)

15. (0, −3)

16. (−6, 0)

17. (4, 5)

18. (−3, −4)

19. (5, 5)

20. (−5, −5)

ANSWERS

In Exercises 21–32, indicate which quadrant contains the given point. If a point lies on one of the coordinate axes, indicate which one.

21. (2, 6)   **22.** (−2, 6)

21. _____

22. _____

23. (−2, −6)   **24.** (2, −6)

23. _____

25. (0, 3)   **26.** (3, 0)

24. _____

25. _____

27. (−2.6, 4.9)   **28.** (0, −1.6)

26. _____

29. (0, 0)   **30.** (8, 47)

27. _____

28. _____

31. (489, −16)   **32.** (−586, 0)

29. _____

QUESTIONS FOR THOUGHT

1. Plot the points in the following set: $\{(x, y) \mid y = x + 2, \text{ and } x = -3, 0, 4\}$.

2. Plot the points in the following set: $\{(x, y) \mid x = y + 2, \text{ and } x = -3, 0, 4\}$.

3. Given the equation $3x + 2y = 12$, complete the given ordered pairs:

30. _____

$$(-2, \quad) \quad (0, \quad) \quad (\quad, -3) \quad (\quad, 0)$$

4. Given the equation $4x - y = 8$, complete the given ordered pairs:

31. _____

$$(-2, \quad) \quad (0, \quad) \quad (\quad, 4) \quad (\quad, 0)$$

5. Why is the point (x, y) called an *ordered* pair?

32. _____

6. What is the difference between (x, y) and $\{x, y\}$?

Graphing a Linear Equation in Two Variables

8.2

In the last section we saw that a solution to the equation $2x + y = 6$ consists of a *pair* of numbers, and that there are infinitely many pairs of numbers which will work. We also saw that every ordered *pair* of real numbers can be thought of as a *point* in a rectangular coordinate system.

We are going to limit our discussion here to first-degree equations involving two variables.

If a pair of numbers (x, y) makes an equation true, we say that *the point (x, y) satisfies the equation.*

EXAMPLE 1 In each of the following, determine whether the given point satisfies the given equation.

(a) Point: $(10, 1)$;   equation: $y - x = 9$

(b) Point: $(-4, -5)$;   equation: $-3x = 7 - y$

Solution   To determine whether a particular point satisfies a given equation, we substitute the first coordinate (the x value) for x, and the second coordinate (the y value) for y to see whether we obtain a true equation.

(a) In order to check the point $(10, 1)$, we substitute $x = 10$, $y = 1$ into the equation $y - x = 9$. Be careful not to confuse the order of x and y in the equation with the order of x and y in the ordered pair.

$$y - x = 9$$
$$1 - 10 \overset{?}{=} 9$$
$$-9 \neq 9 \qquad \text{The point } (10, 1) \text{ does \textit{not} satisfy the equation.}$$

(b) To check the point $(-4, -5)$, we substitute $x = -4$, $y = -5$ into the equation $-3x = 7 - y$.

$$-3x = 7 - y$$
$$-3(-4) \overset{?}{=} 7 - (-5)$$
$$12 \overset{\checkmark}{=} 12 \qquad \text{The point } (-4, -5) \text{ \textit{does} satisfy the equation.} \quad \blacksquare$$

We might also be given an equation and *one* of the two coordinates in an ordered pair, and be asked to find the other coordinate which makes the point satisfy the equation.

EXAMPLE 2 Complete the following ordered pairs so that they satisfy the equation
$$3x - 2y = 12$$

(a) $(\ \ , 0)$ **(b)** $(-6, \ \ )$ **(c)** $(0, \ \ )$

Solution

(a) In order to complete the ordered pair $(\ \ , 0)$, we substitute the value $y = 0$ into the equation and then solve for x.

$$3x - 2y = 12$$
$$3x - 2(0) = 12$$
$$3x = 12$$
$$x = 4 \qquad \text{Thus, the ordered pair is } \boxed{(4, 0)}.$$

✔ **LEARNING CHECKS**

1. (a) Point: $(8, -3)$; equation: $x - y = 5$

(b) Point: $(-2, 1)$; equation: $4y - 3x = 10$

2. Complete each ordered pair so that it satisfies the equation $5y = 10 - 2x$.

(a) $(0, \ \ )$ **(b)** $(\ \ , -4)$
(c) $(\ \ , 0)$

(b) In order to complete the ordered pair $(-6, \ \ )$, we substitute the value $x = -6$ into the equation and solve for y.

$$3x - 2y = 12$$
$$3(-6) - 2y = 12$$
$$-18 - 2y = 12$$
$$-2y = 30$$
$$y = -15 \qquad \text{Thus, the ordered pair is} \ \boxed{(-6, \ -15)}.$$

(c) We substitute $x = 0$ and solve for y.

$$3x - 2y = 12$$
$$3(0) - 2y = 12$$
$$-2y = 12$$
$$y = -6 \qquad \text{Thus, the ordered pair is} \ \boxed{(0, \ -6)}.$$

As is often the case, completing an ordered pair in which one of the coordinates is 0 [as in parts **(a)** and **(c)**] is particularly easy. We will bring this idea up again in a moment. ∎

As we saw in the last section when we plotted some solutions to the equation $2x + y = 6$, the points which satisfy the equation seem to lie on a straight line. While we are not going to prove this fact, we will accept its truth.

DEFINITION The set of all points which satisfy an equation is called the *graph* of the equation.

In mathematics, the statement of an important fact *which can be proven* is called a *theorem*.

THEOREM The graph of an equation of the form $Ax + By = C$ is a straight line (provided A and B are not both equal to 0).

It is for this reason that a first-degree equation in two variables is called a *linear equation*. When a linear equation is written in the form $Ax + By = C$, we say that the equation is in *standard form*.

Keep in mind that not every linear equation is given in standard form. It is enough for us to know that we can put it into standard form if necessary.

When we say that "the solution set to a first-degree equation in two variables is a straight line," we mean to say *two* things. The first is that if we plot all the solutions to such an equation, they will all fall on a straight line. The second is that if we pick any point on this straight line, the point will satisfy the equation.

One suggestive way to think of an equation is to think of it as a condition. For example, the equation $2x + y = 6$ places the following condition on any point which is a member of the solution set:

"Two times the x-coordinate plus the y-coordinate must be equal to 6."

Since we know that the graph of a linear equation is a straight line, we would like to be able to sketch its graph. We know that a straight line is determined by *two* points, and therefore we need to find two points which satisfy the equation. But which two points should we try to find in order to draw the line? We could find *any* two points (pairs of numbers) which satisfy the equation by arbitrarily choosing a number for one variable and solving for the other. However, we just saw in Example 2 that given an equation, completing ordered pairs where one of the coordinates is 0 is quite easy. Why not use those points to draw the line?

Let's illustrate this approach with several examples.

Suppose we want to sketch the graph of the equation $2x - 3y = 12$. We begin by completing the ordered pairs $(0, \quad)$ and $(\quad , 0)$. To complete $(0, \quad)$, we substitute $x = 0$ and solve for y.

$$2x - 3y = 12$$
$$2(0) - 3y = 12$$
$$-3y = 12$$
$$y = -4 \qquad \text{Thus, the first point is } \boxed{(0, -4)}.$$

To complete $(\quad , 0)$, we substitute $y = 0$ and solve for x.

$$2x - 3y = 12$$
$$2x - 3(0) = 12$$
$$2x = 12$$
$$x = 6 \qquad \text{Thus, the second point is } \boxed{(6, 0)}.$$

Having these two points, we can draw our line (see Figure 8.8).

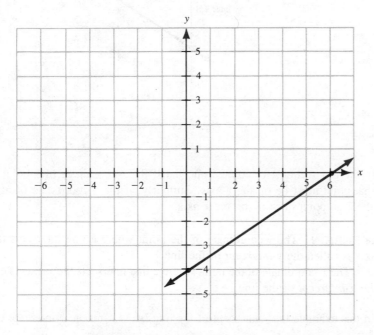

Figure 8.8 *The graph of* $2x - 3y = 12$

However, we must be careful here. Since any two points determine a line, how do we know we have the correct line? It is a very good idea to get one more point on the line just to check that we have not made an error. What additional point shall we find? We simply look at the equation and try to pick a "convenient" value for either x or y, and then solve for the other variable.

Let's pick $x = 3$; we substitute $x = 3$ into the equation and solve for y.

$$2x - 3y = 12$$
$$2(3) - 3y = 12$$
$$6 - 3y = 12$$
$$-3y = 6$$
$$y = -2 \quad \text{Thus, the third point is } \boxed{(3, -2)}.$$

If we now plot this point we see that it falls on the line we have already drawn (see Figure 8.9).

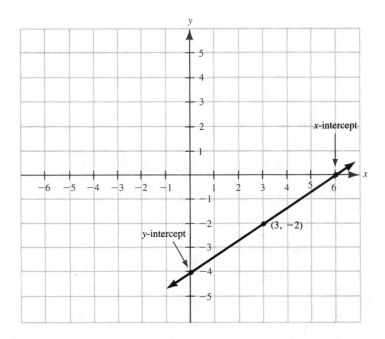

Figure 8.9

In the last section, we pointed out that if the x-coordinate of a point is 0, then that point must lie on the y-axis. Similarly, if the y-coordinate of a point is 0, then the point must lie on the x-axis.

DEFINITION   The y-coordinate of the point where a line intersects the y-axis is called the *y-intercept* of the line.

The x-coordinate of the point where a line intersects the x-axis is called the *x-intercept* of the line.

In this example the y-intercept of the line is -4, and the x-intercept of the line is 6.

Keep in mind that we have drawn only a representative portion of the line. The line extends indefinitely in both directions.

The method we have just outlined is called the *intercept method* for graphing a straight line. In most cases it is the preferred method to follow because the points are the easiest to find and the simplest to plot.

3. Sketch $2x = 3y - 12$.

EXAMPLE 3   Sketch the graph of $5y = 2x + 10$.

Solution   Since this is a linear equation, its graph is a straight line. We will find the x- and y-intercepts as the two points we need to draw the line.

To find the x-intercept, we set $y = 0$ and solve for x.

$$5y = 2x + 10$$
$$5(0) = 2x + 10$$
$$0 = 2x + 10$$
$$-5 = x \qquad \text{Therefore, the } x\text{-intercept is } -5. \text{ We plot the point } (-5, 0).$$

To find the y-intercept, we set $x = 0$, and solve for y.

$$5y = 2x + 10$$
$$5y = 2(0) + 10$$
$$5y = 10$$
$$y = 2 \qquad \text{Therefore, the } y\text{-intercept is 2. We plot the point } (0, 2).$$

For our "check" point we choose $x = 5$. (We choose $x = 5$ so that we will hopefully get a nice y value. The reason we hope for a nice answer is that in order to solve for y we need to divide by 5. Therefore, we pick an x value which is divisible by 5.)

$$5y = 2x + 10$$
$$5y = 2(5) + 10$$
$$5y = 20$$
$$y = 4 \qquad \text{We have as our third point } (5, 4)$$

We can now draw the graph of the line $5y = 2x + 10$ (Figure 8.10).   ∎

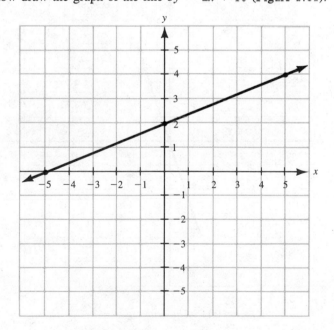

Figure 8.10   *The graph of* $5y = 2x + 10$

There is one difficulty with the intercept method. As we will see in the next example, if a line passes through the origin $(0, 0)$, both the x- and y-intercepts are the same. Rather than getting the necessary two points we get only one.

4. Sketch $y = -2x$.

EXAMPLE 4 Sketch the graph of $y = 3x$.

Solution We proceed as before and try to find the x- and y-intercepts.
To find the x-intercept, we set $y = 0$ and solve for x.

$$y = 3x$$
$$0 = 3x$$
$$0 = x \qquad \text{Thus, the } x\text{-intercept is } 0.$$

If a line crosses the x-axis at the origin, then it must cross the y-axis there as well. That is, the x- and y-intercepts of this line coincide. (Verify for yourself that if you substitute $x = 0$, you will get $y = 0$.) Since we get only one point from the intercept method, we must find another point. We are free to choose any convenient value for either x or y, and then solve for the other variable.
This time we choose $y = 3$, and solve for x.

$$y = 3x$$
$$3 = 3x$$
$$1 = x \qquad \text{Therefore, our second point is } (1, 3).$$

For our check point we choose $x = -1$, and solve for y.

$$y = 3x$$
$$y = 3(-1) = -3 \qquad \text{We have as our check point } (-1, -3).$$

The graph appears in Figure 8.11.

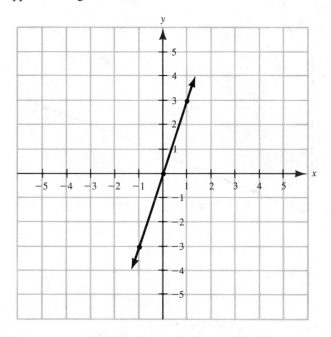

Figure 8.11 *The graph of $y = 3x$* ■

There are two other special types of lines where the intercept method fails. We illustrate these in the next example.

5. Sketch.
   (a) $x = -4$
   (b) $y = 1$

EXAMPLE 5 Sketch the graphs of the following equations in a rectangular coordinate system.
   (a) $x = 3$ **(b)** $y = -2$

Solution   At first glance both the wording of the example and the appearance of the given equations may seem peculiar. Why the extra words "in a rectangular coordinate system"? Why do the equations not have two variables as all the other equations in this chapter have had?

The answers to these questions are closely related. If we are simply asked to graph the solution set of the equation $x = 3$, we might quite reasonably answer as follows:

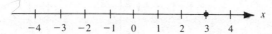

The fact that the question specifies "in a rectangular coordinate system" means that we have to think of the equations $x = 3$ and $y = -2$ in the context of a first-degree equation in *two* variables. As we pointed out earlier, it is often helpful to think of an equation as a condition that x and y must satisfy. If only one of the variables appears, then there is *no* condition on the other variable.

(a) In the equation $x = 3$, y does not appear. Therefore, the equation $x = 3$ places a condition on x only. That condition is that the x-coordinate of the point be 3. In other words, all points which satisfy this equation must be located 3 units to the right. (Think of the standard form $Ax + By = C$. This is just a case where $B = 0$.) That is, the equation $x = 3$ can be thought of as $1 \cdot x + 0 \cdot y = 3$. This equation implies that x must be equal to 3 but that y can be anything!

Our graph is going to be a line parallel to the y-axis and 3 units to the right. The graph appears in Figure 8.12.

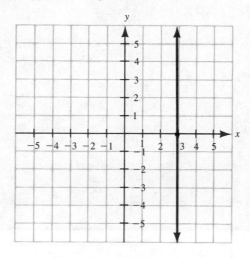

Figure 8.12   *The graph of $x = 3$*

(b) Similarly, in the equation $y = -2$, x does not appear. (This is the special case of $Ax + By = C$ with $A = 0$.) This means that the condition on our point is that it always be located 2 units down. On the other hand, x can be anything!

Our graph is going to be a line parallel to the x-axis and 2 units down. The graph appears in Figure 8.13 on page 450.   ∎

In general, the graph of an equation of the form $x =$ constant is a line parallel to the y-axis. The graph of an equation of the form $y =$ constant is a line parallel to the x-axis.

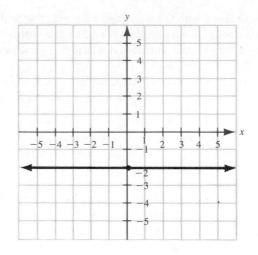

Figure 8.13   *The graph of $y = -2$*

Let's summarize the intercept method for graphing a straight line.

THE INTERCEPT METHOD

1. Locate the *x*-intercept, which is the *x*-coordinate of the point where the line crosses the *x*-axis.

   Do this by substituting $y = 0$ into the equation and solving for *x*.

2. Locate the *y*-intercept, which is the *y*-coordinate of the point where the line crosses the *y*-axis.

   Do this by substituting $x = 0$ into the equation and solving for *y*.

3. Locate a third "check" point by choosing a convenient value for either *x* or *y*, and then solving for the other variable.

4. Draw the line passing through all three points.

Sometimes we obtain fractional values for our points. Since we are just "sketching" the graph, we simply plot the points in a reasonably accurate location.

6. Sketch $5x - 2y = 9$.

EXAMPLE 6   Sketch the graph of $3x + 4y = 10$.

Solution   We sketch the graph using the intercept method. To find the *x*-intercept, set $y = 0$ and solve for *x*.

$$3x + 4y = 10$$
$$3x + 4(0) = 10$$
$$3x = 10$$
$$x = \frac{10}{3} = 3\frac{1}{3} \qquad \text{The } x\text{-intercept is } 3\frac{1}{3}.$$

To find the *y*-intercept, set $x = 0$ and solve for *y*.

$$3x + 4y = 10$$
$$3(0) + 4y = 10$$
$$4y = 10$$
$$y = \frac{10}{4} = 2\frac{1}{2} \qquad \text{The } y\text{-intercept is } 2\frac{1}{2}$$

The graph appears in Figure 8.14. We leave it to the student to check this graph by finding a third point and verifying that it lies on the line.

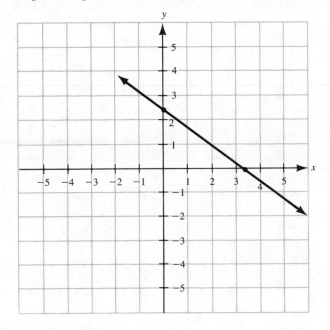

Figure 8.14   *The graph of* $3x + 4y = 10$

✔ *Answers to Learning Checks in Section 8.2*

1. (a) No   (b) Yes   **2.** (a) (0, 2)   (b) (15, −4)   (c) (5, 0)

3.
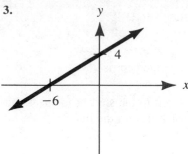

4.
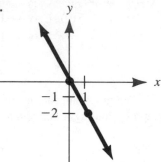

5. (a)
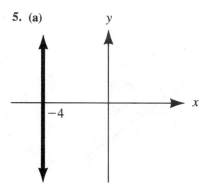

(b)
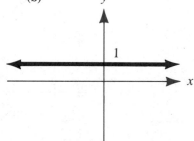

6.
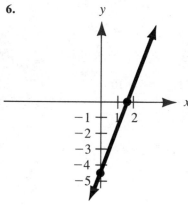

Exercises 8.2

In Exercises 1–14, find the x- and y-intercepts of the equation.

1. $x - y = 7$

2. $x + y = 9$

3. $y - x = -4$

4. $y + 3x = -6$

5. $2x + 3y = 12$

6. $3x + 2y = 12$

7. $y = 4x$

8. $y = -3x$

9. $2x = 5y$

10. $3x = -7y$

11. $x = 5$

12. $y = 4$

13. $x - 4 = 3y$

14. $3x - 8 = 4y$

In Exercises 15–36, use the intercept method to sketch the graph of the given equation in a rectangular coordinate system.

15. $x + y = -5$ **16.** $x - y = 4$

17. $x - y = 8$ **18.** $x + y = -2$

19. $3x - 4y = 12$ **20.** $3x - 7y = 21$

21. $y = 2x - 10$ **22.** $y = 4x + 12$

23. $y = -4x$ **24.** $y = 3x$

25. $y + 7 = x - 5$ **26.** $x + 3 = y - 4$

27. $y = 5$ **28.** $x = 6$

29. $x = -4$ **30.** $y = -1$

31. $y = \dfrac{x - 3}{2}$

32. $x = \dfrac{y + 2}{3}$

33. $2(x - 3) = 4(y + 2)$

34. $6(y - 1) = 3(x + 2)$

35. $\dfrac{2x + 1}{2} = \dfrac{y - 3}{4}$

36. $\dfrac{y + 6}{3} = \dfrac{x - 4}{4}$

QUESTIONS FOR THOUGHT

7. Sketch the graph of the line whose x-intercept is 3, and whose y-intercept is 5.

8. Sketch the graph of the line whose x- and y-coordinates are the same. What would the equation of this line be?

9. Sketch the graph of the line whose x- and y-coordinates are negatives of each other. What would the equation of this line be?

10. Define x- and y-intercepts in two ways:

   (a) In terms of the graph of an equation

   (b) In terms of an algebraic solution to the equation

The Equation of a Line

8.3

In the last section we discussed the procedure for sketching the graph of a line whose equation we are given. We now turn our attention to the reverse situation: How can we obtain the equation of a line given its graph?

In order to answer this question we need to introduce the idea of the *slope* of a line. As we shall see in a moment, the slope of a line is a number which indicates a line's "steepness" and its direction.

DEFINITION   Let (x_1, y_1) and (x_2, y_2) be any two points on a nonvertical line L. The *slope* of L, denoted by m, is

$$m = \frac{\text{Change in } y}{\text{Change in } x} = \frac{y_2 - y_1}{x_2 - x_1}$$

EXAMPLE 1   Compute the slope of the line passing through each of the following pairs of points. Sketch the graph of each line.

(a) $(2, -1)$ and $(5, 3)$ (b) $(-1, 2)$ and $(1, -3)$

(c) $(4, 5)$ and $(-2, 5)$

Solution

(a) Using the definition of slope that we have just given, we can let the "first" point be $(x_1, y_1) = (2, -1)$ and the "second" point be $(x_2, y_2) = (5, 3)$. We then get

$$m = \frac{y_2 - y_1}{x_2 - x_1} = \frac{3 - (-1)}{5 - 2} = \boxed{\frac{4}{3}}$$

Notice that the order in which we choose the points is irrelevant. We could just as well have let $(x_1, y_1) = (5, 3)$ and $(x_2, y_2) = (2, -1)$. We then get

$$m = \frac{y_2 - y_1}{x_2 - x_1} = \frac{-1 - 3}{2 - 5} = \frac{-4}{-3} = \boxed{\frac{4}{3}}$$

What is important is that we subtract the x- and y- coordinates in the *same* order.

The graph appears in Figure 8.15. Note that to get from the point $(2, -1)$ to the point $(5, 3)$ we move right 3 units (the change in x) and up 4 units (the change in y).

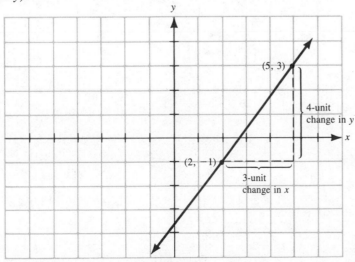

Figure 8.15

(b) The given points are $(-1, 2)$ and $(1, -3)$. We get

$$m = \frac{\text{Change in } y}{\text{Change in } x} = \frac{y_2 - y_1}{x_2 - x_1} = \frac{-3 - 2}{1 - (-1)} = \boxed{\frac{-5}{2}}$$

The graph appears in Figure 8.16. Note that to get from the point $(-1, 2)$ to the point $(1, -3)$ we move right 2 units (the change in x) and down 5 units (the change in y).

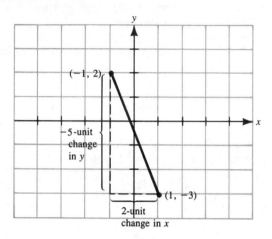

Figure 8.16

(c) The given points are $(4, 5)$ and $(-2, 5)$. We get

$$m = \frac{\text{Change in } y}{\text{Change in } x} = \frac{y_2 - y_1}{x_2 - x_1} = \frac{5 - 5}{4 - (-2)} = \frac{0}{6} = \boxed{0}$$

The graph appears in Figure 8.17. Note that this line which has zero slope is *horizontal*.

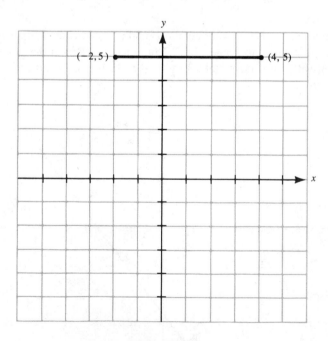

Figure 8.17

It is important to keep in mind that the slope of a line is constant. Basic geometry tells us that regardless of which two points we choose on a particular line, the slope determined by them is the same.

As indicated earlier, the slope of a line is a number which indicates the steepness and direction of the line. Figure 8.18 illustrates lines with various slopes through the point (3, 1). Note that lines with positive slope rise (go up) as we move from left to right, while lines with negative slope fall (go down) as we move from left to right. The larger the slope in absolute value, the steeper the line.

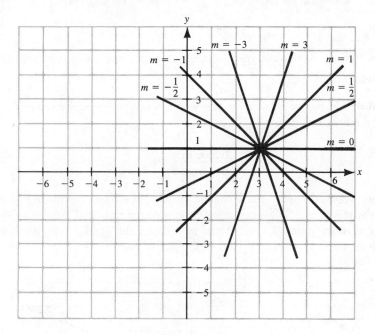

Figure 8.18

If you look back at our definition of slope you will notice that we specified the line be nonvertical. Why? If we try to compute the slope of a vertical line, such as the one passing through the points (1, 3) and (1, 5), we would get

$$m = \frac{5 - 3}{1 - 1} = \frac{2}{0}, \quad \text{which is } \textit{undefined.}$$

Thus, *a vertical line has **no** slope*.

Do not confuse a horizontal line which has 0 slope with a vertical line which has no slope (see Figure 8.19 on page 460).

We summarize this discussion in the accompanying box.

SUMMARY OF SLOPES

A line with positive slope is rising (goes up from left to right).

A line with negative slope is falling (goes down from left to right).

A line with zero slope is horizontal.

A line with no slope (slope undefined) is vertical.

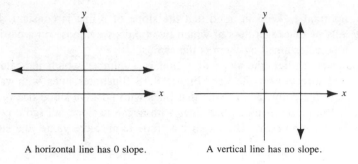

A horizontal line has 0 slope.   A vertical line has no slope.

Figure 8.19

2. Sketch the graph of the line with slope $-\dfrac{4}{3}$ which passes through $(2, 5)$.

EXAMPLE 2   Sketch the graph of the line with slope $\frac{2}{5}$ which passes through the point $(-3, 1)$.

Solution   In order to sketch the line we need *two* points on the line. The slope being $\frac{2}{5}$ means that

$$\frac{\text{Change in } y}{\text{Change in } x} = \frac{2}{5}$$

In other words, for every 5-unit change in x we get a 2-unit change in y. Starting from the point $(-3, 1)$ we can move 5 units to the right and 2 units up, which brings us to the point $(2, 3)$. We can now sketch the graph of the line through the points $(-3, 1)$ and $(2, 3)$ (see Figure 8.20).

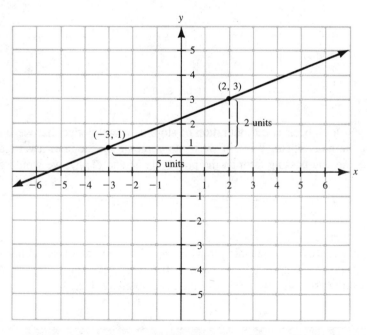

Figure 8.20

As this example illustrates, if we know the slope of the line and a single point on the line, the line is completely determined. (There is only one such line.)   ■

We are now prepared to answer the question raised at the beginning of this section. Suppose we are given a line. That is, suppose we are given the slope of a line and a point on the line: What is the equation of the line?

Let the given slope be m and the given point be (x_1, y_1). If (x, y) is any other point on the line, then the slope determined by the two points (x_1, y_1) and (x, y) must be m. We must have

$$\frac{y - y_1}{x - x_1} = m \qquad \textit{Multiply both sides by } x - x_1.$$

$$y - y_1 = m(x - x_1)$$

We have just derived the following:

POINT–SLOPE FORM OF THE EQUATION OF A LINE

An equation of the line with slope m which passes through the point (x_1, y_1) is

$$y - y_1 = m(x - x_1)$$

EXAMPLE 3   Write an equation of the line with slope 4 passing through the point $(-2, 1)$.

Solution   We use $m = 4$ and $(x_1, y_1) = (-2, 1)$ in the point–slope form.

$$y - y_1 = m(x - x_1)$$
$$y - 1 = 4(x - (-2))$$
$$\boxed{y - 1 = 4(x + 2)}$$

∎

3. Write an equation of the line with slope -3 which passes through $(4, -1)$.

REMEMBER

x_1 and y_1 are the coordinates of the *given* point.

x and y are the variables that appear in the equation.

EXAMPLE 4   Write an equation of the line with slope m and y-intercept b.

Solution   We are given the slope m. Remember that the y-intercept is the y-coordinate of the point where the line crosses the y-axis. Therefore, the fact that the y-intercept is b means that the line passes through the point $(0, b)$.
    Applying the *point–slope form*, we get

$$y - y_1 = m(x - x_1) \qquad \textit{Substitute } (0, b) \textit{ for } (x_1, y_1).$$
$$y - b = m(x - 0)$$
$$y - b = mx$$
$$\boxed{y = mx + b}$$

    This last equation is called the **slope–intercept form** of the equation of a line.

∎

4. Write an equation of the line with slope 4 and y-intercept 7.

> ### SLOPE–INTERCEPT FORM OF THE EQUATION OF A LINE
>
> An equation of the line with slope m and y-intercept b is given by
>
> $$y = mx + b$$

One of the most useful features of the slope–intercept form is that it makes it very easy to determine the slope of a line from its equation.

5. Determine the slopes of the lines whose equations are given.

(a) $y = -2x + 5$

(b) $5x - 3y = 20$

EXAMPLE 5   Determine the slopes of the lines whose equations are given.

(a) $y = 3x - 7$ (b) $2y + 3x = 9$

Solution   The slope–intercept form shows us that when an equation of a line is solved *explicitly* for y, the coefficient of x is the slope.

(a) We simply "read off" the slope from the equation.

$$y = mx + b$$
$$\updownarrow$$
$$y = 3x - 7$$

Thus, $\boxed{m = 3}$. We can also see that the y-intercept is -7.

(b) We must first solve the equation explicitly for y.

$$2y + 3x = 9$$
$$2y = -3x + 9 \qquad \textit{Divide both sides of the equation by 2.}$$
$$y = \frac{-3}{2}x + \frac{9}{2}$$
$$\updownarrow$$
$$y = mx + b$$

Thus, $\boxed{m = -\dfrac{3}{2}}$ Again we can also see that the y-intercept is $\dfrac{9}{2}$.

Keep in mind that if we were asked to find the y-intercept we could just as easily substitute $x = 0$ into the equation and solve for y. ∎

6. Write an equation of the line passing through $(-2, 3)$ and $(4, 8)$.

EXAMPLE 6   Write an equation of the line passing through the points $(3, -4)$ and $(0, 2)$.

Solution   Whether we choose to write an equation of this line using the point–slope form or the slope–intercept form, we need to find the slope of the line.

$$m = \frac{y_2 - y_1}{x_2 - x_1} = \frac{2 - (-4)}{0 - 3} = \frac{6}{-3} = -2$$

Now we can choose to use either form.

Using the *point–slope form*, we can choose to use either $(3, -4)$ or $(0, 2)$ as our given point. If we use $(3, -4)$ we get

$$y - (-4) = -2(x - 3)$$
$$\boxed{y + 4 = -2(x - 3)}$$

Or we can use the *slope–intercept form* by noticing that the line passing through the point (0, 2) means that the y-intercept is 2. (Why?) Thus, we can write

$$y = mx + b \qquad \textit{Substitute } m = -2 \textit{ and } b = 2.$$

$$y = -2x + 2$$

While these two answers may look different, if we solve our first answer for y we get

$$y + 4 = -2(x - 3)$$

$$y + 4 = -2x + 6$$

$$y = -2x + 2 \qquad \textit{This agrees with our second answer.}$$

In conclusion then, when asked to write an equation of a line we are free to use either the point–slope form or the slope–intercept form. Our choice should be guided by which form makes best use of the given information. ∎

✔ *Answers to Learning Checks in Section 8.3*

1. (a) $m = \dfrac{3}{5}$ **(b)** $m = -\dfrac{7}{6}$ **(c)** $m = 0$

2.

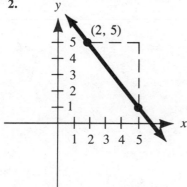

3. $y + 1 = -3(x - 4)$ **4.** $y = 4x + 7$ **5. (a)** $m = -2$ **(b)** $m = \dfrac{5}{3}$

6. $y - 3 = \dfrac{5}{6}(x + 2)$

NOTE TO THE STUDENT

Use the space on this page to write down any questions you have or points you want to review with your instructor.

Exercises 8.3

In Exercises 1–16, find the slope of the line passing through the given points.

1. (3, 5) and (6, 9)

2. (1, 4) and (7, 6)

3. (−1, 4) and (3, −2)

4. (2, −5) and (−4, 3)

5. (−1, −2) and (−3, −4)

6. (−4, −3) and (−2, −5)

7. (0, 2) and (3, 0)

8. (2, 0) and (0, 3)

9. (4, 7) and (−3, 7)

10. (−1, 5) and (3, 5)

11. (2, 6) and (2, 9)

12. (3, −1) and (3, 2)

In Exercises 13–18, sketch the graph of the line satisfying the given conditions.

13. Passing through (2, 1) with slope $\frac{3}{2}$

14. Passing through (2, 1) with slope $\frac{2}{3}$

15. Passing through (−1, 0) with slope −4

16. Passing through (1, −2) with slope $\frac{-2}{3}$

17. Passing through (4, 3) with 0 slope

18. Passing through (4, 3) with no slope

ANSWERS

1. _____

2. _____

3. _____

4. _____

5. _____

6. _____

7. _____

8. _____

9. _____

10. _____

11. _____

12. _____

In Exercises 19–38, write an equation of the line satisfying the given conditions.

19. Passing through (1, 5) with slope 3

20. Passing through (2, 7) with slope −3

21. Passing through (−1, 4) with slope $\dfrac{1}{2}$

22. Passing through (5, −3) with slope $\dfrac{3}{4}$

23. Passing through (0, 6) with slope 5

24. Passing through (0, 2) with slope 4

25. Passing through (−2, 0) with slope $-\dfrac{3}{4}$

26. Passing through (0, −4) with slope $-\dfrac{2}{7}$

27. Passing through (4, −2) with slope 1

28. Passing through (−3, 2) with slope −1

29. Passing through (5, 6) with slope 0

30. Passing through (4, 7) with no slope

31. Passing through (2, 3) and (5, 9)

32. Passing through (1, 5) and (3, 11)

33. Passing through (−1, 4) and (2, −2)

34. Passing through (1, 5) and (3, −4)

35. Vertical line passing through (4, −3)

36. Horizontal line passing through (4, −3)

37. Line has *x*-intercept 5 and *y*-intercept 2

38. Line has *y*-intercept 3 and *x*-intercept 7

In Exercises 39–42, determine the slope of the line from its equation.

39. $y = 5x + 7$

40. $y = -3x - 1$

41. $2y + 3x = 6$

42. $3y - 5x = 12$

CALCULATOR EXERCISES

Find the slope of the line passing through the given pair of points. Round off your answers to the nearest hundredth.

43. (.8, 2.65)   and   (1.3, 4.72)

44. (12.63, 10.44)   and   (9.48, 7.96)

45. (3.7, −1.05)   and   (−2.16, 4.9)

46. (−8.65, 2.8)   and   (12.5, −3.72)

QUESTION FOR THOUGHT

11. What do you think can be said about the slopes of parallel lines?

Solving Systems of Linear Equations: Graphical Method

8.4

In the previous sections we have seen that a linear equation in two variables has infinitely many solutions. If we have two such linear equations, a reasonable question to ask is: "Are there, and can we find solutions (that is, ordered pairs) which satisfy both equations?" Such an ordered pair is called a *simultaneous solution* to the two equations. For example, is there and can we find an ordered pair which satisfies the two equations:

$$\begin{cases} x + y = 6 \\ x - y = 2 \end{cases}$$

This is called a *system of linear equations.* The large brace on the left indicates that the two equations are to be solved simultaneously.

With a little bit of trial and error we might find that the ordered pair (4, 2) works in both equations. (Check it!) However, before we proceed to discuss a systematic method for finding this solution, let's see what we can expect from such a system of equations.

If we keep in mind that the graph of a linear equation is a straight line, then asking for a solution to a system of two linear equations in two variables is the same as asking: "What point do the two lines have in common?" In other words, where do the two lines intersect? Basically, there are three possibilities.

The first possibility is that the two lines intersect in exactly one point. This point is the unique solution to the system of equations. Such a system is called *consistent* (see Figure 8.21).

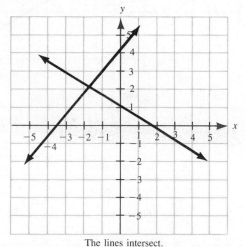

The lines intersect.
There is one solution.
The system is consistent.

Figure 8.21

The second possibility is that the two lines are parallel and so they never intersect. In this case there are no solutions to the system of equations. Such a system is called *inconsistent* (see Figure 8.22 on page 468).

The third possibility is that the graphs of the two equations coincide (they are both the same line). In this case there are infinitely many solutions. All the points which satisfy one equation also satisfy the other. Such as system is called *dependent* (see Figure 8.23).

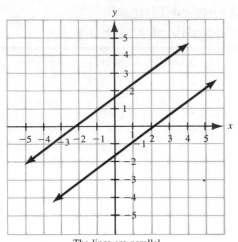

The lines are parallel.
There are no solutions.
The system is inconsistent.

Figure 8.22

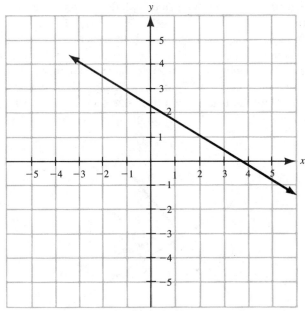

The lines coincide.
There are infinitely many solutions.
The system is dependent.

Figure 8.23

This analysis leads us to the method of solution we will discuss in this section. It is called the ***graphical method.*** In order to solve a system of linear equations we will graph the two equations on the same coordinate axes (by the method we learned in Section 8.2), and then "read off" the point of intersection from our picture. Once we have what we think is the solution, we will check it by substituting the values into both equations.

In Example 1 we will apply this method to the system of equations we mentioned at the beginning of this section.

✔ **LEARNING CHECKS**

1. $\begin{cases} -x + y = 7 \\ x + y = 1 \end{cases}$

——————
EXAMPLE 1   *Solve the system of equations.* $\begin{cases} x + y = 6 \\ x - y = 2 \end{cases}$

Solution We graph both lines by the intercept method. (In order to shorten the solution a bit, we leave the finding of a check point to the student.)

$x + y = 6$ To get the x-intercept, we set $y = 0$ and solve for x.
We get 6 as the x-intercept. Therefore, the line passes through (6, 0).
To get the y-intercept, we set $x = 0$ and solve for y.
We get 6 as the y-intercept. Therefore, the line passes through (0, 6).

$x - y = 2$ Similarly, for the second equation we get 2 as the x-intercept and -2 as the y-intercept.

With this information we graph each of the straight lines (see Figure 8.24). We can see from the picture that the lines cross at the point (4, 2). As we verified before, this point does satisfy both equations.

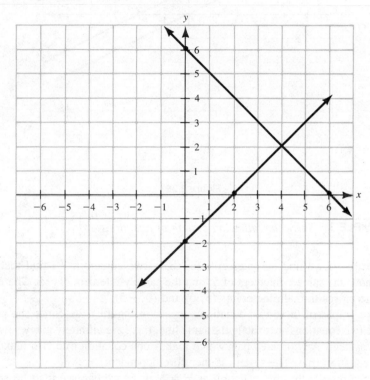

Figure 8.24   ■

EXAMPLE 2 *Solve the following system of equations.* $\begin{cases} x + 2y = 2 \\ 2x + 4y = 0 \end{cases}$ **2.** $\begin{cases} -2x + y = 4 \\ 4x - 2y = 8 \end{cases}$

Solution Again we graph both lines by the intercept method.

$x + 2y = 2$ To get the x-intercept, we set $y = 0$ and solve for x.
Thus, the x-intercept is 2.
To get the y-intercept, we set $x = 0$ and solve for y.
Thus, the y-intercept is 1.

$2x + 4y = 0$ To get the x-intercept, we set $y = 0$ and solve for x.
Thus, the x-intercept is 0. This is also the y-intercept since the line passes through the origin.
Consequently, we need to find a second point on the line.
We choose $x = 4$ and solve for y. We get $y = -2$.
So our second point is (4, -2).

With this information we graph each of the straight lines, as shown in Figure 8.25. From the graphs we can see that the two lines seem to be parallel, which means that the system of equations has no solutions, and is therefore inconsistent. Using the algebraic methods we will discuss in the next section, we can show this is the case.

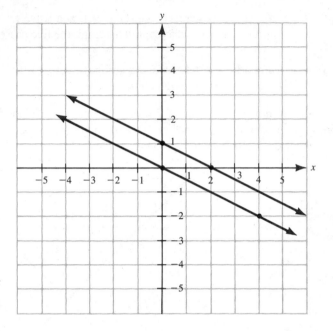

Figure 8.25   ∎

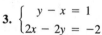

3. $\begin{cases} y - x = 1 \\ 2x - 2y = -2 \end{cases}$

_____

EXAMPLE 3   *Solve the following system of equations.* $\begin{cases} x - y = 5 \\ 3x = 3y + 15 \end{cases}$

Solution   If we find the intercepts for these two equations we find that both lines have the same x-intercept of 5, and the same y-intercept of -5. Therefore, the line passes through the points $(5, 0)$ and $(0, -5)$.

The graphs of these two equations appear in Figure 8.26. Since the graphs for the two equations are exactly the same line, there are infinitely many solutions to the system. Any ordered pair which makes one equation true also makes the other equation true. The system is therefore dependent.

If we rewrite the first equation as $x = y + 5$, we can see that the second equation is just 3 times the first equation. Naturally then, any ordered pair which satisfies one equation must also satisfy the other one. Therefore, the solution set is

$$\boxed{\{(x, y) \mid x - y = 5\}}$$ ∎

Some questions concerning solving a system of equations by the graphical method may have occurred to you by now.

What happens if the actual solution to a system is the point $(2\frac{1}{5}, 4\frac{1}{3})$? Can we be expected to "read" such an answer from the graphs?

In Example 2 we said that we could see that the lines are parallel. Are we sure? What if they intersect at the point $(485, -257)$? These difficulties force us to conclude that the graphical method of solution can be quite imprecise.

Consequently, we need to find another method which will allow us to solve a system of linear equations algebraically—that is, by manipulating the two equations. We will do this in the next section.

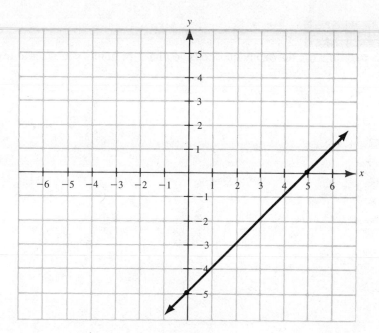

Figure 8.26

✔ *Answers to Learning Checks in Section 8.4*

1. $(-3, 4)$

2. No solutions. Lines are parallel.

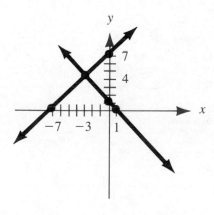

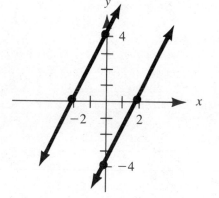

3. $\{(x, y) \mid y - x = 1\}$

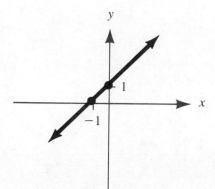

NOTE TO THE STUDENT

Use the space on this page to write down any questions you have or points you want to review with your instructor.

Exercises 8.4

In each of the following exercises, use the graphical method to solve the given system of equations for x and y.

1. $\begin{cases} x + y = 4 \\ x - y = 2 \end{cases}$

2. $\begin{cases} x - y = 5 \\ x + y = 3 \end{cases}$

3. $\begin{cases} 3x + y = 3 \\ x - y = 1 \end{cases}$

4. $\begin{cases} 4x - y = 4 \\ x - y = -2 \end{cases}$

5. $\begin{cases} 3x + y = 6 \\ 6x + 2y = 12 \end{cases}$

6. $\begin{cases} 5x + 2y = 10 \\ 10x + 4y = 40 \end{cases}$

7. $\begin{cases} x + y = 0 \\ x - y = 0 \end{cases}$

8. $\begin{cases} 2x - y = 0 \\ x + 2y = 0 \end{cases}$

9. $\begin{cases} 3x - 2y = 6 \\ x + y = 2 \end{cases}$

10. $\begin{cases} 5x + y = 10 \\ x + y = 6 \end{cases}$

ANSWERS

1. _____

2. _____

3. _____

4. _____

5. _____

6. _____

7. _____

8. _____

9. _____

10. _____

ANSWERS

11. $\begin{cases} 5x - 3y = 15 \\ 2x - y = 4 \end{cases}$

12. $\begin{cases} 3x - 2y = 6 \\ x - 2y = -2 \end{cases}$

11. _____

13. $\begin{cases} y = 2x + 6 \\ y = x + 1 \end{cases}$

14. $\begin{cases} y = 3x - 9 \\ y = x - 5 \end{cases}$

12. _____

13. _____

15. $\begin{cases} y = x - 3 \\ y = x + 4 \end{cases}$

16. $\begin{cases} y = x + 5 \\ y = x - 2 \end{cases}$

14. _____

15. _____

17. $\begin{cases} 2x = y - 8 \\ 3x = y + 6 \end{cases}$

18. $\begin{cases} x = 3y - 12 \\ x = 4y - 8 \end{cases}$

16. _____

17. _____

QUESTIONS FOR THOUGHT

12. Discuss what is meant by a solution to a system of equations.

13. Is it possible for a system of two linear equations to have *exactly* two solutions? Why or why not?

18. _____

Solving Systems of Linear Equations: Algebraic Methods

8.5

As we saw in the last section, the graphical method for solving a system of linear equations has some severe limitations. Consequently, we would like to develop an algebraic method for solving such a system.

By an algebraic method, we mean a method in which we manipulate the equations according to the rules of algebra in order to arrive at the solution to the system. Since we know how to solve a first-degree equation in *one* variable, our strategy is going to be to transform our system of two equations in two variables into one equation in one variable.

The Elimination Method

Let's again consider the system of equations with which we began Section 8.4.

EXAMPLE 1 *Solve the system of equations.* $\begin{cases} x + y = 6 \\ x - y = 2 \end{cases}$

LEARNING CHECKS

1. $\begin{cases} -2x + y = 9 \\ 2x + 3y = 3 \end{cases}$

Solution We know that we can add the same quantity to both sides of an equation and obtain an equivalent equation (an equation with the same solution set). Suppose we add 2 to both sides of the first equation in this system. We obtain

$$\begin{array}{r} x + y = 6 \\ +2 +2 \\ \hline x + y + 2 = 8 \end{array}$$ *This does not seem to help much.*

Now for the slight twist. Since we are seeking a *simultaneous* solution to the system, any solution we find must satisfy *both* equations in the system. The second equation in the system says that the quantity $x - y$ is equal to 2. Therefore, instead of adding 2 to both sides of the first equation, we can add $x - y$ to the left side and 2 to the right side (because $x - y$ and 2 are equal!). This is called "adding the two equations." In this way we still obtain an equivalent equation.

Let's see what happens when we do this.

$$\begin{array}{r} x + y = 6 \\ x - y = 2 \\ \hline 2x = 8 \\ x = 4 \end{array}$$ *We get an equation in **one** variable only. We can solve this equation for x.*

Now we can substitute $x = 4$ into the first equation to solve for y.

$$\begin{array}{r} x + y = 6 \\ 4 + y = 6 \\ y = 2 \end{array}$$ Thus, our solution is the point $\boxed{(4,2)}$.

Finally, we should check this answer in the second equation.

CHECK: $x - y = 2$
$$4 - 2 \overset{\checkmark}{=} 2$$

■

The procedure we used in Example 1 *eliminated* one of the variables from our system of equations, and thereby allowed us to solve for the two variables one at a time. This procedure is called the **elimination method.**

In the system we solved in Example 1, what made the procedure work was that we were fortunate to have the variable *y* appearing with *opposite* coefficients. When we added the two equations, the *y*'s dropped out. Naturally, we cannot expect to have such a system in all cases.

2. $\begin{cases} -3x + 2y = 11 \\ 3x - 3y = -9 \end{cases}$

EXAMPLE 2   *Solve the system of equations.* $\begin{cases} 4x - y = 8 \\ 3x - y = 9 \end{cases}$

Solution   While we could subtract the two equations in this system in order to eliminate *y*, subtracting equations often leads to sign errors; therefore, we will always use *addition* to eliminate one of the variables.

If we simply add the two equations in our system we will get an equivalent equation, *but* we will not eliminate anything.

$$\begin{array}{r} 4x - y = 8 \\ 3x - y = 9 \\ \hline 7x - 2y = 17 \end{array}$$ *This is still an equation in two variables.*

Before we add up our equations, we need to have one of the variables appearing with *opposite* coefficients. Then, when we add up the two equations, that variable will drop out. In order to accomplish this we will multiply both sides of the second equation by -1.

Keep in mind it is only the sign of the *y* coefficient we care about, but the multiplication property of equality requires us to multiply *both sides* of the equation by -1.

$$\begin{cases} 4x - y = 8 \\ 3x - y = 9 \end{cases} \quad \xrightarrow[\text{Multiply by } -1]{\text{As is}} \quad \begin{array}{r} 4x - y = 8 \\ -3x + y = -9 \\ \hline x = -1 \end{array} \quad \begin{array}{l}\textit{Now add the} \\ \textit{resulting equations.}\end{array}$$

Now we can substitute $x = -1$ into either one of the *original* equations to solve for *y*. We will substitute $x = -1$ into the first equation.

$$\begin{aligned} 4x - y &= 8 \\ 4(-1) - y &= 8 \\ -4 - y &= 8 \\ -y &= 12 \\ y &= -12 \end{aligned}$$ Thus, our solution is the point $\boxed{(-1, -12)}$.

We check this solution by substituting $x = -1$, $y = -12$ into the second equation.

CHECK:
$$\begin{aligned} 3x - y &= 9 \\ 3(-1) - (-12) &\overset{?}{=} 9 \\ -3 + 12 &\overset{\checkmark}{=} 9 \end{aligned}$$ ∎

In order to avoid multiplying the equations by fractions, sometimes it is necessary to multiply *both* equations by an appropriate constant in order to eliminate one of the variables.

EXAMPLE 3 *Solve the system of equations.* $\begin{cases} 6x - 7y = 26 \\ -4x - 5y = 2 \end{cases}$

3. $\begin{cases} 5x - 9y = 11 \\ 9x - 6y = 30 \end{cases}$

Solution First we must decide which variable we want to eliminate. In order to convert either $6x$ into $4x$ or $-4x$ into $-6x$ so that the x's will drop out, we would need to multiply by fractions; similarly for the y coefficients.

Instead, we can convert the coefficients of one of the variables in both equations into opposite coefficients. The easiest approach is to convert both coefficients of one of the variables into their least common multiple (LCM). Recall that the least common multiple of two numbers is the smallest number exactly divisible by each of the numbers. For example, the LCM for 4 and 6 is 12; the LCM for 5 and 7 is 35. Since the arithmetic is a little easier, we choose to eliminate x.

We are going to convert the coefficients of x into 12 and -12 so that the x terms will drop out when we add. Remember that we must multiply both sides of the equation in each case.

$$\begin{cases} 6x - 7y = 26 \quad \xrightarrow{\text{Multiply by 2}} \quad 12x - 14y = 52 \\ -4x - 5y = 2 \quad \xrightarrow{\text{Multiply by 3}} \quad \underline{-12x - 15y = 6} \end{cases}$$

$$-29y = 58 \qquad \begin{array}{l}\textit{Add the}\\ \textit{resulting}\\ \textit{equations.}\end{array}$$

$$y = -2$$

Now we substitute $y = -2$ into one of the original equations (we will use the first one) and solve for x.

$$6x - 7y = 26$$
$$6x - 7(-2) = 26$$
$$6x + 14 = 26$$
$$6x = 12$$
$$x = 2 \qquad \text{Thus, our solution is the point } \boxed{(2, -2)}.$$

CHECK: We begin by substituting $x = 2, y = -2$ into the other original equation (in this case, the second one).

$$-4x - 5y = 2$$
$$-4(2) - 5(-2) \stackrel{?}{=} 2$$
$$-8 + 10 \stackrel{\checkmark}{=} 2 \qquad\qquad\qquad ■$$

The choice of which variable to eliminate is yours. In many instances one of the variables offers some advantages. As you do more exercises, you will learn to recognize how to make the easier choice. However, even if you choose the "harder" variable to eliminate, the procedure is still the same.

Let's pause to summarize the elimination method.

THE ELIMINATION METHOD

1. Decide which variable you want to eliminate.
2. Multiply both sides of one or both equations by appropriate constants so that the variable you have chosen to eliminate appears with opposite coefficients.

3. Add the two resulting equations.

4. Solve the equation in *one* variable obtained in step 3.

5. Substitute the value of the variable obtained in the previous step into one of the original equations, and solve for the other variable.

6. Check your solution in the *other* original equation.

4. $\begin{cases} 3x + 5y = 15 \\ 4x + 3y = 9 \end{cases}$

EXAMPLE 4   *Solve the system of equations.* $\begin{cases} 6x - 5y = 6 \\ 7x - 2y = 7 \end{cases}$

Solution   Since the LCM for 6 and 7 is 42, while the LCM for 5 and 2 is 10, we choose to eliminate the y terms. In addition, since the y coefficients have the same sign, we will also need to change the sign of one of them. We will change the sign in the first equation. We do this by multiplying by -2 instead of $+2$.

$\begin{cases} 6x - 5y = 6 \\ 7x - 2y = 7 \end{cases}$ $\begin{array}{l} \xrightarrow{\textit{Multiply by } -2} \\ \xrightarrow{\textit{Multiply by } 5} \end{array}$ $\begin{array}{r} -12x + 10y = -12 \\ 35x - 10y = 35 \\ \hline 23x = 23 \\ x = 1 \end{array}$ *Add the resulting equations.*

Now we substitute $x = 1$ into one of the original equations. This time we will use the second one, and solve for y.

$$7x - 2y = 7$$
$$7(1) - 2y = 7$$
$$7 - 2y = 7$$
$$-2y = 0$$
$$y = 0$$ Thus, the solution is the point $\boxed{(1, 0)}$.

CHECK: We check by substituting $x = 1$, $y = 0$ into the other original equation (in this case, the first one).

$$6x - 5y = 6$$
$$6(1) - 5(0) \overset{?}{=} 6$$
$$6 - 0 \overset{\checkmark}{=} 6$$ ∎

In order for the elimination method to work most easily, it is preferable that the two equations have the like variables lined up in the same column, with the constant on the other side of the equation. Additionally, we would certainly prefer to work with equations without fractional expressions. Consequently, we may have preliminary steps to go through before we are actually ready for the elimination procedure.

5. $\begin{cases} y = -x + 2 \\ \dfrac{x}{4} - \dfrac{y}{5} = 5 \end{cases}$

EXAMPLE 5   *Solve the system of equations.* $\begin{cases} \dfrac{x}{2} + \dfrac{y}{3} = 5 \\ y = 2x + 1 \end{cases}$

Solution   We begin by multiplying both sides of the first equation by 6 in order to clear the denominators, and moving the $2x$ in the second equation from the right side to the left. We do this by subtracting $2x$ from both sides of the second equation.

$$\begin{cases} \dfrac{x}{2} + \dfrac{y}{3} = 5 \\ y = 2x + 1 \end{cases} \xrightarrow[\text{from both sides}]{\substack{\textit{Multiply by 6} \\ \textit{Subtract 2x}}} \begin{array}{l} 3x + 2y = 30 \\ -2x + y = 1 \end{array}$$

Now we can proceed to the elimination process. We choose to eliminate y because it requires multiplying only one of the equations.

$$\begin{cases} 3x + 2y = 30 \\ -2x + y = 1 \end{cases} \xrightarrow[\text{Multiply by } -2]{\textit{As is}} \begin{array}{l} 3x + 2y = 30 \\ 4x - 2y = -2 \\ \hline 7x = 28 \\ x = 4 \end{array} \quad \begin{array}{l} \textit{Now we add the} \\ \textit{resulting equations.} \end{array}$$

Now we substitute $x = 4$ into one of the original equations. This time we use the second one (because it is simpler), and solve for y.

$$y = 2x + 1$$
$$y = 2(4) + 1$$
$$y = 9 \qquad \text{Thus, our solution is the point } \boxed{(4, 9)}.$$

CHECK: We check by substituting $x = 4$, $y = 9$ into the other original equation (in this case, the first one).

$$\frac{x}{2} + \frac{y}{3} = 5$$
$$\frac{4}{2} + \frac{9}{3} \overset{?}{=} 5$$
$$2 + 3 \overset{\checkmark}{=} 5 \qquad\qquad\qquad\qquad \blacksquare$$

We should mention that once we get the value of one of our variables we can substitute that value into any equation which contains both variables, in order to solve for the missing variable. However, if we use equations other than the original ones for this purpose, then we should check our solution in *both* of the original equations. We illustrate this point in the next example.

EXAMPLE 6   *Solve the system of equations.* $\begin{cases} x + \dfrac{y}{2} = 3 \\ \dfrac{2x}{5} - y = -2 \end{cases}$

6. $\begin{cases} x - \dfrac{y}{4} = 1 \\ \dfrac{x}{6} + y = 1 \end{cases}$

Solution   In order to clear the fractions, we multiply both sides of the first equation by 2, and both sides of the second equation by 5.

$$\begin{cases} x + \dfrac{y}{2} = 3 \\ \dfrac{2x}{5} - y = -2 \end{cases} \begin{array}{l} \xrightarrow{\textit{Multiply by 2}} \\ \xrightarrow{\textit{Multiply by 5}} \end{array} \begin{array}{l} 2x + y = 6 \\ 2x - 5y = -10 \end{array}$$

Next we proceed to the elimination process. We choose to eliminate y.

$$\begin{cases} 2x + y = 6 \\ 2x - 5y = -10 \end{cases} \xrightarrow[\textit{As is}]{\textit{Multiply by 5}} \begin{array}{l} 10x + 5y = 30 \\ 2x - 5y = -10 \\ \hline 12x = 20 \\ x = \dfrac{20}{12} = \dfrac{5}{3} \end{array} \quad \begin{array}{l} \textit{Now we add the} \\ \textit{resulting} \\ \textit{equations.} \end{array}$$

Instead of substituting $x = \frac{5}{3}$ into one of the original equations, we might prefer to substitute it into one of the equations with the denominators already cleared. If we substitute $x = \frac{5}{3}$ into the equation $2x + y = 6$, we can solve for y.

$$2x + y = 6$$

$$2\left(\frac{5}{3}\right) + y = 6$$

$$\frac{10}{3} + y = 6 \qquad \textit{Multiply by 3 to clear the fractions.}$$

$$10 + 3y = 18$$

$$3y = 8$$

$$y = \frac{8}{3} \qquad \text{Thus, our solution is the point } \left(\frac{5}{3}, \frac{8}{3}\right).$$

Now we should check the solution in *both* original equations.

CHECK:
$$x + \frac{y}{2} = 3 \qquad\qquad\qquad \frac{2x}{5} - y = -2$$

$$\frac{5}{3} + \frac{\frac{8}{3}}{2} \overset{?}{=} 3 \qquad\qquad\qquad \frac{2\left(\frac{5}{3}\right)}{5} - \frac{8}{3} \overset{?}{=} -2$$

$$\frac{5}{3} + \frac{4}{3} \overset{?}{=} 3 \qquad\qquad\qquad 2 \cdot \frac{5}{3} \cdot \frac{1}{5} - \frac{8}{3} \overset{?}{=} -2$$

$$\frac{9}{3} \overset{\checkmark}{=} 3 \qquad\qquad\qquad \frac{2}{3} - \frac{8}{3} \overset{\checkmark}{=} -2$$

∎

The Substitution Method

There is another method for solving a system of equations, called the ***substitution method.*** In the substitution method we solve for one of the variables in one of the equations, and then substitute for that variable in the other equation. For example, to solve the system

$$\begin{cases} x + 2y = 1 \\ x - y = 4 \end{cases}$$

we proceed as follows. We first solve the second equation for x, obtaining $x = y + 4$, and then substitute the quantity $x = y + 4$ into the first equation.

$$x + 2y = 1 \qquad \textit{Substitute } y + 4 \textit{ for } x.$$

$$\overbrace{y + 4} + 2y = 1$$

$$3y + 4 = 1$$

$$3y = -3$$

$$y = -1$$

Once we know the y value we can proceed to find x as we would in the elimination method. Substituting $y = -1$ into the equation $x = y + 4$, we get

$$x = y + 4$$

$$x = -1 + 4$$

$$x = 3 \qquad \text{Thus, the solution to the system is } \boxed{(3, -1)}.$$

The check is left to the student.

Let's summarize the substitution method.

> **THE SUBSTITUTION METHOD**
>
> 1. Use one of the equations to solve explicitly for one of the variables.
> 2. Substitute the expression obtained in step 1 into the *other* equation.
> 3. Solve the resulting equation in one variable.
> 4. Substitute the value obtained into one of the equations containing both variables (usually the one solved explicitly that was found in step 1) and solve for the other variable.
> 5. Check the solution in the original equations.

While both the elimination and substitution methods will work for all systems of equations, unless you can easily solve explicitly for one of the variables, you will probably find the elimination method simpler to use.

EXAMPLE 7   *Solve the system of equations.* $\begin{cases} 8x + 4y = 10 \\ 4x + 2y = 5 \end{cases}$

7. $\begin{cases} x + 2y = 4 \\ 5x + 10y = 20 \end{cases}$

Solution   Looking at this system of equations, we can see that neither variable looks particularly easy to solve for. This suggests that we use the elimination method rather than the substitution method.

If we choose to use the elimination method there really is no preference here as to which variable to eliminate. In fact, whichever variable we choose to eliminate, we will multiply the second equation by -2.

$$
\begin{array}{lll}
8x + 4y = 10 & \xrightarrow{\quad \textit{As is} \quad} & 8x + 4y = 10 \\
4x + 2y = 5 & \xrightarrow{\textit{Multiply by } -2} & \underline{-8x - 4y = -10} \quad \textit{Add the resulting} \\
& & 0 = 0 \quad \textit{equations.}
\end{array}
$$

Both variables have dropped out entirely, and we are left with an identity. The implication is that the two equations are identical. In fact, if we look back at the two original equations, we can see that the first equation is 2 times the second equation. Thus, both equations have exactly the same solution set. Their graphs are exactly the same line, and so there are infinitely many solutions to this system of equations. The system is dependent.   ■

EXAMPLE 8   *Solve the system of equations.* $\begin{cases} \dfrac{x}{2} - 2y = 3 \\ -x + 4y = 8 \end{cases}$

8. $\begin{cases} -x + \dfrac{y}{3} = 5 \\ 3x - y = 1 \end{cases}$

Solution   We solve this system by the substitution method. Solving the second equation explicitly for x, we get

$$x = 4y - 8 \qquad \textit{Now we substitute } 4y - 8 \textit{ for } x \textit{ in the first equation.}$$

$$\frac{x}{2} - 2y = 3$$

$$\frac{4y - 8}{2} - 2y = 3$$

$$\frac{\overset{2}{\cancel{4}}(y - 2)}{\cancel{2}} - 2y = 3$$

$$2y - 4 - 2y = 3$$

$$-4 = 3 \qquad \textit{This is a contradiction.}$$

Both variables have dropped out entirely, and we are left with a contradiction. The implication is that our assumption that there is a solution to the system of equations has led to a contradiction. Therefore, this assumption must be false, and there are no solutions to this sytem of equations. The system is inconsistent. If we graphed these two lines they would be parallel. ∎

Examples 7 and 8 illustrate that if we eliminate both variables from our system, then there are two possibilities:

If the resulting equation is an identity, then the two equations have the same solution set and the same line for their graphs; the equations are dependent.

If the resulting equation is a contradiction, then the two equations have no simultaneous solution, and their graphs are parallel lines; the equations are inconsistent.

Answers to Learning Checks in Section 8.5

1. $x = -3, y = 3$ **2.** $x = -5, y = -2$ **3.** $x = 4, y = 1$

4. $x = 0, y = 3$ **5.** $x = 12, y = -10$ **6.** $x = \frac{6}{5}, y = \frac{4}{5}$

7. $\{(x, y) \mid x + 2y = 4\}$ **8.** No solutions

Exercises 8.5

Solve each of the following systems of equations using the elimination method.

1. $\begin{cases} x + y = 1 \\ x - y = 3 \end{cases}$

2. $\begin{cases} x - y = 4 \\ x + y = 6 \end{cases}$

3. $\begin{cases} 2x + y = 5 \\ x - y = 4 \end{cases}$

4. $\begin{cases} -x + 3y = 8 \\ x - 2y = -6 \end{cases}$

5. $\begin{cases} 7x + 2y = 15 \\ 3x - 2y = -5 \end{cases}$

6. $\begin{cases} a - 5b = 30 \\ a + 5b = -40 \end{cases}$

7. $\begin{cases} 2x + y = 15 \\ x - 2y = 0 \end{cases}$

8. $\begin{cases} x - 3y = 1 \\ 2x + y = 9 \end{cases}$

1. _____

2. _____

3. _____

4. _____

5. _____

6. _____

7. _____

8. _____

ANSWERS

9. $\begin{cases} 3x + 2y = -11 \\ x + 3y = 1 \end{cases}$

10. $\begin{cases} 5x - y = 18 \\ x + 2y = -3 \end{cases}$

9. _____

10. _____

11. $\begin{cases} r + 2t = 10 \\ 3r + t = -15 \end{cases}$

12. $\begin{cases} 5m + n = 5 \\ m = 2n + 12 \end{cases}$

11. _____

12. _____

13. $\begin{cases} 6x + y = 6 \\ 4x + 1 = y \end{cases}$

14. $\begin{cases} 3x + 6y = 2 \\ -3x - 3y = 1 \end{cases}$

13. _____

14. _____

15. $\begin{cases} 8a + 6b = -3 \\ 12a + 9b = -5 \end{cases}$

16. $\begin{cases} 8a + 6b = 6 \\ 12a - 9b = 3 \end{cases}$

15. _____

16. _____

17. $\begin{cases} 11a - 2b = 30 \\ 3a + 3b = -6 \end{cases}$

18. $\begin{cases} 7a - 5b = 17 \\ 3a - 2b = 7 \end{cases}$

19. $\begin{cases} 2x + 3 = 4y \\ \quad 6x = 9 - 12y \end{cases}$

20. $\begin{cases} \quad 3y = 24 - 9x \\ 3x + y = 8 \end{cases}$

21. $\begin{cases} 5x + 2y = 4y + 9 \\ \quad y = x - 3 \end{cases}$

22. $\begin{cases} 4x + 3 = 2y - 5 \\ \quad x = y - 4 \end{cases}$

23. $\begin{cases} \dfrac{x}{2} + \dfrac{y}{3} = 1 \\[2mm] \dfrac{x}{4} - y = 11 \end{cases}$

24. $\begin{cases} \dfrac{x}{3} - \dfrac{y}{4} = 2 \\[2mm] x + \dfrac{y}{3} = -7 \end{cases}$

ANSWERS

17. _____

18. _____

19. _____

20. _____

21. _____

22. _____

23. _____

24. _____

25. $\begin{cases} \dfrac{x + y}{2} = 4 \\ \quad 3x = 5 - 3y \end{cases}$

26. $\begin{cases} \dfrac{x - y}{3} = 1 \\ \quad \dfrac{x}{5} = y - 1 \end{cases}$

25. _____

27. $\begin{cases} .4x + .2y = 8 \\ .7x - .3y = 1 \end{cases}$

28. $\begin{cases} .2x + .02y = 8 \\ \quad 4x - y = 20 \end{cases}$

26. _____

29. $\begin{cases} 5.2x + 3y = 14 \\ \quad .3x - 2y = 9.5 \end{cases}$

30. $\begin{cases} 8x - 2.8y = 4 \\ .3x + y = 11.2 \end{cases}$

27. _____

CALCULATOR EXERCISES

28. _____

Solve each of the following systems of equations.

31. $\begin{cases} 3.6x + 2.9y = 23.71 \\ 1.7x + 4.5y = 14.64 \end{cases}$

32. $\begin{cases} 2.8x - 4.6y = 3.5 \\ 5.2x - 3.4y = 10.1 \end{cases}$

29. _____

QUESTIONS FOR THOUGHT

14. A student began his solution to a system of equations as follows:

$$\begin{cases} 3x + 2y = 19 \\ 2x - 3y = 4 \end{cases} \quad \xrightarrow[\text{Multiply by 2}]{\text{Multiply by 3}} \quad \begin{array}{r} 9x + 6y = 57 \\ 4x - 6y = 4 \\ \hline 13x = 61 \\ x = \dfrac{61}{13} \end{array}$$

30. _____

What error has the student already made?

15. A student began her solution to a system of equations as follows:

31. _____

$$\begin{cases} 7x - 2y = 3 \\ 5x + y = 0 \end{cases} \quad \xrightarrow[\text{Multiply by 2}]{\text{As is}} \quad \begin{array}{r} 7x - 2y = 3 \\ 10x + 2y = 2 \\ \hline 17x = 5 \\ x = \dfrac{5}{17} \end{array}$$

32. _____

What error has the student already made?

Verbal Problems

8.6

Many of the verbal problems we discuss in this section are very similar to those we have discussed previously. In fact, we will continue to use most of the outline we first gave in Chapter 3 (page 159) for solving verbal problems. (You might find it useful to review that outline before you proceed with this section.)

However, since we now know how to solve a system of two linear equations in two variables, we are no longer restricted to solutions which involve one variable only. As we shall soon see, there are situations where using two variables makes it easier to formulate the solution to a problem.

Let's review an example of the type we have done previously and then illustrate how we would solve the same problem using a two-variable approach.

EXAMPLE 1   Marleen has a collection of fifty coins with a total value of $4.10. If the collection consists entirely of nickels and dimes, how many of each type of coin does she have?

Solution   We recognize that there are two unknown quantities: the number of nickels and the number of dimes.

First we will solve this problem using the approach we outlined in Chapters 3 and 4. From now on we will call this the ***one-variable approach.***

One-variable approach: We begin by letting x represent one of the unknown quantities in the problem.

Let x = the number of *nickels* in the collection.

Then we look for a relationship in the problem between the two unknown quantities. The problem says that there are fifty coins in all.

Therefore, let $50 - x$ = the number of *dimes* in the collection.

Since the value of the collection is $4.10, we write an equation which represents the value of the nickels plus the value of the dimes (in cents):

$$5x + 10(50 - x) = 410$$

Once we have the equation we can solve it rather easily.

$$5x + 10(50 - x) = 410$$
$$5x + 500 - 10x = 410$$
$$-5x + 500 = 410$$
$$-5x = -90$$
$$x = 18$$

Once we know that $x = 18$, then $50 - x = 32$.

Thus, there are $\boxed{18 \text{ nickels and } 32 \text{ dimes}}$.

The check is left to the student.

The second approach is to label the two unknown quantities with two different variables.

 LEARNING CHECKS

1. A total of 40 theater tickets cost $154. Some of the tickets cost $3 each and the rest cost $5 each. How many of each were bought?

Two-variable approach:

> Let n = the number of nickels in the collection.
>
> Let d = the number of dimes in the collection.

Since we are using two variables, we must use the given information in the problem to write *two* equations relating the variables. The two relationships we have are:

> The total number of coins is fifty.
>
> The total value of the coins is $4.10.

Or, equivalently,

> # of nickels + # of dimes = 50
>
> value of the nickels + value of the dimes = 410 cents

If we translate each of these factors into an equation, we obtain the following system:

$$\begin{cases} n + d = 50 \\ 5n + 10d = 410 \end{cases}$$

We can now proceed to solve this system by the method of elimination.

$$\begin{cases} n + d = 50 \\ 5n + 10d = 410 \end{cases} \xrightarrow[\text{As is}]{\text{Multiply by } -5}$$

$$\begin{array}{r} -5n - 5d = -250 \\ 5n + 10d = 410 \\ \hline 5d = 160 \\ d = 32 \end{array}$$

Now we add the resulting equations.

Now we substitute $d = 32$ into the first of the original equations, and solve for n.

$$n + d = 50$$
$$n + 32 = 50$$
$$n = 18$$

Thus, our solution is the same: 18 nickels and 32 dimes. ∎

In some examples the one-variable approach may be easier, while in others the two-variable approach may offer some advantages. The two procedures are similar in that they both require two relationships between the variables. They differ in the way these relationships are used. In general, you should use the method you can adapt most easily to the particular problem you are trying to solve.

In the next few examples we will illustrate the two-variable approach in several different types of situations.

2. The ratio of two numbers is 9 to 4. If their difference is 8, find the numbers.

EXAMPLE 2   The ratio of two numbers is 5 to 2. If the sum of the two numbers is 50, what are the two numbers?

Solution

> Let x = one of the numbers.
>
> Let y = the other number.

The first relationship is that the ratio of the two numbers is 5 to 2. We can translate this as

$$\frac{x}{y} = \frac{5}{2}$$

The second relationship is that the sum of the two numbers is 50. We translate this as

$$x + y = 50$$

We now proceed to solve this system of equations. Notice that our first step is to clear the fractions in the first equation (by multiplying both sides by $2y$) and to bring the y term to the left-hand side.

$$\frac{x}{y} = \frac{5}{2} \xrightarrow[\text{fractions}]{Clear} 2x = 5y \xrightarrow[\text{both sides}]{\substack{Subtract \\ 5y\ from}} 2x - 5y = 0$$

$$x + y = 50 \xrightarrow{As\ is} x + y = 50 \xrightarrow{As\ is} x + y = 50$$

Now we are ready to eliminate one of the variables. We choose to eliminate x.

$$\begin{aligned} 2x - 5y &= 0 \\ x + y &= 50 \end{aligned} \xrightarrow[\text{Multiply by } -2]{As\ is}$$

$$\begin{aligned} 2x - 5y &= 0 \\ -2x - 2y &= -100 \\ \hline -7y &= -100 \end{aligned} \quad \text{Add the two equations.}$$

$$y = \frac{-100}{-7} = \frac{100}{7}$$

We substitute $y = \frac{100}{7}$ into the second of the original equations and solve for x.

$$x + y = 50$$

$$x + \frac{100}{7} = 50 \quad \text{Multiply both sides of the equation by 7 to clear the fractions.}$$

$$7x + 100 = 350$$

$$7x = 250$$

$$x = \frac{250}{7}$$

Thus, our answer to the example is that the two numbers are $\boxed{\dfrac{250}{7} \text{ and } \dfrac{100}{7}}$.

CHECK: $\dfrac{\frac{250}{7}}{\frac{100}{7}} = \dfrac{250}{7} \cdot \dfrac{7}{100} = \dfrac{250}{100} \overset{\checkmark}{=} \dfrac{5}{2}$

$$\frac{250}{7} + \frac{100}{7} = \frac{350}{7} \overset{\checkmark}{=} 50 \qquad\blacksquare$$

EXAMPLE 3 Hubert invests \$12,000 in two savings certificates. One certificate yields 9% yearly interest, and the other yields 10% yearly interest. If Hubert's yearly interest from both certificates is \$1,145, how much was invested in each certificate?

3. $8,000 is split into two investments. Part is invested at 7% and the remainder at 8%. If the total interest from the two investments is $615, how much is invested at each rate?

Solution

Let x = amount invested at 9%   and y = amount invested at 10%.

The first relationship is that the total *amount* invested was $12,000. This translates to

$$x + y = 12,000$$

The second relationship is that the total *interest* from the two investments is $1,145. Keeping in mind that interest is equal to principal times rate, we multiply each principal (the x and the y) by its rate. This tranlates to

$$.09x + .10y = 1,145$$

We can now proceed to solve this system of equations. We first clear the decimals in the second equation (by multiplying both sides of the second equation by 100). Looking ahead, we see that we will get $10y$ in the second equation, so in the same step we multiply both sides of the first equation by -10.

$$
\begin{cases}
x + y = 12,000 \xrightarrow[\text{by } -10]{\text{Multiply}} -10x - 10y = -120,000 \\
.09x + .10y = 1,145 \xrightarrow[\text{by } 100]{\text{Multiply}} \quad 9x + 10y = \quad 114,500 \\
\hline
\qquad\qquad\qquad\qquad\qquad\qquad -x = -5,500 \\
\qquad\qquad\qquad\qquad\qquad\quad \boxed{x = 5,500}
\end{cases}
$$

Add the two equations.

Substituting $x = 5,500$ into the first of our original equations, we can solve for y.

$$x + y = 12,000$$
$$5,500 + y = 12,000$$
$$\boxed{y = 6,500}$$

CHECK:

$$5,500 + 6,500 \overset{\checkmark}{=} 12,000 \quad \text{(The amount invested is \$12,000.)}$$
$$.09(5,500) + .10(6,500) = 495 + 650 \overset{\checkmark}{=} 1,145 \quad \text{(The yearly interest is}$$
$$\text{\$1,145.)} \qquad \blacksquare$$

As we mentioned in Chapters 3 and 4, in order to solve "value"-type problems we must distinguish between two different relationships.

Coin problems: We must distinguish quantity from value.

Interest problems: We must distinguish between amount invested and interest earned.

Mixture problems: We must distinguish the amount of the mixture from the amount of pure substance.

Being able to distinguish between the two relationships allows us to write the two equations we need.

Sometimes a problem is worded in such a way that the one-variable approach is difficult, while the two-variable approach is quite straightforward.

EXAMPLE 4 If x ounces of a 60% alcohol solution are mixed together with y ounces of a 40% alcohol solution, the resulting mixture contains 32 ounces of alcohol. If x ounces of a 50% alcohol solution are mixed together with y ounces of a 30% alcohol solution, the resulting mixture contains 25 ounces of alcohol. Find x and y.

4. If 3 apples and 5 oranges cost $1.70, while 5 apples and 3 oranges cost $1.50, what are the costs per apple and per orange?

Solution We translate the given information about each mixture into an equation. Remember that the amount of actual alcohol contained in a solution is obtained by multiplying the volume (in this case, the number of ounces) by the percentage of the solution (usually called the *concentration*).

$$\begin{cases} .60x + .40y = 32 \\ .50x + .30y = 25 \end{cases}$$

We first clear the decimals from both equations by multiplying both sides of each equation by 10. Then we proceed to eliminate y.

$$\begin{cases} .60x + .40y = 32 \\ .50x + .30y = 25 \end{cases} \xrightarrow[\text{Multiply by 10}]{\text{Multiply by 10}} \begin{array}{l} 6x + 4y = 320 \\ 5x + 3y = 250 \end{array}$$

$$\begin{cases} 6x + 4y = 320 \\ 5x + 3y = 250 \end{cases} \xrightarrow[\text{Multiply by 4}]{\text{Multiply by} -3}$$

$$\begin{array}{l} -18x - 12y = -960 \\ \underline{20x + 12y = 1{,}000} \quad \text{Add the two} \\ \qquad\qquad\qquad\qquad\ \ \text{equations.} \\ \quad\ \ 2x = 40 \\ \quad\ \ \ x = 20 \end{array}$$

Now we can substitute $x = 20$ into the first of our second group of equations which do not involve decimals.

$$\begin{array}{rl} 6x + 4y &= 320 \\ 6(20) + 4y &= 320 \\ 120 + 4y &= 320 \\ 4y &= 200 \\ y &= 50 \end{array} \quad \text{Thus, our answer is } \boxed{x = 20, y = 50}.$$

It is left for the student to check this answer in the words of the problem. ∎

✔ *Answers to Learning Checks in Section 8.6*

1. Twenty-three $3 tickets and seventeen $5 tickets   **2.** $\dfrac{72}{5}$ and $\dfrac{32}{5}$

3. $5,500 at 8% and $2,500 at 7%   **4.** 15¢ per apple and 25¢ per orange

NOTE TO THE STUDENT

Use the space on this page to write down any questions you have or points you want to review with your instructor.

Exercises 8.6

Solve each of the following verbal problems algebraically. You may use either the one- or two-variable approach.

1. The sum of two numbers is 130. If their difference is 28, find the two numbers.

2. The sum of twice one number and another number is 48. If their difference is 3, find the numbers.

3. Sam has eighty coins consisting of nickels and quarters. If the total value of the coins is $13.60, how many of each type of coin are there?

4. Susan has ninety-two packages in her truck. Some of the packages weigh 32 pounds each, and the rest weigh 12 pounds each. If the total weight of all the packages is 1,604 pounds, how many of the lighter packages are there on the truck?

5. Two cars start at the same place and time, and travel in opposite directions. One car is travelling 15 kph faster than the other. After 5 hours, the two cars are 275 kilometers apart. Find the speed of each car.

6. The ratio of two numbers is 3 to 4. If one of the numbers is 5 more than the other, what are the two numbers?

7. Carmen invests a total of $1,700 in two stocks. One stock pays a yearly dividend of 7%, while the other pays 6%. If Carmen received $110 in combined dividends from the two stocks, how much did she invest in each?

8. A storekeeper is preparing a mixture of peanuts and raisins. If peanuts cost 60¢ per pound and raisins cost 85¢ per pound, how many pounds of each should be used to prepare fifty pounds of a mixture selling at 70¢ per pound?

1. _____

2. _____

3. _____

4. _____

5. _____

6. _____

7. _____

8. _____

493

9. Tim and Marge go into a record shop. Tim buys four cassettes and six LPs for a total of $48.80, while Marge buys five cassettes and three LPs for a total of $32.65. What are the prices of an individual cassette and an individual LP?

9. _____

10. A discount building supplies store sells both first-quality and second-quality floor tiles. Robin buys three cases of first-quality tiles and one case of second-quality tiles for a total of $66, while Gene buys one case of first-quality tiles and three cases of second-quality tiles for a total of $54. What is the cost per case for each type of tile?

10. _____

11. The length of a rectangle is twice its width. If the perimeter of the rectangle is 28 inches, what are its dimensions?

11. _____

12. How many 20¢ and 15¢ stamps did Joe buy, if he bought forty-eight stamps and paid $8.10 for them?

12. _____

13. Pat and Carlos both belong to the same book club. Pat orders two regular selections and three specially discounted ones for a total of $23. Carlos orders three regular selections and four specially discounted ones for a total of $32.50. What are the prices of a regular and a specially discounted selection?

13. _____

14. On Monday John walks for 1 hour, jogs for 2 hours, and covers a total of 28 km. On Tuesday he walks for 2 hours, jogs for 1 hour, and covers a total of 23 km. What are his rate walking and his rate jogging?

14. _____

15. The ratio of two numbers is 6 to 5. If the difference between the two numbers is 8, what are the numbers?

15. _____

16. Seven thousand tickets worth $26,875 were sold to a concert. General admission tickets cost $4 each and "standing-room-only" tickets cost $3.50 each. How many of each type were sold?

16. _____

17. Two airplanes leave an airport at the same time flying in opposite directions. One plane is flying at twice the speed of the other. If after 4 hours they are 1,800 miles apart, find the rate of speed of each plane.

18. Repeat Exercise 17 if the planes are flying in the *same* direction.

17. _____

19. Two retailers are ordering from the same source. One retailer orders eight stereo receivers and four turntables at a total cost of $2,060. A second retailer orders five of the same receivers and six of the same turntables at a total cost of $1,690. What are the costs of an individual receiver and an individual turntable?

18. _____

20. A 25% iodine solution is to be mixed with a 75% iodine solution to produce 5 gallons of a 70% iodine solution. How many gallons of each solution are needed?

19. _____

21. John goes into a donut shop and buys ten plain donuts and five cream-filled donuts for $3.70. Jane goes into the same shop and buys five plain donuts and ten cream-filled donuts for $4.10. What is the cost of a plain donut?

20. _____

22. In a recent school election, 571 votes were cast for class president. If the winner received 89 more votes than the loser, how many votes did each receive?

21. _____

23. A bookstore buys thirty-five books for $271. Some of the books cost $7 each and the remainder cost $9 each. How many of each type were bought?

22. _____

24. An artist's supply store sold a total of twenty canvases for $172. If some of the canvases cost $7.50 each, and the remainder were $10.25 each, how many of each type were sold?

23. _____

24. _____

25. Two cars start travelling directly toward each other at the same time from positions 480 kilometers apart. They meet after 4 hours. If one car travels 40 kph faster than the other car, find the speed of each car.

25. _____

26. Repeat Exercise 25 if the slower car leaves 2 hours ahead of the faster car, and they meet 4 hours after the faster car leaves.

26. _____

27. A small single-engine plane travels 150 miles per hour with a tailwind, and 90 miles per hour with a headwind. Find the speed of the wind, and the speed of the plane in still air.

27. _____

28. A 60% acid solution is to be mixed with an 80% acid solution to produce 20 liters of a 65% acid solution. How many liters of each solution are needed?

28. _____

29. A coffee wholesaler wishes to produce sixty pounds of a coffee blend selling at $3.10 per pound. How many pounds of coffee blends selling at $3.35 per pound and $2.75 per pound should be mixed to produce such a mixture?

29. _____

30. Mrs. Thomas has $15,000 to invest. She will invest part of it in a corporate bond which pays 8% interest per year, and the rest in a stock which pays 11% interest per year. How should she divide up the $15,000 so that her total yearly interest will be 10% of her investments?

30. _____

CHAPTER 8 SUMMARY

After having completed this chapter you should be able to:

1. Determine whether an ordered pair satisfies a given equation (Sections 8.1, 8.2).

   For example: Does the point $(3, -6)$ satisfy the equation $2x - 10 = y - 3$?

   To check, we substitute $x = 3$ and $y = -6$ into the equation.

   $$2x - 10 = y - 3$$
   $$2(3) - 10 \overset{?}{=} -6 - 3$$
   $$-4 \neq -9 \quad \cdot \quad \text{Therefore, } (3, -6) \text{ does not satisfy the equation.}$$

2. Complete an ordered pair for an equation, given one of the coordinates (Sections 8.1, 8.2).

   For example: Complete the ordered pair $(-2, \ \ )$ for the equation $x - 5y = 3$.

   We substitute the value $x = -2$ into the equation, and solve for y.

   $$x - 5y = 3$$
   $$-2 - 5y = 3$$
   $$-5y = 5$$
   $$y = -1 \quad \text{Therefore, the ordered pair is } \boxed{(-2, -1)}.$$

3. Sketch the graph of a linear equation using the intercept method (Section 8.2).

   For example: Sketch the graph of $3x - 2y = 6$.

   To find the x-intercept, we set $y = 0$ and solve for x.

   $$3x - 2y = 6$$
   $$3x - 2(0) = 6$$
   $$3x = 6$$
   $$x = 2 \quad \text{Therefore, the } x\text{-intercept is } (2, 0).$$

   To find the y-intercept, we set $x = 0$ and solve for y.

   $$3x - 2y = 6$$
   $$3(0) - 2y = 6$$
   $$-2y = 6$$
   $$y = -3 \quad \text{Therefore, the } y\text{-intercept is } (0, -3).$$

   For our "check point" we choose $x = 4$ and solve for y.

   $$3x - 2y = 6$$
   $$3(4) - 2y = 6$$
   $$12 - 2y = 6$$
   $$-2y = -6$$
   $$y = 3 \quad \text{Therefore, our check point is } (4, 3).$$

We now plot the three points and draw the line passing through them (see Figure 8.27).

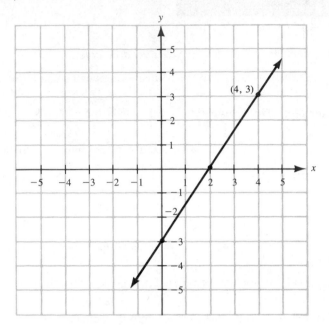

Figure 8.27   *The graph of* $3x - 2y = 6$

4. Use the definition of *slope* to find the slope of a line (Section 8.3).

For example: Find the slope of the line passing through the points $(2, -3)$ and $(-1, 4)$.

$$m = \frac{y_2 - y_1}{x_2 - x_1} = \frac{4 - (-3)}{-1 - 2} = \frac{7}{-3} = \boxed{-\frac{7}{3}}$$

5. Write an equation of a specified line (Section 8.3).

For example: Write an equation of the line with slope $\frac{1}{2}$ which passes through the point $(-2, 4)$.

Using the *point–slope form* for the equation of a straight line, we get

$$y - y_1 = m(x - x_1)$$

$$y - 4 = \frac{1}{2}(x - (-2))$$

$$\boxed{y - 4 = \frac{1}{2}(x + 2)}$$

6. Solve a system of equations by the graphical method (Section 8.4).

For example: Solve the following system by the graphical method:

$$\begin{cases} x - y = 5 \\ 2x + y = 4 \end{cases}$$

The first equation has intercepts $(5, 0)$ and $(0, -5)$.

The second equation has intercepts $(2, 0)$ and $(0, 4)$.

We draw the two lines (see Figure 8.28) and see that they cross at the point $(3, -2)$. We then check that this point satisfies both of the equations.

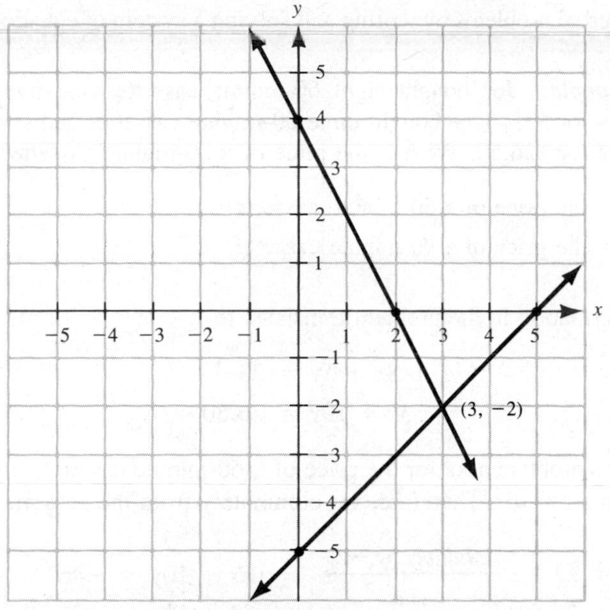

Figure 8.28

CHECK: $3 - (-2) \overset{\checkmark}{=} 5$

$2(3) + (-2) \overset{\checkmark}{=} 4$

7. Solve a system of equations by the elimination or substitution methods (Section 8.5).

For example: Solve the system

$$\begin{cases} 3x - 5y = 1 \\ 7x - 2y = 12 \end{cases}$$

We choose to eliminate y.

$$\begin{cases} 3x - 5y = 1 \\ 7x - 2y = 12 \end{cases} \xrightarrow[\text{Multiply by 5}]{\text{Multiply by } -2} \begin{aligned} -6x + 10y &= -2 \\ 35x - 10y &= 60 \\ \hline 29x &= 58 \\ x &= 2 \end{aligned}$$ *Add the resulting equations.*

Now we substitute $x = 2$ into one of the original equations (we will use the first one) and solve for y.

$$3x - 5y = 1$$
$$3(2) - 5y = 1$$
$$6 - 5y = 1$$
$$-5y = -5$$
$$y = 1$$ Thus, our solution is the ordered pair $\boxed{(2, 1)}$.

CHECK: We substitute $x = 2$, $y = 1$ into the second equation.

$$7x - 2y = 12$$
$$7(2) - 2(1) \overset{?}{=} 12$$
$$14 - 2 \overset{\checkmark}{=} 12$$

8. Solve verbal problems by writing and solving a system of equations (Section 8.6).

For example: Joe bought eight 60-minute cassettes and five 90-minute cassettes for $23. Jane bought three 60-minute cassettes and ten 90-minute cassettes for $26.50. What is the price of a 60-minute cassette?

Let x = the price of a 60-minute cassette.
Let y = the price of a 90-minute cassette.

The information in the problem translates to

$$\begin{cases} 8x + 5y = 23 \\ 3x + 10y = 26.50 \end{cases}$$

Since the problem asks for the price of a 60-minute cassette, it is x we are trying to solve for. Therefore, we eliminate y from the system.

$$\begin{cases} 8x + 5y = 23 \\ 3x + 10y = 26.50 \end{cases}$$

$\xrightarrow{\text{Multiply by } -2}$ $-16x - 10y = -46$

$\xrightarrow{\text{As is}}$ $\underline{3x + 10y = 26.50}$

$-13x = -19.50$ *Add the resulting equations.*

$x = 1.50$

The cost of a 60-minute cassette is $1.50.

ANSWERS TO QUESTIONS FOR THOUGHT

1. $\{(x, y) \mid y = x + 2 \text{ and } x = -3, 0, 4\}$

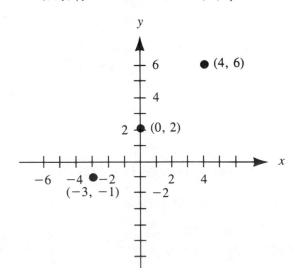

x	y	(x, y)
-3	-1	$(-3, -1)$
0	2	$(0, 2)$
4	6	$(4, 6)$

2. $\{(x, y) \mid x = y + 2 \text{ and } x = -3, 0, 4\}$

x	y	(x, y)
-3	-5	$(-3, -5)$
0	-2	$(0, -2)$
4	2	$(4, 2)$

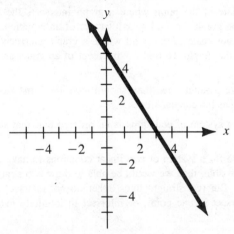

3. $(2, 3), (0, 6), (6, -3), (4, 0)$

4. $(-2, -16), (0, -8), (3, 4), (2, 0)$

5. An ordered pair (x, y) is a pair of numbers in which the order of appearance matters. For instance, $(3, 4)$ and $(4, 3)$ would be different as ordered pairs, even though both involve the same two numbers.

6. The notation $\{x, y\}$ stands for the set containing the two elements x and y. Here, order does not matter. That is, $\{3, 4\}$ and $\{4, 3\}$ are equal as sets. When we write (x, y), we mean that x comes first and y second.

7.

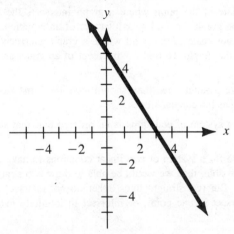

8. Some points on this line are $(-4, -4), (-3, -3), (0, 0), (2, 2),$ and $(5, 5)$. Since the y-coordinate of any such point is equal to its x-coordinate, an equation of this line would be $y = x$.

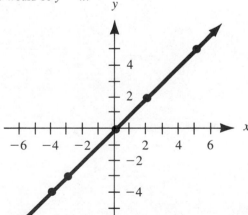

The graph of $y = x$

9. Some points on this line are $(-4, 4)$, $(-3, 3)$, $(0, 0)$, $(2, -2)$, and $(5, -5)$. Since the y-coordinate of any such point is equal to the negative of its x-coordinate, an equation of this line would be $y = -x$.

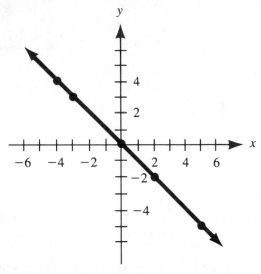

The graph of $y = -x$

10. The x-coordinate of the point where a graph intersects the x-axis is called an x-intercept of the graph. To find an x-intercept of an equation, set $y = 0$ and solve for x. The y-coordinate of the point where a graph intersects the y-axis is called a y-intercept of the graph. To find a y-intercept of an equation, set $x = 0$ and solve for y.

11. If two lines are parallel, then they should have the same steepness. This suggests that parallel lines have equal slopes.

12. A solution to a system of equations is an ordered pair that satisfies both equations simultaneously.

13. It is impossible for a system of two linear equations to have exactly two solutions. If this were possible, then we would be able to draw two straight lines that intersect exactly twice. But two straight lines either do not intersect at all (when they are parallel), intersect in one point, or intersect in infinitely many points (when they coincide).

14. The student forgot to multiply the right-hand side of the second equation by 2.

15. The student mistakenly multiplied 0 times 2 in the second equation and got a product of 2 rather than 0.

CHAPTER 8 REVIEW EXERCISES

In Exercises 1–6, plot the given points on a single rectangular coordinate system.

1. $(-4, 3)$ **2.** $(8, 0)$ **3.** $(6, -2)$

4. $(-7, -9)$ **5.** $(0, 0)$ **6.** $(0, -4)$

In Exercises 7–10, fill in the missing coordinate for the ordered pairs of the given equations.

7. $2x + 4y = 14$; $(3, \ )$ **8.** $2x - 3y = 8$; $(0, \ )$

9. $x - y = 0$; $(0, \ )$ **10.** $x - 2y = 7$; $(\ , -1)$

7. _____

8. _____

9. _____

10. _____

In Exercises 11–18, graph the given equation in a rectangular coordinate system, using the intercept method.

11. $3x + 7y = 21$ **12.** $2y - 6x = 18$

13. $y = 3x + 2$ **14.** $3x + 2y = 5$

15. $x - 2y = 4$ **16.** $x = y$

17. $x = 4$ **18.** $y = -3$

ANSWERS

19. _____

20. _____

21. _____

22. _____

23. _____

24. _____

25. _____

26. _____

27. _____

28. _____

29. _____

30. _____

31. _____

32. _____

33. _____

34. _____

In Exercises 19–22, find the slope of the line passing through the given points.

19. $(2, 5)$ and $(1, 7)$ **20.** $(-1, 4)$ and $(2, -3)$

21. $(1, 8)$ and $(-1, 8)$ **22.** $(6, 2)$ and $(6, 5)$

In Exercises 23–28, write an equation of the line satisfying the given conditions.

23. Passing through $(3, 2)$ with slope 4   **24.** Passing through $(-5, -1)$ with slope $\frac{3}{4}$

25. Passing through the points $(2, -6)$ and $(1, 0)$

26. Passing through the points $(-7, 2)$ and $(0, 3)$

27. Horizontal line passing through $(3, 8)$ **28.** Vertical line passing through $(3, 8)$

In Exercises 29–30, solve the system of equations by the graphical method.

29. $\begin{cases} x + y = 4 \\ x - y = 0 \end{cases}$ **30.** $\begin{cases} 2x + y = 4 \\ 3x - 2y = 6 \end{cases}$

In Exercises 31–36, solve the system of equations algebraically.

31. $\begin{cases} x - 5y = 1 \\ 3x - 2y = 3 \end{cases}$ **32.** $\begin{cases} 2x + 3y = 2 \\ x - 2y = -6 \end{cases}$

33. $\begin{cases} 4x - 3y = 10 \\ 9x + 2y = 5 \end{cases}$ **34.** $\begin{cases} 2x - 4y = 3 \\ 4x = 8y + 6 \end{cases}$

35. $\begin{cases} \dfrac{x}{2} + y = 5 \\ 2y = 8 - x \end{cases}$

36. $\begin{cases} \dfrac{x + y}{2} = 6 \\ x - y = -4 \end{cases}$

35. _____

Write a system of equations, and solve each of the following:

37. Pure water is to be mixed with a 30% solution of alcohol to produce 30 gallons of a 25% solution. How much water should be added?

36. _____

38. A total of 4,300 tickets valued at $18,800 were sold for a concert. Adults' tickets cost $4.50 each, and children's tickets cost $4 each. How many of each type were sold?

37. _____

39. Traci and Gene walk at the same rate and jog at the same rate. Traci walks for 1 hour and jogs for 1 hour, and covers a distance of 17 kilometers. Gene walks for 2 hours and jogs for half an hour, and covers a distance of 16 kilometers. Find the rate at which they both walk, and the rate at which they both jog.

38. _____

39. _____

CHAPTER 8 PRACTICE TEST

1. _____

1. Determine whether the point $(-3, -4)$ satisfies the equation

$$\frac{2y - x}{5} = x - y$$

2. a. _____

2. Find the missing coordinates for the ordered pair of the given equation.
 (a) $3x - y = 12$; $(5, \ )$ **(b)** $x - 3y = 10$; $(\ , -2)$
 (c) $4x + 5y = 15$; $(0, \ )$

b. _____

c. _____

3. Sketch the graphs of each of the following in a rectangular coordinate system using the intercept method.

    (a) $3x - 5y = 15$ **(b)** $y + 2x = 0$ **(c)** $x = 4$

4. _____

5. _____

4. Find the slope of the line passing through the points $(-1, -5)$ and $(2, -3)$.

5. Write an equation of the line with slope $\frac{4}{3}$ which passes through the point $(4, -1)$.

6. _____

6. Write an equation of the line passing through the points $(2, 5)$ and $(4, 0)$.

7. _____

7. Write the equations of the horizontal and vertical lines which pass through the point $(3, 5)$.

8. Solve the following system of equations by the graphical method: $\begin{cases} 2x - y = -8 \\ x + 2y = 6 \end{cases}$

8. _____

9. Solve each of the following systems by the elimination method:

9. a. _____

    (a) $\begin{cases} 4x - 3y = 11 \\ 3x + y = 5 \end{cases}$ **(b)** $\begin{cases} 2x - 7y = 4 \\ 5x - 4y = 10 \end{cases}$ **(c)** $\begin{cases} \dfrac{3x}{2} - y = 6 \\ x - \dfrac{2y}{3} = 5 \end{cases}$

    b. _____

    c. _____

10. *Solve the following problem algebraically.*   Tim ordered five orchestra tickets and three balcony tickets to a certain show for a total of $60. Tom ordered six orchestra tickets and two balcony tickets to the same show for a total of $64. What were the prices per orchestra and balcony ticket?

10. _____

Radical Expressions

Definitions and Basic Notation

9.1

Back in Section 1.6 we introduced the idea of an *irrational* number. Recall that an irrational number is a number which is associated with a point on the real number line, but which cannot be expressed as the quotient of two integers. We also mentioned that the decimal representation of an irrational number is a nonterminating, nonrepeating decimal. As a result of this fact, if we want to discuss the exact value of an irrational number, we can never write down the answer precisely as a decimal.

For example, we indicated in Section 1.6 (without proof) that a number whose square is equal to 2 is an irrational number. Thus, its decimal representation is an infinite nonrepeating decimal. This does not mean that every square root is an irrational number. If we ask for a square root of 100, we can give an exact answer of 10, because $10^2 = 100$. Certainly 10 is not an irrational number!

As we progress through this section, we will elaborate much more on the similarities and differences between square roots which are irrational numbers and square roots which are rational numbers.

Nevertheless, there are times when we want to talk about the "exact value" of an irrational number, such as a square root of 2. In order to do that we must have a symbolic way of referring to it. Let us begin with a definition.

DEFINITION   If $a^2 = b$ then a is called a **square root of b**.

For example:

$$3^2 = 9 \quad \text{so 3 is a square root of 9.}$$
$$(-3)^2 = 9 \quad \text{so } -3 \text{ is also a square root of 9.}$$

Thus, we see that 9 has two square roots—one positive, the other negative.

Similarly, 25 has two square roots, 5 and -5. In general, every positive number has two square roots. However, a negative number does *not* have any square roots in the real number system. For example, the number -16 has no square roots because it is impossible to find a real number which when squared gives *negative* 16. When you square any real number, whether is it positive, negative, or 0, the result cannot be negative.

REMEMBER

Square roots of negative numbers are not defined in the real number system.

Numbers such as 9 and 25, whose square roots are integers, are called **perfect squares.** If, however, b is an integer that is not a perfect square, such as 7, then its square roots are irrational numbers.

Let's now introduce a symbol with which we can represent the square roots of any nonnegative number.

We will write $\sqrt{b}$ for the square root of b. The symbol $\sqrt{}$ is called a **radical sign.**

However, a difficulty still remains. When we see the symbol $\sqrt{9}$, does this mean $+3$ or -3? Both $+3$ and -3 when squared are equal to 9. We certainly do not want the same symbol to stand for two different numbers. Therefore, we give the following definition.

> **DEFINITION** For any positive real number b, $\sqrt{b}$ stands for the *nonnegative* quantity which when squared gives b.
>
> $$\sqrt{b} \quad \text{is read} \quad \textit{the square root of } b.$$

This is called the **radical notation** for a square root.

For instance: $\sqrt{25} = 5$ *because* $5 \cdot 5 = 25$.

Other examples:

$\sqrt{64} = 8$	*because* $8 \cdot 8 = 64$
$-\sqrt{4} = -2$	The minus sign is attached after you compute $\sqrt{4}$.
$\sqrt{-4}$ does not exist.	There is no real number which when squared is equal to -4.
$\sqrt{0} = 0$	*because* $0 \cdot 0 = 0$

Note that it is incorrect to write $\sqrt{25} = -5$, even though $(-5) \cdot (-5) = 25$. Part of the definition of the expression $\sqrt{25}$ is that it be nonnegative.

REMEMBER

The symbol $\sqrt{b}$ stands for the *nonnegative* square root of b.

$\sqrt{25}$ is read as "*the* square root of 25."

$-\sqrt{25}$ is read as "the *negative* square root of 25."

Also note that 0, which is neither positive nor negative, has only one square root, which is 0.

Keep in mind that the square root is defined in terms of its inverse operation—squaring, just as subtraction and division are defined in terms of their inverse operations—addition and multiplication. Although you learned addition and multiplication tables, you never learned a subtraction or a division table. When you subtracted 3 from 8 you thought to yourself, "What do I *add* to 3 to get 8?"; when you divided 28 by 4 you thought, "What do I *multiply* by 4 to get 28?"

To find square roots you have to think in terms of squares. To find $\sqrt{144}$ you think, "What number squared is equal to 144?" We know the answer is 12 because $12 \cdot 12 = 144$. This is often a trial-and-error process.

As we continue with our discussion of square roots, keep the following in mind: Consider the two numbers $\sqrt{4}$ and $\sqrt{7}$. Even though they are numerically different, *conceptually* $\sqrt{4}$ and $\sqrt{7}$ are the same: $\sqrt{4}$ asks for a positive number whose square is 4, while $\sqrt{7}$ asks for a positive number whose square is 7. The only difference is that for $\sqrt{4}$ we get a nice neat numerical answer, while for $\sqrt{7}$ we do not.

On occasion, however, we need to do computations involving square roots which are irrational numbers. In such situations you may see statements such as "$\sqrt{7} = 2.65$" or "$\sqrt{7} = 2.645$." These are only approximate values (remember that $\sqrt{7}$ is an irrational number), and should really be written as $\sqrt{7} \approx 2.65$ or $\sqrt{7} \approx 2.645$. where the symbol "$\approx$" means approximately equal.

When we write $\sqrt{7} = 2.65$, we mean that 2.65 is an approximate value for $\sqrt{7}$ which is correct to one decimal place. Alternatively, we may say that 2.65 is $\sqrt{7}$ rounded off to the nearest hundredth. In other words, in using 2.65 as an approximate value for $\sqrt{7}$, the digits 2 and 6 are accurate; the 5 is obtained by rounding off.

Similarly, if we write $\sqrt{7} = 2.654$, we mean that 2.645 is an approximate value for $\sqrt{7}$ which is correct to two decimal places. That is, it is rounded off to the nearest thousandth, so that the digits 2, 6, and 4 are accurate; the 5 is obtained by rounding off. (The last "rounded" digit might, in fact, be accurate, but usually we do not know.)

If we use a calculator to obtain a value for $\sqrt{7}$, we would typically see the value 2.6457513, which is correct to six decimal places.

In the absence of a calculator with the capacity to compute square roots, you may use Appendix A in the back of the book, which contains a table of square roots and instructions for using the table.

Let's now return to our discussion of radical notation.

✔ LEARNING CHECKS

1. (a) $\sqrt{64}$
   (b) $-\sqrt{64}$
   (c) $\sqrt{-64}$

EXAMPLE 1   Evaluate each of the following.

(a) $\sqrt{49}$ (b) $-\sqrt{49}$ (c) $\sqrt{-49}$

Solution

(a) $\sqrt{49} = \boxed{7}$ because $7 \cdot 7 = 49$.

(b) $-\sqrt{49} = \boxed{-7}$ First we evaluate $\sqrt{49}$, and then we attach the minus sign in front.

(c) $\sqrt{-49}$ does not exist in the real number system. There is *no* real number which when squared is equal to -49. ∎

2. (a) $(\sqrt{10})^6$
   (b) $(\sqrt{11})^3$

EXAMPLE 2   *Evaluate.*   (a) $(\sqrt{7})^4$ (b) $(\sqrt{6})^5$

Solution

(a) You might ask how we can raise $\sqrt{7}$ to the 4th power when we do not know the exact value of $\sqrt{7}$. The point is that even though we do not know its exact value, we do know exactly the kind of number it is. We know that $\sqrt{7}$ is a number which when squared equals 7. That is, $\sqrt{7} \cdot \sqrt{7} = 7$. Therefore,

$$(\sqrt{7})^4 = \underbrace{\sqrt{7} \cdot \sqrt{7}}_{7} \cdot \underbrace{\sqrt{7} \cdot \sqrt{7}}_{7}$$
$$= \quad 7 \quad \cdot \quad 7$$
$$= \boxed{49}$$

(b) We proceed in a similar fashion.

$$(\sqrt{6})^5 = \underbrace{\sqrt{6} \cdot \sqrt{6}}_{6} \cdot \underbrace{\sqrt{6} \cdot \sqrt{6}}_{6} \cdot \sqrt{6}$$
$$= \quad 6 \quad \cdot \quad 6 \quad \cdot \sqrt{6}$$
$$= \boxed{36\sqrt{6}}$$

∎

The basic fact we used in both parts of Example 2, and which we will use over and over again, is:

For all nonnegative numbers a, $\boxed{\sqrt{a} \cdot \sqrt{a} = a}$

This is simply the definition of the square root written symbolically.

EXAMPLE 3   *Evaluate.*   **(a)** $\sqrt{36 + 64}$ **(b)** $\sqrt{36}\sqrt{64}$

3. (a) $\sqrt{25 - 16}$
   (b) $\sqrt{25}\sqrt{16}$

Solution

(a) $\sqrt{36 + 64} = \sqrt{100} = \boxed{10}$

Note that the answer is *not* obtained by computing the square root of each number under the radical sign separately! $\sqrt{36 + 64} \neq 6 + 8 = 14$!

(b) As with other algebraic expressions, when we write two radicals next to each other it automatically means multiplication.

$$\sqrt{36}\sqrt{64} = 6 \cdot 8 = \boxed{48}$$ ∎

✔   *Answers to Learning Checks in Section 9.1*

1. (a) 8   **(b)** −8   **(c)** No such real number

2. (a) 1,000   **(b)** $11\sqrt{11}$ **3. (a)** 3   **(b)** 20

NOTE TO THE STUDENT

Use the space on this page to write down any questions you have or points you want to review with your instructor.

Exercises 9.1

Evaluate or simplify each of the following. Assume that all variables represent nonnegative quantities.

1. $\sqrt{4}$

2. $\sqrt{9}$

3. $-\sqrt{4}$

4. $-\sqrt{9}$

5. $\sqrt{-4}$

6. $\sqrt{-9}$

7. $\sqrt{25}$

8. $\sqrt{49}$

9. $-\sqrt{100}$

10. $-\sqrt{36}$

11. $\sqrt{64}$

12. $\sqrt{81}$

13. $\sqrt{121}$

14. $\sqrt{144}$

15. $\sqrt{169}$

16. $\sqrt{196}$

17. $\sqrt{225}$

18. $\sqrt{256}$

19. $\sqrt{289}$

20. $\sqrt{324}$

21. $\sqrt{361}$

22. $\sqrt{400}$

23. $\sqrt{3}\sqrt{3}$

24. $\sqrt{10}\sqrt{10}$

25. $\sqrt{29}\sqrt{29}$

26. $\sqrt{67}\sqrt{67}$

27. $(\sqrt{11})^2$

28. $(\sqrt{5})^2$

29. $(\sqrt{33})^2$

30. $(\sqrt{41})^2$

1. _____
2. _____
3. _____
4. _____
5. _____
6. _____
7. _____
8. _____
9. _____
10. _____
11. _____
12. _____
13. _____
14. _____
15. _____
16. _____
17. _____
18. _____
19. _____
20. _____
21. _____
22. _____
23. _____
24. _____
25. _____
26. _____
27. _____
28. _____
29. _____
30. _____

31. _____

32. _____

33. _____

34. _____

35. _____

36. _____

37. _____

38. _____

39. _____

40. _____

41. _____

42. _____

43. _____

44. _____

45. _____

46. _____

47. _____

48. _____

49. _____

50. _____

51. _____

52. _____

53. _____

54. _____

31. $\sqrt{x}\sqrt{x}$

32. $(\sqrt{x})^2$

33. $(\sqrt{7})^4$

34. $(\sqrt{3})^6$

35. $(\sqrt{7})^5$

36. $(\sqrt{3})^7$

37. $\sqrt{25-9}$

38. $\sqrt{100}-\sqrt{36}$

39. $\sqrt{25}-\sqrt{9}$

40. $\sqrt{100-36}$

41. $(\sqrt{25}-\sqrt{9})^2$

42. $(\sqrt{100-36})^2$

43. $(\sqrt{25-9})^2$

44. $(\sqrt{100}-\sqrt{36})^2$

CALCULATOR EXERCISES

Use a calculator to find the following square roots. Round off your answers to the nearest hundredth.

45. $\sqrt{425}$

46. $\sqrt{983}$

47. $\sqrt{637}$

48. $\sqrt{730}$

49. $\sqrt{73.6}$

50. $\sqrt{58.9}$

51. $\sqrt{2.09}$

52. $\sqrt{7.85}$

53. $\sqrt{.037}$

54. $\sqrt{.00049}$

QUESTIONS FOR THOUGHT

1. Determining whether a number is a perfect square is not as hard as it might at first seem. For instance, suppose we wanted to know whether the number 648 is a perfect square. Do you see why the square root of 648 must be between 20 and 30? Can you think of an easy way to check if 648 can possibly be a perfect square? [*Hint:* Think about the possible final digit a perfect square can have.] Is 648 a perfect square?

2. Use the result of the previous question to check if 841 can possibly be a perfect square. Is it?

3. Discuss the correctness of the following steps:

$$\sqrt{2} \overset{?}{=} \sqrt{1+1}$$
$$\overset{?}{=} \sqrt{1} + \sqrt{1}$$
$$\overset{?}{=} 1+1$$
$$\overset{?}{=} 2$$

4. Describe the following statement in words: $\sqrt{a}\sqrt{a} = a$

Properties of Radicals and Simplest Radical Form

9.2

Just as we did with all the different types of algebraic expressions we have studied up to now, such as polynomials and rational expressions, we will discuss how to add, subtract, multiply, and divide radical expressions.

It should come as no surprise that we will want our radical expressions to be in "simplified" form. Consequently, we need to describe what it means for a radical expression to be in its simplest form.

Properties of Radicals

Let's look at a few numerical examples to help us recognize some of the properties of radicals that we will use in our simplifying process.

$$\sqrt{4 \cdot 100} = \sqrt{400} = 20 \quad \text{and} \quad \sqrt{4}\sqrt{100} = 2 \cdot 10 = 20$$

Therefore, we see that $\sqrt{4 \cdot 100} = \sqrt{4}\sqrt{100}$.

$$\sqrt{9 \cdot 16} = \sqrt{144} = 12 \quad \text{and} \quad \sqrt{9}\sqrt{16} = 3 \cdot 4 = 12$$

Therefore, we see that $\sqrt{9 \cdot 16} = \sqrt{9}\sqrt{16}$.

Similarly,

$$\frac{\sqrt{100}}{\sqrt{4}} = \frac{10}{2} = 5 \quad \text{and} \quad \sqrt{\frac{100}{4}} = \sqrt{25} = 5$$

Therefore, we see that $\dfrac{\sqrt{100}}{\sqrt{4}} = \sqrt{\dfrac{100}{4}}$

$$\frac{\sqrt{9}}{\sqrt{16}} = \frac{3}{4} \quad \text{and} \quad \sqrt{\frac{9}{16}} = \frac{3}{4} \quad \text{because} \quad \frac{3}{4} \cdot \frac{3}{4} = \frac{9}{16}$$

Therefore, we see that $\dfrac{\sqrt{9}}{\sqrt{16}} = \sqrt{\dfrac{9}{16}}$.

These examples suggest the following properties of radicals:

PROPERTIES OF RADICALS

For all nonnegative numbers a and b,

Property 1 $\sqrt{ab} = \sqrt{a}\sqrt{b}$

Property 2 $\sqrt{\dfrac{a}{b}} = \dfrac{\sqrt{a}}{\sqrt{b}}, \quad b \neq 0$

In words, these properties say that the square root of a *product* (*quotient*) is the *product* (*quotient*) of the square roots. *But* these properties say nothing about the square root of a *sum* or *difference*.

Keep in mind that we have not proved these properties. We have merely observed some numerical evidence which suggests that they are true. Let's prove property 1.

We want to show that $\sqrt{a}\sqrt{b} = \sqrt{ab}$.

$\sqrt{ab}$ means the quantity which when squared is equal to ab. Let's see what happens when we square $\sqrt{a}\sqrt{b}$.

$$(\sqrt{a}\sqrt{b})^2 = (\sqrt{a}\sqrt{b})(\sqrt{a}\sqrt{b}) \qquad \textit{Reorder and regroup the factors.}$$
$$= \underbrace{\sqrt{a}\sqrt{a}}_{a} \underbrace{\sqrt{b}\sqrt{b}}_{b}$$
$$= \quad a \quad \cdot \quad b$$
$$= ab$$

Thus, we see that when we square $\sqrt{a}\sqrt{b}$ we get ab. This is precisely the definition of $\sqrt{ab}$, and we conclude that $\sqrt{ab} = \sqrt{a}\sqrt{b}$. Property 2 is proven in an analogous way.

 **LEARNING CHECKS**

1. $\sqrt{25 \cdot 16} - \sqrt{25 - 16}$

EXAMPLE 1 *Evaluate.* $\sqrt{9 \cdot 16} + \sqrt{9 + 16}$

Solution *Be careful!* The misuse of the properties of radicals is a very common error.

$$\sqrt{9 \cdot 16} = \sqrt{9}\sqrt{16} \qquad \textit{By property 1}$$
$$\textit{but}$$
$$\sqrt{9 + 16} \neq \sqrt{9} + \sqrt{16} \quad \textit{because} \quad \sqrt{25} \neq 3 + 4$$

Returning to our example, we have

$$\sqrt{9 \cdot 16} + \sqrt{9 + 16} = \sqrt{9}\sqrt{16} + \sqrt{25}$$
$$= (3)(4)$$
$$= 12 + 5$$
$$= \boxed{17} \qquad \blacksquare$$

REMEMBER

The square root of a *sum* is not equal to the sum of the square roots.

Given an example such as $\sqrt{3}\sqrt{3}$, students often use propety 1 mentally to think

$$\sqrt{3}\sqrt{3} = \sqrt{3 \cdot 3} = \sqrt{9} = 3$$

While this is, of course, perfectly correct, it is severely missing the point.

$\sqrt{3}\sqrt{3} = 3$ because that is what $\sqrt{3}$ *means*. $\sqrt{3}$ is the positive number which when multiplied by itself gives 3.

To emphasize this point, think about how you would compute

$$\sqrt{4,583}\sqrt{4,583}$$

Hopefully, you do not want to do the following:

$$\sqrt{4{,}583}\sqrt{4{,}583} = \sqrt{4{,}583 \cdot 4{,}583} = \sqrt{21{,}003{,}889} = 4{,}583$$

Understanding what the radical sign *means* allows us to do computations such as $\sqrt{4{,}583}\sqrt{4{,}583} = 4{,}583$. It is just a specific example of the basic fact about square roots, that for all nonnegative numbers a,

$$\boxed{\sqrt{a}\sqrt{a} = a}$$

We need to clarify one last detail before we move on to define what we mean by the simplest radical form.

There is a natural tendency to accept the statement

$$\sqrt{a^2} = a \quad \text{because} \quad a \cdot a = a^2$$

However, this is not quite true, as the following example illustrates.

$$\sqrt{4^2} = \sqrt{16} = 4$$
$$\textit{but} \quad \sqrt{(-4)^2} = \sqrt{16} = 4 \neq -4$$

So we see that $\sqrt{a^2}$ is not equal to a when a is negative.

In order to avoid this complication, we will agree from here on to assume that all variables appearing under square root signs are nonnegative, unless otherwise specified.

Up to this point we have used the phrase *algebraic expression* without having formally defined it. Now that we have defined radicals, we can give the following definition.

> **DEFINITION**   An *algebraic expression* is an expression obtained by adding, subtracting, multiplying, dividing, and taking radicals of constants and/or variables.

Some examples of algebraic expressions that we have had, or soon will have, are:

$$-\sqrt{17}, \qquad 3x^2 - 5x + 3, \qquad \frac{4x - 3}{x^2 - 5}, \qquad \sqrt{2x - 7}, \qquad \frac{3\sqrt{x} - 4}{\sqrt{x + 9}}$$

Simplest Radical Form

As with the other types of expressions we have worked with, we want algebraic expressions involving radicals to be in simplest form. Consequently, we must define what we mean by simplest radical form.

> **DEFINITION**   An expression is said to be in *simplest radical form* if it satisfies the following three conditions:
>
> 1. The expression under the radical sign does not contain any perfect square factors. In other words, we want the expression under the radical sign to be as "small" as possible. As we shall soon see, expressions such as $\sqrt{12}$ and $\sqrt{a^7}$ violate this condition.
>
> 2. There are no fractions under the radical sign. For example, $\sqrt{\dfrac{3}{5}}$ violates this condition.
>
> 3. There are no radicals in denominators. For example, $\dfrac{2}{\sqrt{3}}$ violates this condition.

Radicals which satisfy these three conditions are generally easier to work with.

Let's see how the two properties of radicals help us transform radical expressions into simplest radical form.

2. (a) $\sqrt{18}$

(b) $\sqrt{48}$

EXAMPLE 2   Express in simplest radical form.

(a) $\sqrt{12}$ **(b)** $\sqrt{72}$

Solution

(a) If the expression appearing under the radical sign (which is called the *radicand*) has a factor which is a perfect square, we can use property 1 of radicals to simplify it.

$$\sqrt{12} = \sqrt{4 \cdot 3} = \sqrt{4}\sqrt{3} = \boxed{2\sqrt{3}}$$

Note that it would not have helped us to write $\sqrt{12} = \sqrt{6 \cdot 2}$, because neither 6 nor 2 is a perfect square.

(b) We will work out the solution via two slightly different paths.

First path:

$$\sqrt{72} = \sqrt{9 \cdot 8} \quad \leftarrow \textit{Use radical property 1.} \rightarrow$$
$$= \sqrt{9}\sqrt{8}$$
$$= 3\sqrt{8}$$
$$= 3\sqrt{4 \cdot 2} \quad \leftarrow \textit{Use radical property 1 again.}$$
$$= 3\sqrt{4}\sqrt{2}$$
$$= 3(2)\sqrt{2}$$
$$= \boxed{6\sqrt{2}}$$

Second path:

$$\sqrt{72} = \sqrt{36 \cdot 2}$$
$$= \sqrt{36}\sqrt{2}$$
$$= \boxed{6\sqrt{2}}$$

Clearly, it is much more efficient to try to find the *largest* perfect square you can at the beginning. However, even if you do not, keep applying radical property 1 until the quantity under the radical sign has no perfect square factors remaining.

■

As we pointed out in the last section, in order to avoid complications, we will assume that all variables appearing under radical signs are nonnegative.

3. (a) $\sqrt{a^8}$

(b) $\sqrt{a^{10}}$

(c) $\sqrt{a^{12}}$

(d) $\sqrt{a^{36}}$

EXAMPLE 3   *Simplify.*

(a) $\sqrt{x^2}$ **(b)** $\sqrt{x^4}$ **(c)** $\sqrt{x^6}$ **(d)** $\sqrt{x^{16}}$

Solution

(a) $\sqrt{x^2} = \boxed{x}$ *Because* $x \cdot x = x^2$

(b) $\sqrt{x^4} = \boxed{x^2}$ *Because* $x^2 \cdot x^2 = x^4$

(c) $\sqrt{x^6} = \boxed{x^3}$ *Because* $x^3 \cdot x^3 = x^6$

(d) $\sqrt{x^{16}} = \boxed{x^8}$ *Because* $x^8 \cdot x^8 = x^{16}$

Watch out! We are looking for $\sqrt{x^{16}}$, not $\sqrt{16}$.   ■

In Example 3, we notice that in each case finding the square root of a power involves looking for two equal numbers which *add* up to the power. We can always do this when the power is *even*.

> Even powers are always perfect squares.

Note that when we square an expression in exponential form we end up multiplying the exponent by 2 according to Exponent Rule 2.

$$(x^5)^2 = x^5 \cdot x^5 = x^{2 \cdot 5} = x^{10}$$
$$(x^n)^2 = x^n \cdot x^n = x^{2n}$$

Thus, finding the square root of a power requires dividing the exponent by 2, as we just saw in Example 3. This is fine for even powers, but what about odd powers?

EXAMPLE 4 Simplify as completely as possible.

(a) $\sqrt{x^7}$ **(b)** $\sqrt{18x^6}$ **(c)** $\sqrt{20x^7y^{10}}$

Solution

(a) Since 7 is not even, x^7 is *not* a perfect square. In order to satisfy condition 1 for the simplest radical form, we factor out the biggest perfect square (meaning even power) in x^7, which is x^6. If we factor out only x^2 or x^4, which are also perfect squares, we will not have simplified *completely*, since the expression under the radical sign will still have a perfect square factor.

$$\sqrt{x^7} = \sqrt{x^6 \cdot x} \qquad \text{\textit{This step can be thought of as a mental step.}}$$
$$= \sqrt{x^6}\sqrt{x} \qquad \text{\textit{We use radical property} 1.}$$
$$= \boxed{x^3\sqrt{x}}$$

(b) Let's visualize the radical in two factors. The first part consists of the perfect square factors, and the second part consists of whatever factors remain.

$$\sqrt{18x^6} = \sqrt{9x^6 \cdot 2}$$
$$= \underbrace{\sqrt{9x^6}}_{\substack{\text{Perfect square} \\ \text{factors}}} \cdot \underbrace{\sqrt{2}}_{\substack{\text{Remaining} \\ \text{factors}}}$$

Now use radical property 1.

$$= \sqrt{9}\sqrt{x^6}\sqrt{2}$$
$$= \boxed{3x^3\sqrt{2}}$$

(c) Let's show how we factor the radical in steps.

$$\sqrt{20x^7y^{10}} = \sqrt{\quad}\ \sqrt{\quad} \qquad \text{\textit{We want two radicals: the first for the perfect}}$$

square factors, the second for what is left. First we factor 20.

$$= \sqrt{4}\ \ \sqrt{5} \qquad \text{\textit{Next we factor} } x^7.$$
$$= \sqrt{4x^6}\ \ \sqrt{5x} \qquad \text{\textit{The } } y^{10} \text{ \textit{is a perfect square, so it goes in the}}$$

first radical.

$$= \sqrt{4x^6y^{10}}\sqrt{5x} \qquad \leftarrow \text{\textit{This radical contains the remaining factors.}}$$

↑

This radical contains perfect square factors.

$$= \sqrt{4}\sqrt{x^6}\sqrt{y^{10}}\sqrt{5x}$$
$$= \boxed{2x^3y^5\sqrt{5x}}$$

■

4. (a) $\sqrt{a^9}$

(b) $\sqrt{12a^{12}}$

(c) $\sqrt{24a^8b^{13}}$

5. (a) $\sqrt{a^4 b^4}$

   (b) $\sqrt{a^4 + b^4}$

6. (a) $\sqrt{\dfrac{7}{9}}$

   (b) $\sqrt{\dfrac{7}{10}}$

   (c) $\dfrac{12}{\sqrt{14}}$

EXAMPLE 5   Simplify as completely as possible.

(a) $\sqrt{x^2 y^2}$ (b) $\sqrt{x^2 + y^2}$

Solution

(a) Using radical property 1, we get:

$$\sqrt{x^2 y^2} = \sqrt{x^2}\sqrt{y^2} = \boxed{xy}$$

(b) $\sqrt{x^2 + y^2}$ *cannot be simplified!*

$$\sqrt{x^2 + y^2} \neq x + y \quad \text{because} \quad (x + y)^2 = (x + y)(x + y)$$
$$= x^2 + 2xy + y^2$$

Watch out! This is a very common error.   ∎

EXAMPLE 6   Express in simplest radical form.

(a) $\sqrt{\dfrac{3}{4}}$ (b) $\sqrt{\dfrac{3}{5}}$ (c) $\dfrac{10}{\sqrt{6}}$

Solution

(a) This is not in simplified form because the fraction inside the radical violates condition 2 of simplest radical form. Using radical property 2, we get:

$$\sqrt{\dfrac{3}{4}} = \dfrac{\sqrt{3}}{\sqrt{4}} = \boxed{\dfrac{\sqrt{3}}{2}} \qquad \text{This is now in simplest radical form.}$$

(b) This also violates condition 2. Again we use radical property 2.

$$\sqrt{\dfrac{3}{5}} = \dfrac{\sqrt{3}}{\sqrt{5}}$$

However, the radical in the denominator still violates condition 3 of simplest radical form. In order to get rid of the radical sign in the denominator, we apply the Fundamental Principle of Fractions in a clever way.

$$\sqrt{\dfrac{3}{5}} = \dfrac{\sqrt{3}}{\sqrt{5}} = \dfrac{\sqrt{3}}{\sqrt{5}} \cdot \dfrac{\sqrt{5}}{\sqrt{5}}$$

$$= \boxed{\dfrac{\sqrt{15}}{5}}$$

We want to multiply the denominator by $\sqrt{5}$ to get rid of the radical ($\sqrt{5}\sqrt{5} = 5$), but then we must also multiply the numerator by $\sqrt{5}$. $\sqrt{3}\sqrt{5} = \sqrt{3 \cdot 5} = \sqrt{15}$

This process is called *rationalizing the denominator,* and $\dfrac{\sqrt{5}}{\sqrt{5}}$ is called the *rationalizing factor.*

(c) Condition 3 for simplest radical form is violated because of the radical in the denominator. We rationalize the denominator using the rationalizing factor of $\dfrac{\sqrt{6}}{\sqrt{6}}$.

$$\dfrac{10}{\sqrt{6}} = \dfrac{10}{\sqrt{6}} \cdot \dfrac{\sqrt{6}}{\sqrt{6}}$$

$$= \dfrac{10\sqrt{6}}{6} \qquad \text{Because } \sqrt{6} \cdot \sqrt{6} = 6; \text{ now reduce } \dfrac{10}{6}.$$

$$= \frac{\overset{5}{\cancel{10}}\sqrt{6}}{\underset{3}{\cancel{6}}}$$ *We can reduce the 10 with the 6 because they are **factors** of the numerator and denominator, respectively.*

$$= \boxed{\frac{5\sqrt{6}}{3}}$$ ∎

EXAMPLE 7   *Simplify as completely as possible.* $\dfrac{21}{2\sqrt{3}}$

7. $\dfrac{8}{3\sqrt{2}}$

Solution   Condition 3 of simplest radical form is again violated.

What is the rationalizing factor? You may use $\dfrac{2\sqrt{3}}{2\sqrt{3}}$ (try it!), but that is really not necessary since the 2 in the denominator is not bothering us. It is the $\sqrt{3}$ we want to rationalize. The simplest rationalizing factor is $\dfrac{\sqrt{3}}{\sqrt{3}}$.

$$\frac{21}{2\sqrt{3}} = \frac{21}{2\sqrt{3}} \cdot \frac{\sqrt{3}}{\sqrt{3}}$$

$$= \frac{21\sqrt{3}}{2 \cdot 3}$$ *Because* $\sqrt{3} \cdot \sqrt{3} = 3$; *now reduce the 3 and the 21.*

$$= \boxed{\frac{7\sqrt{3}}{2}}$$ ∎

In the next section we will begin our discussion of how we perform the various arithmetic operations with radical expressions.

✔   *Answers to Learning Checks in Section 9.2*

1. 17   **2. (a)** $3\sqrt{2}$ **(b)** $4\sqrt{3}$ **3. (a)** a^4 **(b)** a^5 **(c)** a^6 **(d)** a^{18}

4. (a) $a^4\sqrt{a}$ **(b)** $2a^6\sqrt{3}$ **(c)** $2a^4b^6\sqrt{6b}$

5. (a) a^2b^2 **(b)** Cannot be simplified

6. (a) $\dfrac{\sqrt{7}}{3}$ **(b)** $\dfrac{\sqrt{70}}{10}$ **(c)** $\dfrac{6\sqrt{14}}{7}$ **7.** $\dfrac{4\sqrt{2}}{3}$

NOTE TO THE STUDENT

Use the space on this page to write down any questions you have or points you want to review with your instructor.

Exercises 9.2

Simplify each of the following expressions as completely as possible. Be sure your answers are in simplest radical form. Assume that all variables appearing under radical signs are nonnegative.

1. $\sqrt{64}$ **2.** $\sqrt{144}$ **3.** $\sqrt{18}$

4. $\sqrt{20}$ **5.** $\sqrt{32}$ **6.** $\sqrt{48}$

7. $\sqrt{50}$ **8.** $\sqrt{54}$ **9.** $\sqrt{400}$

10. $\sqrt{900}$ **11.** $\sqrt{x^6}$ **12.** $\sqrt{x^{10}}$

13. $\sqrt{x^7}$ **14.** $\sqrt{x^{11}}$ **15.** $\sqrt{16x^{16}}$

16. $\sqrt{36x^{36}}$ **17.** $\sqrt{9x^9}$ **18.** $\sqrt{25x^{25}}$

19. $\sqrt{40x^8}$ **20.** $\sqrt{24a^{12}}$ **21.** $\sqrt{25x^7}$

22. $\sqrt{16a^{11}}$ **23.** $\sqrt{12x^5}$ **24.** $\sqrt{20y^3}$

25. $\sqrt{a^2b^4}$ **26.** $\sqrt{a^2 + b^4}$ **27.** $\sqrt{x^6 + y^8}$

28. $\sqrt{x^6y^8}$ **29.** $\sqrt{49a^8b^{12}}$ **30.** $\sqrt{36m^6n^{18}}$

31. $\sqrt{28x^9y^6}$ **32.** $\sqrt{4x^{10}y^7}$ **33.** $\sqrt{50m^7n^{11}}$

1. _____
2. _____
3. _____
4. _____
5. _____
6. _____
7. _____
8. _____
9. _____
10. _____
11. _____
12. _____
13. _____
14. _____
15. _____
16. _____
17. _____
18. _____
19. _____
20. _____
21. _____
22. _____
23. _____
24. _____
25. _____
26. _____
27. _____
28. _____
29. _____
30. _____
31. _____
32. _____
33. _____

ANSWERS

34. _____

35. _____

36. _____

37. _____

38. _____

39. _____

40. _____

41. _____

42. _____

43. _____

44. _____

45. _____

46. _____

47. _____

48. _____

49. _____

50. _____

51. _____

52. _____

53. _____

54. _____

55. _____

56. _____

34. $\sqrt{60r^{13}t^5}$ **35.** $\sqrt{48x^6y^8z^9}$ **36.** $\sqrt{54x^7y^{14}z^{14}}$

37. $\sqrt{\dfrac{4}{9}}$ **38.** $\sqrt{\dfrac{16}{49}}$ **39.** $\sqrt{\dfrac{7}{25}}$

40. $\sqrt{\dfrac{13}{64}}$ **41.** $\sqrt{\dfrac{5}{6}}$ **42.** $\sqrt{\dfrac{3}{10}}$

43. $\dfrac{1}{\sqrt{2}}$ **44.** $\dfrac{1}{\sqrt{3}}$ **45.** $\dfrac{18}{\sqrt{10}}$

46. $\dfrac{12}{\sqrt{14}}$ **47.** $\dfrac{3}{\sqrt{x}}$ **48.** $\dfrac{4}{\sqrt{y}}$

49. $\dfrac{12}{5\sqrt{6}}$ **50.** $\dfrac{24}{7\sqrt{3}}$ **51.** $\dfrac{8x}{\sqrt{2x}}$

52. $\dfrac{12y}{\sqrt{3y}}$ **53.** $\dfrac{x^2}{\sqrt{xy}}$ **54.** $\dfrac{a^3}{\sqrt{2a}}$

55. $\dfrac{\sqrt{8}}{\sqrt{6}}$ **56.** $\dfrac{\sqrt{12}}{\sqrt{18}}$

QUESTIONS FOR THOUGHT

5. A student suggested the following procedure for rationalizing the denominator:

To rationalize the denominator in the fraction $\dfrac{2}{\sqrt{5}}$ "square the numerator and denominator" to get

$$\frac{2}{\sqrt{5}} = \frac{2^2}{(\sqrt{5})^2} = \frac{4}{5}$$

What is wrong with this "procedure?

6. Historically, the conditions for simplest radical form were motivated by arithmetic considerations. Use Appendix A to compute $\dfrac{1}{\sqrt{5}}$.

Now rationalize the denominator in $\dfrac{1}{\sqrt{5}}$ and then do the computation again. Which computation was easier? Why?

7. Repeat the previous question for $\dfrac{\sqrt{3}}{\sqrt{7}}$.

Adding and Subtracting Radical Expressions

9.3

It is very important to keep in mind that the procedures and principles we are going to use for combining radical expressions are the same as those we have used for other types of expressions—only the objects we are working with are different.

For example, we know that $5x + 3x = 8x$. We called this addition process *combining like terms*, and it is derived from the Distributive Law:

$5x + 3x = (5 + 3)x = 8x$
Similarly, $5\sqrt{2} + 3\sqrt{2} = 8\sqrt{2}$.

On the other hand, neither $5x + 3y$ nor $5\sqrt{2} + 3\sqrt{7}$ can be combined since in both cases we are dealing with unlike terms.

What about an expression such as $5x + \sqrt{2}x$? According to our definition of like terms, these are like terms since their variable parts are identical. However, to combine them we would have to write

$$5x + \sqrt{2}x = (5 + \sqrt{2})x$$

which is not of much help. Consequently, in such a case we will allow ourselves to leave the answer in the form $5x + \sqrt{2}x$.

EXAMPLE 1   Simplify as completely as possible.
(a) $5\sqrt{3} - 7\sqrt{3}$ (b) $4\sqrt{7} + 2\sqrt{7} - \sqrt{7}$
(c) $3\sqrt{2} - 4\sqrt{3} + 6\sqrt{2} + 2\sqrt{3} + \sqrt{5}$

Solution
(a) $5\sqrt{3} - 7\sqrt{3}$ *Think: "This is just like $5x - 7x$."*

$= (5 - 7)\sqrt{3}$

$= -2\sqrt{3}$

(b) $4\sqrt{7} + 2\sqrt{7} - \sqrt{7} = (4 + 2 - 1)\sqrt{7} = 5\sqrt{7}$

(c) $3\sqrt{2} - 4\sqrt{3} + 6\sqrt{2} + 2\sqrt{3} + \sqrt{5}$ *We look for and combine like terms.*

$= (3 + 6)\sqrt{2} + (-4 + 2)\sqrt{3} + \sqrt{5}$

$= 9\sqrt{2} - 2\sqrt{3} + \sqrt{5}$ ∎

EXAMPLE 2   *Simplify.*
(a) $\sqrt{3}\sqrt{3}$ (b) $\sqrt{3} + \sqrt{3}$

Solution   Read the example carefully. Do not confuse multiplication with addition.

(a) $\sqrt{3}\sqrt{3} = 3$ *By the definition of square root*

(b) $\sqrt{3} + \sqrt{3} = 2\sqrt{3}$ *This is just like $x + x = 2x$.*   ∎

Often, like terms involving radicals are not immediately apparent.

✔ **LEARNING CHECKS**

1. (a) $8\sqrt{2} - 3\sqrt{2}$
   (b) $7\sqrt{5} - 2\sqrt{5} + \sqrt{5}$
   (c) $3\sqrt{6} - 5\sqrt{3} + 4\sqrt{6} - \sqrt{3}$

2. (a) $\sqrt{5}\sqrt{5}$
   (b) $\sqrt{5} + \sqrt{5}$

3. (a) $\sqrt{54} - \sqrt{24}$
   (b) $\sqrt{20} + \sqrt{50}$
   (c) $\sqrt{25} + \sqrt{27}$

EXAMPLE 3 *Simplify.*

(a) $\sqrt{75} - \sqrt{12}$ (b) $\sqrt{40} + \sqrt{60}$ (c) $\sqrt{16} + \sqrt{18}$

Solution At first glance we do not see any like terms that can be combined. However, these expressions are *not* in simplest radical form. Let's see what happens if we first put each term into simplest radical form.

(a) Using radical property 1, we get:

$$\sqrt{75} - \sqrt{12} = \sqrt{25}\sqrt{3} - \sqrt{4}\sqrt{3}$$
$$= 5\sqrt{3} - 2\sqrt{3} \quad \text{Now combine like terms.}$$
$$= \boxed{3\sqrt{3}}$$

Actually, we had like terms all along; they were just not obvious.

(b) Using radical property 1, we get

$$\sqrt{40} + \sqrt{60} = \sqrt{4}\sqrt{10} + \sqrt{4}\sqrt{15}$$
$$= \boxed{2\sqrt{10} + 2\sqrt{15}}$$

There are no like terms, and so we cannot simplify any further.

(c) We evaluate $\sqrt{16}$ and simplify $\sqrt{18}$.

$$\sqrt{16} + \sqrt{18} = 4 + \sqrt{9}\sqrt{2}$$
$$= \boxed{4 + 3\sqrt{2}}$$

Note that $4 + 3\sqrt{2} \neq 7\sqrt{2}$; $4 + 3\sqrt{2}$ is just like $4 + 3x$. Since these are not like terms, we cannot combine them. ∎

The next few examples illustrate several more variations on the same theme.

4. $5\sqrt{90a^5} + a\sqrt{40a^3}$

EXAMPLE 4 *Simplify as completely as possible.* $3\sqrt{28y^3} + y\sqrt{63y}$

Solution We begin by using radical property 1 to get each radical into simplest radical form.

$$3\sqrt{28y^3} + y\sqrt{63y} = 3\sqrt{4y^2}\sqrt{7y} + y\sqrt{9}\sqrt{7y}$$
$$= 3(2y)\sqrt{7y} + y(3)\sqrt{7y}$$
$$= 6y\sqrt{7y} + 3y\sqrt{7y} \quad \text{Combine like terms.}$$
$$= \boxed{9y\sqrt{7y}} \quad ∎$$

5. $4(\sqrt{5} - \sqrt{3}) - 6(\sqrt{3} - 2\sqrt{5})$

EXAMPLE 5 *Multiply and simplify.* $3(\sqrt{2} - \sqrt{7}) - 5(2\sqrt{7} - \sqrt{2})$

Solution We work out this example as if it were "$3(x - y) - 5(2y - x)$." We begin by multiplying out using the Distributive Law.

$$3(\sqrt{2} - \sqrt{7}) - 5(2\sqrt{7} - \sqrt{2}) = 3\sqrt{2} - 3\sqrt{7} - 5(2)\sqrt{7} - 5(-\sqrt{2})$$
$$= 3\sqrt{2} - 3\sqrt{7} - 10\sqrt{7} + 5\sqrt{2}$$

Combine like terms.

$$= \boxed{8\sqrt{2} - 13\sqrt{7}} \quad ∎$$

EXAMPLE 6 *Combine.* $\sqrt{12} + \dfrac{1}{\sqrt{3}}$

6. $\sqrt{18} + \dfrac{1}{\sqrt{2}}$

Solution We first put each term into simplest radical form, which means something different for each term. In the first term we factor the $\sqrt{12}$, and in the second term we rationalize the denominator.

$$\sqrt{12} + \frac{1}{\sqrt{3}} = \sqrt{4}\sqrt{3} + \frac{1}{\sqrt{3}} \cdot \frac{\sqrt{3}}{\sqrt{3}}$$

$$= 2\sqrt{3} + \frac{\sqrt{3}}{3} \qquad \textit{Now we combine; the LCD is 3.}$$

$$= \frac{3 \cdot 2\sqrt{3}}{3} + \frac{\sqrt{3}}{3}$$

$$= \frac{6\sqrt{3} + \sqrt{3}}{3} \qquad \textit{Combine like terms.}$$

$$= \boxed{\frac{7\sqrt{3}}{3}} \qquad\qquad\qquad ■$$

EXAMPLE 7 *Combine.* $\sqrt{\dfrac{2}{3}} + \sqrt{\dfrac{3}{2}}$

7. $\sqrt{\dfrac{3}{5}} + \sqrt{\dfrac{5}{3}}$

Solution There are a number of reasonable ways to proceed, all of which will lead us to the correct answer. We offer the one that we think is the most "natural." We begin by using radical property 2.

$$\sqrt{\frac{2}{3}} + \sqrt{\frac{3}{2}} = \frac{\sqrt{2}}{\sqrt{3}} + \frac{\sqrt{3}}{\sqrt{2}} \qquad \textit{Next we rationalize each denominator.}$$

$$= \frac{\sqrt{2}}{\sqrt{3}} \cdot \frac{\sqrt{3}}{\sqrt{3}} + \frac{\sqrt{3}}{\sqrt{2}} \cdot \frac{\sqrt{2}}{\sqrt{2}}$$

$$= \frac{\sqrt{6}}{3} + \frac{\sqrt{6}}{2} \qquad \textit{Add the fractions; the LCD is 6.}$$

$$= \frac{2\sqrt{6}}{2 \cdot 3} + \frac{3\sqrt{6}}{2 \cdot 3}$$

$$= \frac{2\sqrt{6}}{6} + \frac{3\sqrt{6}}{6}$$

$$= \boxed{\frac{5\sqrt{6}}{6}} \qquad\qquad\qquad ■$$

Keep in mind that our basic approach to these problems has been to first put each term into simplest radical form before adding and/or subtracting.

NOTE TO THE STUDENT

Use the space on this page to write down any questions you have or points you want to review with your instructor.

Exercises 9.3

In each of the following, perform the indicated operations and simplify as completely as possible. Assume all variables appearing under radical signs are nonnegative.

1. $x + 2x + 3x$

2. $3a + a + 5a$

3. $\sqrt{5} + 2\sqrt{5} + 3\sqrt{5}$

4. $3\sqrt{7} + \sqrt{7} + 5\sqrt{7}$

5. $4x - x$

6. $7a - a$

7. $4\sqrt{6} - \sqrt{6}$

8. $7\sqrt{11} - \sqrt{11}$

9. $3x + 5y$

10. $6a + b$

11. $3\sqrt{5} + 5\sqrt{3}$

12. $6\sqrt{2} + \sqrt{5}$

13. $x \cdot x$

14. $a \cdot a$

15. $\sqrt{5} \cdot \sqrt{5}$

16. $\sqrt{13} \cdot \sqrt{13}$

17. $x + x$

18. $a + a$

19. $\sqrt{5} + \sqrt{5}$

20. $\sqrt{13} + \sqrt{13}$

21. $\sqrt{5} + 3\sqrt{7} - 4\sqrt{5} - 5\sqrt{7}$

22. $\sqrt{11} - 8\sqrt{2} - \sqrt{2} - 3\sqrt{11}$

23. $3\sqrt{3} + 4\sqrt{3} - \sqrt{2}$

24. $5\sqrt{5} + 2\sqrt{5} - 4\sqrt{3}$

25. $2(x - y) + 3(y - x)$

26. $4(a - b) + 5(b - a)$

1. _____
2. _____
3. _____
4. _____
5. _____
6. _____
7. _____
8. _____
9. _____
10. _____
11. _____
12. _____
13. _____
14. _____
15. _____
16. _____
17. _____
18. _____
19. _____
20. _____
21. _____
22. _____
23. _____
24. _____
25. _____
26. _____

27. $2(\sqrt{5} - \sqrt{3}) + 3(\sqrt{3} - \sqrt{5})$

28. $4(\sqrt{11} - \sqrt{7}) + 5(\sqrt{7} - \sqrt{11})$

29. $5(3\sqrt{6} + 2\sqrt{7}) - 4(\sqrt{7} - 2\sqrt{6})$

30. $3(2\sqrt{5} + 4\sqrt{10}) - 5(\sqrt{5} - 3\sqrt{10})$

31. $6(\sqrt{m} - \sqrt{n}) - (3\sqrt{m} + 6\sqrt{n})$

32. $2(\sqrt{x} - \sqrt{y}) - (4\sqrt{x} - 2\sqrt{y})$

33. $\sqrt{8} + \sqrt{18}$

34. $\sqrt{12} + \sqrt{27}$

35. $\sqrt{25} + \sqrt{24}$

36. $\sqrt{49} + \sqrt{50}$

37. $\sqrt{20} - \sqrt{5}$

38. $\sqrt{28} - \sqrt{7}$

39. $4\sqrt{12} - \sqrt{75}$

40. $6\sqrt{8} - \sqrt{98}$

41. $\sqrt{20} + \sqrt{40} + \sqrt{60}$

42. $\sqrt{18} + \sqrt{27} + \sqrt{45}$

43. $3\sqrt{72} - 5\sqrt{32}$ **44.** $4\sqrt{96} - 5\sqrt{24}$

45. $5\sqrt{36} + 4\sqrt{30}$ **46.** $8\sqrt{25} - 3\sqrt{21}$

43. _____

44. _____

45. _____

47. $\sqrt{25x} + \sqrt{36x}$ **48.** $\sqrt{16m} - \sqrt{64m}$

46. _____

47. _____

49. $\sqrt{12w} + \sqrt{27w}$ **50.** $\sqrt{18z} + \sqrt{8z}$

48. _____

49. _____

51. $\sqrt{45x} - \sqrt{20x}$ **52.** $\sqrt{54x} - \sqrt{24x}$

50. _____

51. _____

53. $\sqrt{20y^3} - \sqrt{45y^3}$ **54.** $\sqrt{50t^7} - \sqrt{32t^7}$

52. _____

53. _____

55. $x\sqrt{28xy^3} + y\sqrt{63x^3y}$ **56.** $b^2\sqrt{40a^5b^2} + a\sqrt{90a^3b^6}$

54. _____

55. _____

56. _____

57. $\dfrac{\sqrt{32x^3y^2}}{2xy} - \sqrt{8x}$

58. $\dfrac{\sqrt{48x^5y^4}}{4x^2y^2} - \sqrt{12x}$

59. $\sqrt{2} + \sqrt{\dfrac{1}{2}}$

60. $\sqrt{5} + \sqrt{\dfrac{1}{5}}$

61. $\sqrt{27} + \dfrac{4}{\sqrt{3}}$

62. $\sqrt{24} + \dfrac{5}{\sqrt{6}}$

63. $\dfrac{\sqrt{8}}{7} + \dfrac{7}{\sqrt{2}}$

64. $\dfrac{\sqrt{12}}{5} + \dfrac{5}{\sqrt{3}}$

65. $\sqrt{\dfrac{2}{7}} + \sqrt{\dfrac{7}{2}}$

66. $3\sqrt{\dfrac{x}{3}} + x\sqrt{\dfrac{3}{x}}$

57. _____

58. _____

59. _____

60. _____

61. _____

62. _____

63. _____

64. _____

65. _____

66. _____

QUESTIONS FOR THOUGHT

8. Use Appendix A to compute $\sqrt{80}$.
   Then simplify $\sqrt{80}$ and again use the table to compute the value of the simplified form. Are the results the same? Should they be?
   Now compute $\sqrt{80}$ in both forms using a calculator. Are the results the same? Should they be?

9. Repeat the previous question for $\sqrt{150}$.

Multiplying and Dividing Radical Expressions

9.4

Just as we did in the last section in the case of adding and subtracting radical expressions, we will multiply and divide them by applying the same basic procedures that we use for polynomials. We will constantly be using the properties of radicals, which we repeat in the box for easy reference.

Property 1 $\sqrt{ab} = \sqrt{a}\sqrt{b}$ Property 2 $\sqrt{\dfrac{a}{b}} = \dfrac{\sqrt{a}}{\sqrt{b}}$

EXAMPLE 1   *Multiply.* $\sqrt{2}\sqrt{3}\sqrt{5}$

Solution   In the same way that we use radical property 1 to "break up" a radical (over multiplication!), so too we can use it to write the product of two or more radicals as a single radical.

$$\sqrt{2}\sqrt{3}\sqrt{5} = \sqrt{2 \cdot 3 \cdot 5}$$
$$= \boxed{\sqrt{30}}$$
■

EXAMPLE 2   *Multiply.* $\sqrt{12} \cdot \sqrt{18}$

Solution   We offer two approaches.

First approach:

$$\sqrt{12} \cdot \sqrt{18} = \sqrt{12 \cdot 18}$$ *By radical property 1*
$$= \sqrt{216}$$ *We look for the largest perfect square factor of 216, which is 36.*
$$= \sqrt{36}\sqrt{6}$$
$$= \boxed{6\sqrt{6}}$$

Second approach: We first put each radical into simplest radical form.

$$\sqrt{12} \cdot \sqrt{18} = \sqrt{4}\sqrt{3} \cdot \sqrt{9}\sqrt{2}$$
$$= 2\sqrt{3} \cdot 3\sqrt{2}$$ *Rearrange the factors.*
$$= 2 \cdot 3\sqrt{3}\sqrt{2}$$
$$= \boxed{6\sqrt{6}}$$
■

Note that the second approach used in Example 2 kept the numbers much smaller. The arithmetic was easier when we simplified the radical first. Since this is usually the case, we will generally use the following outline for working with radicals.

LEARNING CHECKS

1. $\sqrt{3}\sqrt{5}\sqrt{7}$

2. $\sqrt{15}\sqrt{20}$

3. (a) $a(3a - b)$
   (b) $\sqrt{2}(3 - 4\sqrt{2})$

EXAMPLE 3 *Multiply.*

(a) $x(2x + y)$ (b) $\sqrt{3}(2\sqrt{3} + \sqrt{5})$

Solution   At first glance, part (b) may look a bit difficult. However, it is built along exactly the same lines as part (a), which should be quite familiar by now.

(a) $x(2x + y) = \boxed{x \cdot 2x + x \cdot y} = \boxed{2xx + xy}$

$= \boxed{2x^2 + xy}$

(b) $\sqrt{3}(2\sqrt{3} + \sqrt{5}) = \sqrt{3} \cdot 2\sqrt{3} + \sqrt{3}\sqrt{5} = 2\underset{3}{\underline{\sqrt{3}\sqrt{3}}} + \sqrt{3 \cdot 5}$

$= 2 \cdot 3 + \sqrt{15} = \boxed{6 + \sqrt{15}}$

While the first two steps in part (a) are surely mental steps by now, in part (b) they may not be. In fact, because the objects we are working with in part (b) are still somewhat new, it might be a good idea to write the "mental steps" for as long as you feel it is necessary.   ∎

4. $5\sqrt{a}(\sqrt{a} - 4) - 3(2 - 4\sqrt{a})$

EXAMPLE 4   *Multiply and simplify.* $2\sqrt{x}(\sqrt{x} - 3) - 4(3 - 5\sqrt{x})$

Solution   We proceed as we would if there were no radicals—by using the Distributive Law to remove the parentheses.

$2\sqrt{x}(\sqrt{x} - 3) - 4(3 - 5\sqrt{x}) = \underline{2\sqrt{x}\sqrt{x}} - 6\sqrt{x} - 12 + 20\sqrt{x}$

$= 2x - 6\sqrt{x} - 12 + 20\sqrt{x}$

Combine like terms.

$= \boxed{2x + 14\sqrt{x} - 12}$ ∎

5. $(\sqrt{t} + 6)(\sqrt{t} - 2)$

EXAMPLE 5   *Multiply and simplify.* $(\sqrt{x} - 3)(\sqrt{x} + 5)$

Solution   We handle this example just as we would $(a - 3)(a + 5)$. Each term in the first set of parentheses multiplies each term in the second set. In the case of two binomials, this process was called the FOIL method.

$(\sqrt{x} - 3)(\sqrt{x} + 5) = \underline{\sqrt{x}\sqrt{x}} + 5\sqrt{x} - 3\sqrt{x} - 15$

$= x + 5\sqrt{x} - 3\sqrt{x} - 15$ *Combine like terms.*

$= \boxed{x + 2\sqrt{x} - 15}$ ∎

EXAMPLE 6 *Multiply and simplify.* $(\sqrt{7} - \sqrt{3})^2$

Solution *Watch out!* Avoid the temptation to square each term separately.

$$\boxed{\text{Remember: } (a + b)^2 \neq a^2 + b^2}$$

$$
\begin{aligned}
(\sqrt{7} - \sqrt{3})^2 &= (\sqrt{7} - \sqrt{3})(\sqrt{7} - \sqrt{3}) \\
&= \sqrt{7}\sqrt{7} - \sqrt{7}\sqrt{3} - \sqrt{7}\sqrt{3} + \sqrt{3}\sqrt{3} \\
&= 7 - \sqrt{21} - \sqrt{21} + 3 \qquad \textit{Combine like terms.} \\
&= \boxed{10 - 2\sqrt{21}}
\end{aligned}
$$

■

6. $(\sqrt{6} + 2)^2$

EXAMPLE 7 *Multiply and simplify.* $(\sqrt{a} - 3)^2 - (\sqrt{a - 3})^2$

Solution Note the difference between the two expressions being squared. The first is binomial; the second is not.

$$
\begin{aligned}
(\sqrt{a} - 3)^2 - (\sqrt{a - 3})^2 &= (\sqrt{a} - 3)(\sqrt{a} - 3) - \underbrace{\sqrt{a - 3}\sqrt{a - 3}} \\
&= \sqrt{a}\sqrt{a} - 3\sqrt{a} - 3\sqrt{a} + 9 - (a - 3)
\end{aligned}
$$

Note that the parentheses around $a - 3$ are essential.

$$
\begin{aligned}
&= a - 6\sqrt{a} + 9 - a + 3 \\
&= \boxed{-6\sqrt{a} + 12}
\end{aligned}
$$

■

7. $(\sqrt{u} + 5)^2 - (\sqrt{u + 5})^2$

EXAMPLE 8 *Simplify.* **(a)** $\dfrac{\sqrt{72}}{\sqrt{6}}$ **(b)** $\dfrac{\sqrt{6b^7}}{\sqrt{30ab}}$

Solution

(a) We offer two solutions to illustrate a point.

Solution 1
We first simplify $\sqrt{72}$.

$$
\begin{aligned}
\frac{\sqrt{72}}{\sqrt{6}} &= \frac{\sqrt{36}\sqrt{2}}{\sqrt{6}} \\
&= \frac{6\sqrt{2}}{\sqrt{6}} \qquad \textit{Rationalize.} \\
&= \frac{6\sqrt{2}}{\sqrt{6}} \cdot \frac{\sqrt{6}}{\sqrt{6}} \\
&= \frac{6\sqrt{12}}{6} \\
&= \sqrt{12} \\
&= \sqrt{4}\sqrt{3} \\
&= \boxed{2\sqrt{3}}
\end{aligned}
$$

Solution 2
We first make one radical using property 2.

$$
\begin{aligned}
\frac{\sqrt{72}}{\sqrt{6}} &= \sqrt{\frac{72}{6}} \\
&= \sqrt{12} \\
&= \sqrt{4}\sqrt{3} \\
&= \boxed{2\sqrt{3}}
\end{aligned}
$$

8. **(a)** $\dfrac{\sqrt{48}}{\sqrt{2}}$

(b) $\dfrac{\sqrt{5r^3}}{\sqrt{20rs}}$

Clearly, the second method is more efficient. If you have the quotient of two radical expressions and you see that there are common factors which can be reduced, it is usually a better strategy to use property 2 first to make a single radical and reduce the fraction within the radical sign. Then proceed to simplify the remaining expression.

(b) Since we see that there are common factors (the 6 and the 30 will reduce, etc.), we begin by using property 2.

$$\frac{\sqrt{6b^7}}{\sqrt{30ab}} = \sqrt{\frac{6b^7}{30ab}} \qquad \textit{Reduce.}$$

$$= \sqrt{\frac{b^6}{5a}} \qquad \textit{Use property 2 again.}$$

$$= \frac{\sqrt{b^6}}{\sqrt{5a}} \qquad \textit{Simplify } \sqrt{b^6}.$$

$$= \frac{b^3}{\sqrt{5a}} \qquad \textit{Rationalizing, we get}$$

$$= \frac{b^3}{\sqrt{5a}} \cdot \frac{\sqrt{5a}}{\sqrt{5a}}$$

$$= \boxed{\frac{b^3\sqrt{5a}}{5a}} \qquad\qquad\qquad \blacksquare$$

9. $(\sqrt{7} - 1)(\sqrt{7} + 1)$

EXAMPLE 9   *Multiply and simplify.* $(\sqrt{13} - 3)(\sqrt{13} + 3)$

Solution   Multiply out using FOIL.

$$(\sqrt{13} - 3)(\sqrt{13} + 3) = \sqrt{13}\sqrt{13} + 3\sqrt{13} - 3\sqrt{13} - 9 \qquad \begin{array}{l}\textit{The cross terms} \\ \textit{combine to 0.}\end{array}$$

$$= 13 - 9$$

$$= \boxed{4} \qquad \textit{This answer does not involve radicals.}$$

We will soon make good use of this fact. $\blacksquare$

In the last section we saw that in order to rationalize the denominator of an expression such as $\dfrac{5}{\sqrt{13}}$, we multiply the fraction by the rationalizing factor of $\dfrac{\sqrt{13}}{\sqrt{13}}$.

$$\frac{5}{\sqrt{13}} = \frac{5}{\sqrt{13}} \cdot \frac{\sqrt{13}}{\sqrt{13}} = \frac{5\sqrt{13}}{13}$$

In this way we have removed the radical from the denominator, and the expression now satisfies *all* the conditions of simplest radical form.

But what if we had to simplify $\dfrac{5}{\sqrt{13} - 3}$? If we again multiply the numerator and denominator by $\sqrt{13}$ (which is mathematically legal), we do not get the job done. We still end up with a radical in the denominator. *Try it!*

If we look back to Example 9, we can see what does work. The rationalizing factor is $\dfrac{\sqrt{13} + 3}{\sqrt{13} + 3}$.

$$\frac{5}{\sqrt{13} - 3} = \frac{5}{\sqrt{13} - 3} \cdot \frac{\sqrt{13} + 3}{\sqrt{13} + 3} \qquad \begin{array}{l}\textit{In Example 9 we saw that} \\ (\sqrt{13} - 3)(\sqrt{13} + 3) = 4.\end{array}$$

$$= \boxed{\frac{5(\sqrt{13} + 3)}{4}}$$

The expressions $\sqrt{13} - 3$ and $\sqrt{13} + 3$ are called ***conjugates*** of each other. To get a conjugate of a binomial just change the sign of the second term. Thus, a conjugate of $\sqrt{7} + 2$ is $\sqrt{7} - 2$.

EXAMPLE 10 *Rationalize the denominator,* $\dfrac{20}{\sqrt{10} + \sqrt{6}}$

10. $\dfrac{16}{\sqrt{11} - \sqrt{3}}$

Solution The rationalizing factor for this expression is $\dfrac{\sqrt{10} - \sqrt{6}}{\sqrt{10} - \sqrt{6}}$.

$$\frac{20}{\sqrt{10} + \sqrt{6}} = \frac{20}{\sqrt{10} + \sqrt{6}} \cdot \frac{\sqrt{10} - \sqrt{6}}{\sqrt{10} - \sqrt{6}}$$

In the denominators, the conjugates multiply like the difference of two squares. We get the square of the first term minus the square of the last term. The cross terms drop out.

$$= \frac{20(\sqrt{10} - \sqrt{6})}{10 - 6}$$

$$= \frac{20(\sqrt{10} - \sqrt{6})}{4}$$

Note that we do not multiply out the top; we keep the factored form to see if we can reduce the fraction. We reduce the 20 and the 4.

$$= \frac{\overset{5}{\cancel{20}}(\sqrt{10} - \sqrt{6})}{\underset{}{\cancel{4}}}$$

$$= \boxed{5(\sqrt{10} - \sqrt{6})} \quad \text{or} \quad \boxed{5\sqrt{10} - 5\sqrt{6}} \qquad \blacksquare$$

EXAMPLE 11 *Simplify as completely as possible.* $\dfrac{8}{3 - \sqrt{5}} - \dfrac{10}{\sqrt{5}}$

11. $\dfrac{21}{3 - \sqrt{2}} - \dfrac{6}{\sqrt{2}}$

Solution We begin by rationalizing each denominator. Keep in mind that each fraction has its own rationalizing factor.

$$\frac{8}{3 - \sqrt{5}} - \frac{10}{\sqrt{5}} = \frac{8}{3 - \sqrt{5}} \cdot \frac{3 + \sqrt{5}}{3 + \sqrt{5}} - \frac{10}{\sqrt{5}} \cdot \frac{\sqrt{5}}{\sqrt{5}}$$

$$= \frac{8(3 + \sqrt{5})}{9 - 5} - \frac{10\sqrt{5}}{5}$$

$$= \frac{8(3 + \sqrt{5})}{4} - \frac{10\sqrt{5}}{5} \qquad \text{\textit{Reduce each fraction.}}$$

$$= \frac{\overset{2}{\cancel{8}}(3 + \sqrt{5})}{\cancel{4}} - \frac{\overset{2}{\cancel{10}}\sqrt{5}}{\cancel{5}}$$

$$= 2(3 + \sqrt{5}) - 2\sqrt{5}$$

$$= 6 + 2\sqrt{5} - 2\sqrt{5}$$

$$= \boxed{6} \qquad \blacksquare$$

EXAMPLE 12 *Simplify as completely as possible.* $\dfrac{12 + \sqrt{18}}{6}$

12. $\dfrac{10 + \sqrt{20}}{8}$

Solution We begin by simplifying the radical.

$$\frac{12 + \sqrt{18}}{6} = \frac{12 + \sqrt{9}\sqrt{2}}{6}$$

$$= \frac{12 + 3\sqrt{2}}{6} \qquad \text{\textit{Factor out the common factor of 3 in the numerator.}}$$

$$= \frac{3(4 + \sqrt{2})}{6} \qquad \text{\textit{Now reduce the 3 and the 6.}}$$

$$= \frac{\cancel{3}(4 + \sqrt{2})}{\underset{2}{\cancel{6}}}$$

$$= \boxed{\frac{4 + \sqrt{2}}{2}} \qquad \text{\textit{Note that we \textbf{cannot} reduce the 4 and the 2.}} \qquad \blacksquare$$

Being able to work with radical expressions allows us to broaden the scope of the types of questions we can answer.

13. Verify that $3 - \sqrt{2}$ is a solution to $x^2 - 6x + 7 = 0$.

EXAMPLE 13   Verify that $3 + \sqrt{2}$ is a solution to the equation

$$x^2 - 6x + 7 = 0$$

Solution   We must replace each x in the equation by the value $3 + \sqrt{2}$, and show that it satisfies the equation.

$$x^2 - 6x + 7 = 0$$
$$(3 + \sqrt{2})^2 - 6(3 + \sqrt{2}) + 7 \overset{?}{=} 0$$
$$(3 + \sqrt{2})(3 + \sqrt{2}) - 6(3 + \sqrt{2}) + 7 \overset{?}{=} 0$$
$$9 + 3\sqrt{2} + 3\sqrt{2} + 2 - 18 - 6\sqrt{2} + 7 \overset{?}{=} 0 \qquad \textit{Combine like terms.}$$
$$0 \overset{\checkmark}{=} 0 \qquad\blacksquare$$

✔

1. $\sqrt{105}$ **2.** $10\sqrt{3}$ **3. (a)** $3a^2 - ab$ **(b)** $3\sqrt{2} - 8$

4. $5a - 8\sqrt{a} - 6$ **5.** $t + 4\sqrt{t} - 12$ **6.** $10 + 4\sqrt{6}$ **7.** $-10\sqrt{u} - 20$

8. (a) $2\sqrt{6}$ **(b)** $\dfrac{r\sqrt{s}}{2s}$ **9.** 6   **10.** $2(\sqrt{11} + \sqrt{3})$ **11.** 9

12. $\dfrac{5 + \sqrt{5}}{4}$

Exercises 9.4

Perform the indicated operations. Simplify all answers as completely as possible. Assume that all variables appearing under radical signs are nonnegative.

1. $\sqrt{3}\sqrt{11}$

2. $\sqrt{5}\sqrt{7}$

3. $\sqrt{3}\sqrt{5}\sqrt{13}$

4. $\sqrt{2}\sqrt{7}\sqrt{11}$

5. $\sqrt{6} + \sqrt{24}$

6. $\sqrt{5} + \sqrt{45}$

7. $\sqrt{6}\sqrt{24}$

8. $\sqrt{5}\sqrt{45}$

9. $\sqrt{3}\sqrt{5}\sqrt{6}$

10. $\sqrt{2}\sqrt{6}\sqrt{10}$

11. $\sqrt{3}(\sqrt{5} + \sqrt{6})$

12. $\sqrt{2}(\sqrt{6} + \sqrt{10})$

13. $\sqrt{18}\sqrt{32}$

14. $\sqrt{24}\sqrt{28}$

15. $\sqrt{3}(2\sqrt{3} - 3\sqrt{2})$

16. $\sqrt{5}(3\sqrt{5} - 4\sqrt{3})$

17. $5\sqrt{x}(\sqrt{x} - 2\sqrt{5})$

18. $4\sqrt{a}(2\sqrt{a} + 3\sqrt{7})$

19. $3\sqrt{2}(\sqrt{2} - 4) + \sqrt{2}(5 - \sqrt{2})$

20. $5\sqrt{3}(\sqrt{3} - 2) + \sqrt{3}(7 - \sqrt{3})$

1. _____

2. _____

3. _____

4. _____

5. _____

6. _____

7. _____

8. _____

9. _____

10. _____

11. _____

12. _____

13. _____

14. _____

15. _____

16. _____

17. _____

18. _____

19. _____

20. _____

539

21. $4\sqrt{x}(\sqrt{x} - \sqrt{2}) - \sqrt{x}(3\sqrt{x} - 2\sqrt{2})$

22. $2\sqrt{y}(\sqrt{y} - \sqrt{3}) - \sqrt{y}(3\sqrt{y} - 4\sqrt{3})$

21. _____

22. _____

23. $(\sqrt{11} + 3)(\sqrt{11} - 6)$

24. $(\sqrt{10} - 5)(\sqrt{10} + 2)$

23. _____

24. _____

25. $(\sqrt{x} + \sqrt{3})^2$

26. $(\sqrt{a} - \sqrt{5})^2$

25. _____

26. _____

27. $(\sqrt{x} + \sqrt{3})(\sqrt{x} - \sqrt{3})$

28. $(\sqrt{a} - \sqrt{5})(\sqrt{a} + \sqrt{5})$

27. _____

28. _____

29. $(3\sqrt{2} - 2\sqrt{5})^2$

30. $(4 + 5\sqrt{3})^2$

29. _____

30. _____

31. $(3\sqrt{x} - \sqrt{7})(3\sqrt{x} + \sqrt{7})$

32. $(2\sqrt{y} + 3\sqrt{6})(2\sqrt{y} - 3\sqrt{6})$

31. _____

32. _____

33. $(2\sqrt{x} - 3)(3\sqrt{x} + 4)$

34. $(4\sqrt{a} + 1)(3\sqrt{a} - 1)$

33. _____

34. _____

35. $(\sqrt{28} - \sqrt{24})(\sqrt{7} - \sqrt{6})$

36. $(\sqrt{32} - \sqrt{20})(\sqrt{2} + \sqrt{5})$

35. _____

36. _____

37. $(\sqrt{t + 9})^2 + (\sqrt{t} + 9)^2$

38. $(\sqrt{z} - 1)^2 + (\sqrt{z - 1})^2$

37. _____

38. _____

39. $(\sqrt{x} + 2)^2 - (\sqrt{x + 2})^2$

40. $(\sqrt{m} - 4)^2 - (\sqrt{m - 4})^2$

41. $\dfrac{10}{\sqrt{11}}$

42. $\dfrac{12}{\sqrt{5}}$

43. $\dfrac{\sqrt{54}}{\sqrt{3}}$

44. $\dfrac{\sqrt{200}}{\sqrt{5}}$

45. $\dfrac{\sqrt{xy^3}}{\sqrt{x^3y}}$

46. $\dfrac{\sqrt{3a^7}}{\sqrt{27a^5}}$

47. $\dfrac{\sqrt{a^2b^5}}{\sqrt{ab^8}}$

48. $\dfrac{\sqrt{12a^2b^3}}{\sqrt{3a^3b}}$

49. $\dfrac{10}{4 - \sqrt{11}}$

50. $\dfrac{12}{3 + \sqrt{5}}$

51. $\dfrac{6}{\sqrt{x} + \sqrt{3}}$

52. $\dfrac{8}{\sqrt{a} - \sqrt{2}}$

53. $\dfrac{\sqrt{3}}{2 + \sqrt{3}}$

54. $\dfrac{\sqrt{5}}{\sqrt{6} - \sqrt{5}}$

55. $\dfrac{\sqrt{5} + \sqrt{3}}{\sqrt{5} - \sqrt{3}}$

56. $\dfrac{\sqrt{11} - \sqrt{7}}{\sqrt{11} + \sqrt{7}}$

ANSWERS

39. _____

40. _____

41. _____

42. _____

43. _____

44. _____

45. _____

46. _____

47. _____

48. _____

49. _____

50. _____

51. _____

52. _____

53. _____

54. _____

55. _____

56. _____

In Exercises 57–62, rationalize the denominators and simplify.

57. $\dfrac{8}{\sqrt{5} - \sqrt{3}} - \dfrac{12}{\sqrt{3}}$

58. $\dfrac{16}{\sqrt{6} + \sqrt{2}} + \dfrac{8}{\sqrt{2}}$

57. _____

58. _____

59. $\dfrac{6}{3 - \sqrt{7}} - \dfrac{21}{\sqrt{7}}$

60. $\dfrac{12}{\sqrt{15} + \sqrt{3}} - \dfrac{30}{\sqrt{15}}$

59. _____

61. $(3 + \sqrt{5})^2 + \dfrac{8}{3 - \sqrt{5}}$

62. $(4 - \sqrt{7})^2 - \dfrac{18}{4 + \sqrt{7}}$

60. _____

In Exercises 63–66, reduce to lowest terms.

63. $\dfrac{4 + \sqrt{8}}{6}$

64. $\dfrac{6 + \sqrt{2}}{8}$

61. _____

62. _____

65. $\dfrac{12 - \sqrt{20}}{10}$

66. $\dfrac{10 - \sqrt{24}}{12}$

63. _____

67. Verify that $2 + \sqrt{10}$ is a solution to the equation $x^2 - 4x - 6 = 0$.

64. _____

65. _____

68. Determine whether $3 - \sqrt{13}$ is a solution to the equation $x^2 - 6x = 3$.

66. _____

QUESTION FOR THOUGHT

67. _____

10. Discuss what (if anything) is *wrong* with the following:

    (a) $\dfrac{3\sqrt{10}}{6} = \dfrac{\cancel{3}\sqrt{10}}{\underset{2}{\cancel{6}}} = \dfrac{\sqrt{10}}{2} = \dfrac{\sqrt{5 \cdot 2}}{2} = \dfrac{\sqrt{5} \cdot \cancel{2}}{\cancel{2}} = \sqrt{5}$

    (b) $\dfrac{2 + \sqrt{5}}{4} = \dfrac{\cancel{2} + \sqrt{5}}{\underset{2}{\cancel{4}}} = \dfrac{1 + \sqrt{5}}{2}$

68. _____

CHAPTER 9 SUMMARY

After having completed this chapter you should be able to:

1. Recognize and understand basic radical notation (Section 9.1).

   For example:

   (a) $\sqrt{36}$ means the nonnegative number which when squared is equal to 36. Therefore, $\sqrt{36} = 6$ because $6 \cdot 6 = 36$.

   (b) $\sqrt{x^8} = x^4$ because $x^4 \cdot x^4 = x^8$.

2. Apply the basic properties of radicals to obtain an expression in *simplest radical form* (Section 9.2).

   For example:

   (a) $\sqrt{18x^9} = \sqrt{9x^8}\sqrt{2x} = \boxed{3x^4\sqrt{2x}}$

   (b) $\dfrac{5}{\sqrt{x}} = \dfrac{5}{\sqrt{x}} \cdot \dfrac{\sqrt{x}}{\sqrt{x}} = \boxed{\dfrac{5\sqrt{x}}{x}}$

3. Add and subtract radical expressions (Section 9.3).

   For example:

   (a) $3\sqrt{7} - 8\sqrt{7} = \boxed{-5\sqrt{7}}$

   (b) $\sqrt{40} + \sqrt{50} + \sqrt{90} = \sqrt{4}\sqrt{10} + \sqrt{25}\sqrt{2} + \sqrt{9}\sqrt{10}$

   $\qquad\qquad\qquad\qquad = 2\sqrt{10} + 5\sqrt{2} + 3\sqrt{10}$

   $\qquad\qquad\qquad\qquad = \boxed{5\sqrt{10} + 5\sqrt{2}}$

4. Multiply radical expressions (Section 9.4).

   For example:

   (a) $2\sqrt{3}(\sqrt{3} - 4\sqrt{5}) = 2\sqrt{3}\sqrt{3} - 8\sqrt{3}\sqrt{5} = \boxed{6 - 8\sqrt{15}}$

   (b) $(2\sqrt{x} - \sqrt{3})(3\sqrt{x} - 4\sqrt{3})$

   $\qquad = 6\sqrt{x}\sqrt{x} - 8\sqrt{3x} - 3\sqrt{3x} + 4\sqrt{3}\sqrt{3}$

   $\qquad = 6x - 11\sqrt{3x} + 4 \cdot 3$

   $\qquad = \boxed{6x - 11\sqrt{3x} + 12}$

5. Divide radical expressions, which frequently involves rationalizing denominators (Section 9.4).

   For example:

   $$\dfrac{12}{\sqrt{5} - \sqrt{2}} = \dfrac{12}{\sqrt{5} - \sqrt{2}} \cdot \dfrac{\sqrt{5} + \sqrt{2}}{\sqrt{5} + \sqrt{2}}$$

   $$= \dfrac{12(\sqrt{5} + \sqrt{2})}{5 - 2}$$

   $$= \dfrac{12(\sqrt{5} + \sqrt{2})}{3} \qquad \textit{Reduce.}$$

   $$= \boxed{4(\sqrt{5} + \sqrt{2})}$$

1. Since $20^2 = 400$, $30^2 = 900$, and 648 is between 400 and 900, $\sqrt{648}$ must be between 20 and 30. Consider the following table:

$$0^2 = \underline{0} \qquad 5^2 = 2\underline{5}$$
$$1^2 = \underline{1} \qquad 6^2 = 3\underline{6}$$
$$2^2 = \underline{4} \qquad 7^2 = 4\underline{9}$$
$$3^2 = \underline{9} \qquad 8^2 = 6\underline{4}$$
$$4^2 = 1\underline{6} \qquad 9^2 = 8\underline{1}$$

This implies that any perfect square must end in one of the digits underlined: 0, 1, 4, 9, 6, or 5. Therefore, any number that ends in either 2, 3, 7, or 8 cannot be a perfect square. So 648 cannot be a perfect square.

2. Using the same argument as in (1), $\sqrt{841}$ must be between 20 and 30. Since 841 ends in the digit 1, there are only two possibilities, if 841 is a perfect square: $(21)^2 = 841$ or $(29)^2 = 841$. Since 841 is closer to 900 than it is to 400, try $(29)^2$ first, since 29 is closer to 30 than it is to 20. Since $29 \cdot 29 = 841$, conclude that 841 *is* a perfect square, and that $\sqrt{841} = 29$.

3. It is incorrect to claim that $\sqrt{1 + 1} = \sqrt{1} + \sqrt{1}$. In fact, if a and b are positive, it is *never* true that $\sqrt{a + b} = \sqrt{a} + \sqrt{b}$. That is, the square root of the sum of two positive numbers is never equal to the sum of the square roots of those numbers.

4. For any nonnegative number a, $\sqrt{a}\sqrt{a} = a$ tells us that $\sqrt{a}$ is the nonnegative quantity whose square is equal to a.

5. When we square a real number other than 0 or 1, we get an answer that is different from the original number. So it is incorrect to say that $\dfrac{2}{\sqrt{5}} = \dfrac{2^2}{(\sqrt{5})^2}$. When we rationalize the denominator properly we multiply $\dfrac{2}{\sqrt{5}}$ by $\dfrac{\sqrt{5}}{\sqrt{5}}$. This means that we multiply by 1, which does *not* change the value of the original number.

6. $\dfrac{1}{\sqrt{5}} = \dfrac{1 \cdot \sqrt{5}}{\sqrt{5} \cdot \sqrt{5}} = \dfrac{\sqrt{5}}{5}$

$\dfrac{1}{\sqrt{5}} = \dfrac{1}{2.2361} = .4472$ correct to 3 places

$\dfrac{\sqrt{5}}{5} = \dfrac{2.2361}{5} = .4472$ correct to 3 places

It is easier to compute $\dfrac{\sqrt{5}}{5}$ than to compute $\dfrac{1}{\sqrt{5}}$, since $\dfrac{1}{\sqrt{5}}$ involves division by a decimal quantity, whereas $\dfrac{\sqrt{5}}{5}$ does not.

7. $\dfrac{\sqrt{3}}{\sqrt{7}} = \dfrac{1.7321}{2.6458} = .6547$ correct to 3 places

$\dfrac{\sqrt{3}}{\sqrt{7}} = \dfrac{\sqrt{3} \cdot \sqrt{7}}{\sqrt{7} \cdot \sqrt{7}} = \dfrac{\sqrt{21}}{7} = \dfrac{4.5826}{7} = .6547$ correct to 3 places

8. $\sqrt{80} = \sqrt{16 \cdot 5} = \sqrt{16}\sqrt{5} = 4\sqrt{5}$

$\sqrt{80} = 8.9442719$

$4\sqrt{5} = 4(2.2360679) = 8.9442716$

The two results are the same, to 6 places. These results should be equal, and only appear to differ because of rounding off.

9. $\sqrt{150} = \sqrt{25 \cdot 6} = \sqrt{25}\sqrt{6} = 5\sqrt{6}$

$\sqrt{150} = 12.247449$

$5\sqrt{6} = 5(2.4494897) = 12.247449$

10. (a) The "2" in the numerator is under the square root and is thus $\sqrt{2}$. This cannot be cancelled with the "2" in the denominator.

(b) The cancellation is not valid since 2 is not a common factor of the numerator. Remember that terms cannot be cancelled.

NOTE TO THE STUDENT

Use the space on this page to write down any questions you have or points you want to review with your instructor.

CHAPTER 9 REVIEW EXERCISES

In Exercises 1–12, simplify the given expression as completely as possible. Assume that all variables appearing under radical signs are nonnegative.

1. $\sqrt{49}$

2. $-\sqrt{100}$

3. $\sqrt{-16}$

4. $\sqrt{90}$

5. $\sqrt{96}$

6. $\sqrt{16x^{16}}$

7. $\sqrt{9x^9}$

8. $\sqrt{20x^7y^{10}}$

9. $\sqrt{\dfrac{4}{9}}$

10. $\sqrt{\dfrac{3}{4}}$

11. $\sqrt{\dfrac{3}{5}}$

12. $\sqrt{\dfrac{4}{5}}$

ANSWERS

13. _____

14. _____

15. _____

16. _____

17. _____

18. _____

19. _____

20. _____

21. _____

22. _____

23. _____

24. _____

25. _____

26. _____

In Exercises 13–37, perform the indicated operations. Be sure to express your answer in simplest radical form. Assume that all variables appearing under radical signs are nonnegative.

13. $8\sqrt{7} - 5\sqrt{7} - \sqrt{7}$

14. $3\sqrt{5} - 4\sqrt{3} + 3\sqrt{3} - 7\sqrt{5}$

15. $\sqrt{45} - \sqrt{20}$

16. $8\sqrt{32} - 5\sqrt{18} - \sqrt{8}$

17. $\sqrt{75x} + \sqrt{12x}$

18. $x\sqrt{54x^3} - \sqrt{24x^5}$

19. $\dfrac{\sqrt{12x^3y^2}}{xy} + \sqrt{27x}$

20. $\sqrt{\dfrac{5}{7}} + \sqrt{\dfrac{7}{5}}$

21. $\sqrt{5}(3\sqrt{5} + \sqrt{2})$

22. $(4\sqrt{x} - \sqrt{3})(\sqrt{x} - 2\sqrt{3})$

23. $(3\sqrt{7} - 2\sqrt{3})(2\sqrt{7} + 5\sqrt{3})$

24. $(\sqrt{36} - \sqrt{16})^2$

25. $(\sqrt{x} - 3)^2$

26. $(\sqrt{a} - \sqrt{10})(\sqrt{a} + \sqrt{10})$

27. $\dfrac{7}{\sqrt{3}}$

28. $\dfrac{10}{\sqrt{6}}$

29. $\dfrac{x^2}{\sqrt{x}}$

30. $\dfrac{7}{3\sqrt{6}}$

31. $\dfrac{18}{\sqrt{12}}$

32. $\dfrac{\sqrt{8m^5}}{\sqrt{18m}}$

33. $\dfrac{14}{3-\sqrt{2}}$

34. $\dfrac{14}{3\sqrt{2}}$

35. $\dfrac{2+\sqrt{5}}{6+\sqrt{5}}$

36. $\dfrac{12}{\sqrt{10}-\sqrt{6}} - \dfrac{18}{\sqrt{6}}$

37. $(\sqrt{x+7})^2 - (\sqrt{x}+\sqrt{7})^2$

38. Show that $2+\sqrt{3}$ is a solution to the equation $x^2 - 4x + 1 = 0$.

27. _____
28. _____
29. _____
30. _____
31. _____
32. _____
33. _____
34. _____
35. _____
36. _____
37. _____

CHAPTER 9 PRACTICE TEST

Perform the indicated operations. Make sure your final answer is in simplest radical form. Assume that all variables appearing under radical signs are nonnegative.

1. _____

1. $\sqrt{25x^{16}y^6}$ **2.** $2\sqrt{27} - 3\sqrt{12} + \sqrt{300}$

2. _____

3. _____

3. $\sqrt{50x^3} - x\sqrt{32x}$ **4.** $\dfrac{\sqrt{48x^5y^{12}}}{\sqrt{8xy^2}}$

4. _____

5. $\sqrt{20x^8y^9} + 3x^4y^4\sqrt{5y}$ **6.** $\dfrac{3x^2}{\sqrt{6x}}$

5. _____

6. _____

7. $(2\sqrt{x} - \sqrt{5})(\sqrt{x} + 3\sqrt{5})$ **8.** $(\sqrt{100} - \sqrt{36})^2$

7. _____

9. $(\sqrt{x} - 4)^2 - (\sqrt{x-4})^2$ **10.** $\dfrac{10}{\sqrt{7} - \sqrt{3}}$

8. _____

9. _____

11. Determine whether $1 - \sqrt{2}$ is a solution to the equation $x^2 - 2x = 1$.

10. _____

11. _____

CUMULATIVE REVIEW
Chapters 7–9

In Exercises 1–10, *simplify the expression as completely as possible. Fractions should be reduced to lowest terms.*

1. $\sqrt{36x^{16}y^{12}}$

2. $\dfrac{x^2 - 4x}{x^2 - 16}$

3. $\dfrac{t^2 - 5t + 6}{t^2 - 6t + 9}$

4. $\sqrt{\dfrac{45x^9y^5}{5x^2y}}$

5. $\dfrac{7}{\sqrt{6}}$

6. $\dfrac{9}{\sqrt{6} - 2}$

7. $\dfrac{20}{3 - \sqrt{5}}$

8. $\dfrac{15}{\sqrt{5}}$

9. $\sqrt{120}$

10. $\sqrt{18x^3y^8}$

In Exercises 11–28, *perform the indicated operations and simplify as completely as possible.*

11. $\dfrac{3}{4x} + \dfrac{5}{x + 4}$

12. $\sqrt{45t} - \sqrt{20t}$

13. $\dfrac{5}{6xy^3} - \dfrac{7}{4x^2}$

14. $3\sqrt{2}(\sqrt{3} - \sqrt{6}) - 5(\sqrt{12} - \sqrt{6})$

15. $\sqrt{\dfrac{3}{7}} + \sqrt{21}$

16. $\dfrac{x^2 - 5x}{10x} \cdot \dfrac{x^2}{x^2 - 25}$

17. $(2\sqrt{x} - 3)(\sqrt{x} + 5)$

18. $\dfrac{6rt}{r^2 - 2rt + t^2} \div \dfrac{t^2}{r^2 - t^2}$

ANSWERS

1. _____

2. _____

3. _____

4. _____

5. _____

6. _____

7. _____

8. _____

9. _____

10. _____

11. _____

12. _____

13. _____

14. _____

15. _____

16. _____

17. _____

18. _____

ANSWERS

19. _____

20. _____

21. _____

22. _____

23. _____

24. _____

25. _____

26. _____

27. _____

28. _____

29. _____

30. _____

31. _____

32. _____

33. _____

34. _____

35. _____

36. _____

19. $\dfrac{6}{x^2 + 2x} - \dfrac{4}{x^2 - 2x}$

20. $\sqrt{27x^6 y^5} - xy\sqrt{12x^4 y^3}$

21. $\dfrac{6}{\sqrt{5}} + \dfrac{3}{\sqrt{20}}$

22. $\dfrac{6}{u + 3} - \dfrac{4}{3u} + \dfrac{1}{2u + 6}$

23. $(3\sqrt{2} - 4\sqrt{5})(4\sqrt{2} - 2\sqrt{5})$

24. $\dfrac{\dfrac{a}{2} - \dfrac{8}{a}}{\dfrac{a^2 - 8a + 16}{4}}$

25. $8 \cdot \dfrac{t}{t + 2} - \dfrac{5}{3t^2 + 6t} \cdot (6t^2 + 18t)$

26. $(\sqrt{x} + 3)^2 + (\sqrt{x + 3})^2$

27. $\dfrac{15}{\sqrt{7} - \sqrt{2}} - \dfrac{10}{\sqrt{2}}$

28. $\dfrac{2z + 9}{4z + 12} - \dfrac{5z + 8}{4z + 12} + \dfrac{3z + 1}{4z + 12}$

In Exercises 29–36, solve the given equation. If it contains more than one variable, solve for the indicated variable.

29. $\dfrac{11}{x} - \dfrac{2}{3} = \dfrac{25}{3x}$

30. $\dfrac{4}{3a + 6} - \dfrac{3}{2} = \dfrac{5}{6a + 12}$

31. $5x - 7t = 9x - 4t + 12$ for t

32. $\dfrac{3y}{5} - a = 4y + 3a - 6$ for y

33. $\dfrac{8}{z - 2} + 5 = \dfrac{4z}{z - 2}$

34. $\dfrac{x + 8}{8} - \dfrac{x + 6}{6} = \dfrac{x + 3}{3} - \dfrac{x + 4}{4}$

35. $\dfrac{3}{4}(x - 6) + \dfrac{2}{5}(x - 7) = 1 - (x + 4)$

36. $\dfrac{3c + 1}{3c - 2} = \dfrac{3c}{3c - 1}$

In Exercises 37–44, sketch the graph of the given equation in a rectangular coordinate system. Label the intercepts.

37. $y = 2x - 6$

38. $3x - 5y = 15$

39. $3y - 6x = 12$

40. $4x + 3y = 10$

41. $x - 2 = 0$

42. $y + 3 = 0$

43. $y = 5x$

44. $3y = x$

In Exercises 45–48, find the slope of the line passing through the given pair of points.

45. $(2, -1)$ and $(-3, 4)$

46. $(2, 0)$ and $(4, 5)$

47. $(2, 4)$ and $(-1, 4)$

48. $(3, 1)$ and $(3, -2)$

In Exercises 49–52, write an equation of the line with the given slope which passes through the given point.

49. $m = 4$; $(2, 3)$

50. $m = \dfrac{1}{2}$; $(-4, -1)$

51. $m = -\dfrac{3}{4}$; $(0, 3)$

52. m is undefined; $(0, 3)$

53. Write an equation of the line passing through the points $(-3, 5)$ and $(2, -2)$.

54. Write an equation of the line passing through the points $(0, 4)$ and $(4, 0)$.

ANSWERS

45. _____

46. _____

47. _____

48. _____

49. _____

50. _____

51. _____

52. _____

53. _____

54. _____

ANSWERS

55. _____

56. _____

57. _____

58. _____

59. _____

60. _____

61. _____

62. _____

63. _____

64. _____

65. _____

66. _____

67. _____

68. _____

In Exercises 55–56, solve the given system of equations graphically.

55. $\begin{cases} 2x - y = 7 \\ x + 2y = 6 \end{cases}$ **56.** $\begin{cases} 2x + 3y = 3 \\ 5x - 4y = 19 \end{cases}$

In Exercises 57–64, solve each system of equations algebraically.

57. $\begin{cases} 2x - y = 7 \\ x + 2y = 6 \end{cases}$ **58.** $\begin{cases} 2x + 3y = 3 \\ 5x - 4y = 19 \end{cases}$

59. $\begin{cases} 4x - 3y = 0 \\ 2x - y = \dfrac{1}{3} \end{cases}$ **60.** $\begin{cases} 5x - 7y = 13 \\ 6x - 4y = 20 \end{cases}$

61. $\begin{cases} y = 5x - 4 \\ x = 3y + 12 \end{cases}$ **62.** $\begin{cases} x - 8y = -4 \\ 4y - x = 1 \end{cases}$

63. $\begin{cases} x + \dfrac{y}{2} = 5 \\ 2x + y = 10 \end{cases}$ **64.** $\begin{cases} \dfrac{x}{3} - y = 2 \\ x - 3y = 5 \end{cases}$

65. If Bob can plow a field in 6 days and Martha can plow the same field in 4 days, how long will it take them to plow the field working together?

66. If Roger can overhaul an engine in 8 hours working alone or in 5 hours working with Pat, how long will it take Pat to overhaul the engine working alone?

67. The numerator of a fraction is 4 less than the denominator. If the numerator is increased by 3 and the denominator is increased by 1, the value of the fraction is $\frac{1}{2}$. Find the original fraction.

68. John goes into a clothing store and buys six shirts and two ties for $88.68. Bob buys four of the same priced shirts and three of the same priced ties for $68.27. Find the prices of a single shirt and a single tie.

CUMULATIVE PRACTICE TEST
Chapters 7–9

ANSWERS

1. _____

2. _____

3. _____

4. _____

5. _____

6. _____

7. _____

8. _____

9. _____

10. _____

11. _____

12. _____

13. _____

14. _____

15. _____

16. _____

In Problems 1–6, simplify the expression as completely as possible. Reduce fractions to lowest terms and express radicals in simplest radical form.

1. $\sqrt{64x^{16}}$

2. $\dfrac{3x^2 - 12x}{x^2 - x - 12}$

3. $\sqrt{40x^7y^{10}}$

4. $\dfrac{20}{\sqrt{6}}$

5. $\dfrac{t^2 - t - 6}{t^2 + t - 6}$

6. $\dfrac{12}{4 + \sqrt{7}}$

In Problems 7–16, perform the indicated operations and simplify as completely as possible.

7. $\dfrac{5}{x + 5} - \dfrac{4}{x + 4}$

8. $2\sqrt{5}(\sqrt{3} - \sqrt{2}) - (\sqrt{60} - \sqrt{40})$

9. $(\sqrt{x} + \sqrt{y})^2$

10. $\dfrac{w^2 - 3w - 10}{4w^2 + 8w} \cdot \dfrac{w^2}{w^2 - 10w + 25}$

11. $\dfrac{6}{a^2 + 3a} - \dfrac{3}{a^2 - 3a}$

12. $\sqrt{12x^7} + 3x\sqrt{3x^5}$

13. $(2\sqrt{z} - 3\sqrt{7})(3\sqrt{z} + \sqrt{7})$

14. $\dfrac{u^2 - 9u}{u^2} \div (u^2 - 81)$

15. $\dfrac{2 + \dfrac{1}{x}}{4 - \dfrac{1}{x^2}}$

16. $\dfrac{15}{\sqrt{7} - 2} - \dfrac{35}{\sqrt{7}}$

ANSWERS

17. _____

18. _____

19. _____

20. _____

21. _____

22. _____

23. _____

24. _____

25. _____

26. _____

27. _____

28. _____

In Problems 17–19, solve the given equation.

17. $\dfrac{9}{4t - 12} - \dfrac{2}{3} = \dfrac{11}{12t - 36}$

18. *Solve for u.* $\dfrac{2}{5}u - 4x = au - x + 7$

19. $\dfrac{10}{x + 4} + \dfrac{3}{5} = \dfrac{6 - x}{x + 4}$

In Problems 20–22, sketch the graph of the equation in a rectangular coordinate system. Label the intercepts.

20. $x - 3y = 0$ **21.** $x - 3y = 6$ **22.** $x - 3 = 6$

23. Find the slope of the line passing through the points $(-2, 2)$ and $(3, 6)$.

24. Write an equation of the line passing through the points $(2, -4)$ and $(-1, 3)$.

In Exercises 25–26, solve the system of equations algebraically.

25. $\begin{cases} 3x - 2y = 7 \\ 4x + y = -9 \end{cases}$ **26.** $\begin{cases} \dfrac{x}{2} - y = 5 \\ -x + 2y = 8 \end{cases}$

27. Jim can paint a house alone in 12 days while Susan can paint the same house alone in 9 days. If Jim paints alone for 5 days and then stops, how long will it take Susan to finish painting the house alone?

28. Sylvia buys four blouses at regular price and three blouses on special sale for a total of $55.30. Barbara buys two blouses at regular price and five on special sale for a total of $50.40. What are the prices of an individual blouse regularly and on special sale?

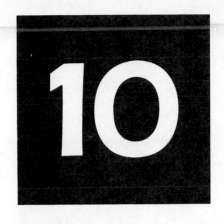

Quadratic Equations

In this chapter we will discuss solving second-degree equations in one variable. That is, we will learn how to solve equations in which the variable appears to the second power, as for example in the equation $x^2 + x = 30$. Such equations are called *quadratic equations*. We will see that, depending on the particular equation we are trying to solve, there are several methods of solution available to us. The goals of this chapter are not only to learn how to solve quadratic equations, but also to learn to choose the most efficient method for doing so.

The Factoring Method

In Chapter 3 we learned how to solve first-degree equations in one variable. Recall that for a first-degree equation there are three possibilities: Either the equation has no solutions (it is a contradiction), or one solution, or infinitely many solutions (it is an identity).

Quadratic equations give rise to one more possibility. For instance, the equation $x^2 + x = 30$ has two solutions since both $x = 5$ and $x = -6$ satisfy the equation.

$$x^2 + x = 30 \qquad\qquad x^2 + x = 30$$
$$5^2 + 5 \overset{?}{=} 30 \qquad (-6)^2 + (-6) \overset{?}{=} 30$$
$$25 + 5 \overset{?}{=} 30 \qquad\qquad 36 - 6 \overset{?}{=} 30$$
$$30 \overset{\checkmark}{=} 30 \qquad\qquad\qquad 30 \overset{\checkmark}{=} 30$$

Many of the quadratic equations we will solve will have two distinct solutions. The question we need to address is, "How do we *find* the solutions to a quadratic equation?"

As we proceed through this chapter we will learn several methods for solving a quadratic equation. For the most part, we will want to start with our equation in a particular form.

> **DEFINITION** A quadratic equation is said to be in *standard form* if it is written in the form $ax^2 + bx + c = 0$, with $a > 0$.

Note that standard form for a second-degree *equation* is very similar to standard form for a second-degree *polynomial*. However, do not confuse a polynomial, which is an expression, with an equation.

Also, do not take this definition too literally. It does not mean to say that all the terms are on the left-hand side of the equation and 0 is on the right-hand side. It does mean to say that for a quadratic equation to be in standard form, all the terms must be on one side of the equation with the coefficient of x^2 positive, and 0 on the other side.

EXAMPLE 1 Put each of the following quadratic equations into standard form.
(a) $x^2 + 3x - 5 = x - 3$ **(b)** $3x - 5 - x^2 = 0$
(c) $(x + 3)^2 = 2x(x - 5)$

Solution
(a) In order to get this equation into standard form, we can subtract x and add 3 to both sides of the equation.

$$\begin{array}{rcl} x^2 + 3x - 5 &=& x - 3 \\ -x + 3 && -x + 3 \\ \hline \boxed{x^2 + 2x - 2 = 0} \end{array}$$

Factor each of the following as completely as possible. If you get fewer than 4 out of 5 correct, you should review Sections 6.3 and 6.4 before proceeding to this section.

1. $x^2 - 6x$

2. $x^2 - 6x - 16$

3. $x^2 - 16$

4. $2x^2 + 3x - 20$

5. $4x^2 - 24x + 36$

✔ **LEARNING CHECKS**

1. **(a)** $3x^2 - 5x + 4 = x^2 + x$

   (b) $8 - 5x - x^2 = 0$

   (c) $(x - 5)^2 = 4x(x + 3)$

ANSWERS TO PROGRESS CHECK

1. $x(x - 6)$ **2.** $(x - 8)(x + 2)$

3. $(x + 4)(x - 4)$

4. $(2x - 5)(x + 4)$

5. $4(x - 3)(x - 3)$

(b) This equation does have all the terms on one side and 0 on the other side, but the coefficient of x^2 is not positive. In order to get the x^2 term to have a positive coefficient, we can multiply both sides of the equation by -1. (While we are at it, we might as well rearrange the terms in the order that we usually prefer—from highest power to lowest.)

$$3x - 5 - x^2 = 0 \qquad \textit{Rearrange the terms to get}$$
$$-x^2 + 3x - 5 = 0 \qquad \textit{Now multiply both sides by } -1.$$
$$-1(-x^2 + 3x - 5) = -1 \cdot 0$$
$$\boxed{x^2 - 3x + 5 = 0} \qquad \textit{This equation is now in standard form.}$$

(c) We begin by multiplying out each side of the equation.

$$(x + 3)^2 = 2x(x - 5)$$
$$(x + 3)(x + 3) = 2x(x - 5)$$
$$x^2 + 6x + 9 = 2x^2 - 10x$$

Keeping in mind that we want the coefficient of x^2 to be positive, we get all the terms on the right-hand side and 0 on the left-hand side. (If we did it the other way around, we would end up with a negative coefficient for the x^2 term, and we would then have the additional step of multiplying both sides of the equation by -1.)

$$x^2 + 6x + 9 = 2x^2 - 10x$$
$$\underline{-x^2 - 6x - 9 \qquad -x^2 - 6x - 9}$$
$$\boxed{0 = x^2 - 16x - 9} \qquad \blacksquare$$

The first method of solution for quadratic equations that we will discuss is based on the following basic fact about real numbers.

ZERO-PRODUCT RULE

If $a \cdot b = 0$, then either $a = 0$, or $b = 0$, or both a and b are equal to 0.

Thus, if we want to solve the equation

$$(x - 3)(x + 4) = 0$$

the zero-product rule tells us that if the *product* of two factors is equal to 0, then at least one of the factors must be equal to 0. Thus, we can conclude that

$$x - 3 = 0 \quad \text{or} \quad x + 4 = 0 \qquad \textit{We can now solve these two simple first-degree equations.}$$

$$\boxed{x = 3} \quad \text{or} \quad \boxed{x = -4}$$

We can easily see that both of these values are solutions of the original equation, $(x - 3)(x + 4) = 0$.

It is worthwhile noting that this same idea extends to more than two factors. For example, if we want to solve the equation

$$4z(z + 5)(z - 7) = 0,$$

we use the fact that in order for the product to be equal to 0, at least one of the factors must be equal to 0. There are four factors: 4, z, $z + 5$, and $z - 7$. We need to find the values of z which make the individual factors equal to 0.

Since the first factor is 4, and 4 can never be equal to 0, we simply ignore it. If you like, you can think of dividing both sides of the equation by 4: 0 divided by 4 is still 0. (We can always ignore *constant* factors when we are interested in values of the variable which make a product equal to 0.)

This now leaves us with three possibilities:

$$z = 0 \quad \text{or} \quad z + 5 = 0 \quad \text{or} \quad z - 7 = 0$$

$$\boxed{z = 0} \quad \text{or} \quad \boxed{z = -5} \quad \text{or} \quad \boxed{z = 7}$$

As before, we can see that each of these values satisfies the original equation.

Suppose we want to solve the equation $x^2 + x = 30$, which we mentioned at the beginning of this section. We cannot apply the zero-product rule to this equation because we do not have the two necessary ingredients to use it. We do not have a product and we do not have 0.

The difficulty of not having 0 on one side of the equation can be overcome by putting the quadratic equation into standard form. The other difficulty of not having a product can be overcome if we can *factor* the nonzero side of our equation. This is exactly what the *factoring* process does: It changes a sum into a product. Thus, our solution is as follows:

$$x^2 + x = 30 \qquad \textit{We begin by putting the equation in standard form}$$
$$x^2 + x - 30 = 0 \qquad \textit{Now we try to factor the left-hand side.}$$
$$(x + 6)(x - 5) = 0$$
$$x + 6 = 0 \qquad \text{or} \quad x - 5 = 0$$
$$\boxed{x = -6} \quad \text{or} \quad \boxed{x = 5}$$

which are the solutions we saw earlier.

The method of solution that we have just illustrated is called the ***factoring method*** for obvious reasons.

There were two basic steps in this solution. The first was to get 0 on one side of the equation, which we can *always* do. The second was to factor the resulting second-degree polynomial that we get on one side of the equation, which we know from experience we cannot always do. Thus, the factoring method has the serious weakness that it does not always work.

In the next few sections we will discuss other methods which do not depend on being able to factor, but for the remainder of this section we will look at several more quadratic equations that can be solved by the factoring method.

2. $5t^2 = 4t$

EXAMPLE 2   *Solve for y.* $3y^2 = 7y$

Solution   In order to solve by the factoring method (which is the only method we have so far), we begin by putting the equation into standard form.

$$3y^2 = 7y \qquad \textit{Subtract 7y from both sides.}$$
$$3y^2 - 7y = 0 \qquad \textit{Factor; remember to look for any common factors first.}$$
$$y(3y - 7) = 0 \qquad \textit{Now we can use the zero-product rule.}$$

$$y = 0 \quad \text{or} \quad 3y - 7 = 0$$

$$\boxed{y = 0} \qquad 3y = 7$$

$$\boxed{y = \frac{7}{3}}$$

Some students attempt to solve this example by dividing both sides of the equation by y. This will lead to an incomplete solution (try it!). The reason is that $y = 0$ may be one of the solutions (as it is in this example), in which case you are dividing by 0, which is not allowed.

CHECK: We substitute the values $y = 0$ and $y = \frac{7}{3}$ into the original equation.

$$y = 0: \qquad 3y^2 = 7y \qquad y = \frac{7}{3}: \qquad 3y^2 = 7y$$

$$3(0)^2 \overset{?}{=} 7(0) \qquad\qquad 3\left(\frac{7}{3}\right)^2 \overset{?}{=} 7\left(\frac{7}{3}\right)$$

$$3(0) \overset{?}{=} 0 \qquad\qquad 3\left(\frac{49}{9}\right) \overset{?}{=} 7\left(\frac{7}{3}\right)$$

$$0 \overset{\checkmark}{=} 0 \qquad\qquad \frac{\cancel{3}}{1} \cdot \frac{49}{\underset{3}{\cancel{9}}} \overset{?}{=} \frac{7}{1} \cdot \frac{7}{3}$$

$$\frac{49}{3} \overset{\checkmark}{=} \frac{49}{3} \qquad \blacksquare$$

Let's pause to summarize the factoring method.

SOLVING QUADRATIC EQUATIONS BY THE FACTORING METHOD

1. Get the equation into standard form. This may require several steps.
2. Factor the nonzero side of the equation.
3. Use the zero-product rule to set each of the factors equal to 0.
4. Check each of the answers in the original equation.

EXAMPLE 3   *Solve for a.* $(2a + 1)(a - 1) = (3a - 2)(2a - 4)$

Solution   Following our outline, we begin by multiplying out both sides of the equation, and then get it into standard form.

$$(2a + 1)(a - 1) = (3a - 2)(2a - 4)$$

$$\begin{array}{rl}
2a^2 - a - 1 = & 6a^2 - 16a + 8 \\
\underline{-2a^2 + a + 1 \quad\; -2a^2 +\;\;\; a + 1} \\
0 = & 4a^2 - 15a + 9 \\
0 = & (4a - 3)(a - 3)
\end{array}$$

We can most easily get standard form by collecting terms on the right-hand side.

Factoring this requires a bit of trial and error.

$$4a - 3 = 0 \quad \text{or} \quad a - 3 = 0$$

$$\boxed{a = \frac{3}{4}} \qquad \boxed{a = 3}$$

3. $(3u - 2)(u + 4) = (u + 12)(u + 1)$

CHECK: We substitute $a = 3$ into the original equation.

$$(2a + 1)(a - 1) = (3a - 2)(2a - 4)$$
$$(2(3) + 1)(3 - 1) \stackrel{?}{=} (3(3) - 2)(2(3) - 4)$$
$$(6 + 1)(2) \stackrel{?}{=} (9 - 2)(6 - 4)$$
$$(7)(2) \stackrel{\checkmark}{=} (7)(2)$$

The fact that $a = 3$ checks gives us a great deal of confidence that our other solution is correct as well. Nevertheless, in order to be absolutely sure we must also substitute $a = \frac{3}{4}$ into the original equation. This check is left to the student. ∎

4. $(t + 5)(t - 3) = 9$

EXAMPLE 4   *Solve for y.* $(y - 6)(y + 1) = 8$

Solution   Be careful not to misinterpret the zero-product rule. We *cannot* set each of the factors $y - 6$ and $y + 1$ equal to 8. The zero-product rule requires that the product of the factors be equal to *zero*.

Therefore, we must again begin by getting the equation into standard form.

$$(y - 6)(y + 1) = 8$$
$$y^2 - 5y - 6 = 8 \qquad \textit{Subtract 8 from both sides.}$$
$$y^2 - 5y - 14 = 0 \qquad \textit{Factor.}$$
$$(y - 7)(y + 2) = 0$$
$$y - 7 = 0 \quad \text{or} \quad y + 2 = 0$$
$$\boxed{y = 7} \quad \text{or} \quad \boxed{y = -2}$$

CHECK: $y = 7$: $(y - 6)(y + 1) = 8$
$$(7 - 6)(7 + 1) \stackrel{?}{=} 8$$
$$(1)(8) \stackrel{\checkmark}{=} 8$$

$y = -2$: $(y - 6)(y + 1) = 8$
$$(-2 - 6)(-2 + 1) \stackrel{?}{=} 8$$
$$(-8)(-1) \stackrel{\checkmark}{=} 8 \qquad ∎$$

As the next example shows, fractional equations can give rise to quadratic equations.

5. $a + \dfrac{5a}{a - 2} = \dfrac{10}{a - 2}$

EXAMPLE 5   *Solve for x.* $x + \dfrac{4x}{x - 1} = \dfrac{4}{x - 1}$

Solution   We begin as we would any fractional equation—by clearing the fractions.

$$x + \frac{4x}{x - 1} = \frac{4}{x - 1} \qquad \textit{Multiply both sides of the equation by the LCD, which is } x - 1.$$
$$x(x - 1) + \frac{\cancel{x - 1}}{1} \cdot \frac{4x}{\cancel{x - 1}} = \frac{\cancel{x - 1}}{1} \cdot \frac{4}{\cancel{x - 1}}$$
$$x(x - 1) + 4x = 4$$
$$x^2 - x + 4x = 4$$

Since this is a quadratic equation, we get it into standard form.

$$x^2 + 3x - 4 = 0 \qquad \textit{Factor.}$$

$$(x + 4)(x - 1) = 0$$

$$x + 4 = 0 \quad \text{or} \quad x - 1 = 0$$

$$x = -4 \quad \text{or} \qquad x = 1 \qquad \textit{Notice that we have not put a box}$$
$$\textit{around our answers.}$$

Since this is a fractional equation, we *must* check to make sure that we do not have any extraneous solutions (ones that do not satisfy the *original* equation).

CHECK: We check in the *original* equation.

$$x = -4: \qquad x + \frac{4x}{x - 1} = \frac{4}{x - 1} \qquad\qquad x = 1: \quad x + \frac{4x}{x - 1} = \frac{4}{x - 1}$$

$$-4 + \frac{4(-4)}{-4 - 1} \stackrel{?}{=} \frac{4}{-4 - 1} \qquad\qquad 1 + \frac{4(1)}{1 - 1} \stackrel{?}{=} \frac{4}{1 - 1}$$

$$-4 + \frac{-16}{-5} \stackrel{?}{=} \frac{4}{-5} \qquad\qquad\qquad 1 + \frac{4}{0} \stackrel{?}{=} \frac{4}{0}$$

$$\frac{-20}{5} + \frac{16}{5} \stackrel{\checkmark}{=} -\frac{4}{5} \qquad\qquad \textit{x = 1 does not work.}$$
$$\textit{We cannot divide by 0.}$$

Therefore, we have only one solution: $\boxed{x = -4}$ ■

In Example 5 we made an important remark. We said that because the equation we obtained as a result of simplifying was a quadratic equation, "we get it into standard form." It is important to keep in mind that if the equation we obtain from our simplifying process is a *first-degree* equation, then we solve it by simply isolating the variable. There would be no need to put the equation into standard form. Standard form is the form we prefer for *quadratic* equations only.

✔ *Answers to Learning Checks in Section 10.1*

1. (a) $2x^2 - 6x + 4 = 0$ **(b)** $x^2 + 5x - 8 = 0$ **(c)** $3x^2 + 22x + 25 = 0$

2. $t = 0, \dfrac{4}{5}$ **3.** $u = -\dfrac{5}{2}, 4$ **4.** $t = 4, -6$ **5.** $a = -5$

NOTE TO THE STUDENT

Use the space on this page to write down any questions you have or points you want to review with your instructor.

Exercises 10.1

Solve each of the following equations. If the equation is quadratic, use the factoring method. If the equation cannot be factored, say so.

1. $(x - 2)(x + 3) = 0$

2. $(a + 4)(a - 3) = 0$

3. $(x - 2)(x + 3) = 6$

4. $(a + 4)(a - 3) = 8$

5. $y(y - 4) = 0$

6. $w(w - 6) = 0$

7. $y(y - 4) = 12$

8. $w(w - 6) = 27$

9. $x^2 - x - 6 = 0$

10. $t^2 + 2t - 8 = 0$

11. $x^2 - 3x = 10$

12. $t^2 + 10 = 7t$

13. $-m^2 + 2m + 8 = 0$

14. $-w^2 + 9w - 20 = 0$

15. $-m^2 = 8 - 9m$

16. $-w^2 = -w - 20$

17. $p^2 + 3p = p(p + 4)$

18. $2n^2 - 6n = n(2n + 4) + 5$

19. $2a^2 = 11a - 12$

20. $14b = 3b^2 - 5$

21. $2a(a + 3) = 0$

22. $3z(z - 5) = 0$

23. $2a(a + 3) = 20$

24. $3z(z - 5) = -12$

1. _____
2. _____
3. _____
4. _____
5. _____
6. _____
7. _____
8. _____
9. _____
10. _____
11. _____
12. _____
13. _____
14. _____
15. _____
16. _____
17. _____
18. _____
19. _____
20. _____
21. _____
22. _____
23. _____
24. _____

ANSWERS

25. **25.** $2x^2 + 5x - 4 = x^2 + 3x - 7$ **26.** $3x^2 - 2x - 6 = 2x^2 - 6x - 3$

26. _____

27. $5x^2 = 45$ **28.** $3x^2 = 48$

27. _____

29. $(x + 3)^2 = 3x^2 - 10$ **30.** $(x - 4)^2 = 2x^2 - 11x - 2$

28. _____

31. $4y = 4y^2 + 1$ **32.** $9y^2 + 4 = 12y$

29. _____

33. $(x + 2)^2 = 25$ **34.** $(z - 3)^2 = 16$

30. _____

35. $(x - 4)(x + 1) = (x - 3)(x - 2)$ **36.** $(y + 2)(y + 5) = (y - 1)(y + 6)$

31. _____

32. _____   **37.** $x + \dfrac{1}{x} = 2$ **38.** $x + \dfrac{3}{x} = \dfrac{7}{2}$

33. _____

39. $\dfrac{x - 1}{x + 1} = \dfrac{x}{x + 3}$ **40.** $\dfrac{x - 1}{x} = \dfrac{3x}{x - 2}$

34. _____

35. _____

41. $a - \dfrac{5a}{a + 1} = \dfrac{5}{a + 1}$ **42.** $a + \dfrac{3a}{a - 3} = \dfrac{9}{a - 3}$

36. _____

37. _____

QUESTIONS FOR THOUGHT

38. _____

1. Explain what is wrong (and why) with the following "solutions."

39. _____
    (a) $(x - 3)(x - 4) = 7$
    $$x - 3 = 7 \quad \text{or} \quad x - 4 = 7$$
    $$x = 10 \quad \text{or} \quad x = 11$$

40. _____
    (b) $3x(x - 2) = 0$
    $$x = 3 \quad \text{or} \quad x = 0 \quad \text{or} \quad x - 2 = 0$$
    $$x = 3 \quad \text{or} \quad x = 0 \quad \text{or} \quad x = 2$$

41. _____

2. Consider the equation $x^2 + 4 = 0$. Is it possible for this equation to have any solutions in the real number system? Why or why not?

42. _____

The Square Root Method

10.2

As we mentioned in Section 10.1, the factoring method has the drawback of being a "special" method, in that it works only when we can factor.

The next method we will discuss may also seem somewhat special at first glance, but as we shall see in this section and the next, we will be able to generalize it to a method which works for all equations.

If asked to solve the equation $x^2 = 9$, we could proceed as follows (using the factoring method):

$$x^2 = 9 \qquad \textit{Get the equation into standard form by subtracting 9 from both sides.}$$

$$x^2 - 9 = 0 \qquad \textit{Factor.}$$

$$(x - 3)(x + 3) = 0$$

$$x - 3 = 0 \quad \text{or} \quad x + 3 = 0$$

$$\boxed{x = 3} \quad \text{or} \quad \boxed{x = -3}$$

If we think a moment about the equation $x^2 = 9$, it should certainly come as no surprise that the solutions are $+3$ and -3. After all, the equation $x^2 = 9$ is asking for a number whose square is equal to 9. In other words, the equation requires that x be a square root of 9. As we saw in Chapter 9, every positive number has two square roots, one positive and the other negative. Thus, solving this equation involves finding the square roots of 9, which are $\sqrt{9}$ and $-\sqrt{9}$, or 3 and -3.

More generally, we can state the following theorem.

> **THEOREM** If $u^2 = d$ then $u = \sqrt{d}$ or $u = -\sqrt{d}$, for $d \geq 0$.

A shorter way of writing the two solutions $u = \sqrt{d}$ and $u = -\sqrt{d}$ is to write

$$u = \pm\sqrt{d} \quad \text{(read "plus or minus } \sqrt{d}\text{")}$$

We use the symbol "$\pm$" instead of writing the equation twice, once with a plus sign and once with a minus sign. The reason we must insist on $d \geq 0$ is that otherwise we will get the square root of a negative number, which is not defined in the real number system.

When we invoke this theorem we will say that we are "taking square roots" of the equation. This procedure of taking square roots is called the *square root method*.

It is important to remember that the symbol $\sqrt{d}$ still stands for the *positive* square root of d. However, when we are looking for a solution to the equation $u^2 = d$, we have no prior knowledge as to whether u is positive or negative, and so we must take both square roots in our solution set.

While the square root method is often shorter than the factoring method, if that were its only advantage we would not bother with "another" method. In fact, the square root method works in some situations where we cannot factor.

EXAMPLE 1 *Solve for a.* $a^2 = 11$

Solution

$$a^2 = 11 \qquad \textit{This equation is in exactly the form that our theorem requires.}$$
$$\textit{Take square roots.}$$
$$\boxed{a = \pm\sqrt{11}}$$

✔ **LEARNING CHECKS**

1. $t^2 = 15$

We can see that these solutions satisfy the equation. (Throughout the remainder of this chapter, whenever the solutions to an equation are not checked, the check is left to the student.)

Note that had we tried to solve this equation by the factoring method, we would have subtracted 11 from both sides, and gotten $a^2 - 11 = 0$. However, this does not factor (with integers), and so the factoring method fails. ∎

Recall that the standard form for a quadratic equation is $ax^2 + bx + c = 0$. If we analyze the square root method we see that it works because there is no x term. Thus, we will normally use it when $b = 0$, since we can then isolate the x^2 term and take square roots.

2. (a) $3t^2 + 8 = 15 - 6t^2$
(b) $5t^2 + 4 = 6$

EXAMPLE 2 Solve for x.

(a) $x^2 + 2 = 5 - 3x^2$ **(b)** $3x^2 - 2 = 8$

Solution

(a) Since we see no x term we will use the square root method.

$$x^2 + 2 = 5 - 3x^2 \qquad \textit{Isolate the } x^2 \textit{ term.}$$
$$4x^2 = 3$$
$$x^2 = \frac{3}{4} \qquad \textit{Take square roots.}$$
$$x = \pm\sqrt{\frac{3}{4}} = \pm\frac{\sqrt{3}}{\sqrt{4}} = \boxed{\pm\frac{\sqrt{3}}{2}}$$

Notice that we always put our final answer in simplest radical form.

(b) Again we see no x term, so we use the square root method.

$$3x^2 - 2 = 8$$
$$3x^2 = 10$$
$$x^2 = \frac{10}{3} \qquad \textit{Take square roots.}$$
$$x = \pm\sqrt{\frac{10}{3}} \qquad \textit{Simplify the radical.}$$
$$x = \pm\frac{\sqrt{10}}{\sqrt{3}} = \pm\frac{\sqrt{10}}{\sqrt{3}} \cdot \frac{\sqrt{3}}{\sqrt{3}} = \boxed{\pm\frac{\sqrt{30}}{3}}$$ ∎

3. (a) $(u + 5)(u - 3) = 2u - 1$
(b) $3z^2 + 15 = 11$

EXAMPLE 3 Solve for z.

(a) $(z - 3)(z - 2) = 13 - 5z$ **(b)** $2z^2 + 7 = 1$

Solution

(a) We begin by getting the equation into standard form.

$$(z - 3)(z - 2) = 13 - 5z$$
$$z^2 - 5z + 6 = 13 - 5z$$
$$\underline{+5z - 13 \qquad -13 + 5z}$$
$$z^2 - 7 = 0$$
$$z^2 = 7 \qquad \textit{Take square roots.}$$
$$\boxed{z = \pm\sqrt{7}}$$

Of course, if you notice that the z term is going to drop out, you would not bother subtracting 13 from both sides. Rather, you would subtract 6 from both sides in order to isolate the second-degree term more quickly. Thus, we see that if we are going to use the square root method, it is not important that the equation be in standard form.

(b) $2z^2 + 7 = 1$

$$2z^2 = -6$$
$$z^2 = -3 \qquad \text{\textit{Take square roots.}}$$
$$z = \pm\sqrt{-3} \qquad \text{There are \textbf{\textit{no real solutions.}}}$$

The $\sqrt{-3}$ does not exist in the real number system. Therefore, our equation has *no real solutions*. ∎

We mentioned above that the square root method is used primarily in those cases when the standard form of the equation has no x term. There is one very important exception to this rule. If we look carefully at the content of the theorem quoted earlier in this section, we can paraphrase it as follows:

$$\text{If } (\textit{something})^2 = d, \text{ then } \textit{something} = \pm\sqrt{d}.$$

The theorem gives us the solutions to an equation which is in the form

$$\textit{Perfect square} = \textit{Nonnegative number}$$

In other words, it is not necessary that the *something* be a single letter. In fact, the *something* can itself be an expression. In particular, we are interested in the case when it is a binomial.

EXAMPLE 4 *Solve for y.* $(y - 5)^2 = 9$

4. $(a + 3)^2 = 4$

Solution This equation is in the form $()^2 = 9$ and so according to our theorem $() = \pm\sqrt{9}$. In other words, the *something* in this problem is the binomial $y - 5$.

$$(y - 5)^2 = 9 \qquad \text{\textit{Take square roots.}}$$
$$y - 5 = \pm\sqrt{9}$$
$$y - 5 = \pm 3 \qquad \text{\textit{We have two equations.}}$$
$$y - 5 = 3 \quad \text{or} \quad y - 5 = -3$$
$$\boxed{y = 8} \quad \text{or} \quad \boxed{y = 2}$$

CHECK: $y = 8$: $(y - 5)^2 = 9$
$$(8 - 5)^2 \overset{?}{=} 9$$
$$3^2 \overset{\checkmark}{=} 9$$

$y = 2$: $(y - 5)^2 = 9$
$$(2 - 5)^2 \overset{?}{=} 9$$
$$(-3)^2 \overset{\checkmark}{=} 9$$
∎

You might be asking yourself, "Could we have solved the equation in Example 4 by the factoring method as well?" Let's try it and see.

$$(y - 5)^2 = 9 \qquad \textit{Multiply out.}$$
$$(y - 5)(y - 5) = 9$$
$$y^2 - 10y + 25 = 9$$
$$y^2 - 10y + 16 = 0$$
$$(y - 8)(y - 2) = 0$$
$$y - 8 = 0 \quad \text{or} \quad y - 2 = 0$$
$$\boxed{y = 8} \quad \text{or} \quad \boxed{y = 2}$$

So we see that the factoring method works as well, although the square root method was a bit simpler.

It is worth repeating that if the only advantage of the square root method were that it made some solutions a bit simpler, we would not bother ourselves with learning yet another method. However, as the next example further illustrates, the square root method offers us an approach which works for cases in which the factoring method fails totally.

5. $(t - 2)^2 = 6$

EXAMPLE 5   *Solve for x.* $(x + 3)^2 = 7$

Solution   Suppose we try to solve this equation by the factoring method.

$$(x + 3)^2 = 7$$
$$(x + 3)(x + 3) = 7$$
$$x^2 + 6x + 9 = 7$$
$$x^2 + 6x + 2 = 0 \qquad \textit{Try to factor.}$$

We are stuck! However, since this equation is in the form $(\textit{binomial})^2 = $ number, let's try the square root method.

$$(x + 3)^2 = 7 \qquad \textit{Take square roots.}$$
$$x + 3 = \pm\sqrt{7} \qquad \textit{We get two equations.}$$
$$x + 3 = \sqrt{7} \quad \text{or} \quad x + 3 = -\sqrt{7} \qquad \textit{Subtract 3 from both sides of each equation.}$$
$$\boxed{x = -3 + \sqrt{7}} \quad \text{or} \quad \boxed{x = -3 - \sqrt{7}} \qquad\qquad ■$$

Note that unlike Example 4, separating the equation $x + 3 = \pm\sqrt{7}$ into two equations in Example 5 did not yield any simplification of our answers. In such a case we might as well write

$$x + 3 = \pm \sqrt{7} \qquad \textit{Subtract 3 from both sides.}$$
$$\underline{-3 \quad -3}$$
$$x = -3 \pm \sqrt{7}$$

Whenever we have a radical that is an irrational number we may leave our answer in this form. However, if the radical is not an irrational number, then we are expected to give each answer individually, as in the following example.

__EXAMPLE 6__ *Solve for x.* $\left(x - \dfrac{2}{3}\right)^2 = \dfrac{1}{4}$

6. $\left(u - \dfrac{3}{2}\right)^2 = \dfrac{4}{9}$

__Solution__ Using the square root method, we get

$$\left(x - \frac{2}{3}\right)^2 = \frac{1}{4} \qquad \textit{Take square roots.}$$

$$x - \frac{2}{3} = \pm\sqrt{\frac{1}{4}} = \pm\frac{\sqrt{1}}{\sqrt{4}}$$

$$x - \frac{2}{3} = \pm\frac{1}{2} \qquad \textit{Add } \frac{2}{3} \textit{ to both sides.}$$

$$\underline{+\frac{2}{3} \quad +\frac{2}{3}}$$

$$x = \frac{2}{3} \pm \frac{1}{2} \qquad \textit{We do not leave the answer in this form.}$$

$$x = \frac{2}{3} + \frac{1}{2} \quad \text{or} \quad x = \frac{2}{3} - \frac{1}{2}$$

$$\boxed{x = \frac{7}{6}} \quad \text{or} \quad \boxed{x = \frac{1}{6}}$$

■

To summarize then, the square root method allows us to find the solutions (if any) of quadratic equations which are in the form

$$(x + p)^2 = d$$

where p and d are numbers.

In the next section we will carry the square root method one step further. We will learn how to adapt it to situations in which the "x term is not missing"— that is, to situations in which we do not have a perfect square.

✔ *Answers to Learning Checks in Section 10.2*

1. $t = \pm\sqrt{15}$ **2. (a)** $t = \pm\dfrac{\sqrt{7}}{3}$ **(b)** $t = \pm\dfrac{\sqrt{10}}{5}$

3. (a) $u = \pm\sqrt{14}$ **(b)** No real solutions

4. $a = -5, -1$ **5.** $t = 2\pm\sqrt{6}$ **6.** $u = \dfrac{13}{6}, \dfrac{5}{6}$

NOTE TO THE STUDENT

Use the space on this page to write down any questions you have or points you want to review with your instructor.

Exercises 10.2

In Exercises 1–36, solve the given equation using the square root method. If the equation has no real solutions, say so.

1. $x^2 = 25$

2. $a^2 = 81$

3. $b^2 - 16 = 0$

4. $y^2 - 36 = 0$

5. $9b^2 - 16 = 0$

6. $4y^2 - 36 = 0$

7. $b^2 + 16 = 0$

8. $y^2 + 36 = 0$

9. $25x^2 = 4$

10. $49x^2 = 100$

11. $36x^2 - 15 = 0$

12. $64x^2 - 30 = 0$

13. $3b^2 = 11$

14. $11w^2 - 7 = 0$

15. $3b^2 = 12$

16. $11w^2 - 11 = 0$

17. $9a^2 = 20$

18. $4c^2 = 27$

19. $3y^2 = 32$

20. $5z^2 = 48$

21. $7y^2 - 4 = 5y^2 + 6$

22. $8x^2 - 7 = 3x^2 + 3$

23. $5a^2 - 3a + 4 = 2a^2 - 3a + 13$

24. $2a^2 - 4a + 1 = 1 - 4a$

25. $3a^2 - 18 = 5a^2 - 10$

26. $1 - 4w^2 = w^2 + 11$

1. _____
2. _____
3. _____
4. _____
5. _____
6. _____
7. _____
8. _____
9. _____
10. _____
11. _____
12. _____
13. _____
14. _____
15. _____
16. _____
17. _____
18. _____
19. _____
20. _____
21. _____
22. _____
23. _____
24. _____
25. _____
26. _____

27. $(x + 2)^2 = 4(x + 7)$

28. $(x - 4)^2 = 4(9 - 2x)$

29. $(t - 2)^2 = 9$

30. $(t + 2)^2 = 25$

31. $(a + 5)^2 = 7$

32. $(y - 4)^2 = 13$

33. $(x - 6)^2 = 12$

34. $(x + 1)^2 = 18$

35. $\left(x + \dfrac{2}{5}\right)^2 = \dfrac{3}{25}$

36. $\left(x - \dfrac{3}{7}\right)^2 = \dfrac{5}{49}$

In Exercises 37–44, solve the given equation. For quadratic equations, choose either the factoring method or the square root method, whichever you think is the easier to use.

37. $2x^2 + 7x - 5 = 3x^2 + 9x - 4$

38. $x^2 + 4x + 9 = 3x^2 + 4x + 1$

39. $(y - 2)(y + 3) = y + 10$

40. $(3y - 1)(2y + 3) = 7(y^2 + y - 2)$

41. $(y - 2)(y + 3) = (2y - 7)(y + 4)$

42. $(y + 3)(y - 5) = (2y + 5)(y - 3)$

43. $4(x + 1) = \dfrac{9}{x + 1}$

44. $\dfrac{1}{x} + \dfrac{6}{x + 2} = 2$

CALCULATOR EXERCISES

Solve the following equations using the square root method. Round off your answers to the nearest hundredth.

45. $x^2 = 7$

46. $t^2 = 21$

47. $3k^2 = 20$

48. $5u^2 = 18$

49. $4x^2 = 19.7$

50. $6w^2 = 1.6$

51. $(2a - .3)^2 = 7.5$

52. $(3x - .5)^2 = 10.4$

Method of Completing the Square

10.3

Thus far we have learned two methods for solving quadratic equations: the factoring method, which is limited to those equations which can be factored; and the square root method, which is limited to equations of the special form $(x + p)^2 = d$.

In this section we will see how to convert any quadratic equation into one which is of this special form and can therefore be solved by the square root method. Let's begin by analyzing an example. (All checks in the next few sections are left to the student.)

EXAMPLE 1 *Solve for x.* $(x - 3)^2 = 17$

Solution In the last section we saw that an equation in this form can be solved quite easily by the square root method.

$$(x - 3)^2 = 17 \qquad \textit{Take square roots.}$$
$$x - 3 = \pm\sqrt{17}$$
$$\boxed{x = 3 \pm \sqrt{17}}$$

But what if this equation had been given in any of the following alternate *equivalent* forms?

$$(x - 3)(x - 3) = 17 \qquad \textit{Written out}$$
$$x^2 - 6x + 9 = 17 \qquad \textit{Multiplied out}$$
$$x^2 - 6x - 8 = 0 \qquad \textit{In standard form}$$

It is possible that if we were looking for it, we might recognize the first two of these alternate forms as perfect squares. However, it is unreasonable for us to be expected to recognize the last of these as being equivalent to a perfect square. Keep these three forms of the equation $(x - 3)^2 = 17$ in mind, as we will refer to them again. ∎

How do we proceed with an equation such as $x^2 - 6x - 8 = 0$, which we cannot factor and is not a perfect square? Before we proceed let us analyze the structure of an expression of the form $(x + p)^2$.

$$(x + p)^2 = (x + p)(x + p)$$
$$= x^2 + px + px + p^2$$
$$= x^2 + 2px + p^2$$

Do not get hung up on the letters. In most examples, p will have some numerical value. The key thing to keep in mind is that the coefficient of the x (the first-degree term) is $2p$, twice the number in the binomial; the numerical term is p^2, the square of the number in the binomial.

We know we can solve a quadratic equation of the form $(x + p)^2 = d$ by the square root method. So let's use $x^2 + 2px + p^2$ as a model for our equations, and see if we can transform them into perfect squares.

✔ LEARNING CHECKS

1. $(t + 4)^2 = 13$

MODEL OF A PERFECT SQUARE

$$(x + p)^2 = x^2 + 2px + p^2$$

An example should make this idea clear. Suppose we want to solve the equation $x^2 - 6x - 8 = 0$. We cannot factor it, so the factoring method fails, and it is not a perfect square, so the square root method fails. Let's see if looking at our model of a perfect square can help us.

Our first step is to add 8 to both sides of the equation. Remember that having an equation in standard form is important for the factoring method where we need 0 on one side of the equation. Since we *cannot* factor here, 0 is no longer important.

$$x^2 - 6x - 8 = 0$$
$$x^2 - 6x \quad = 8 \qquad \textit{Notice that we left a space where the } -8 \textit{ used to be.}$$

Now we compare $x^2 - 6x$ to our model perfect square $x^2 + 2px + p^2$. If our model and example are to match *exactly*, then they must match up term by term.

- The x^2 in our model matches the x^2 in our example exactly.

- If the x terms are to match as well, then their coefficients must be the same.

  The coefficient of x in our model is $2p$ (it will always be the same in the model), while the coefficient of x in this example is -6.

  Thus, for the x coefficients to be the same we must have $2p = -6$, which implies that $p = -3$.

- Finally, our model contains the quantity p^2.

  We have just determined that $p = -3$, therefore $p^2 = (-3)^2 = 9$.

  In order to have a perfect square as in the model we want our example to have $x^2 - 6x + 9$ on the left-hand side of the equation. In order to get this we add 9 to *both* sides of the equation.

Thus far, our solution looks as follows:

$$x^2 - 6x - 8 = 0$$
$$x^2 - 6x \quad = 8 \qquad \textit{By looking at the model we determine that } 2p = -6;$$
$$\textit{therefore, } p = -3, \textit{ and so the ``missing'' term is } p^2 = 9.$$
$$\textit{Therefore, we add 9 to both sides of the equation.}$$

$$\underline{\qquad +9 \quad +9}$$
$$x^2 - 6x + 9 = 17 \qquad \textit{The left-hand side of this equation is now a perfect square.}$$
$$(x - 3)^2 = 17 \qquad \textit{Notice that the left-hand side of the equation conforms to}$$
$$(x - 3)^2 = 17 \qquad \textit{the model. Since } p = -3 \textit{ we have}$$
$$\uparrow \qquad (x + p)^2 = (x + (-3))^2 = (x - 3)^2.$$

p goes in here according to our model.

If you look back now at Example 1 you will see that we have taken one of the "unrecognizable" forms of that equation, and reconstructed the "nice" perfect square form of that equation. We have not finished yet because we have not solved the equation. From here on we proceed exactly as we did in Example 1 by taking square roots, and solving for x.

The process we have just carried out is called the ***method of completing the square.*** It consists of determining the number which is needed to make a perfect square on one side of the equation.

Our analysis of the perfect square form $(x + p)^2 = x^2 + 2px + p^2$, and our use of $2p$, p, and p^2 were just devices to help us determine what the "missing" term was. Since $2p$ is always the coefficient of x in our model, and p^2 is always the constant term in our model, we will always be computing p (which is one-half of $2p$), and then squaring that to get p^2. In other words, the term needed to complete the square is always *the square of one-half the coefficient of x (the first-degree term)*.

Let's use the method of completing the square in the next example.

EXAMPLE 2 *Solve by completing the square.* $y^2 + 8y - 5 = 0$

2. $u^2 - 8u - 3 = 0$

Solution We first add 5 to both sides of the equation to get $y^2 + 8y = 5$. Now we compare the left-hand side of this equation to our model. Do not get hung up on the letters. Instead of the variable x in our model, this equation happens to have the variable y.

- The y^2 in this example matches the x^2 in our model.
- The coefficient of x in our model is $2p$ and that must match the coefficient of the first-degree term in this example, which is 8.

  Thus, we have $2p = 8$ and so $p = 4$.

- Therefore, the missing p^2 term is $p^2 = 4^2 = 16$, and so we add 16 to both sides of the equation. Thus far our solution looks as follows:

$$y^2 + 8y - 5 = 0 \qquad \text{\textit{Add 5 to both sides.}}$$
$$y^2 + 8y = 5 \qquad \text{\textit{$2p = 8$; $p = 4$; $p^2 = 16$}}$$
$$\text{\textit{or}} \ \left(\frac{1}{2} \cdot 8\right)^2 = (4)^2 = 16$$

Add 16 to both sides of the equation to complete the square.

$$y^2 + 8y + 16 = 5 + 16 \qquad \text{\textit{The left-hand side is now a perfect square. It matches our model with $p = 4$.}}$$
$$(y + p)^2 = (y + 4)^2$$

$$(y + 4)^2 = 21 \qquad \text{\textit{Now we can take square roots.}}$$
$$y + 4 = \pm\sqrt{21} \qquad \text{\textit{Subtract 4 from both sides.}}$$
$$\boxed{y = -4 \pm \sqrt{21}} \qquad\qquad\qquad\qquad\quad ■$$

Remember that the whole purpose of completing the square is to be able to solve the equation by the square root method. Do not forget to solve the equation after you have completed the square.

EXAMPLE 3 *Solve by completing the square.* $2x^2 + 94 = 24x$

3. $3t^2 = 18t - 30$

Solution In order to complete the square we want to start with the equation having the constant (numerical) term isolated, and of course with the leading coefficient being positive. Therefore, we rewrite the equation as

$$2x^2 + 94 = 24x$$
$$2x^2 - 24x = -94 \qquad \text{\textit{Comparing this to our model we see that we have $2x^2$ instead of x^2. Therefore, we divide both sides of the equation by 2.}}$$
$$x^2 - 12x = -47 \qquad \text{\textit{We compute }} \left[\frac{1}{2}(-12)\right]^2 = (-6)^2 = 36. \text{\textit{ To}}$$
$$x^2 - 12x + 36 = -47 + 36 \qquad \text{\textit{complete the square we add 36 to both sides.}}$$
$$(x - 6)^2 = -11 \qquad \text{\textit{Now take square roots.}}$$
$$x - 6 = \pm\sqrt{-11} \qquad \textbf{\textit{No real solutions}} \text{ \textit{due to the square root of a negative number}}$$

Notice that the method of completing the square also tells us when an equation has no solutions. ∎

Let's outline the method of completing the square.

METHOD OF COMPLETING THE SQUARE

1. Isolate the constant (numerical term), making sure that the second-degree term has a positive coefficient.
2. If the coefficient of the second-degree term is not 1, divide both sides of the equation by that coefficient.
3. Take one-half the coefficient of the first-degree term and square it. Add that quantity to both sides of the equation.
4. Now having a perfect square on one side of the equation, we can solve the equation by the square root method.

Keep in mind that the computation we have been doing with $2p$, p, and p^2 has simply been a device to help us understand and carry out step 3 in the outline.

4. $u^2 + 3u - 4 = 0$

EXAMPLE 4 *Solve by completing the square.* $x^2 - 5x + 4 = 0$

Solution Following our outline we proceed as follows:

$x^2 - 5x + 4 = 0$ *We isolate the constant.*

$x^2 - 5x \quad = -4$ *We compute* $\left[\frac{1}{2}(-5)\right]^2 = \left(\frac{-5}{2}\right)^2 = \frac{25}{4}$.

To complete the square we must add $\frac{25}{4}$ *to both sides of the equation.*

$x^2 - 5x + \frac{25}{4} = -4 + \frac{25}{4}$ *Because we built it, we know that the left-hand side is a perfect square. We know it is* $(x + p)^2 =$

$$\left[x + \left(-\frac{5}{2}\right)\right]^2 = \left(x - \frac{5}{2}\right)^2$$

Thus, the left-hand side becomes

$\left(x - \frac{5}{2}\right)^2 = -4 + \frac{25}{4}$ *Combine* -4 *and* $\frac{25}{4}$; *the LCD is 4.*

$\left(x - \frac{5}{2}\right)^2 = -\frac{16}{4} + \frac{25}{4}$ *Combine fractions.*

$\left(x - \frac{5}{2}\right)^2 = \frac{9}{4}$ *Take square roots.*

$x - \frac{5}{2} = \pm\sqrt{\frac{9}{4}} = \pm\frac{3}{2}$ *Add* $\frac{5}{2}$ *to both sides.*

$x = \frac{5}{2} \pm \frac{3}{2}$

$x = \frac{5}{2} + \frac{3}{2} = \frac{8}{2} = 4$ or $x = \frac{5}{2} - \frac{3}{2} = \frac{2}{2} = 1$

$\boxed{x = 4}$ or $\boxed{x = 1}$

The fact that we get rational answers tells us that we could have used the factoring method in the original equation.

$$x^2 - 5x + 4 = 0$$
$$(x - 4)(x - 1) = 0$$
$$x - 4 = 0 \quad \text{or} \quad x - 1 = 0$$
$$\boxed{x = 4} \quad \text{or} \quad \boxed{x = 1}$$

Comparing the two methods of solution, it is quite obvious that the factoring method is easier and quicker for this equation. This example simply gave us an opportunity to practice the method of completing the square.   ∎

At this point you can probably see that while the method of completing the square has the advantage of *always working*, it also has the potential for creating some messy arithmetic.

The method of completing the square has numerous applications in mathematics. In the next section we will see how we can use the method of completing the square to derive a formula which also works for all quadratic equations, but which is generally much easier to use.

✔ *Answers to Learning Checks in Section 10.3*

1. $t = -4 \pm \sqrt{13}$ **2.** $u = 4 \pm \sqrt{19}$ **3.** No real solutions

4. $u = -4, u = 1$

NOTE TO THE STUDENT

Use the space on this page to write down any questions you have or points you want to review with your instructor.

Exercises 10.3

In Exercises 1–22, solve the given equation by the method of completing the square.

1. $x^2 + 8x + 6 = 0$

2. $x^2 + 10x + 20 = 0$

3. $x^2 - 4x - 3 = 0$

4. $x^2 - 2x - 5 = 0$

5. $x^2 - 10x = 15$

6. $x^2 - 6x = 19$

7. $a^2 - 8a - 20 = 0$

8. $a^2 - 4a - 5 = 0$

9. $-x^2 - 12x = 6$

10. $-x^2 + 10x = 5$

11. $2z^2 - 12z + 4 = 0$

12. $3z^2 + 6z - 18 = 0$

13. $10 = 5y^2 + 20y$

14. $24 = 4y^2 - 8y$

15. $u^2 + 5u - 2 = 0$

16. $u^2 - 3u - 1 = 0$

1. _____

2. _____

3. _____

4. _____

5. _____

6. _____

7. _____

8. _____

9. _____

10. _____

11. _____

12. _____

13. _____

14. _____

15. _____

16. _____

ANSWERS

17. _____

18. _____

19. _____

20. _____

21. _____

22. _____

23. _____

24. _____

25. _____

26. _____

27. _____

28. _____

29. _____

30. _____

17. $x^2 + 4x + 5 = 2x - 3$

18. $x^2 + 8x + 10 = 2x - 2$

19. $(x - 4)(x + 3) = 1 - x$

20. $(y + 5)(y - 2) = 2 + 3y$

21. $2x^2 + 3 = 6x$

22. $3x^2 + 1 = 6x$

In Exercises 23–30, solve the given equation either by the factoring method or the square root method (completing the square where necessary). Choose whichever method you think is more appropriate.

23. $(x + 3)^2 = 6$

24. $(x - 2)^2 = 10x$

25. $(x + 3)^2 = 6x$

26. $(x - 2)^2 = 10$

27. $x^2 + 8x - 9 = 0$

28. $x^2 + 8x - 7 = 0$

29. $3x^2 + 4 = 8x$

30. $5x^2 + 6x = 8$

QUESTIONS FOR THOUGHT

3. State the relationship between the coefficient of the middle term and the constant of a perfect square.

4. Solve this equation and check one of the solutions:

$$\frac{x}{x - 1} = \frac{2}{x - 2}$$

The Quadratic Formula

10.4

Completing the square is a useful algebraic technique which will be needed again in several places in intermediate algebra, precalculus, and calculus. It is the most powerful of the methods that we have learned so far, because unlike the others it can be applied to *all* quadratic equations. However, as we mentioned in the last section, it is also potentially the messiest and most tedious to use.

Let us begin by solving a quadratic equation by the method of completing the square.

EXAMPLE 1 *Solve by completing the square.* $2x^2 + 7x + 4 = 0$

Solution We will follow the outline we presented in the last section, with one slight addition. We will number the basic steps in the process. These numbers appear at the right-hand side of the page. We will use these numbers to make it easier to refer back to the steps in the solution.

$2x^2 + 7x + 4 = 0$ *Subtract 4 from both sides.* **(1)**

$2x^2 + 7x \quad = -4$ *Divide both sides of the equation by 2.* **(2)**

$$\frac{2x^2}{2} + \frac{7x}{2} = \frac{-4}{2}$$

$$x^2 + \frac{7}{2}x = -2$$

We want to take one-half of the coefficient of x and square it.

$$\left[\frac{1}{2}\left(\frac{7}{2}\right)\right]^2 = \left(\frac{7}{4}\right)^2 = \frac{49}{16}$$

Note that this means $p = \frac{7}{4}$.

We add $\frac{49}{16}$ *to both sides.* **(3)**

$$x^2 + \frac{7}{2}x + \frac{49}{16} = -2 + \frac{49}{16}$$

On the left-hand side we know that we have $(x + p)^2$ *with* $p = \frac{7}{4}$.

On the right-hand side we combine -2 *and* $\frac{49}{16}$; *the LCD is 16.*

$$\left(x + \frac{7}{4}\right)^2 = \frac{-32}{16} + \frac{49}{16}$$ **(4)**

$$\left(x + \frac{7}{4}\right)^2 = \frac{17}{16}$$ *Take square roots.* **(5)**

$$x + \frac{7}{4} = \pm\sqrt{\frac{17}{16}} = \pm\frac{\sqrt{17}}{\sqrt{16}} = \pm\frac{\sqrt{17}}{4}$$ *Subtract* $\frac{7}{4}$ *from both sides.* **(6)**

$$x = -\frac{7}{4} \pm \frac{\sqrt{17}}{4}$$ *Since the denominators are the same we can combine.*

$$\boxed{x = \frac{-7 \pm \sqrt{17}}{4}}$$ **(7)**

∎

We certainly do not relish the thought of making this our usual method for solving quadratic equations. Instead, we can *use* algebra to carry out the process of completing the square *once* and obtain a formula which we can apply to all quadratic equations.

✔ **LEARNING CHECKS**

1. $3x^2 - 5x + 1 = 0$

We begin with the general quadratic equation in standard form. That is, we start with the equation $ax^2 + bx + c = 0$, where $a > 0$, and carry out the process of completing the square for this general equation. We will number the steps just as we did in Example 1, so that you can see that the process is the same. The only difference is that we will be working with letters instead of numbers.

$$ax^2 + bx + c = 0 \qquad \text{Subtract } c \text{ from both sides.} \qquad \textbf{(1)}$$

$$ax^2 + bx \quad\;\; = -c \qquad \text{Divide both sides of the equation by } a. \qquad \textbf{(2)}$$

$$\frac{ax^2}{a} + \frac{bx}{a} = \frac{-c}{a}$$

$$x^2 + \frac{b}{a}x = -\frac{c}{a} \qquad \text{We want to take one-half the coefficient of } x \text{ and square it.}$$

$$\left[\frac{1}{2}\left(\frac{b}{a}\right)\right]^2 = \left(\frac{b}{2a}\right)^2 = \frac{b^2}{4a^2}$$

Note this means that $p = \dfrac{b}{2a}$.

We add $\dfrac{b^2}{4a^2}$ to both sides. $\qquad$ **(3)**

$$x^2 + \frac{b}{a}x + \frac{b^2}{4a^2} = -\frac{c}{a} + \frac{b^2}{4a^2}$$

On the left-hand side we know that we have $(x + p)^2$ with $p = \dfrac{b}{2a}$.

On the right-hand side we combine $-\dfrac{c}{a}$ and $\dfrac{b^2}{4a^2}$; the LCD is $4a^2$.

$$\left(x + \frac{b}{2a}\right)^2 = -\frac{4ac}{4a^2} + \frac{b^2}{4a^2} \qquad\qquad \textbf{(4)}$$

$$\left(x + \frac{b}{2a}\right)^2 = \frac{-4ac + b^2}{4a^2} \qquad \text{Take square roots.} \qquad \textbf{(5)}$$

$$x + \frac{b}{2a} = \pm\sqrt{\frac{b^2 - 4ac}{4a^2}} = \pm\frac{\sqrt{b^2 - 4ac}}{\sqrt{4a^2}} = \pm\frac{\sqrt{b^2 - 4ac}}{2a}$$

Subtract $\dfrac{b}{2a}$ from both sides. $\qquad$ **(6)**

$$x = -\frac{b}{2a} \pm \frac{\sqrt{b^2 - 4ac}}{2a} \qquad \text{Since the denominators are the same we can combine.}$$

$$\boxed{x = \frac{-b \pm \sqrt{b^2 - 4ac}}{2a}} \qquad\qquad \textbf{(7)}$$

Thus, we have derived what is called the **quadratic formula.**

THE QUADRATIC FORMULA

The solutions to the quadratic equation $ax^2 + bx + c = 0$ are given by the formula

$$x = \frac{-b \pm \sqrt{b^2 - 4ac}}{2a}$$

In words, the quadratic formula tells us that if we have a quadratic equation in standard form, then all we have to do is substitute the values of a, b, and c into the formula to get the solutions (if any).

EXAMPLE 2 *Solve by using the quadratic formula.* $x^2 - 3x - 5 = 0$

2. $t^2 - 7t - 3 = 0$

Solution This equation is already in standard form. In order to use the formula we must identify a, b, and c. Remember that a is the coefficient of the second-degree term, b is the coefficient of the first-degree term, and c is the constant. Therefore, we have $a = 1$, $b = -3$, and $c = -5$. Substituting these values into the formula, we get

$$x = \frac{-b \pm \sqrt{b^2 - 4ac}}{2a} \qquad \textit{Our first step is just substituting the values.}$$

$$x = \frac{-(-3) \pm \sqrt{(-3)^2 - 4(1)(-5)}}{2(1)}$$

$$x = \frac{3 \pm \sqrt{9 + 20}}{2}$$

$$\boxed{x = \frac{3 \pm \sqrt{29}}{2}}$$

   ■

Notice that the quadratic formula is much easier to use than the method of completing the square. This is because the formula contains within it all the steps that we would have to do if we were completing the square.

Here are some things to watch out for when using the quadratic formula:

1. If b is a negative number, then $-b$ will be positive.

2. If a is positive (as it will be if the equation is in standard form), and c is negative, then you will end up *adding* the numbers under the radical sign because $-4ac$ is positive.

3. Do not forget that the quantity $2a$ is the denominator of the *entire* expression $-b \pm \sqrt{b^2 - 4ac}$.

EXAMPLE 3 *Solve by using the quadratic formula.* $3u^2 - 4u = 5$

3. $2y^2 = 10y - 3$

Solution We begin by putting the equation into standard form.

$$3u^2 - 4u = 5$$

$3u^2 - 4u - 5 = 0$ *We substitute into the formula $a = 3$, $b = -4$, $c = -5$. Thus, from the formula we get the solutions*

$$u = \frac{-(-4) \pm \sqrt{(-4)^2 - 4(3)(-5)}}{2(3)}$$

$$u = \frac{4 \pm \sqrt{16 + 60}}{6}$$

$$u = \frac{4 \pm \sqrt{76}}{6} \qquad \textit{Simplify the radical.}$$

$$u = \frac{4 \pm \sqrt{4}\sqrt{19}}{6} = \frac{4 \pm 2\sqrt{19}}{6} \qquad \textit{Now factor out a 2 in the numerator.}$$

$$u = \frac{2(2 \pm \sqrt{19})}{6} \qquad \textit{Reduce.}$$

$$u = \frac{\cancel{2}(2 \pm \sqrt{19})}{\cancel{6}_{3}}$$

$$\boxed{u = \frac{2 \pm \sqrt{19}}{3}}$$

   ■

4. $5 + 4u = u^2$

EXAMPLE 4   *Solve by using the quadratic formula.* $3 - 2t = t^2$

Solution   We put the equation into standard form and identify a, b, and c.

$$3 - 2t = t^2$$
$$0 = t^2 + 2t - 3 \qquad a = 1, b = 2, \text{ and } c = -3$$
$$t = \frac{-2 \pm \sqrt{2^2 - 4(1)(-3)}}{2(1)}$$
$$t = \frac{-2 \pm \sqrt{4 + 12}}{2} = \frac{-2 \pm \sqrt{16}}{2} = \frac{-2 \pm 4}{2}$$
$$t = \frac{-2 + 4}{2} = \frac{2}{2} = 1 \quad \text{or} \quad t = \frac{-2 - 4}{2} = \frac{-6}{2} = -3$$
$$\boxed{t = 1} \qquad\qquad \text{or} \quad \boxed{t = -3}$$

As with completing the square, when we get rational solutions it means that we could have solved the equation by the factoring method. *Try it!*   ∎

5. $(t + 6)(t - 4) = (t + 3)(t + 2)$

EXAMPLE 5   *Solve for t.* $2t^2 - 5t + 7 = t(2t - 3)$

Solution   *Be careful!* Do not automatically assume that just because there is a second-degree term in the initial equation that this must be a quadratic equation.
We begin by multiplying out the right-hand side.

$$2t^2 - 5t + 7 = t(2t - 3)$$

$2t^2 - 5t + 7 = \quad 2t^2 - 3t$	*Put the equation in standard form.*
$\underline{-2t^2 + 3t \qquad\quad -2t^2 + 3t}$	
$\qquad -2t + 7 = 0$	*This is not a quadratic equation at all! This is a*
$\qquad\quad -2t = -7$	*first-degree equation. We simply isolate t.*

$$\boxed{t = \frac{7}{2}}$$
∎

As Example 5 clearly shows, the method of solution we choose for an equation depends on the *type* of equation we are dealing with. Look carefully at the equation before deciding on a method of solution.

6. $t^2 + 2t + 5 = 0$

EXAMPLE 6   *Solve by using the quadratic formula.* $x^2 + 3x + 4 = 0$

Solution   Since the equation is already in standard form, we begin by identifying a, b, and c.

$$a = 1, \quad b = 3, \quad \text{and} \quad c = 4$$
$$x = \frac{-b \pm \sqrt{b^2 - 4ac}}{2a}$$
$$x = \frac{-3 \pm \sqrt{3^2 - 4(1)(4)}}{2(1)}$$
$$x = \frac{-3 \pm \sqrt{9 - 16}}{2}$$
$$x = \frac{-3 \pm \sqrt{-7}}{2} \qquad \textbf{\textit{No real solutions}}$$

As soon as we see that the answer involves the square root of a negative number, we can stop and say that the equation has *no real solutions*.   ∎

In the next section we will discuss how to decide which method to choose for solving a particular quadratic equation.

✔ *Answers to Learning Checks in Section 10.4*

1. $x = \dfrac{5 \pm \sqrt{13}}{6}$ **2.** $t = \dfrac{7 \pm \sqrt{61}}{2}$ **3.** $y = \dfrac{5 \pm \sqrt{19}}{2}$

4. $u = 5, u = -1$ **5.** $t = -10$ **6.** No real solutions

NOTE TO THE STUDENT

Use the space on this page to write down any questions you have or points you want to review with your instructor.

Exercises 10.4

In Exercises 1–8, identify a, b, and c as used in the quadratic formula.

1. $x^2 + 3x - 5 = 0$ **2.** $x^2 + 5x - 2 = 0$

3. $t^2 - 7t = 6$ **4.** $t^2 + 7 = 6t$

5. $2u^2 = 8u$ **6.** $5u = 10u^2$

7. $3x^2 - 11 = 0$ **8.** $4z^2 = 7$

In Exercises 9–36, solve the equation by using the quadratic formula.

9. $x^2 + 3x - 5 = 0$ **10.** $x^2 + 5x - 2 = 0$

11. $y^2 + 4y - 6 = 0$ **12.** $y^2 + 2y - 5 = 0$

13. $u^2 - 2u + 3 = 0$ **14.** $u^2 - 3u + 3 = 0$

15. $t^2 - 7t = 6$ **16.** $t^2 + 6 = 6t$

1. _____

2. _____

3. _____

4. _____

5. _____

6. _____

7. _____

8. _____

9. _____

10. _____

11. _____

12. _____

13. _____

14. _____

15. _____

16. _____

ANSWERS

17. _____

18. _____

19. _____

20. _____

21. _____

22. _____

23. _____

24. _____

25. _____

26. _____

17. $2x^2 - 3x - 1 = 0$

18. $3x^2 + 5x + 2 = 0$

19. $5x^2 - x = 2$

20. $7x^2 - 3 = x$

21. $t^2 - 3t + 4 = 2t^2 + 4t - 3$

22. $2t^2 + 4t + 1 = 3t^2 + 7t + 4$

23. $(5w + 2)(w - 1) = 3w + 1$

24. $(3w - 1)(2w - 3) = w$

25. $(x - 1)^2 = x(x - 2)$

26. $(x + 2)^2 = x(x + 4)$

27. $3x^2 - 5x + 7 = 2x(x - 5) + 9x + 5$ **28.** $t^2 + 4t = t(2t - 2) + 13$

27. _____

29. $2u^2 = 6u + 3$ **30.** $8r - 2 = 5r^2$

28. _____

29. _____

31. $x^2(x - 1) = (x - 1)^3$ **32.** $y^2(y + 2) = (y + 2)^3$

30. _____

31. _____

33. $4x = 9x^2$ **34.** $5u^2 = 3u$

32. _____

33. _____

35. $4 = 9x^2$ **36.** $5u^2 = 3$

34. _____

35. _____

36. _____

37. _____

38. _____

39. _____

40. _____

41. _____

42. _____

43. _____

44. _____

45. _____

46. _____

37. $\dfrac{w}{2} = \dfrac{3}{w+2}$

38. $\dfrac{t}{6} = \dfrac{3}{t+4}$

39. $\dfrac{y}{y+1} = \dfrac{y+2}{3y}$

40. $\dfrac{z}{2} + \dfrac{3}{z} = z$

CALCULATOR EXERCISES

Solve the following equations. Round off your answers to the nearest hundredth.

41. $t^2 - 7t = 10$

42. $u^2 - 5u + 5 = 0$

43. $3w^2 + 7w = 21$

44. $2z^2 = 13 - 31z$

45. $1.7x^2 - 3.2x = 6.1$

46. $3.4x^2 + 7.3x + 2.05 = 0$

QUESTIONS FOR THOUGHT

5. Solve Example 1 of this section, which was solved by the method of completing the square, by the quadratic formula. Which method of solution do you think is easier?

6. Compare and contrast the various methods we have discussed for solving quadratic equations.

7. Discuss what is wrong with each of the following "solutions."

(a) $x^2 - 3x - 1 = 0$

$$x = \frac{-3 \pm \sqrt{9 - 4(1)}}{2}$$

$$x = \frac{-3 \pm \sqrt{5}}{2}$$

(b) $x^2 - 5x - 3 = 0$

$$x = \frac{5 \pm \sqrt{25 - 12}}{2}$$

$$x = \frac{5 \pm \sqrt{13}}{2}$$

(c) $x^2 - 5x + 3 = 0$

$$x = 5 \pm \frac{\sqrt{25 - 12}}{2}$$

$$x = 5 \pm \frac{\sqrt{13}}{2}$$

(d) $x^2 - 6x - 3 = 0$

$$x = \frac{6 \pm \sqrt{36 + 12}}{2} = \frac{6 \pm \sqrt{48}}{2}$$

$$x = \frac{6 \pm \sqrt{16}\sqrt{3}}{2} = \frac{6 \pm 4\sqrt{3}}{2}$$

$$x = 6 \pm 2\sqrt{3}$$

Choosing a Method

At this point we have three methods for solving a quadratic equation:

1. The factoring method
2. The square root method

   Within this method we include the process of completing the square, which is often necessary before we can take square roots.

3. The quadratic formula

We know that both completing the square and the quadratic formula work for *all* quadratic equations. On the other hand, we have seen that the factoring method, when it works, is often the easiest method. How then should we decide on which method to choose for a particular equation?

Based on what we have seen up to now, it is fairly safe to say that unless an equation is given in the form of a perfect square, it is generally easier to solve a quadratic equation by the factoring method or by using the quadratic formula, than it is to solve by completing the square. Consequently, our first step in the process of solving a quadratic equation will usually be to get the equation into standard form, since standard form is required for both the factoring method and the quadratic formula.

Let's look at several examples and analyze why we choose a particular method in each case. Then based on this analysis we can try to offer some general guidelines as to how to choose the "best method."

EXAMPLE 1   Solve for x.

(a) $3x^2 = 5x$ **(b)** $x^2 = 3x + 28$

Solution
(a) We begin by putting the equation into standard form.

$$3x^2 = 5x$$

$$3x^2 - 5x = 0 \qquad \textit{We \textbf{should} notice a common factor of x; the factoring method is the most convenient.}$$

$$x(3x - 5) = 0$$

$$x = 0 \quad \text{or} \quad 3x - 5 = 0$$

$$\boxed{x = 0} \quad \text{or} \quad \boxed{x = \frac{5}{3}}$$

Clearly then, in such a situation where there is no constant term ($c = 0$) but there is a first-degree term ($b \neq 0$), there will be a common variable factor, and so the factoring method will work.

(b) Again we begin by putting the equation in standard form.

$$x^2 = 3x + 28$$

$$x^2 - 3x - 28 = 0 \qquad \textit{Since the leading coefficient is 1, we should feel fairly confident that if the left-hand side factors we will find the}$$

$$(x - 7)(x + 4) = 0 \qquad \textit{correct factorization quickly.}$$

$$x - 7 = 0 \quad \text{or} \quad x + 4 = 0$$

$$\boxed{x = 7} \quad \text{or} \quad \boxed{x = -4}$$

This example is further evidence of the fact that the factoring method, when it works, is usually the easiest method to use, particularly if there is a common factor of the variable as in part **(a)**. ∎

✔ LEARNING CHECKS

1. **(a)** $5y^2 = 8y$
   (b) $t^2 = 5t + 14$

2. **(a)** $z(z + 7) = z - 3$

(b) $4x^2 - 8x + 50 = 2x^2 - 5x + 70$

EXAMPLE 2 Solve for x.

(a) $x(x - 2) = 6x - 2$ **(b)** $10x^2 + 39x - 100 = 20 - 5x - 14x^2$

Solution

(a) We begin by multiplying out the left-hand side, and putting the equation into standard form.

$$x(x - 2) = 6x - 2$$

$$x^2 - 2x = 6x - 2$$

$x^2 - 8x + 2 = 0$ *At a glance we should be able to see that this cannot be factored. Therefore, we use the quadratic formula with* $a = 1, b = -8,$ *and* $c = 2$.

$$x = \frac{-b \pm \sqrt{b^2 - 4ac}}{2a}$$

$$x = \frac{-(-8) \pm \sqrt{(-8)^2 - 4(1)(2)}}{2(1)}$$

$$x = \frac{8 \pm \sqrt{64 - 8}}{2} = \frac{8 \pm \sqrt{56}}{2} \qquad \textit{Simplify the radical.}$$

$$x = \frac{8 \pm \sqrt{4}\sqrt{14}}{2} = \frac{8 \pm 2\sqrt{14}}{2} \qquad \textit{Factor and reduce.}$$

$$x = \frac{2(4 \pm \sqrt{14})}{2} = \frac{\cancel{2}(4 \pm \sqrt{14})}{\cancel{2}}$$

$$\boxed{x = 4 \pm \sqrt{14}}$$

(b) We begin by getting the equation into standard form.

$$10x^2 + 39x - 100 = 20 - 5x - 14x^2$$

$24x^2 + 44x - 120 = 0$ *Factor out the common factor of 4.*

$4(6x^2 + 11x - 30) = 0$ *Divide both sides of the equation by 4.*

$$\frac{\cancel{4}(6x^2 + 11x - 30)}{\cancel{4}} = \frac{0}{4}$$

$6x^2 + 11x - 30 = 0$ *This looks like it may be difficult to factor, or may not factor at all. Let's use the formula.* $a = 6, b = 11,$ *and* $c = -30$

$$x = \frac{-11 \pm \sqrt{11^2 - 4(6)(-30)}}{2(6)}$$

$$x = \frac{-11 \pm \sqrt{121 + 720}}{12} = \frac{-11 \pm \sqrt{841}}{12}$$

$$\sqrt{841} = 29^*$$

$$x = \frac{-11 \pm 29}{12}$$

$$x = \frac{-11 + 29}{12} = \frac{18}{12} = \frac{3}{2} \quad \text{or} \quad x = \frac{-11 - 29}{12} = \frac{-40}{12} = -\frac{10}{3}$$

$$\boxed{x = \frac{3}{2}} \qquad \text{or} \qquad \boxed{x = -\frac{10}{3}}$$

---

*We have previously mentioned how we can quickly determine whether a reasonably small number is a perfect square. Because 841 ends in 1, the only possible candidates for its square roots are 21 and 29.

This example illustrates several things. First, even though using the quadratic formula is a straightforward mechanical procedure, the arithmetic can get messy and there are numerous opportunities to make careless errors.

Second, factoring out the common factor of 4 was useful even though we used the quadratic formula and not the factoring method. (Imagine what the arithmetic would have been like if we had used the formula with $a = 24$, $b = 44$, and $c = -120$.)

Third, the fact that we obtained rational solutions tells us that we could have used the factoring method. (*Try it!*) Nevertheless, it is not clear whether the factoring method (which requires a fair amount of trial and error) or the quadratic formula (which is a bit messy) is the "better" method. ∎

Basically, if you are good at factoring, it probably pays to invest a minute or two in trying to use the factoring method. If that fails, then generally you would use the quadratic formula.

EXAMPLE 3 *Solve for x.* $(3x - 5)^2 = 17$

3. $(5t - 2)^2 = 10$

Solution   Since this equation is given in perfect square form, we might as well take advantage of that fact and use the square root method.

$$(3x - 5)^2 = 17 \qquad \textit{Take square roots.}$$
$$3x - 5 = \pm\sqrt{17}$$
$$3x = 5 \pm \sqrt{17}$$
$$\boxed{x = \frac{5 \pm \sqrt{17}}{3}}$$

Had we wanted to, we could have multiplied out $(3x - 5)^2$, put the equation into standard form, and then used the formula. (*Try it!*) However, if an equation is given in the form of a perfect square, it seems foolish not to take advantage of the square root method. ∎

Unless you are given specific instructions as to which method of solution to use, you are free to choose whichever method *you* find the most efficient.

EXAMPLE 4 *Solve for t.* $3t^2 + 2 = 5t^2 - 4$

4. $u^2 + 8 = 2u^2 - 3$

Solution   We can begin by getting this equation into standard form.

$$3t^2 + 2 = 5t^2 - 4$$
$$0 = 2t^2 - 6 \qquad \textit{We note that the square root method will work.}$$
$$6 = 2t^2 \qquad \textit{Divide both sides by 2.}$$
$$3 = t^2 \qquad \textit{Take square roots.}$$
$$\boxed{\pm\sqrt{3} = t}$$

As soon as we see that there is no first-degree term ($b = 0$), we know that we can use the square root method. ∎

Based on our discussion thus far we offer the following outline for solving quadratic equations.

OUTLINE FOR SOLVING QUADRATIC EQUATIONS

1. Simplify both sides of the equation as completely as possible, and put the equation into standard form.

2. Factor out any common factors, and divide both sides of the equation by any common *numerical* factor to eliminate it.

3. If there is a common variable factor, use the factoring method. In other words, the factoring method works when $c = 0$.

4. If there is no first-degree term ($b = 0$), then the square root method will work.

5. If it looks like the nonzero side of the equation can be factored fairly easily, then try the factoring method.

6. If it looks too complicated to factor, or does not factor, then use the quadratic formula.

7. Check the equation either by substituting into the original equation, or by solving the equation using a different method.

Answers to Learning Checks in Section 10.5

1. (a) $y = 0, \frac{8}{5}$ (b) $t = 7, -2$ **2.** (a) $z = -3 \pm \sqrt{6}$ (b) $x = -\frac{5}{2}, 4$

3. $t = \dfrac{2 \pm \sqrt{10}}{5}$ **4.** $u = \pm\sqrt{11}$

Exercises 10.5

*In each of the following exercises use the method **you** think is the most appropriate to solve the given equation. Check your answers by using a different method.*

1. $x^2 + 6x + 5 = 0$

2. $y^2 + 4y + 3 = 0$

3. $x^2 + 6x - 5 = 0$

4. $y^2 + 4y - 3 = 0$

5. $2r^2 + 1 = 3r$

6. $2r^2 + 1 = 4r$

7. $w^2 = 4w + 5$

8. $w^2 = 4w$

9. $(x + 1)(x + 2) = (x + 3)(x + 4)$.

10. $(z + 3)(z - 2) = (z - 3)(z + 2)$

11. $(a + 1)^2 - (a + 3)(a - 2) = a + 6$

12. $y(y + 6) = (y + 4)(y + 2)$

1. _____

2. _____

3. _____

4. _____

5. _____

6. _____

7. _____

8. _____

9. _____

10. _____

11. _____

12. _____

597

ANSWERS

13. $4x^2 = 16x - 28$

14. $6x^2 = 24x - 30$

13. _____

14. _____

15. $(x - 1)^2 = 5$

16. $(m - 3)^2 = 7m.$

15. _____

17. $(x - 1)^2 = 5x$

18. $(m - 3)^2 = 7$

16. _____

17. _____

19. $(u + 2)^2 = 4(u + 5)$

20. $(u + 2)^2 = u(u + 5)$

18. _____

19. _____

21. $y^2 - 4y + 10 = 5(y + 2)$

22. $3y^2 - 7y - 12 = 4(2y^2 - 3)$

20. _____

21. _____

22. _____

23. $(t + 4)(t - 8) = 13$ **24.** $(c - 3)(c + 6) = 22$

25. $(n + 2)(n + 1) = 3$ **26.** $(n + 2)(n + 1) = 3n$

27. $z^2 - 3z = 3z - 9$ **28.** $4z + 5 = z^2 + 12z + 21$

29. $16z + 12 = 3z^2$ **30.** $17z - 12 = 5z^2$

31. $x + \dfrac{1}{x} = 2$ **32.** $w + \dfrac{2}{w} = 1$

33. $x^2 + 1 = \dfrac{5}{2}x$ **34.** $x^2 + \dfrac{2}{3}x = 4$

23. _____

24. _____

25. _____

26. _____

27. _____

28. _____

29. _____

30. _____

31. _____

32. _____

33. _____

34. _____

35. $\dfrac{x}{x+1} = \dfrac{4}{x+4}$

36. $\dfrac{x+2}{x} = \dfrac{5}{x-4}$

35. _____

37. $\dfrac{x}{x+1} = \dfrac{x+2}{x+3}$

38. $\dfrac{x+2}{x+4} = \dfrac{x+3}{x+6}$

36. _____

37. _____

39. $\dfrac{3x}{x+1} + \dfrac{2}{x-1} = 4$

40. $\dfrac{x+4}{x+1} - \dfrac{4}{x+2} = 2$

38. _____

39. _____

QUESTIONS FOR THOUGHT

8. Solve the equation $2x^2 + 3x = 20$ by all three methods. Which was the easiest? The most difficult? Why?

9. Solve the equation $3x^2 - 5x - 1 = 0$. Check your answer by substituting one of your solutions back into the original equation. Also check your answer by solving the equation using a different method.

   Which method of checking was easier? Why? Which check do you have more confidence in? Why?

40. _____

Verbal Problems

10.6

Now that we have the ability to solve quadratic equations, we can solve an even wider variety of problems. Keep in mind that when you are solving a verbal problem you do not know whether the resulting equation will be linear or quadratic. Be sure to look carefully at the equation before deciding on a method of solution.

EXAMPLE 1 The sum of a number and its reciprocal is $\frac{29}{10}$. Find the number.

Solution We translate the words in the problem as follows:

Let $x = $ the number; then $\dfrac{1}{x} = $ its reciprocal.

The sum of a number and its reciprocal is $\frac{29}{10}$.

$$x + \frac{1}{x} = \frac{29}{10}$$

Thus, our equation is

$$x + \frac{1}{x} = \frac{29}{10}$$

Multiply both sides of the equation by the LCD, which is $10x$, to clear the fractions.

$$\frac{10x}{1}\left(x + \frac{1}{x}\right) = \frac{10x}{1}\left(\frac{29}{10}\right)$$

$$10x \cdot x + \frac{10x}{1} \cdot \frac{1}{x} = \frac{10x}{1} \cdot \frac{29}{10}$$

$$10x^2 + \frac{10x}{1} \cdot \frac{1}{x} = \frac{10x}{1} \cdot \frac{29}{10}$$

$$10x^2 + 10 = 29x \qquad \textit{Put the equation in standard form.}$$
$$10x^2 - 29x + 10 = 0 \qquad \textit{The factoring method works.}$$
$$(5x - 2)(2x - 5) = 0$$

$$5x - 2 = 0 \quad \text{or} \quad 2x - 5 = 0$$

$$x = \frac{2}{5} \quad \text{or} \quad x = \frac{5}{2}$$

We leave it to the student to check that both solutions satisfy the statement of the problem. Notice that the two solutions are reciprocals of each other. ∎

EXAMPLE 2 The length of a rectangle is three less than twice the width, and its area is equal to 44 square centimeters. Find the dimensions of the rectangle.

Solution We draw a diagram of the rectangle and label the sides according to the information in the problem (see Figure 10.1 on page 602).

LEARNING CHECKS

1. The difference between a number and its reciprocal is $\frac{5}{6}$. Find the number.

2. The length of a rectangle is 1 more than 3 times the width. If the area is 52 square centimeters, find the dimensions of the rectangle.

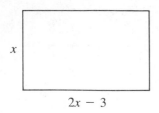

x

2x − 3

Figure 10.1   *Rectangle for Example 2*

The area of a rectangle is equal to its length times its width. Therefore, our equation is

$$x(2x - 3) = 44$$ *Multiply out and put the equation in standard form.*

$$2x^2 - 3x - 44 = 0$$ *The factoring method works.*

$$(2x - 11)(x + 4) = 0$$

$$2x - 11 = 0 \quad \text{or} \quad x + 4 = 0$$

$$x = \frac{11}{2} \quad \text{or} \quad x = -4$$

Since it makes no sense for the width of a rectangle to have a negative dimension, we reject the negative answer. Thus, the answer to the problem is that the

$$\text{width} = \frac{11}{2} \text{ cm.}$$

and the length $= 2(\frac{11}{2}) - 3 = 11 - 3 = 8$,

$$\text{length} = 8 \text{ cm.}$$

CHECK: We check to see that the area is equal to 44 square centimeters.

$$\frac{11}{2} \cdot 8 = 11 \cdot 4 = 44$$ ∎

3. A salesman drives a distance of 300 km. at a certain rate of speed, and then makes the return trip driving 25 kph faster. If the total driving time was 10 hours, find his rate going.

EXAMPLE 3   Samantha and Jenna are both driving to a tennis tournament 200 miles away. They both leave at the same time, but Samantha arrives 1 hour ahead of Jenna because she was driving 10 miles per hour faster than Jenna. Find the rate at which each drove to the tournament.

Solution   We should recognize that our starting point in such a problem is the relationship $d = rt$. How do we use it here? The problem asks us to find the rates at which they drove.

Let r = Jenna's rate.

Then $r + 10$ = Samantha's rate   (she was driving 10 mph faster than Jenna).

The problem also tells us that Samantha arrived 1 hour ahead of Jenna, which means that (since they left at the same time) Samantha's driving time was 1 hour *less*. Thus, we have the following "time relationship":

$$t_{\text{Samantha}} = t_{\text{Jenna}} - 1$$

We use the fact that $d = rt$ in the equivalent form of $t = \dfrac{d}{r}$, by substituting into the time relationship:

$$\frac{d_{\text{Samantha}}}{r_{\text{Samantha}}} = \frac{d_{\text{Jenna}}}{r_{\text{Jenna}}} - 1$$

However, we know that they both travelled the same distance, which is given to be 200 miles, and we have represented the rates above. Let's substitute these quantities into the last equation.

$$\frac{200}{r + 10} = \frac{200}{r} - 1$$

We multiply both sides of the equation by the LCD of $r(r + 10)$, to clear fractions.

$$\frac{r(r + 10)}{1} \cdot \frac{200}{r + 10} = \frac{r(r + 10)}{1} \cdot \frac{200}{r} - r(r + 10)$$

$$200r = 200(r + 10) - r(r + 10)$$

$$200r = 200r + 2,000 - r^2 - 10r$$

Put the equation in standard form.

$$r^2 + 10r - 2,000 = 0$$

$$(r + 50)(r - 40) = 0$$

$$r + 50 = 0 \quad \text{or} \quad r - 40 = 0$$

$$r = -50 \quad \text{or} \quad r = 40$$

Since it makes no sense for a rate to be negative, the answer to the problem is that $r = 40$.

Thus, we have $\boxed{\text{Jenna's rate is 40 mph}}$ and $\boxed{\text{Samantha's rate is 50 mph}}$.

The check is left to the student. ■

A **right triangle** is a triangle with a right (90°) angle. The sides which form the right angle are called the **legs**. The side opposite the right angle (that is, the longest side of the triangle) is called the **hypotenuse**.

There is a famous theorem of Pythagoras which says that the sum of the squares of the legs of a right triangle is equal to the square of the hypotenuse. Symbolically, we draw our right triangle and label the lengths of the legs a and b, and the length of the hypotenuse c, as indicated in the accompanying figure.

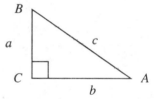

PYTHAGOREAN THEOREM

In a right triangle with legs a and b, and hypotenuse c,

$$a^2 + b^2 = c^2$$

EXAMPLE 4 The diagonal of a square is 10 cm. long. Find the length of a side of the square.

Solution A square has the properties that all its sides have the same length and that all its angles are right angles. If we let s represent the length of a side of the square, then we have the situation shown in Figure 10.2.

We see that the diagonal forms two right triangles. We use the Pythagorean Theorem to obtain the equation

$$s^2 + s^2 = 10^2$$

$$2s^2 = 100$$

$$s^2 = 50 \qquad \textit{Take square roots.}$$

$$s = \pm\sqrt{50} = \pm\sqrt{25}\sqrt{2} = \pm 5\sqrt{2}$$

4. Find the length of the diagonal of a rectangle whose dimensions are 5 cm. by 8 cm.

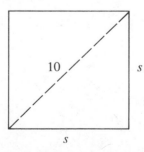

Figure 10.2 *Diagram for Example 4*

Since it makes no sense for a length to be negative, we reject the negative solution.

The answer to the question is that the length of a side of the square is

$$5\sqrt{2} \text{ cm.}$$

∎

Sometimes a verbal problem involving quadratic equations involves nothing more than substituting values into a given formula.

5. The height in feet, S, of a ball above the ground after t seconds is given by the equation

$$S = 100 - 30t - 16t^2$$

Find the height of the ball after 1 second.

EXAMPLE 5   The profit in dollars (P) on each TV set made daily by the AAA Television Company is related to the number of TV sets produced (x) at the AAA factory according to the equation

$$P = -\frac{x^2}{4} + 45x - 1{,}625$$

What is the company's profit on each TV set if it produces sixty TV sets per day?

Solution   Solving this example requires us to recognize that, in terms of the language of the given equation, we want to find P when $x = 60$. Therefore, we simply substitute $x = 60$ into the given equation and compute P.

$$P = -\frac{x^2}{4} + 45x - 1{,}625 \qquad \textit{Substitute } x = 60.$$

$$P = -\frac{(60)^2}{4} + 45(60) - 1{,}625$$

$$P = -\frac{3{,}600}{4} + 2{,}700 - 1{,}625$$

$$P = -900 + 2{,}700 - 1{,}625$$

$$\boxed{P = \$175}$$

The profit is \$175 per TV set.   ∎

✔ *Answers to Learning Checks in Section 10.6*

1. $\frac{3}{2}$ or $-\frac{2}{3}$ **2.** 4 cm. by 13 cm.   **3.** 50 kph   **4.** $\sqrt{89}$ cm.   **5.** 54 feet

Exercises 10.6

Solve each of the following exercises algebraically.

1. The sum of a number and its reciprocal is $\frac{13}{6}$. Find the number.

2. The sum of a number and 3 times its reciprocal is $\frac{79}{10}$. Find the number.

3. The sum of two numbers is 20, and their product is 96. Find the two numbers.

4. One number is 5 more than 3 times a second number. If their product is -2, what are the numbers?

5. The length of a rectangle is 3 more than twice its width, and its area is 90 square meters. Find its dimensions.

6. The width of a rectangle is one-third its length. If the area of the rectangle is 20 square inches, what are the dimensions of the rectangle?

7. One leg of a right triangle is 7 cm. long and the hypotenuse is 15 cm. long. What is the length of the other leg?

8. A rectangle has dimensions 10 mm. by 20 mm. What is the length of the diagonal of the rectangle?

ANSWERS

9. _____

10. _____

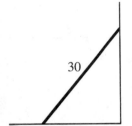

Diagram for Exercise 11

11. _____

12. _____

13. _____

14. _____

15. _____

16. _____

9. Find the length of the diagonal of a square whose side is 8 in.

10. Find the length of the sides of a square whose diagonal is 8 in.

11. Harold leans a 30-foot ladder against a building (see the accompanying figure). The base of the ladder is 8 feet from the building. How high up the building does the ladder reach?

12. Sam leans a 50-foot ladder against a building so that the top of the ladder is 45 feet above the ground. How far is the base of the ladder from the building?

13. The lengths of the sides of a right triangle are three consecutive integers. Find them.

14. The lengths of the sides of a right triangle are three consecutive even integers. Find them.

15. The denominator of a fraction is one more than the numerator. If the numerator is increased by three, the resulting fraction is one more than the original fraction. Find the original fraction.

16. The numerator of a fraction is one less than the denominator. If $\frac{7}{12}$ is added to the fraction the result is the reciprocal of the original fraction. Find the original fraction.

17. A concert hall contains 768 seats. If the number of rows of seats is eight less than the number of seats in each row, how many seats are there in each row?

18. A box contains 800 marbles in small bags. There are half as many bags as there are marbles in each bag. How many bags are there?

17. _____

19. A motorist completes a trip covering 150 kilometers in 2 hours. She covers the first 120 kilometers at a certain rate of speed, and then decreases her speed by 20 kilometers per hour for the remaining 30 kilometers. Find her speed during the first 120 kilometers.

18. _____

20. Arnold travels from town A to town B, which are 300 kilometers apart. His rate going is twice as fast as his rate returning. If his total travelling time was $7\frac{1}{2}$ hours, what was his rate of speed going from A to B?

19. _____

20. _____

21. Suppose the profit (P) made on the sale of theater tickets is related to the price (x) of the tickets according to the following equation:

$$P = 1,000(-x^2 + 15x - 35)$$

How much profit is earned if the price per ticket is $5? $4?

22. Suppose that a ball is thrown up into the air in such a way that its height (H) above the ground t seconds after it is thrown is given by the equation

$$H = -16t^2 + 80t + 10$$

How high is the ball after 3 seconds? In how many seconds will it be 5 feet above the ground?

CALCULATOR EXERCISES

Solve the following problems algebraically. Round off your answers to the nearest tenth.

23. If an object is dropped from a height of 1,000 feet, its height h after t seconds is given by the formula $h = 1,000 - 16t^2$.

   (a) How long will it take for the object to reach a height of 700 feet?

   (b) How long will it take for the object to hit the ground? [*Hint:* When is $h = 0$?]

24. If an object is thrown upward from an initial height of 80 meters with an initial velocity of 10 meters per second, then its height h after t seconds is given by

$$h = 80 + 10t - 9.6t^2$$

   (a) Approximately how long will it take the object to reach a height of 50 meters?

   (b) Approximately how long will it take the object to hit the ground?

25. A manufacturer has determined that his cost per unit, C, for producing x radios is approximately given by the formula $C = x^2 - 27x + 19$.

   (a) How many radios can be produced in order to have a unit cost of $15 per radio?

   (b) How many radios can be produced in order to have a unit cost of $5 per radio?

26. Repeat Exercise 25 if the cost equation is $C = 5x^2 - 12x + 21$.

CHAPTER 10 SUMMARY

After having completed this chapter you should be able to:

1. Solve a quadratic equation by the factoring method (Section 10.1).

   For example: Solve for t.

   $$2t^2 + 10t = 3t - 10 \qquad \text{\textit{Put the equation in standard form.}}$$
   $$t^2 + 7t + 10 = 0 \qquad \text{\textit{Factor.}}$$
   $$(t + 5)(t + 2) = 0$$
   $$t + 5 = 0 \quad \text{or} \quad t + 2 = 0$$
   $$\boxed{t = -5} \text{ or } \boxed{t = -2}$$

2. Solve a quadratic equation by taking square roots (Section 10.2).

   For example: Solve for z.

   $$4z^2 = 11 \qquad \text{\textit{Isolate the } } z^2 \text{ \textit{term.}}$$
   $$z^2 = \frac{11}{4} \qquad \text{\textit{Take square roots.}}$$
   $$z = \pm \sqrt{\frac{11}{4}} = \pm \frac{\sqrt{11}}{\sqrt{4}}$$
   $$\boxed{z = \pm \frac{\sqrt{11}}{2}}$$

3. Solve a quadratic equation by the square root method where it is necessary to first complete the square (Section 10.3).

   For example:

   $$x^2 - 8x + 4 = 0 \qquad \text{\textit{Subtract 4 from both sides.}}$$
   $$x^2 - 8x = -4 \qquad \text{\textit{Take one-half the coefficient of the first-degree term and square it. That is, take one-half of } -8, \text{ \textit{which is } } -4, \text{ \textit{and square it; the result is 16. Therefore, we add 16 to both sides of the equation.}}}$$
   $$x^2 - 8x + 16 = -4 + 16 \qquad \text{\textit{The left-hand side is now a perfect square.}}$$
   $$(x - 4)^2 = 12 \qquad \text{\textit{Take square roots.}}$$
   $$x - 4 = \pm\sqrt{12} = \pm\sqrt{4}\sqrt{3} = \pm2\sqrt{3} \qquad \text{\textit{Add 4 to both sides.}}$$
   $$\boxed{x = 4 \pm 2\sqrt{3}}$$

4. Solve a quadratic equation by using the quadratic formula (Section 10.4) which says:

   The solutions to the equation $ax^2 + bx + c = 0$ are given by the formula

   $$x = \frac{-b \pm \sqrt{b^2 - 4ac}}{2a}$$

For example:

$$3x^2 + 5 = 9x$$

$$3x^2 - 9x + 5 = 0 \qquad a = 3, b = -9, \text{ and } c = 5$$

$$x = \frac{-(-9) \pm \sqrt{(-9)^2 - 4(3)(5)}}{2(3)}$$

$$x = \frac{9 \pm \sqrt{81 - 60}}{6}$$

$$\boxed{x = \frac{9 \pm \sqrt{21}}{6}}$$

5. Solve verbal problems that give rise to quadratic equations (Section 10.6).

   For example: The product of a number and 2 more than itself is 4. Find the number.

   Let x = number.

   Then $x + 2$ = "two more than itself."

   The problem tells us that the product of the two numbers is 4. Therefore, our equation is

   $$\boxed{x(x + 2) = 4} \qquad \textit{We put the equation in standard form.}$$

   $$x^2 + 2x - 4 = 0 \qquad \textit{The left-hand side does not factor; use the formula.}$$

   $$x = \frac{-2 \pm \sqrt{2^2 - 4(1)(-4)}}{2(1)}$$

   $$x = \frac{-2 \pm \sqrt{4 + 16}}{2} = \frac{-2 \pm \sqrt{20}}{2} = \frac{-2 + \sqrt{4}\sqrt{5}}{2} = \frac{-2 \pm 2\sqrt{5}}{2}$$

   $$x = \frac{2(-1 \pm \sqrt{5})}{2} = \frac{\cancel{2}(-1 \pm \sqrt{5})}{\cancel{2}}$$

   $$x = 1 \pm \sqrt{5}$$

   We apparently have two solutions: $x = -1 + \sqrt{5}$ and $x = -1 - \sqrt{5}$. Let's check one of these, say the first one.

   CHECK: If the first number is $-1 + \sqrt{5}$, then "two more than itself" $-1 + \sqrt{5} + 2 = 1 + \sqrt{5}$. The product of the two numbers is supposed be 4:

   $$(-1 + \sqrt{5})(1 + \sqrt{5}) \overset{?}{=} 4 \qquad \textit{Multiply out.}$$

   $$-1 - \sqrt{5} + \sqrt{5} + 5 \overset{?}{=} 4$$

   $$4 \overset{\checkmark}{=} 4$$

   We leave it to the student to verify that $-1 - \sqrt{5}$ also checks.

ANSWERS TO QUESTIONS FOR THOUGHT

1. **(a)** We cannot set each of the factors equal to 7. The zero-product rule requires that the product of the factors be equal to 0.

   (b) $x = 3$ is not a possible solution. The first factor on the left side of the equation can never be equal to zero. We can either ignore its presence or divide both sides of the equation by it. (This logic is valid for constant factors of a zero product, but *not* for variable factors.)

2. Since the square of any real number must be nonnegative, x^2 must be at least zero. Therefore, $x^2 + 4$ must be at least 4, which means that $x^2 + 4$ cannot ever equal 0. Thus, the equation $x^2 + 4 = 0$ cannot have any solution in the real number system.

3. The constant of a perfect square is the square of one-half of the coefficient of the middle term.

4.
$$\frac{x}{x-1} = \frac{2}{x-2}$$

$$\frac{(x-1)(x-2)}{1} \cdot \frac{x}{x-1} = \frac{(x-1)(x-2)}{1} \cdot \frac{2}{x-2}$$

$$x(x-2) = 2(x-1)$$

$$x^2 - 2x = 2x - 2$$

$$x^2 - 4x = -2$$

$$\underline{\ +4\quad +4}$$

$$x^2 - 4x + 4 = 2$$

$$(x-2)^2 = 2$$

$$x - 2 = \pm\sqrt{2}$$

$$x = 2 \pm \sqrt{2}$$

CHECK $x = 2 + \sqrt{2}$:

$$\frac{x}{x-1} = \frac{2}{x-2}$$

$$\frac{2+\sqrt{2}}{2+\sqrt{2}-1} \overset{?}{=} \frac{2}{2+\sqrt{2}-2}$$

$$\frac{2+\sqrt{2}}{1+\sqrt{2}} \overset{?}{=} \frac{2}{\sqrt{2}}$$

$$\frac{2+\sqrt{2}}{1+\sqrt{2}} = \frac{(2+\sqrt{2})(1-\sqrt{2})}{(1+\sqrt{2})(1-\sqrt{2})}$$

$$= \frac{2 - 2\sqrt{2} + \sqrt{2} - 2}{1-2} = \frac{-\sqrt{2}}{-1} = \sqrt{2}$$

$$\frac{2}{\sqrt{2}} = \frac{2\sqrt{2}}{\sqrt{2}\sqrt{2}} = \frac{2\sqrt{2}}{2} = \sqrt{2} \quad \text{so} \quad \frac{2+\sqrt{2}}{1+\sqrt{2}} \overset{\checkmark}{=} \frac{2}{\sqrt{2}}$$

5. $2x^2 + 7x + 4 = 0$ $a = 2, b = 7, c = 4$

$$x = \frac{-b \pm \sqrt{b^2 - 4ac}}{2a}$$

$$x = \frac{-7 \pm \sqrt{7^2 - 4(2)(4)}}{2(2)}$$

$$x = \frac{-7 \pm \sqrt{49 - 32}}{4}$$

$$x = \frac{-7 \pm \sqrt{17}}{4}$$

Using the quadratic formula is easier than the method of completing the square.

6. The factoring method, when it works, is usually the easiest method to use. However, it does not always work. Completing the square and the quadratic formula work for any quadratic equation. Generally, completing the square is the more complicated of these two methods.

7. (a) Since $b = -3$ and $c = -1$,

$$x = \frac{-(-3) \pm \sqrt{9 - 4(-1)}}{2} = \frac{3 \pm \sqrt{9+4}}{2} = \frac{3 \pm \sqrt{13}}{2}$$

(b) The minus sign under the square root should be a plus sign.

(c) The "5" should be divided by 2 as well. That is,

$$x = \frac{5 \pm \sqrt{25 - 12}}{2} = \frac{5 \pm \sqrt{13}}{2}$$

(d) This is correct up until the last step. Then

$$\frac{6 \pm 4\sqrt{3}}{2} = \frac{\cancel{2}(3 \pm 2\sqrt{3})}{\cancel{2}} = 3 \pm 2\sqrt{3}, \quad \text{not } 6 \pm 2\sqrt{3}$$

8. *Method* 1: *Factoring*

$$2x^2 + 3x - 20 = 0$$
$$(2x - 5)(x + 4) = 0$$
$$2x - 5 = 0 \quad \text{or} \quad x + 4 = 0$$
$$2x = 5$$
$$x = \frac{5}{2} \quad \text{or} \qquad x = -4$$

Method 2: Completing the square

$$2x^2 + 3x \qquad = 20$$
$$x^2 + \frac{3}{2}x \qquad = 10$$
$$\underline{\qquad\qquad + \frac{9}{16} \quad + \frac{9}{16}\qquad\qquad}$$
$$x^2 + \frac{3}{2}x + \frac{9}{16} = 10 + \frac{9}{16}$$
$$\left(x + \frac{3}{4}\right)^2 = \frac{160}{16} + \frac{9}{16}$$
$$\left(x + \frac{3}{4}\right)^2 = \frac{169}{16}$$
$$x + \frac{3}{4} = \pm\sqrt{\frac{169}{16}} = \pm\frac{\sqrt{169}}{\sqrt{16}}$$
$$x + \frac{3}{4} = \pm\frac{13}{4}$$
$$x = -\frac{3}{4} \pm \frac{13}{4}$$
$$x = \frac{-3 \pm 13}{4}$$
$$x = \frac{-3 + 13}{4} = \frac{10}{4} = \frac{5}{2} \quad \text{or}$$
$$x = \frac{-3 - 13}{4} = \frac{-16}{4} = -4$$

Method 3: Quadratic formula $2x^2 + 3x - 20 = 0$ $a = 2, b = 3, c = -20$

$$x = \frac{-b \pm \sqrt{b^2 - 4ac}}{2a}$$
$$x = \frac{-3 \pm \sqrt{3^2 - 4(2)(-20)}}{2(2)}$$
$$x = \frac{-3 \pm \sqrt{9 + 160}}{4}$$
$$x = \frac{-3 \pm \sqrt{169}}{4} = \frac{-3 \pm 13}{4}$$
$$x = \frac{-3 + 13}{4} = \frac{10}{4} = \frac{5}{2} \quad \text{or}$$
$$x = \frac{-3 - 13}{4} = \frac{-16}{4} = -4$$

The easiest of these methods is the first, while the second one appears to be the most difficult.

9. $3x^2 - 5x - 1 = 0$ $a = 3, b = -5, c = -1$

$$x = \frac{-b \pm \sqrt{b^2 - 4ac}}{2a}$$

$$x = \frac{-(-5) \pm \sqrt{(-5)^2 - 4(3)(-1)}}{2(3)}$$

$$x = \frac{5 \pm \sqrt{25 + 12}}{6} = \frac{5 \pm \sqrt{37}}{6}$$

CHECK $x = \dfrac{5 + \sqrt{37}}{6}$:

$$3x^2 - 5x - 1 = 0$$

$$3\left(\frac{5 + \sqrt{37}}{6}\right)^2 - 5\left(\frac{5 + \sqrt{37}}{6}\right) - 1 \stackrel{?}{=} 0$$

$$3\left(\frac{5 + \sqrt{37}}{6}\right)\left(\frac{5 + \sqrt{37}}{6}\right) - 5\left(\frac{5 + \sqrt{37}}{6}\right) - 1 \stackrel{?}{=} 0$$

$$\frac{\cancel{3}}{1}\left(\frac{25 + 10\sqrt{37} + 37}{\underset{12}{\cancel{36}}}\right) - \frac{\cancel{5}}{1}\left(\frac{5 + \sqrt{37}}{6}\right) - 1 \stackrel{?}{=} 0$$

$$\frac{62 + 10\sqrt{37}}{12} - \frac{25 + 5\sqrt{37}}{6} - 1 \stackrel{?}{=} 0$$

$$\frac{\cancel{2}(31 + 5\sqrt{37})}{\underset{6}{\cancel{12}}} - \frac{25 + 5\sqrt{37}}{6} - 1 \stackrel{?}{=} 0$$

$$\frac{31 + 5\sqrt{37} - (25 + 5\sqrt{37})}{6} - 1 \stackrel{?}{=} 0$$

$$\frac{6}{6} - 1 \stackrel{?}{=} 0$$

$$1 - 1 \stackrel{?}{=} 0$$

$$0 \stackrel{\checkmark}{=} 0$$

$$3x^2 - 5x - 1 = 0$$
$$\underline{\ + 1 \quad +1}$$
$$3x^2 - 5x = 1$$
$$x^2 - \frac{5}{3}x = \frac{1}{3}$$
$$\underline{\phantom{x^2 - \frac{5}{3}x}\ + \frac{25}{36} \quad +\frac{25}{36}}$$
$$x^2 - \frac{5}{3}x + \frac{25}{36} = \frac{1}{3} + \frac{25}{36}$$

$$\left(x - \frac{5}{6}\right)^2 = \frac{12}{36} + \frac{25}{36}$$

$$\left(x - \frac{5}{6}\right)^2 = \frac{37}{36}$$

$$x - \frac{5}{6} = \pm\sqrt{\frac{37}{36}} = \pm\frac{\sqrt{37}}{\sqrt{36}}$$

$$x - \frac{5}{6} = \pm\frac{\sqrt{37}}{6}$$

$$x = \frac{5}{6} \pm \frac{\sqrt{37}}{6} = \frac{5 \pm \sqrt{37}}{6}$$

Deciding which method of checking is easier is up to you.

NOTE TO THE STUDENT

Use the space on this page to write down any questions you have or points you want to review with your instructor.

CHAPTER 10 REVIEW EXERCISES

Solve each of the following equations.

1. $x^2 - 7x - 6 = 0$

2. $x^2 - 10x - 24 = 0$

3. $x^2 + 5 = 4x$

4. $x^2 + 4x = 5$

5. $(u - 6)^2 = 13$

6. $(z + 3)^2 = -9$

7. $2y^2 + 7y = 15$

8. $2y^2 - 1 = 3y$

9. $18x^2 - 24x + 6 = 0$

10. $20x^2 - 8x - 16 = 0$

11. $(x - 6)^2 = (x + 3)(x - 5)$

12. $2x^2 - 13x + 5 = x(x - 3)$

13. $u^2 + 1 = \dfrac{13u}{6}$

14. $\dfrac{z^2}{9} - \dfrac{z}{3} = 2$

15. $3x(x - 2) = (x - 3)^2$

16. $2w = 4 + \dfrac{3}{w}$

17. $\dfrac{x + 3}{x + 6} = \dfrac{x + 2}{x + 4}$

18. $\dfrac{x}{x - 1} = \dfrac{9}{x} - \dfrac{5}{2}$

Solve the next two equations by completing the square.

19. $x^2 - 7x + 3 = 0$

20. $3x^2 - 12x = 6$

1. _____

2. _____

3. _____

4. _____

5. _____

6. _____

7. _____

8. _____

9. _____

10. _____

11. _____

12. _____

13. _____

14. _____

15. _____

16. _____

17. _____

18. _____

19. _____

20. _____

ANSWERS

1. _____

2. _____

3. _____

4. _____

5. _____

6. _____

7. _____

8. _____

9. _____

10. _____

CHAPTER 10 PRACTICE TEST

Solve each of the following equations. Choose any method you like.

1. $(x + 5)(x - 2) = 18$

2. $2x^2 - 3x - 3 = 0$

3. $x^2 - x + 14 = 2x(x - 3)$

4. $(x + 5)^2 = 10$

5. $(x + 5)^2 = 10x$

6. $4x^2 - 8x + 2 = x^2 - 21x + 12$

7. $5x^2 + 15 = 30x$

8. $\dfrac{x}{2} - \dfrac{8}{x} = x - 4$

9. Solve the following equation by completing the square, and check your answer by using the quadratic formula: $3x^2 - 12x = 7$

10. The length of a rectangle is 7 more than twice its width. If the area of the rectangle is 30 sq. cm., find its dimensions.

Squares and Square Roots

Using the Table

For example, to find $\sqrt{210}$ we look down the column headed n to line 21, then over to the column headed $\sqrt{10n} = \sqrt{10(21)} = \sqrt{210}$. Thus,

$$\sqrt{210} = 14.491$$

Squares and Square Roots

n	n^2	$\sqrt{n}$	$\sqrt{10n}$	n	n^2	$\sqrt{n}$	$\sqrt{10n}$
1	1	1.000	3.162	26	676	5.099	16.125
2	4	1.414	4.472	27	729	5.196	16.432
3	9	1.732	5.477	28	784	5.292	16.733
4	16	2.000	6.325	29	841	5.385	17.029
5	25	2.236	7.071	30	900	5.477	17.321
6	36	2.449	7.746	31	961	5.568	17.607
7	49	2.646	8.367	32	1,024	5.657	17.889
8	64	2.828	8.944	33	1,089	5.745	18.166
9	81	3.000	9.487	34	1,156	5.831	18.439
10	100	3.162	10.000	35	1,225	5.916	18.708
11	121	3.317	10.488	36	1,296	6.000	18.974
12	144	3.464	10.954	37	1,369	6.083	19.235
13	169	3.606	11.402	38	1,444	6.164	19.494
14	196	3.742	11.832	39	1,521	6.245	19.748
15	225	3.873	12.247	40	1,600	6.325	20.000
16	256	4.000	12.649	41	1,681	6.403	20.248
17	289	4.123	13.038	42	1,764	6.481	20.494
18	324	4.243	13.416	43	1,849	6.557	20.736
19	361	4.359	13.784	44	1,936	6.633	20.976
20	400	4.472	14.142	45	2,025	6.708	21.213
21	441	4.583	14.491	46	2,116	6.782	21.448
22	484	4.690	14.832	47	2,209	6.856	21.679
23	529	4.796	15.166	48	2,304	6.928	21.909
24	576	4.899	15.492	49	2,401	7.000	22.136
25	625	5.000	15.811	50	2,500	7.071	22.361

Squares and
Square Roots
(continued)

n	n^2	$\sqrt{n}$	$\sqrt{10n}$	n	n^2	$\sqrt{n}$	$\sqrt{10n}$
51	2,601	7.141	22.583	76	5,776	8.718	27.568
52	2,704	7.211	22.804	77	5,929	8.775	27.749
53	2,809	7.280	23.022	78	6,084	8.832	27.928
54	2,916	7.348	23.238	79	6,241	8.888	28.107
55	3,025	7.416	23.452	80	6,400	8.944	28.284
56	3,136	7.483	23.664	81	6,561	9.000	28.460
57	3,249	7.550	23.875	82	6,724	9.055	28.636
58	3,364	7.616	24.083	83	6,889	9.110	28.810
59	3,481	7.681	24.290	84	7,056	9.165	28.983
60	3,600	7.746	24.495	85	7,225	9.220	29.155
61	3,721	7.810	24.698	86	7,396	9.274	29.326
62	3,844	7.874	24.900	87	7,569	9.327	29.496
63	3,969	7.937	25.100	88	7,744	9.381	29.665
64	4,096	8.000	25.298	89	7,921	9.434	29.833
65	4,225	8.062	25.495	90	8,100	9.487	30.000
66	4,356	8.124	25.690	91	8,281	9.539	30.166
67	4,489	8.185	25.884	92	8,464	9.592	30.332
68	4,624	8.246	26.077	93	8,649	9.644	30.496
69	4,761	8.307	26.268	94	8,836	9.695	30.659
70	4,900	8.367	26.458	95	9,025	9.747	30.822
71	5,041	8.426	26.646	96	9,216	9.798	30.984
72	5,184	8.485	26.833	97	9,409	9.849	31.145
73	5,329	8.544	27.019	98	9,604	9.899	31.305
74	5,476	8.602	27.203	99	9,801	9.950	31.464
75	5,625	8.660	27.386	100	10,000	10.000	31.623

Sample Final Exams

SAMPLE FINAL EXAM
Form A

Simplify all answers as completely as possible; reduce all fractions to lowest terms. **Show all work.**

In Problems 1–7, perform the indicated operations and simplify as completely as possible.

1. $-9 - 4 + 5 - (-3) - 1$

2. $\dfrac{-2(-4)(-6)}{-2(-4) - 6}$

3. If $x = -3$ and $y = -2$, evaluate $(x - y)^2 - xy^2$.

4. $5a(a^2 - 4ab) - a^2(a - 3b)$

5. $3mn^2(m^5 - 8m^2n) - 2m(6mn)(-2mn^2)$

6. $x - 4[x - 3(x - 2)]$

7. $(5x - 4)(x + 2) + (x - 3)^2$

8. Use long division to find the quotient:
$$\frac{2x^3 - 17x + 1}{x + 3}$$

9. Evaluate $2^0 + 2^{-1} + 8 \cdot 4^{-2}$.

10. Compute using scientific notation:
$$\frac{(.004)(250)}{.08}$$

11. Simplify as completely as possible; express your answer with positive exponents only.
$$\frac{(3a^{-4})^2}{(a^2)^{-3}}$$

1. _____

2. _____

3. _____

4. _____

5. _____

6. _____

7. _____

8. _____

9. _____

10. _____

11. _____

ANSWERS

In Problems 12–16, factor the given expression as completely as possible.

12. $18m^2n - 6mn + 12mn^2$

13. $x^2 - 5x - 36$

12. _____

14. $4z^2 - 32z + 64$

15. $a^3 - 16a$

13. _____

14. _____

16. $x^2 + 4x + xy + 4y$

17. Reduce to lowest terms. $\dfrac{x^2 - 6x - 27}{x^2 + 2x - 3}$

15. _____

16. _____

18. Evaluate. $\sqrt{25 - 16} + (\sqrt{64} + \sqrt{36})^2$

19. Simplify. $\sqrt{54x^5}$

17. _____

18. _____

In Problems 20–27, perform the indicated operations and simplify as completely as possible.

20. $\dfrac{3x}{4} - \dfrac{5x}{6}$

21. $\dfrac{3x}{4} \cdot \dfrac{5x}{6}$

19. _____

20. _____

22. $\dfrac{24x^3y^2}{5z^4} \div (15xyz)$

23. $\dfrac{4}{x + 4} - \dfrac{2}{x + 2}$

21. _____

22. _____

24. $\dfrac{5}{3x} + \dfrac{4}{x^2 + 3x}$

25. $2\sqrt{27} - 3\sqrt{12}$

23. _____

24. _____

25. _____

26. $\dfrac{\sqrt{2} + \sqrt{8} + \sqrt{18}}{\sqrt{2}}$

26. _____

In Problems 27–32, solve the given equation or inequality for x.

27. $14 - 5x = 17 - 4x$

28. $6x + 2(4 - x) = 20 - (x - 8)$

27. _____

28. _____

29. $\dfrac{4}{x} - \dfrac{3}{x + 1} = 1$

30. $3x^2 - 4x = 1 - x$

29. _____

31. $4x + 5y + 6 = 7x - 2y + 1$

32. Solve and sketch the solution set on a number line.
$9 - 4(x - 3) \geq 29$

30. _____

31. _____

33. Solve for x and y. $\begin{cases} 4x - 3y = 14 \\ 3x - 5y = 5 \end{cases}$

32. _____

33. _____

ANSWERS

34. Sketch the graph of $5x - 2y - 10 = 0$. Label the intercepts.

35. Write an equation of the line which passes through the points $(-1, 4)$ and $(2, -3)$.

34. _____

*Solve each of the following problems **algebraically**.*

36. The length of a rectangle is 5 less than 3 times the width. If the perimeter of the rectangle is 54 cm., find the dimensions of the rectangle.

35. _____

37. If there are 28.4 grams in 1 ounce, how many ounces are there (to the nearest tenth) in 300 grams?

36. _____

38. Jim can mow a lawn alone in 3 hours, while Sarah can mow the same lawn in 4 hours. If Jim mows alone for 1 hour, how long will it take Sarah to finish mowing the lawn alone?

37. _____

39. A certain sum of money is invested at 8% and twice that amount is invested at 10%. If the total interest from the two investments was $672, how much was invested at each rate?

38. _____

40. A total of $162 was spent on 15 books. Some of the books cost $12 each and the remainder cost $9 each. How many of each type were bought?

39. _____

40. _____

SAMPLE FINAL EXAM
Form B

Simplify all answers as completely as possible; reduce all fractions to lowest terms. **Show all work**.

In Problems 1–7, perform the indicated operations and simplify as completely as possible.

1. $-8 - 5 + 4 - (-2) - 3$

2. $\dfrac{-4(-3)(-8)}{-4(-3) - 8}$

3. If $x = -2$ and $y = -3$, evaluate $(y - x)^2 - (xy)^2$.

4. $6a^2(a - 2ab) - a(a^2 - 4a^2b)$

5. $6m^2n(m^4 - 4mn) - 3m(n^2)(-8m^2)$

6. $a - 2[a - 3(a - 4)]$

7. $(3x - 2)(4x^2 + 5x - 6)$

8. Use long division to find the quotient:
$$\frac{3x^3 - 5x^2 - 4}{x - 2}$$

9. Evaluate. $4^0 + 4^{-1} + 6 \cdot 2^{-3}$.

10. Compute using scientific notation:
$$\frac{(.03)(900)}{.006}$$

11. Simplify as completely as possible; express your answer with positive exponents only.
$$\frac{(4a^{-4})^3}{(a^{-3})^2}$$

1. _____

2. _____

3. _____

4. _____

5. _____

6. _____

7. _____

8. _____

9. _____

10. _____

11. _____

ANSWERS

In Problems 12–16, *factor the given expression as completely as possible.*

12. $15uv^2 - 10u^2v^3 + 5uv$

13. $x^2 + 8x - 48$

12. _____

13. _____

14. $6a^2 - 24a + 24$

15. $2c^3 - 50c$

14. _____

16. $x^2 + 6x - xy - 6y$

17. Reduce to lowest terms. $\dfrac{2x^2 + 6x}{x^2 + 6x + 9}$

15. _____

16. _____

18. Evaluate. $\sqrt{100 - 36} + (\sqrt{9} + \sqrt{16})^2$ **19.** Simplify. $\sqrt{24x^9}$

17. _____

18. _____

In Problems 20–26, *perform the indicated operations and simplify as completely as possible.*

20. $\dfrac{3x}{8} - \dfrac{7x}{12}$

21. $\dfrac{3x}{8} \cdot \dfrac{7x}{12}$

19. _____

20. _____

22. $\dfrac{28x^2y^3}{9z^5} \div (12xyz)$

23. $\dfrac{5}{x - 5} - \dfrac{3}{x - 3}$

21. _____

22. _____

24. $\dfrac{3}{x^2 - 3x + 2} + \dfrac{3}{x^2 - x}$

25. $3\sqrt{8x} - 2\sqrt{18x}$

23. _____

24. _____

25. _____

26. $\dfrac{\sqrt{3} + \sqrt{12} + \sqrt{27}}{\sqrt{3}}$

26. _____

In Problems 27–32, solve the given equation or inequality for x.

27. $12 - 7x = 18 - 6x$

28. $4x + 3(8 - 3x) = 16 - 2(x - 1)$

27. _____

28. _____

29. $\dfrac{3}{x - 2} + \dfrac{4}{x} = 1$

30. $5x^2 + 2x - 1 = 3x^2 + 3x + 6$

29. _____

31. $2(x + y) - 3 = 7x - (y - 4)$

32. Solve and sketch the solution set on a number line.
$$6 < 3 - x \le 10$$

30. _____

31. _____

33. Solve for x and y. $\begin{cases} 3x - 4y = -10 \\ 4x - 3y = -11 \end{cases}$

32. _____

33. _____

ANSWERS

34. Sketch the graph of $8y - 3x = 24$. Label the intercepts.

34. _____

35. Write an equation of the line with x-intercept 4 and y-intercept -2.

35. _____

Solve each of the following problems **algebraically**.

36. Three less than 4 times a number is 7 more than twice the number. Find the number.

36. _____

37. The length of a rectangle is 3 more than twice its width. If the area of the rectangle is 65 cm., find the dimensions of the rectangle.

37. _____

38. The ratio of men to women in a certain factory is 8 to 5. If there are 320 men, how many women are there?

38. _____

39. Sam and Joan are supposed to stuff a total of 570 envelopes. Sam works for awhile at the rate of 12 envelopes per minute, and then Joan takes over at the rate of 18 envelopes per minute. If they take a total of 40 minutes, how long did Sam work?

39. _____

40. How much of a 30% salt solution must be mixed with 40 liters of a 50% salt solution to produce a 35% salt solution?

40. _____

Answers to Odd-Numbered Exercises

Exercises 0.1

1. True   **3.** True   **5.** True   **7.** True   **9.** True   **11.** {1, 2, 3, 4, 5, 6, 7}   **13.** {0, 2, 4, 6, 8, 10, 12, 14, 16, 18}
15. {0, 1, 2, 3, 4, 5, 6}   **17.** {0, 1, 2, 3, 4, 5}   **19.** {6, 7, 8, . . .}   **21.** {7, 8, 9, . . .}   **23.** {3, 4, 5}   **25.** {0, 4, 8, 12, . . .}
27. {0, 12, 24, 36, . . .}   **29.** $\varnothing$ **31.** $4 > 2$ **33.** $7 = 7$ **35.** $19 > 14$ **37.** $72 = 72$ **39.** $0 = 0$ **41.** $<, \leq, \neq$
43. $=, \leq, \geq$ **45.** $>, \geq, \neq$ **47.** $<, \leq, \neq$ **49.** $>, \geq, \neq$ **51.** $2 \cdot 7$ **53.** $3 \cdot 11$ **55.** $2 \cdot 3 \cdot 5$ **57.** Prime
59. $2 \cdot 2 \cdot 2 \cdot 2 \cdot 2 \cdot 2$ **61.** $2 \cdot 2 \cdot 2 \cdot 2 \cdot 2 \cdot 3$ **63.** $3 \cdot 29$

Exercises 0.2

1. $\frac{2}{3}, \frac{12}{18}$ **3.** $\frac{2}{7}, \frac{28}{70}$ **5.** $\frac{3}{4}, \frac{6}{8}$ **7.** $\frac{10}{2}, \frac{15}{3}$ **9.** $\frac{2}{2}, \frac{3}{3}$ **11.** $\frac{4}{5}$ **13.** $\frac{1}{7}$ **15.** $\frac{1}{2}$ **17.** Cannot be reduced   **19.** $\frac{7}{18}$ **21.** 6
23. $\frac{11}{4}$ **25.** Cannot be reduced   **27.** 20   **29.** 21   **31.** 64   **33.** 6   **35.** $\frac{4}{5}$ on the computer; $\frac{1}{5}$ on the software

Exercises 0.3

1. $\frac{10}{21}$ **3.** $\frac{1}{36}$ **5.** $\frac{2}{3}$ **7.** $\frac{2}{9}$ **9.** $\frac{25}{8}$ **11.** $\frac{4}{5}$ **13.** $\frac{125}{144}$ **15.** 1   **17.** $\frac{9}{25}$ **19.** 27   **21.** 12   **23.** $\frac{1}{12}$ **25.** $\frac{49}{50}$ **27.** $\frac{8}{7}$
29. 1   **31.** $\frac{9}{5}$ **33.** $\frac{7}{4}$

Exercises 0.4

1. $\frac{5}{7}$ **3.** $\frac{3}{8}$ **5.** $\frac{1}{2}$ **7.** 1   **9.** $\frac{3}{4}$ **11.** $\frac{11}{24}$ **13.** $\frac{5}{16}$ **15.** $\frac{11}{12}$ **17.** $\frac{13}{12}$ **19.** $\frac{37}{60}$ **21.** $\frac{23}{140}$ **23.** $\frac{28}{5}$ **25.** $\frac{25}{3}$ **27.** $\frac{15}{4}$
29. $\frac{64}{5}$ **31.** $4\frac{3}{5}$ **33.** $7\frac{1}{6}$ **35.** $\frac{2}{3}$ **37.** $\frac{5}{12}$

Exercises 0.5

1. 29.9   **3.** 9.52   **5.** 82.956   **7.** 8.583   **9.** 1.42   **11.** 22.62   **13.** 5.97   **15.** 34.3215   **17.** 1.9092   **19.** 7.92
21. 420   **23.** 20.5   **25.** .0802   **27.** .0016   **29.** 7   **31.** .000006   **33.** 260

Exercises 0.6

1. .25   **3.** 78%   **5.** 5%   **7.** .09   **9.** 1.5   **11.** .28   **13.** 67%   **15.** 200%   **17.** 1.37   **19.** .7%   **21.** .624   **23.** .086
25. 21   **27.** 2.52   **29.** .04   **31.** 25%   **33.** 62.5%   **35.** 40%

Chapter 0 Review Exercises

1. {2, 4, 6, 8, 10, 12, 14, 16, 18}   **2.** {1, 3, 5, 7, 9, 11, 13, 15, 17, 19}   **3.** {2, 3, 5, 7, 11, 13, 17, 19}
4. {4, 6, 8, 9, 10, 12, 14, 15, 16, 18}   **5.** {23, 29}   **6.** $\varnothing$ **7.** $2 \cdot 3 \cdot 5$ **8.** $2 \cdot 2 \cdot 7$ **9.** Prime   **10.** $2 \cdot 2 \cdot 2 \cdot 3 \cdot 3$
11. $2 \cdot 2 \cdot 5 \cdot 5$ **12.** $3 \cdot 19$ **13.** $\frac{23}{20}$ **14.** $\frac{3}{10}$ **15.** $\frac{8}{15}$ **16.** $\frac{15}{8}$ **17.** 36.118   **18.** 15.092   **19.** 4,000   **20.** 174.05
21. .000216   **22.** 148.6   **23.** $\frac{31}{24}$ **24.** $\frac{37}{36}$ **25.** $\frac{9}{4}$ **26.** $\frac{1}{32}$ **27.** .8   **28.** .92   **29.** 47%   **30.** .417   **31.** .063   **32.** .5%

Chapter 0 Practice Test

1. (a) True   **(b)** False   **(c)** False   **(d)** True   **(e)** $\varnothing$ **2.** $2 \cdot 2 \cdot 3 \cdot 7$ **3.** $\frac{29}{30}$ **4.** $\frac{1}{9}$ **5.** $\frac{25}{4}$ **6.** 129.582   **7.** 11.57
8. 61.44   **9.** 81.32   **10.** 84.1%   **11.** 48   **12.** 30%

Exercises 1.1

1. True, Commutative Law of Addition   **3.** True, Commutative Law of Addition   **5.** False   **7.** False
9. True, Commutative Law of Multiplication   **11.** True, Commutative Law of Multiplication   **13.** False

15. False   **17.** False   **19.** Associative Law of Addition   **21.** Associative Law of Multiplication
23. Commutative Law of Addition; Associative Law of Addition   **25.** Commutative Law of Multiplication; Associative Law of Multiplication
27. -4 **29.** 4   **31.** 4   **33.** 4   **35.** -4 **37.** -4 **39.** 5   **41.** 5   **43.** 19   **45.** 0   **47.** 26   **49.** 20   **51.** 20
53. 50   **55.** 3   **57.** 29   **59.** 1   **61.** 57   **63.** 2

Exercises 1.2

1. -2 **3.** -12 **5.** 7   **7.** 5   **9.** 4   **11.** -8 **13.** -4 **15.** 8   **17.** -15 **19.** -4 **21.** 0   **23.** -7 **25.** -44
27. 6   **29.** -15 **31.** -5 **33.** -7 **35.** -3 **37.** 6   **39.** -6 **41.** -5 **43.** -29 **45.** -67 **47.** -9 **49.** -1
51. 1   **53.** 12   **55.** 0   **57.** \$33   **59.** 14-yard line

Exercises 1.3

1. -4 **3.** -11 **5.** 9   **7.** -6 **9.** 3   **11.** -3 **13.** 13   **15.** -13 **17.** 3   **19.** -3 **21.** -13 **23.** 13   **25.** 1
27. -11 **29.** -11 **31.** 3   **33.** -3 **35.** 3   **37.** -7 **39.** -13 **41.** 5   **43.** -1 **45.** 0   **47.** -16 **49.** 0
51. -13 **53.** -1 **55.** -9 **57.** 0   **59.** 8   **61.** 0

Exercises 1.4

1. -15 **3.** -15 **5.** 15   **7.** -8 **9.** -2 **11.** -4 **13.** -4 **15.** 4   **17.** 0   **19.** Undefined   **21.** -14 **23.** -4
25. 8   **27.** -32 **29.** -60 **31.** 15   **33.** 17   **35.** -3 **37.** 7   **39.** -15 **41.** 2   **43.** 7   **45.** 3   **47.** 0   **49.** -5
51. 2   **53.** Undefined   **55.** 42   **57.** 19   **59.** 4   **61.** 8

Exercises 1.5

1. 10.1   **3.** 20.4   **5.** 4   **7.** $24.53\overline{3}$ **9.** 70   **11.** -7.3 **13.** -10.6 **15.** 10.6   **17.** -7 **19.** -22 **21.** Between 4 and 5
23. Between 2 and 3   **25.** Between 7 and 8
27.   **29.**   **31.**
33.   **35.**

Chapter 1 Review Exercises

1. -4 **2.** -5 **3.** -12 **4.** -10 **5.** 4   **6.** -4 **7.** -15 **8.** -9 **9.** 0   **10.** -10 **11.** -3 **12.** -5 **13.** -3
14. -8 **15.** 9   **16.** -2 **17.** -8 **18.** 24   **19.** -1 **20.** -7 **21.** 26   **22.** 22   **23.** 11   **24.** -3 **25.** 144   **26.** 72
27. 25   **28.** 37   **29.** -10 **30.** -11 **31.** 4   **32.** 5   **33.** -20 **34.** -12 **35.** 1   **36.** -2 **37.** -1 **38.** 7
39. 4   **40.** -3 **41.** -1 **42.** -1 **43.** 54   **44.** 10   **45.**

46.   **47.**   **48.**

49.   **50.**

Chapter 1 Practice Test

1. -11 **2.** 0   **3.** -4 **4.** -10 **5.** -12 **6.** 23   **7.** -19 **8.** 91   **9.** 29   **10.**

Exercises 2.1

1. $xxxxxx$ **3.** $(-x)(-x)(-x)(-x)$ **5.** $-xxxx$ **7.** $xxyyy$ **9.** $xx + yyy$ **11.** $xyyy$ **13.** a^4 **15.** x^2y^3 **17.** $-r^2s^3$
19. $-x^2(-y)^3$ **21.** x^3x^5 or x^8 **23.** 243   **25.** -8 **27.** 16   **29.** -45 **31.** -14 **33.** 17   **35.** 576   **37.** 243   **39.** x^8
41. a^7 **43.** $60x^3$ **45.** $6r^5$ **47.** $-15x^5$ **49.** $40c^5$ **51.** $32x^4y^7$ **53.** $6x^4y^4$ **55.** $9a^8$ **57.** $-64n^6$ **59.** x^{14} **61.** .073
63. 14.758   **65.** 107.916   **67.** 48.337

Exercises 2.2

1. 3   **3.** -3 **5.** -1 **7.** 5   **9.** 5   **11.** 5   **13.** -1 **15.** -5 **17.** 3   **19.** -2 **21.** 2   **23.** 1   **25.** 18   **27.** 11
29. 13   **31.** 25   **33.** -96 **35.** 2   **37.** -18 **39.** 100   **41.** 63   **43.** 20   **45.** 21   **47.** -113 **49.** 28
51. 2 terms: $3x$, coefficient 3, literal part x; $-4y$, coefficient -4, literal part y **53.** 1 term: coefficient -12, literal part xy
55. 1 term: coefficient 3, literal part $x(z - y)$ **57.** 3 terms: $4x^2$, coefficient 4, literal part x^2; $-3x$, coefficient -3, literal part x; 2
59. 3 terms: $-x^2$, coefficient -1, literal part x^2; y, coefficient 1, literal part y; -13
61. 2 terms: $6x^2$, coefficient 6, literal part x^2; $20y^2$, coefficient 20, literal part y^2 **63.** 2.031   **65.** -9.862 **67.** .0456   **69.** 25.667

Exercises 2.3

1. Essential   **3.** Nonessential   **5.** 1st nonessential, 2nd essential   **7.** x, $2x$; coefficients 1, 2; y, $3y$: coefficients 1, 3
9. $-x$, $-3x$: coefficients -1, -3; $2x^2$, $-x^2$: coefficients 2, -2; $4x^3$: 4   **11.** $-x^2y$, $-2x^2y$: coefficients -1, -2; $2xy^2$, $3xy^2$: coefficients 2, 3; x^2y^2: 1
13. $7x$ **15.** $7x^2$ **17.** $-3a$ **19.** $-4y$ **21.** $-6x$ **23.** $-5x + 2y$ **25.** $3x + 5y + 2z$ **27.** $4x^2 + 10x$ **29.** $2x^2 - 3x$
31. $6x^2y - 4x^2$ **33.** $-2y$ **35.** $3st$ **37.** $a^2b - ab^2$ **39.** $2(x + 5)$ **41.** $5(y - 4)$ **43.** $3(3x + y - 2)$ **45.** $x(x + y)$
47. $3x + 12$ **49.** $5y - 10$ **51.** $-2x - 14$ **53.** $15x + 6$ **55.** $-12x - 4$ **57.** $x^2 + 3x$ **59.** $x^3 + 3x^2$ **61.** $10x^2 - 20x$
63. $7x + 2y$ **65.** $7x - y$ **67.** $3x + 3y - 4xy$ **69.** $7x^3 + 21x$ **71.** $11x^3 + 15x$ **73.** $5.5x$ **75.** $-2.14y^3 + 4.62x^2$

Exercises 2.4

1. $8x + 5y$ **3.** $8m + 7n$ **5.** $-7x + 11y$ **7.** $3x - 1$ **9.** $-3x + 11$ **11.** $2x - 4$ **13.** $12 - 3x$ **15.** $7y - 1$ **17.** $4x - 18y$
19. $4x - 12y$ **21.** $5x - 15y + 3xy$ **23.** $-12xy$ **25.** $-45x^2y^2$ **27.** $5x^3 - 8x^2$ **29.** $13a^2 - 7a$ **31.** $3x^2 + 21x$
33. $3a^3 + 9ab + 4a^2b^2 - 4b^3$ **35.** $-5x^2 - 8x + 1$ **37.** $-4x^3y^2 + 4x^3y$ **39.** $4u^3v - 5u^2v^2 - uv^3$ **41.** $8x + 15y + 12xy$
43. $12m - 14n - 12mn$ **45.** $9t^9 - 13t^5 - t^4$ **47.** 0   **49.** $3a - 8$ **51.** $x^2 - 16x$ **53.** $19x - 18$ **55.** $8y - 22$
57. $13.34x - 2.3y$ **59.** $7.62x^2 + 8.26xy$

Exercises 2.5

1. $n + 4$ **3.** $n - 4$ **5.** $n - 4$ **7.** $5n + 6$ **9.** $2n - 9$ **11.** $n(n + 7)$ **13.** $(n + 2)(n - 6)$ **15.** $2n - 8 = 14$
17. $5n + 4 = n - 2$ **19.** $r + s = rs$ **21.** $2(r + s) = rs - 3$ **23.** Let n = the first integer; $n + (n + 1)$
25. Let n = the first even integer; $n + (n + 2)$ **27.** Let n = the first odd integer; $n(n + 2)(n + 4)$
29. Let n = the even integer; $8n = 7(n + 2) - 4$ **31.** Let n = the first integer; $n^2 + (n + 1)^2 + (n + 2)^2 = 5$
33. (a) 8   **(b)** 40 cents   **(c)** 12   **(d)** 120 cents   **(e)** 9   **(f)** 225 cents   **(g)** 29   **(h)** 385 cents
35. (a) 200   **(b)** 15   **(c)** 3,000   **(d)** 160   **(e)** 20   **(f)** 3,200   **(g)** 6,200
37. (a) 100 m./min.   **(b)** 25 min.   **(c)** 2,500 m.   **(d)** 220 m./min.   **(e)** 35 min.   **(f)** 7,700 m.   **(g)** 10,200 m.

Chapter 2 Review Exercises

1. $xyyy$ **2.** $(xy)(xy)(xy)$ **3.** $-xxxx$ **4.** $(-x)(-x)(-x)(-x)$ **5.** $3xx$ **6.** $(3x)(3x)$ **7.** x^2y^3 **8.** $x^2 + x^3$ **9.** $a^2 - b^3$ **10.** $-a^2b^3$
11. -16 **12.** 16   **13.** -48 **14.** 144   **15.** 25   **16.** 29   **17.** 8   **18.** -2 **19.** 17   **20.** -25 **21.** x^8 **22.** r^9
23. $2a^9$ **24.** $5y^6$ **25.** $4x^5 + 3x^6$ **26.** $5a^8 + 2a^7$ **27.** $-2x^2 - 8x + 4$ **28.** $3m^3 - 4m^2 - 9$ **29.** $6a^3b^5$ **30.** $-10a^3b^4$
31. $6a^3b + 2a^2b^5$ **32.** $-10a^3b - 5ab^4$ **33.** $11x^2 + 12x - 15$ **34.** $10z^2 - 10z + 4$ **35.** $3y^3 - 3y^2$ **36.** $c^4 + 5c^3$
37. $3x^3y - 6xy^2 + 4xy^3 - 4x^2y^2$ **38.** $-3r^2s^2 - 2r^3s^2 - 5r^2s^3$ **39.** $4x - 12y$ **40.** $10a - 8b$ **41.** $-x^7 - 6x^4y^2$ **42.** $6x^3y^5$
43. 3   **44.** $3 - 3x + 3x^2 - x^3$ **45.** $5(x^2 + 2)$ **46.** $2(a^5 + 8)$ **47.** $3(y - 2z + 3)$ **48.** $11(2x - 3y + 1)$
49. $n + 7 = 3n - 4$ **50.** $2n - 5 = 3n + 4$ **51.** $n + (n + 2) = n - 5$ **52.** $n + 4(n + 1) = 3(n + 2) + 8$
53. (a) 12   **(b)** \$2   **(c)** 9   **(d)** \$5   **(e)** \$24   **(f)** \$45   **(g)** \$69
54. (a) n **(b)** 2   **(c)** $2n - 4$ **(d)** \$5   **(e)** $2n$ **(f)** $5(2n - 4)$ **(g)** $2n + 5(2n - 4)$

Chapter 2 Practice Test

1. -81 **2.** 81   **3.** -1 **4.** -41 **5.** -1 **6.** -2 **7.** -24 **8.** 94   **9.** $3x^2y - 8xy - y^2$ **10.** $-24x^6y^3$ **11.** xy
12. $3x^5$ **13.** $3x - 12$ **14. (a)** $2n + 4$ **(b)** $5n - 20 = n$
15. (a) \$3   **(b)** x **(c)** \$2   **(d)** $2x - 5$ **(e)** $3(2x - 5)$ **(f)** $2x$ **(g)** $2x + 3(2x - 5)$

Exercises 3.1

1. Identity   **3.** Contradiction   **5.** Identity   **7.** Contradiction   **9.** Contradiction   **11.** Identity   **13.** Contradiction   **15.** Identity
17. $x = -7$ **19.** None   **21.** $y = -5$ **23.** None   **25.** None   **27.** None   **29.** $x = 2$ **31.** $a = -1, a = 4$ **33.** $y = 2$

Exercises 3.2

1. $x = 5$ **3.** $y = 11$ **5.** $a = -2$ **7.** $a = -3$ **9.** $x = 7$ **11.** $x = 3$ **13.** $x = -2$ **15.** $z = 2$ **17.** $x = 3$
19. $t = -5$ **21.** $a = -1$ **23.** $w = 7$ **25.** $y = 5$ **27.** $a = -4$ **29.** $r = -6$ **31.** $x = -3$ **33.** $u = -1$ **35.** $x = 0$
37. Identity   **39.** No solutions   **41.** $x = 7$ **43.** $t = 5$ **45.** $y = 4$ **47.** $y = 0$ **49.** $a = 0$ **51.** $z = -3$ **53.** $t = -9$
55. $t = -3$ **57.** $y = -1$ **59.** No solutions   **61.** $a = 0$ **63.** $x = -2$ **65.** $z = 6$ **67.** $x = 6.5$ **69.** $t = .03$
71. $t = -8.19$

Exercises 3.3

We include the equations used to solve each exercise along with the answers. Try them on your own before looking here.

1. The numbers are n and $3n + 4$; $n + 3n + 4 = 24$; $\boxed{5, 19}$ **3.** The numbers are x and $4x - 5$; $x + 4x - 5 = 10$; $\boxed{3, 7}$

5. The number is y; $y + 5y + 3 = 27$; $\boxed{4}$ **7.** The number is n; $2n + 4 - n = 12$; $\boxed{8}$

9. The smallest number is x, the largest number is $2x + 10$, the middle number is $2x - 5$; $x + 2x + 10 + 2x - 5 = 80$; $\boxed{15, 25, 40}$

11. The width $= w$, the length $= 2w + 1$; $2w + 2(2w + 1) = 26$; $\boxed{4 \text{ by } 9}$

13. Second side $= x$; first side $= x + 10$; third side $= 3x$; $x + 10 + x + 3x = 45$; $\boxed{7, 17, 21}$

15. $w =$ original width, $3w - 2 =$ original length; $2(3w) + 2(3w - 4) = 2w + 2(3w - 2) + 12$; $\boxed{4 \text{ by } 10}$

17. $x = $ # of 15¢ stamps, $29 - x = $ # of 12¢ stamps; $15x + 12(29 - x) = 399$; $\boxed{\text{seventeen 15¢ stamps, twelve 12¢ stamps}}$

19. $n = $ # of nickels, $2n = $ # of dimes, $2n - 3 = $ # of quarters; $5n + 10(2n) + 25(2n - 3) = 450$; $\boxed{\text{7 nickels, 14 dimes, 11 quarters}}$

21. $x = $ # of books she sells, $80 - x = $ # of magazine subscriptions she sells; $150x + 225(80 - x) = 15{,}750$; $\boxed{\text{30 books}}$

23. $t = $ # of hours until they pass each other; $20t + 40t = 300$; $\boxed{\text{3 P.M.}}$

25. $t = $ # of hours 55-kph person travels; $55t + 45(t - 1) = 355$; $\boxed{\text{6 P.M.}}$

27. $t = $ # of hours to complete running section; $18t + 50(6 - t) = 172$; $\boxed{\text{4 hours; 72 km.}}$

29. $t = $ # of hours trainee works; $7t + 15(t - 2) = 124$; $\boxed{\text{4 P.M.}}$

Exercises 3.4

1. Yes   **3.** No   **5.** Yes   **7.** No   **9.** No   **11.** No   **13.** Yes   **15.** Yes   **17.** Yes

19. $x < 5$;

21. $a > -3$;

23. $x \le 4$;

25. $y > -2$;

27. $x > -2$;

29. $x > 2$;

31. $a > 0$;

33. $a \le 0$;

35. $x > -3$;

37. $x > 3$;

39. $-1 < a \le 6$;

41. $-2 \le x \le 2$;

43. $-2 < y < 1$;

45. $5 > x > 1$;

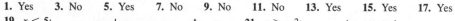

Exercises 3.5

1. $x < -2$ **3.** $a > -1$ **5.** $y < 4$ **7.** $y > -4$ **9.** $y > -4$ **11.** $y < 4$ **13.** $x > -4$ **15.** $1 < y$ **17.** $x \leq 1$
19. $x \geq 12$ **21.** $z \geq -2$ **23.** $w \geq 4$ **25.** $a > 6$ **27.** $y \geq 2$ **29.** Contradiction   **31.** Identity   **33.** $x > -4$
35. $a < -4$ **37.** $y \geq -10$ **39.** $-5 < x < 3$ **41.** $-1 < a < 1$ **43.** $3 < x \leq 5$

45. $x > 5$;

47. $a \leq 4$;

49. $-4 < x < -1$;

51. $-2 \leq t < 1$;

53. $-3 \leq x < 4$;

55. $x \geq 3.2$ **57.** $6.32 < t \leq 11.51$

Chapter 3 Review Exercises

1. Identity   **2.** Contradiction   **3.** Conditional   **4.** Identity   **5.** $x = -1$ **6.** $x = 4$ **7.** $y = -1, y = 1$ **8.** $w = 2$
9. $t = -2$ **10.** $z = -3$ **11.** $x = -5$ **12.** None   **13.** $a = -2$ **14.** None   **15.** $x = -5$ **16.** $x = -4$ **17.** Identity
18. No solutions   **19.** $a = -5$ **20.** $t = 2$ **21.** $x = 28$ **22.** $x = -1$ **23.** $a = 2$ **24.** $w = -1$ **25.** $x > 3$ **26.** $x \leq 3$

27. $x \leq -9$;

28. $x > -5$;

29. $-2 \leq a < 4$;

30. $-4 < t < 3$;

31. Let n = one number, $2n - 3$ = the other number; $n + 2n - 3 = 18$; $\boxed{7, 11}$

32. Let s = smaller number, $3s - 7$ = larger number; $s + 3s - 7 = s + 8$; $\boxed{5, 8}$

33. Let w = width, $5w + 4$ = length; $2w + 2(5w + 4) = 80$; $\boxed{6 \text{ by } 34}$

34. Let L = length, $2L - 8$ = width; $2L + 2(2L - 8) = 5(2L - 8)$; $\boxed{4 \text{ by } 6}$

35. Let f = # of \$12 skirts, $150 - f$ = # of \$7 skirts; $12f + 7(150 - f) = 1,500$; $\boxed{\text{ninety \$12 skirts, sixty \$7 skirts}}$

36. Let t = time going, $7 - t$ = time returning; $45t = 60(7 - t)$; $\boxed{360 \text{ km.}}$

37. Let v = # of overtime hours worked; $6 \cdot 40 + 9v \geq 348$; $\boxed{12 \text{ hours}}$

Chapter 3 Practice Test

1. (a) Contradiction   **(b)** Identity   **(c)** Conditional   **2. (a)** Yes   **(b)** No   **(c)** No
3. (a) $x = 3$ **(b)** $y = 0$ **(c)** $a = -2$ **(d)** $x \leq 3$

(e) $-2 \leq x < 2$

4. Let w = width, $4w - 5$ = length; $2w + 2(4w - 5) = 7w + 11$; $\boxed{7 \text{ by } 23}$

5. Let n = # of new cassettes, $20 - n$ = # of old cassettes; $3n + 20 - n = 46$; $\boxed{13 \text{ new, 7 used}}$ **6.** Let x = number; $2x - 8 = x - 3$; $\boxed{5}$

Cumulative Review: Chapters 1–3

The number appearing in parentheses after the answer indicates the section in which the material is discussed.
1. -20 (§1.4)   **2.** 27 (§1.5)   **3.** 20 (§1.5)   **4.** 5 (§1.5)   **5.** -25 (§2.1)   **6.** 25 (§2.1)   **7.** x^6 (§2.1)   **8.** $2x^9$ (§2.1)
9. $-2x^2y - 3xy^2$ (§2.3)   **10.** $z^2 - 6z - 8$ (§2.3)   **11.** $6x^3 - 8xy$ (§2.3)   **12.** $-24x^3y$ (§2.1)   **13.** $15u^5v$ (§2.1)
14. $-3u^5 + 15u^2v$ (§2.3)   **15.** $10m - 15n$ (§2.4)   **16.** $-t^2 - 16r^3$ (§2.4)   **17.** $-2a^3b + 2a^2b^2$ (§2.4)   **18.** $3x^3yz^2 - 9x^2y^2z$ (§2.4)

19. $2x^2y - 2xy^2$ (§2.4) **20.** $-7y$ (§2.4) **21.** $10x - 60$ (§2.4) **22.** $a - b + ab - a^2$ (§2.4) **23.** $-6xy^2$ (§2.4) **24.** 0 (§2.4)
25. 4 (§2.2) **26.** -4 (§2.2) **27.** -54 (§2.2) **28.** 49 (§2.2) **29.** 4 (§2.2) **30.** 31 (§2.2) **31.** -40 (§2.2) **32.** 11 (§2.2)
33. $x = -7$ (§3.2) **34.** $t = -4$ (§3.2) **35.** $a = -1$ (§3.2) **36.** $w = 3$ (§3.2) **37.** Identity (§3.1)
38. $a \geq 9$ (§3.5);

39. $s > -7$ (§3.5);

40. No solutions (§3.1) **41.** $3 < y \leq 4$ (§3.5);

42. $-2 \leq z \leq 3$ (§3.5); **43.** $d = 0$ (§3.2) **44.** $x = 0$ (§3.2)

45. Let $n = $ one number, $3n - 7 = $ other number; $n + 3n - 7 = 41$; $\boxed{12, 29}$ (§3.3)

46. Let $w = $ width, $3w + 8 = $ length; $2w + 2(3w + 8) = 24$; $\boxed{\text{width} = 1 \text{ cm., length} = 11 \text{ cm.}}$ (§3.3)

47. Let $d = $ # of danishes, $18 - n = $ # of pastries; $40n + 55(18 - n) = 825$; $\boxed{11 \text{ danishes and 7 pastries}}$ (§3.3)

48. Let $t = $ # of minutes old copier works, $t - 15 = $ # of minutes new copier works; $20t + 25(t - 15) = 885$; $\boxed{10{:}28 \text{ A.M.}}$ (§3.3)

Cumulative Practice Test: Chapters 1–3

1. (a) -105 (§2.1) (b) 36 (§2.1) **2.** (a) 13 (§2.2) (b) 5 (§ 2.2)
3. (a) $-2x^2 - 5x + 3$ (§2.3) (b) $-3x - y$ (§2.3) (c) $-2a^3 - ab$ (§2.4) (d) 0 (§2.4) (e) $4x^3y^3$ (§2.4)
(f) $6 - 6a + 6a^2 - a^3$ (§2.4)
4. (a) $x = -3$ (§3.2) (b) $w = -9$ (§3.2) (c) $x \geq 3$ (§3.4) (d) $a = -5$ (§3.2) (e) No solutions (§3.1) (f) Identity (§3.1)
5. (a) $a \leq -2$ (§3.4); (b) $1 \leq x < 5$ (§3.4);

6. (a) Let $n = $ the number, $3n - 5 = $ the other number; $n + 3n - 5 = 27$; $\boxed{8, 19}$ (§3.3)

(b) Let $n = $ # of tires at \$32 each, $40 - n = $ # of tires at \$19 each; $32n + 19(40 - n) = 1{,}124$; $\boxed{28 \text{ new, 12 old}}$ (§3.3)

(c) Let $t = $ # of hours to complete the race at 20 kph, $t - 1 = $ # of hours to complete the race at 25 kph; $20t = 25(t - 1)$;
$\boxed{\text{distance} = 100 \text{ kilometers}}$ (§3.3)

Exercises 4.1

1. $\frac{3}{5}$ **3.** $-\frac{3}{7}$ **5.** $\frac{5}{2}$ **7.** $\frac{1}{2}$ **9.** $\frac{1}{3}$ **11.** x^2 **13.** $\frac{1}{x^2}$ **15.** $\frac{5}{2x}$ **17.** $-\frac{3z^4}{5}$ **19.** $\frac{2}{5t^5}$ **21.** $-\frac{3b^3}{a^2}$ **23.** 1 **25.** $\frac{-r^2}{2t^2}$
27. $-\frac{5b}{a}$ **29.** $\frac{x^2}{2}$ **31.** $\frac{-2}{x^6}$ **33.** $\frac{1}{8x^3y^3}$ **35.** $\frac{1}{2}$ **37.** $\frac{1}{12}$ **39.** $\frac{4s - 3t}{8s - 9t}$ **41.** a **43.** 1

Exercises 4.2

1. $\frac{8}{27}$ **3.** -1 **5.** $\frac{2x}{15y}$ **7.** $\frac{5x^3}{12y^2}$ **9.** $\frac{16}{25}$ **11.** $\frac{12x^3}{y^3}$ **13.** $\frac{w}{4}$ **15.** $\frac{x}{3}$ **17.** $\frac{48}{x}$ **19.** $\frac{x}{48}$ **21.** $\frac{3}{2y}$ **23.** $\frac{3m^2}{n}$
25. $\frac{3v}{2u}$ **27.** $4x^2$ **29.** $9y^2$ **31.** $\frac{1}{9y^2}$ **33.** $-\frac{2x^2}{3y^2}$ **35.** 1 **37.** $\frac{9}{a^2}$ **39.** $\frac{81}{a^4}$ **41.** $\frac{81}{a^4}$ **43.** $\frac{3}{5}$ **45.** $2x$ **47.** $\frac{x^3}{y^3}$
49. $\frac{u^2}{2z^3}$ **51.** $\frac{1}{y}$ **53.** xy

Exercises 4.3

1. 3 **3.** $-\frac{4}{5}$ **5.** $-\frac{2}{3}$ **7.** $\frac{22}{15}$ **9.** $\frac{8}{15}$ **11.** $-\frac{1}{6}$ **13.** $\frac{27}{8}$ **15.** $\frac{4}{x}$ **17.** $\frac{32}{9x^2}$ **19.** $\frac{2y}{7x}$ **21.** 0 **23.** $\frac{2x - 3}{3x}$ **25.** $-\frac{y}{4}$
27. $\frac{2}{5}$ **29.** $\frac{2}{3w}$ **31.** $\frac{3y + 2x}{xy}$ **33.** $\frac{6}{xy}$ **35.** $\frac{10 - 21x}{6x}$ **37.** $\frac{5y + 6x}{4xy}$ **39.** $\frac{8 - 3x}{2x^2}$ **41.** $\frac{6}{x^3}$ **43.** $\frac{35 - 9a}{20a^2}$ **45.** $\frac{1 + 2x}{x}$
47. $\frac{10y + x}{6xy^2}$ **49.** $\frac{14b^2 + 9a}{12a^2b^3}$ **51.** $\frac{7}{8a^3b^4}$ **53.** $\frac{18n^2 - 20m + 3m^2n}{24m^2n^3}$ **55.** $\frac{5x^2 + 2y^2}{2xy}$ **57.** $\frac{t^2 - 3}{t}$ **59.** $\frac{a^2 - 5a + 6}{2a}$

Exercises 4.4

1. 8   **3.** −1   **5.** 8   **7.** $y < 3$ **9.** 4   **11.** $a \le -30$ **13.** 5   **15.** 40   **17.** $\frac{34}{85} = \frac{2}{5}$ **19.** −7   **21.** 5   **23.** 7

25. $y < -6$ **27.** 1   **29.** 0   **31.** −2   **33.** −2   **35.** 4   **37.** 2   **39.** 0   **41.** $-39 \le x \le -21$ **43.** $\frac{31x}{30}$ **45.** $x = 60$

47. $x = -3$ **49.** $\frac{x + 11}{4}$

Exercises 4.5

1. $\frac{7}{5}$ **3.** $\frac{5}{7}$ **5.** $\frac{11}{5}$ **7.** $\frac{1}{3}$ **9.** $\frac{a}{b + c}$ **11.** $x = 20$ **13.** $a = 10$ **15.** $y = 50$ **17.** $y = 12$ **19.** 294   **21.** 9

23. 8 cm.   **25.** $\frac{2}{5}$ **27.** $8\frac{1}{3}$ cm.   **29.** 4.55 kg.   **31.** 108.7 yd.   **33.** $x = .12$ **35.** $t = 11.59$ **37.** $y = -2.05$

Exercises 4.6

1. Let n = the number; $\frac{2}{3}n + 5 = 9$; $\boxed{6}$ **3.** Let n = the number, $\frac{3}{4}n - 2 = \frac{1}{8}n - 7$; $\boxed{-8}$

5. Let L = length, $\frac{L}{2}$ = width; $2L + 2\left(\frac{L}{2}\right) = 36$; $\boxed{\text{length} = 12 \text{ m., width} = 6 \text{ m.}}$

7. Let L = length of longest side, $\frac{3}{4}L$ = length of medium side, $\frac{1}{2} \cdot \frac{3}{4}L = \frac{3}{8}L$ = length of shortest side; $L + \frac{3}{4}L + \frac{3}{8}L = 17$;

$\boxed{\text{short} = 3 \text{ in., medium} = 6 \text{ in., long} = 8 \text{ in.}}$

9. Let c = # of combination tickets sold, $350 - c$ = # of regular tickets sold; $7c + 3(350 - c) = 1{,}990$;

$\boxed{\text{235 combination tickets, 115 regular tickets}}$

11. Let q = # of quarters, $2q + 3$ = # of dimes; $25q + 10(2q + 3) = 255$; $\boxed{\text{5 quarters, 13 dimes}}$

13. Let t = # of minutes new machine works, $t + 15$ = # of minutes old machine works; $250t + 175(t + 15) = 13{,}675$; $\boxed{10{:}41}$

15. Let x = amount invested at 8%, $x + 4{,}000$ = amount invested at 11%; $.08x + .11(x + 4{,}000) = 1{,}390$; $\boxed{\$9{,}000 \text{ at } 11\%}$

17. Let x = amount invested at 9%, $800 - x$ = amount invested at 6%; $.09x + .06(800 - x) = 67.50$; $\boxed{\$650 \text{ at } 9\%, \$150 \text{ at } 6\%}$

19. Let x = amount invested at 8%, $6{,}000 - x$ = amount invested at 12%; $.08x + .12(6{,}000 - x) = .09(6{,}000)$; $\boxed{\$4{,}500 \text{ at } 8\%, \$1{,}500 \text{ at } 12\%}$

21. Let n = # of ml. of 30% solution needed; $.30n + .50(30) = .45(n + 30)$; $\boxed{10 \text{ ml.}}$

23. Let s = # of liters of 25% salt solution, $90 - s$ = # of liters of 55% salt solution; $.25s + .55(90 - s) = .50(90)$;

$\boxed{\text{75 liters of 55\% sol., 15 liters of 25\% sol.}}$

25. Let n = # of gallons of pure antifreeze; $n + .30(10) = .50(10 + n)$; $\boxed{\text{4 gallons}}$

27. Let n = # of pounds of \$3.75 candy; $3.75n + 5(35) = 4.25(n + 35)$; $\boxed{52.5 \text{ lb.}}$

29. Let t = # of hours until they meet; $4t + 8t = 9$; $\boxed{8{:}45}$

31. Let t = # of hours Susan jogs until they meet; $t + \frac{1}{4}$ = # of hours John walks until they meet; $4\left(t + \frac{1}{4}\right) + 8t = 9$; $\boxed{8{:}40}$

33. Let t = time jogging, $2 - t$ = time walking; $9t + 5(2 - t) = 16$; $\boxed{1\frac{1}{2} \text{ hr.}}$

Chapter 4 Review Exercises

1. $-\frac{3}{7}$ **2.** $-\frac{2}{5}$ **3.** $\frac{5x^4}{2}$ **4.** $\frac{2}{7a^6}$ **5.** $-\frac{5x^2}{2y^5}$ **6.** $-\frac{8x^4}{5y^4}$ **7.** $\frac{1}{t}$ **8.** $\frac{w}{z}$ **9.** $\frac{a^2}{16}$ **10.** $\frac{a}{2}$ **11.** $\frac{a}{3}$ **12.** $\frac{7}{5}$ **13.** $\frac{3x - 2}{6x}$

14. $\frac{5x + 1}{5x}$ **15.** $\frac{y^2}{2}$ **16.** 1   **17.** $\frac{3}{4y^3}$ **18.** $\frac{x^2 y}{48}$ **19.** $\frac{a^2}{8}$ **20.** $\frac{3a}{4}$ **21.** $\frac{2x^2}{3}$ **22.** $\frac{x^2}{4}$ **23.** $\frac{3x + 8}{2x^2}$ **24.** $\frac{9y + 10}{6y^2}$

25. $\dfrac{18b^2 - 20ab^2 + 21a^2}{24a^2b^3}$ **26.** $\dfrac{9t + 14t^2 - 40r}{24rt^3}$ **27.** $x = 5$ **28.** $x = 7$ **29.** $t < 0$ **30.** $a = \dfrac{1}{2}$ **31.** $y = 2$ **32.** $z < 8$

33. $x = 1$ **34.** $x = 6$ **35.** $x = 4$ **36.** $x = 20$ **37.** Let $x = $ # of ounces in 1 kg.; $\dfrac{x}{1,000} = \dfrac{1}{28.4}$; $\boxed{35.21 \text{ oz.}}$

38. Let $n = $ the number; $\dfrac{3}{4}n - 1 = n - 4$; $\boxed{12}$

39. Let $x = $ amount invested at 6%, $2x = $ amount invested at 7%, $7,000 - 3x = $ amount invested at 8%;

$.06x + .07(2x) + .08(7,000 - 3x) \geq 500$; $\boxed{\$1,500}$

40. Let $n = $ number of nickels, $2n = $ number of dimes, $30 - 3n = $ number of quarters; $5n + 10(2n) + 25(30 - 3n) = 350$;

$\boxed{8 \text{ nickels, 16 dimes, 6 quarters}}$

41. Let $r = $ Bill's present speed; $5r = 3(r + 20)$; $\boxed{30 \text{ mph}}$

Chapter 4 Practice Test

1. (a) $-\dfrac{5}{12}$ (b) x^8 (c) $2a^3$ (d) $\dfrac{-5t^2}{3r^2}$

2. (a) $\dfrac{3x}{2y}$ (b) $\dfrac{3x^2}{4y}$ (c) $\dfrac{3a^2}{4}$ (d) $\dfrac{2y^2}{81x^2}$ (e) $\dfrac{2a}{5}$ (f) $\dfrac{a^2}{25}$ (g) $\dfrac{3}{x}$ (h) $\dfrac{11a}{24}$ (i) $\dfrac{17}{12x}$ (j) $\dfrac{70xy + 9}{30x^2y}$ (k) $-\dfrac{1}{x}$

3. (a) $x = 15$ (b) $x \geq 15$ (c) $a = 2$ (d) $t = 20$ **4.** Let $x = $ # of miles in 50 kilometers; $\dfrac{x}{50} = \dfrac{1}{1.61}$; $\boxed{31.06 \text{ miles}}$

5. Let $x = $ the number; $x + \dfrac{2}{3}x = 2x - 5$; $\boxed{x = 15}$

6. Let $x = $ # of tickets sold at the door; $9x + 7.50(400 - x) = 3,330$; $\boxed{220 \text{ tickets}}$

7. Let $d = $ amount invested at 13%; $7,000 - d = $ amount invested at 8%; $.13d + .08(7,000 - d) = 750$; $\boxed{\$3,800 \text{ at } 13\%, \$3,200 \text{ at } 8\%}$

8. Let $x = $ # of ounces of 20% solution; $.20x + .65(24) = .50(x + 24)$; $\boxed{12 \text{ ounces}}$

9. Let $t = $ # of hours second person travels until they are 604 kilometers apart, $t + 4 = $ # of hours first person travels until they are

604 kilometers apart; $48(t + 4) + 55t = 604$; $\boxed{7:00 \text{ P.M.}}$

Exercises 5.1

1. x^5 **3.** x^6 **5.** x^9 **7.** 10^9 **9.** 648   **11.** y^2 **13.** $3u^6v^4$ **15.** a^7 **17.** x^8 **19.** x^4y^2 **21.** $x^{10}y^{15}$ **23.** $16r^{12}s^{20}$

25. $-x^9y^3$ **27.** $\dfrac{x^{12}}{y^8}$ **29.** $144x^{16}$ **31.** xy^5 **33.** $\dfrac{32x^{10}}{y^{10}}$ **35.** $-\dfrac{27a^6b^9}{8c^3}$ **37.** -1 **39.** -1 **41.** -7

Exercises 5.2

1. (a) -6 (b) $x^{-1} = \dfrac{1}{x}$ (c) $x^{-6} = \dfrac{1}{x^6}$ (d) $\dfrac{1}{8}$ **3.** 1   **5.** 5   **7.** x **9.** $\dfrac{1}{25}$ **11.** 25   **13.** 1   **15.** $\dfrac{1}{x^{10}}$ **17.** $\dfrac{1}{a^8}$

19. $\dfrac{1}{1,000}$ or .001   **21.** x^4y^4 **23.** $\dfrac{2}{a^3}$ **25.** $\dfrac{-3}{y^2}$ **27.** $\dfrac{x}{y}$ **29.** $\dfrac{a^8}{b^6}$ **31.** $\dfrac{9y^6}{x^4z^8}$ **33.** $\dfrac{4x^3}{y^3}$ **35.** x^3 **37.** $\dfrac{-1}{3a^{12}}$

39. $\dfrac{1}{x^2} + \dfrac{1}{y}$ or $\dfrac{y + x^2}{x^2y}$ **41.** x **43.** $\dfrac{x^6}{y^5}$ **45.** $3(10^4)$ or 30,000   **47.** a^{15} **49.** $\dfrac{y^5}{x^8}$ **51.** $\dfrac{n^4}{4m^8}$ **53.** $\dfrac{3}{xy}$

Exercises 5.3

1. 4.53×10^3 **3.** 4.53×10^{-2} **5.** 7×10^{-5} **7.** 7×10^6 **9.** 8.537×10^4 **11.** 8.537×10^{-3} **13.** 9×10 **15.** 9

17. 9×10^{-1} **19.** 9×10^{-2} **21.** 3×10^{-8} **23.** 2.8×10 **25.** 4.75×10 **27.** 9.7273×10^3 **29.** 28,000   **31.** .00028

33. 42,900,000   **35.** .000000429   **37.** .00352   **39.** 352,860   **41.** .000026   **43.** 1   **45.** 5×10^{-7} **47.** .004   **49.** .03

51. 5.98×10^{27} kg.   **53.** 7.44×10^{-18} gm.   **55.** 10^{-8} cm.   **57.** 6.73×10^{24} tons ,   **59.** 5.87×10^{12} miles

5.4

1. (b) 5 (c) 5 **3.** (a) 2 (b) 1, 0 (c) 1 **5.** (a) 2 (b) 2, 3 (c) 3 **7.** (a) 1 (b) 5 (c) 5
9. (a) 1 (b) 0 (c) 0 **11.** (a) 3 (b) 3, 2, 1 (c) 3 **13.** (a) 2 (b) 3, 5 (c) 5
15. (a) 2, 1, 0 (b) 2 (c) 1, -5, 6
17. (a) 2, 0 (b) 2 (c) 1, 0, 4 **19.** (a) 3, 0 (b) 3 (c) 1, 0, 0, -1 **21.** $-a^3 + 12$ **23.** $-u^3 + 4u^2 - 9u + 7$
25. $3t^3 + 2t^2$ **27.** $4r^2s^3 - 3r^2s^2$ **29.** $9w^3 - 4w^2 - 5w + 18$ **31.** $3x^2 + 7x - 9$ **33.** $x^2 + 4x + 2$ **35.** $a^2b + a^2 + b$
37. 33 **39.** 61 **41.** 22

Exercises 5.5

1. $60x^6$ **3.** $15x^4 + 12x^3$ **5.** $-60x^2y^2z^2$ **7.** $12xy^2z - 20x^2yz$ **9.** $3x^3 + 13x^2y - 12xy^2$ **11.** $3x^2y^3 - 3xy^3$ **13.** $x^3 + x^2 + x + 6$
15. $y^3 - 3y^2 - 16y + 30$ **17.** $3x^3 + 7x^2 - 21x + 10$ **19.** $15z^3 + 16z^2 + 44z + 16$ **21.** $x^3 + y^3$ **23.** $x^4 + 2x^3 + x^2 - 1$
25. $x^2 + 8x + 15$ **27.** $x^2 - 8x + 15$ **29.** $x^2 - 2x - 15$ **31.** $x^2 + 2x - 15$ **33.** $a^2 + 3ab - 40b^2$ **35.** $12x^2 - 19x + 4$
37. $x^4 + 5x^2 + 6$ **39.** $x^2 + 14x + 49$ **41.** $x^2 - 49$ **43.** $x^2 - 8x + 16$ **45.** $x^3 + 6x^2 + 12x + 8$ **47.** $2x^4 - 8x^3 - 64x^2$
49. $45x^3 - 84x^2 + 36x$ **51.** $2x^2 - 7x$ **53.** $-4a + 14$ **55.** $-24x$ **57.** $1.92x^3 - 2.4x^2 + 4.88x$ **59.** $.06x^2 + .01x - .4$
61. $.007x^2 - 1.876x + 31.5$

Chapter 5 Review Exercises

1. $\frac{1}{81}$ **2.** 10 **3.** $\frac{49}{144}$ **4.** $\frac{12}{7}$ **5.** $\frac{y^2}{x^5}$ **6.** y^6 **7.** $\frac{9x^6}{y^4}$ **8.** $\frac{1}{x^5}$ **9.** x^6 **10.** $\frac{x^2}{y^2}$ **11.** $\frac{1}{4x^8}$ **12.** $\frac{y}{x^{10}}$ **13.** 5.87×10^7
14. 5.87×10^{-3} **15.** 2×10^{-6} **16.** 7×10^3 **17.** .00256 **18.** 879,000 **19.** 577,300,000 **20.** .00000007447 **21.** 2,000
22. 1.8 **23.** 40 **24.** 5.7×10^5 **25.** (a) 3 (b) 2, 1, 0 (c) 2 **26.** (a) 4 (b) 3, 2, 1, 0 (c) 3
27. (a) 3 (b) 4, 2, 2 (c) 4 **28.** (a) 4 (b) 5, 6, 2, 1 (c) 6 **29.** (a) 2 (b) 1, 0 (c) 1
30. (a) 2 (b) 0, 1 (c) 1 **31.** (a) 1 (b) 0 (c) 0 **32.** (a) 1 (b) Undefined (c) Undefined
33. (a) 1 (b) 8 (c) 8 **34.** (a) 2 (b) 5, 3 (c) 5 **35.** $2x^3 - 7x^2 + 0x + 4$ **36.** $3t^5 + 0t^4 + 0t^3 - t^2 + 0t - 10$
37. $y^5 + 0y^4 + 0y^3 + y^2 - 2y - 1$ **38.** $-x^4 + 0x^3 + 0x^2 + 0x + 1$ **39.** $2x^2 + 2$ **40.** $4y^4 + y^2 + 8y$ **41.** $4x^2 - 10x + 12$
42. $6y^4 - 3y^2 + 10y$ **43.** $-x^2y + xy^2 - 7x^2y^2$ **44.** $-8m^2 - 6m^2n$ **45.** $5x^2y - 5xy^2 - 8x^2y^2$ **46.** $3r^2s - 7rs^2 + 3r^2s^2$ **47.** $4a^3$
48. $7m^2n - 8m^2n^2 - n^3 + 2mn^2$ **49.** $8x$ **50.** $5a^2 - 5b^2$ **51.** $3x^2$ **52.** $-2a^3 + 2a^2 - 5a - 9$ **53.** $x^2 - 3x - 28$
54. $a^2 - 9a + 20$ **55.** $8x^2 - 22x + 15$ **56.** $30x^2 - 29x + 4$ **57.** $6a^2 + 7ab - 20b^2$ **58.** $28x^2 - 29xy + 6y^2$
59. $x^3 - 7x - 6$ **60.** $x^3 - 9x^2 + 26x - 24$ **61.** $x^2 + 12x + 36$ **62.** $9x^2 - 12xy + 4y^2$ **63.** $x^3 - 15x^2 + 75x - 125$
64. $8x^3 - 12x^2 + 6x - 1$ **65.** $3x^4 - 6x^3 - 24x^2$ **66.** $6y^3 - 28y^2 - 10y$ **67.** $x^2 - 25$ **68.** $9x^2 - 4y^2$ **69.** $x^3 - x^2 - 2x + 8$
70. $x^3 - 2x^2y - 7xy^2 + 12y^3$ **71.** $x^4 + 4x^3 + 4x^2 - 1$ **72.** $y^4 - 6y^3 + 9y^2 - 16$ **73.** $x^2 + 8x - 14$ **74.** $-2x + 5$

Chapter 5 Practice Test

1. $\frac{3}{2}$ **2.** $\frac{x^8}{y^2}$ **3.** $\frac{1}{x^9}$ **4.** x^{20} **5.** $\frac{4y^{19}}{x^6}$ **6.** (a) 4 (b) $-1, 0$ (c) 4 **7.** $24x^7y^2$ **8.** $-12x^4y^2 + 6x^5y$ **9.** xy
10. $12x^3 - 23x^2 + 28x - 12$ **11.** $6x^3 - 4x^2y - xy^2$ **12.** $x^2 + x + 1$ **13.** $-4a$ **14.** (a) 3.16×10^{-3} (b) 3.16×10^4
15. 2×10^5

Exercises 6.1

1. $x^2 + 7x + 12$ **3.** $x^2 - 7x + 12$ **5.** $x^2 + x - 12$ **7.** $x^2 - x - 12$ **9.** $x^2 + 8x + 12$ **11.** $x^2 - 8x + 12$ **13.** $x^2 + 4x - 12$
15. $x^2 - 4x - 12$ **17.** $a^2 + 16a + 64$ (perfect square) **19.** $a^2 - 16a + 64$ (perfect square) **21.** $a^2 - 64$ (difference of two squares)
23. $c^2 - 8c + 16$ (perfect square) **25.** $c^2 + 8c + 16$ (perfect square) **27.** $c^2 - 16$ (difference of two squares) **29.** $3x^2 + 25x + 28$
31. $3x^2 + 19x + 28$ **33.** $3x^2 - 17x - 28$ **35.** $3x^2 + 17x - 28$ **37.** $15x^2 + 41x + 28$ **39.** $15x^2 + 47x + 28$ **41.** $15x^2 - x - 28$
43. $15x^2 + x - 28$ **45.** $4a^2 + 20a + 25$ (perfect square) **47.** $4a^2 - 25$ (difference of two squares) **49.** $9x^2y^2$
51. $x^6 + 2x^3y^2 + y^4$ (perfect square)

Exercises 6.2

1. $5(x + 4)$ **3.** $4(2a - 3)$ **5.** Not factorable **7.** $x(x + 3)$ **9.** $a(a + 1)$ **11.** $x(x - 5 + y)$ **13.** $3c^3(c^3 - 2)$ **15.** $xy(x - y)$
17. $3x(2x + 1)$ **19.** Not factorable **21.** $4c^2d^3(3cd^2 + 1)$ **23.** Not factorable **25.** $2xyz^2(xz + 4 - 5xy)$ **27.** $6u^3v^2(1 + 3v - 2v^3)$
29. $(x - 5)(x + 4)$ **31.** $(y + 6)(y - 3)$ **33.** $(x + 8)(x + y)$ **35.** $(m + n)(m + 9)$ **37.** $(x - y)(x - 4)$ **39.** $(x + 2)(3xy - 5)$

Exercises 6.3

1. $x(x + 3)$ **3.** $(x + 2)(x + 1)$ **5.** $(x - 2)(x - 1)$ **7.** Not factorable **9.** $(x + 2)(x - 1)$ **11.** $(x - 2)(x + 1)$ **13.** $(a + 6)(a + 2)$
15. $(a - 4)(a + 3)$ **17.** Not factorable **19.** $a(a - 12)$ **21.** $(a + 4)(a - 3)$ **23.** $(x - y)(x - 2y)$ **25.** $(a + 6)(a + 4)$
27. $(y + 6)(y + 6)$ **29.** $(y + 6)(y - 6)$ **31.** $(x - 9)(x + 2)$ **33.** $(r - 5s)(r + 2s)$ **35.** $(c - 5)(c - 1)$ **37.** $4(x + 1)(x + 1)$
39. $(x + 6)(x - 5)$ **41.** $2(x + 5)(x - 5)$ **43.** $(x - 5)(x + 4)$ **45.** Not factorable **47.** $(y + 7)(y + 4)$ **49.** $2(y + 7)(y - 6)$
51. $(7 + d)(7 - d)$ **53.** Not factorable

Exercises 6.4

1. $x(x + 3)$ **3.** $(x + 2)(x + 1)$ **5.** Not factorable **7.** $(3x + 2)(x + 2)$ **9.** $(2x + 3)(x + 4)$ **11.** $2(x + 3)(x + 2)$
13. $(5x - 2)(x - 5)$ **15.** $5(x - 2)(x - 1)$ **17.** $(2y + 3)(y - 2)$ **19.** $(5a - 6)(a + 3)$ **21.** $(2t + 3)(t + 2)$ **23.** $2(t^2 + 3t + 3)$
25. $3(w^2 - 2w - 10)$ **27.** Not factorable **29.** $(3x - 5y)(x - 3y)$ **31.** $(6a + 5)(a + 2)$ **33.** $(3a + 10)(2a - 1)$
35. $6(a - 4)(a + 1)$ **37.** $(x - 6y)(x + 6y)$ **39.** $4(x - 3y)(x + 3y)$ **41.** $x(x + 8)(x - 3)$ **43.** $6x(x - 4)$ **45.** $4x^2(x - 4)(x - 2)$
47. $2xy(3x - 4y + 6)$ **49.** $(3x - 16)(x + 3)$ **51.** $8x(x - 4)$ **53.** $-(x - 5)(x + 3)$ **55.** $-4xy(x + 7)(x - 3)$ **57.** $(5 - x)(5 + x)$

Exercises 6.5

1. $\dfrac{x + 4}{2}$ **3.** $\dfrac{t - 6}{6}$ **5.** $x - 3y$ **7.** $\dfrac{x - 3y}{2xy}$ **9.** $2ab - 3c - 4a^2c^2$ **11.** $x - 5, R = 12$ **13.** $t + 2$ **15.** $w + 1, R = -24$
17. $2x - 1, R = 6$ **19.** $y^2 + 3y + 7$ **21.** $2a^2 - a - 2, R = 4$ **23.** $x^2 - 4x + 12$ **25.** $x^3 + 2x^2 + 4x + 8$ **27.** $x^2 + 4x - 2$
29. $2t^2 + 5t - 4, R = 4$

Chapter 6 Review Exercises

1. $x^2 + 12x + 35$ **2.** $a^2 + 9a + 18$ **3.** $x^2 - 12x + 35$ **4.** $a^2 - 9a + 18$ **5.** $x^2 - 2x - 35$ **6.** $a^2 - 3a - 18$
7. $x^2 + 2x - 35$ **8.** $a^2 + 3a - 18$ **9.** $x^2 - 10x + 25$ **10.** $a^2 + 12a + 36$ **11.** $x^2 - 25$ **12.** $a^2 - 36$ **13.** $x^2 - 81y^2$
14. $a^2 - 49b^2$ **15.** $2x^2 - 11x - 21$ **16.** $3a^2 + 14a - 24$ **17.** $15x^2 + 14x - 8$ **18.** $28a^2 + 13ab - 6b^2$ **19.** $(x + 4)(x + 3)$
20. $(x - 4)(x - 3)$ **21.** $x(x + 7)$ **22.** Not factorable **23.** $(x - 12)(x - 1)$ **24.** $(x - 4)(x + 3)$ **25.** $(x - 9y)(x + 3y)$
26. $(r - 6t)(r - 2t)$ **27.** $(x - 8)(x + 8)$ **28.** $16(x - 2)(x + 2)$ **29.** $(2x + 5)(x + 2)$ **30.** $2(x + 5)(x - 1)$ **31.** $3(x - 4)(x + 2)$
32. $(3x + 4)(x - 6)$ **33.** $6(a + 4)(a + 2)$ **34.** $(3a + 16)(2a + 3)$ **35.** $5xy(x - 4y)(x + 4y)$ **36.** $2(3m^2n - 4mr^3 + 4n^2r)$
37. $x(x + 9)$ **38.** Not factorable **39.** $(5t + 1)(5t - 1)$ **40.** $25(t + 2)(t - 2)$ **41.** $-(x - 6)(x + 5)$ **42.** Not factorable
43. $-3(x - 3)(x - 1)$ **44.** $-4(x - 5)(x + 1)$ **45.** $x - y$ **46.** $\dfrac{3r - 2t^2 + 5rt^3}{t}$ **47.** $x - 3, R = -8$ **48.** $y^2 + y + 3, R = 5$
49. $2x^2 + 6x + 14, R = 38$ **50.** $2x^2 - x + 2$ **51.** $x^2 - 2x + 4$ **52.** $16x^3 + 32x^2 + 64x + 128, R = 192$

Chapter 6 Practice Test

1. $3x(2x^2 + 4x - 5)$ **2.** $2xy(2x - 4y - 1)$ **3.** $(x + 8)(x + 1)$ **4.** $(x - 10y)(x + y)$ **5.** $4x(x - 5)$ **6.** $5x(x + 3)(x - 3)$
7. $6(x + 3)(x + 1)$ **8.** $(2x + 3)(x - 5)$ **9.** $(x + 2)(x + 2)$ **10.** $(3x - 2)(2x + 3)$ **11.** $(xy + 3)(xy - 3)$ **12.** $\dfrac{6r - 9 + 10rt^2}{2rt}$
13. $2x^2 + 4x + 3, R = 12$

Cumulative Review: Chapters 4–6

1. $-\dfrac{4}{7}$ (§4.1) **2.** $\dfrac{3}{5x^4}$ (§4.1) **3.** $\dfrac{9t}{5s}$ (§4.1) **4.** $-\dfrac{1}{2a}$ (§4.1) **5.** $\dfrac{12}{5}$ (§4.2) **6.** $\dfrac{3x^2}{125}$ (§4.2) **7.** $\dfrac{6x^2 + 250}{25x}$ (§4.3) **8.** $\dfrac{2}{x}$ (§4.3)
9. $\dfrac{2}{t}$ (§4.3) **10.** $\dfrac{1}{5u^2}$ (§4.3) **11.** $x^2 + 3x - 40$ (§5.5) **12.** $x^3 + 9x^2 + 3x - 40$ (§5.5) **13.** $a^2 + 2ab + b^2 - c^2$ (§5.5)
14. $2z^3 + 6z^2 - 36z$ (§5.5) **15.** $\dfrac{6x^2y}{25z}$ (§4.2) **16.** $\dfrac{2u}{3v^3w^4}$ (§4.2) **17.** $20a^2 + 3ac - 9c^2$ (§5.5) **18.** $t^2 - t$ (§5.5) **19.** $\dfrac{-11}{6x}$ (§4.3)
20. $\dfrac{18z + 11y}{6yz}$ (§4.3) **21.** $\dfrac{25y^2 - 27x}{30x^2y^3}$ (§4.3) **22.** $\dfrac{3}{4x^3y^4}$ (§4.2) **23.** $2x^2 + 3x$ (§5.5) **24.** $-2x$ (§5.5) **25.** $2x$ (§4.2)
26. $\dfrac{a}{3}$ (§4.3) **27.** $4a^3 - 12a^2 - 9a + 27$ (§5.5) **28.** $x^4 - x^2 - x^2y^2 + y^2$ (§5.5) **29.** $\dfrac{2x^2 + 3x - 1}{x^2}$ (§4.3)
30. $\dfrac{120st - 4t^2 + 9s}{24s^2t^2}$ (§4.3) **31.** $-4y^2 - xy$ (§5.4) **32.** $-2a^3 + 3a - 9$ (§5.4) **33.** (a) 4 (b) -3 (§5.4)
34. $3x^3 + 0x^2 - x + 4$ (§5.4) **35.** $\dfrac{x + 8}{2}$ (§6.5) **36.** $\dfrac{2(2r - 3s^2 - r^2s)}{3rs}$ (§6.5) **37.** $y - 1, R = 2$ (§6.5) **38.** $2a^2 + a - 2$ (§6.5)
39. $6x^2 + 8x + 9, R = 8$ (§6.5) **40.** $t^3 - 2t^2 + 3t - 6, R = 6$ (§6.5) **41.** $\dfrac{19}{16}$ (§5.2) **42.** 1 (§5.2) **43.** x (§5.1) **44.** $\dfrac{1}{a}$ (§5.2)

45. $\frac{4}{x^3}$ (§5.1)   **46.** $\frac{x^{13}}{y^{22}}$ (§5.2)   **47.** $\frac{a^{11}}{27t^2}$ (§5.2)   **48.** $\frac{x^3}{125y^{15}}$ (§5.2)   **49.** 4.39×10^{-4} (§5.3)   **50.** 5.78×10^5 (§5.3)

51. 10^5 (§5.3)   **52.** $.002$ (§5.3)   **53.** $x = 8$ (§4.4)   **54.** $t = -3$ (§4.4)   **55.** Identity (§4.4)   **56.** $z = 0$ (§4.4)   **57.** $y = \frac{1}{2}$ (§4.4)

58. $x = -\frac{15}{4}$ (§4.4)   **59.** $x = 80$ (§4.4)   **60.** $x = 17$ (§4.4)   **61.** $(x + 1)(x + 5)$ (§6.3)   **62.** $x(x + 6)$ (§6.2)

63. $(x - 2)(x - 3)$ (§6.3)   **64.** $(x - 6)(x + 1)$ (§6.3)   **65.** $3xy(2x^2 - 4y - 3x)$ (§6.2)   **66.** $5m^2n^3(2mn^2 - 1)$ (§6.2)

67. $(u + 7)(u - 7)$ (§6.4)   **68.** $4(a - 3)(a - 3)$ (§6.3)   **69.** $(2r - 5)(r + 3)$ (§6.4)   **70.** $t^2(t + 6)(t - 6)$ (§6.4)

71. $5(t^2 + 2t + 3)$ (§6.3)   **72.** $xy(x + y)(x - y)$ (§6.4)   **73.** $(2x - 3y)(3x - 4y)$ (§6.4)   **74.** $-(x - 12)(x + 2)$ (§6.3)

75. $x(x + 16)$ (§6.2)   **76.** Not factorable (§6.3)   **77.** $(x + y)(x + a)$ (§6.2)   **78.** $(a - 3)(a + z)$ (§6.2)   **79.** $(x - a)(x - 4)$ (§6.2)

80. $(x^4 + y^4)(x^2 + y^2)(x + y)(x - y)$ (§6.4)

81. Let $n = $ # of \$8.75 tickets, $360 - n = $ # of \$6.25 tickets; $8.75n + 6.25(360 - n) = 2,850$; $\boxed{240 \ \$8.75 \text{ tickets, } 120 \ \$6.25 \text{ tickets}}$ (§4.6)

82. Let $x = $ # of grams in 15 ounces; $\frac{x}{15} = \frac{454}{16}$; $\boxed{425.63 \text{ grams}}$ (§4.5)

83. Let $x = $ # of votes party A received; $\frac{x}{15,700} = \frac{8}{5}$; $\boxed{25,120}$ (§4.5)

84. $x = $ amount invested at 8%, $2x = $ amount invested at 9%, $2x + 1,000 = $ amount invested at 10%;

$.08x + .09(2x) + .10(2x + 1,000) = 3,090$; $\boxed{\$33,500}$ (§4.6)

85. Let $t = $ # of hours to overtake; $80t = 65\left(t + \frac{1}{4}\right)$; $\boxed{1 \text{ hour and 5 minutes}}$ (§4.6)

Cumulative Practice Test: Chapters 4–6

1. $x^3 - 5x^2y + 5xy^2 + 2y^3$ (§5.5)   **2.** $\frac{2y^8}{3x^4}$ (§5.2)   **3.** $\frac{4d^3s}{t}$ (§4.2)   **4.** $7a^2 - 20a + 24$ (§5.5)   **5.** $-20x$ (§5.5)   **6.** $\frac{5a}{3x}$ (§4.3)

7. $\frac{-3x}{y}$ (§4.3)   **8.** $\frac{1}{16x^3}$ (§5.2)   **9.** $\frac{15 + 8ab}{18ab^2}$ (§4.3)   **10.** $\frac{2(3t^3 - 2s)}{3st}$ (§6.5)   **11.** $4x^2 + 5x + 15, R = 10$ (§6.5)

12. $x^3 - 2x^2 + 3x - 6$ (§6.5)   **13.** $\frac{4x^2y^4}{81z^5}$ (§4.2)   **14. (a)** 9.16×10^{-4} **(b)** 9.16×10^5 (§5.3)   **15.** 222 (§5.3)

16. $-2x^2 + 7x + 1$ (§5.4)   **17.** $a = 324$ (§4.4)   **18.** $t = -4$ (§4.4)   **19.** $(x - 12)(x + 2)$ (§6.3)   **20.** $3ab^3(2ab^2 - 1)$ (§6.2)

21. $(2t - 3)(t + 4)$ (§6.4)   **22.** $6(x^2 - 6x + 12)$ (§6.2)   **23.** $3xy(x + 2y)(x - 2y)$ (§6.4)   **24.** $(a + 5)(a - 7)$ (§6.2)

25. $(x - 3)(x - y)$ (§6.2)   **26.** $2(u^2 + 4)(u + 2)(u - 2)$ (§6.4)

27. Let $L = $ length, $\frac{1}{3}L + 5 = $ width; $2L + 2\left(\frac{1}{3}L + 5\right) = 34$; $\boxed{9 \text{ by } 8}$ (§4.6)

28. Let $n = $ # of 22¢ stamps, $28 - n = $ # of 13¢ stamps; $22n + 13(28 - n) = 535$; $\boxed{\text{nineteen 22¢ stamps, nine 13¢ stamps}}$ (§4.6)

29. Let $t = $ # of hours they work together, $t + 4 = $ # of hours the associate works; $40t + 24(t + 4) = 480$; $\boxed{6 \text{ hours}}$ (§4.6)

30. Let $t = $ # of hours Tom walks until they are 9 km. apart, $t + \frac{1}{3} = $ # of hours Terry walks until they are 9 km. apart;

$6\left(t + \frac{1}{3}\right) + 8t = 9$; $\boxed{11:50}$ (§4.6)

Exercises 7.1

1. $\frac{4y^5}{5x^3}$ **3.** $\frac{1}{3x}$ **5.** $\frac{2(x + 4)^4}{3x}$ **7.** $\frac{3}{2}$ **9.** $\frac{3}{5}$ **11.** $\frac{1}{2}$ **13.** $\frac{x - 2}{2(x + 2)}$ **15.** $\frac{y}{2(y + 2)}$ **17.** $\frac{6}{x - 3}$ **19.** Cannot be reduced

21. Cannot be reduced   **23.** $\frac{1}{2x}$ **25.** $\frac{y + 1}{y - 6}$ **27.** $\frac{s + 3}{s - 1}$ **29.** $\frac{x + 3}{x}$ **31.** $\frac{3a - 2}{a - 2}$ **33.** $\frac{4x - 1}{x + 2}$ **35.** $\frac{x - 2}{x}$

37. $\frac{2(x - 3)(x + 1)}{(x - 5)(x + 2)}$ **39.** $\frac{x + 2}{x - 4}$ **41.** Cannot be reduced   **43.** $\frac{y(y + 1)}{6(y + 2)}$ **45.** $\frac{1}{c}$

Exercises 7.2

1. $\frac{16ay^2}{3z^2}$ **3.** $\frac{s^3}{25p}$ **5.** $72x^3y^4$ **7.** $49ab^4z^4$ **9.** $\frac{x + 4}{(x + 5)(x + 1)}$ **11.** $\frac{x(x + 4)}{(x + 2)(x - 2)}$ **13.** $\frac{2r^2(r + 1)}{(r - 2)(r - 1)}$ **15.** $\frac{m + 6}{3(m + 3)}$

17. $\frac{x + 1}{x^2 + 2}$ **19.** $\frac{(y + 1)(y + 2)}{(y - 4)^2}$ **21.** $\frac{x + 2}{x + 1}$ **23.** $\frac{6x + 1}{x - 3}$ **25.** $\frac{4}{5}$ **27.** $\frac{x(x + 4)}{4}$ **29.** $\frac{t^2}{(2t + 3)(t + 1)}$ **31.** $\frac{(x + 5)^2}{x - 5}$

33. $\frac{(x + y)^2}{4(x - y)}$ **35.** $\frac{x}{x - 3}$ **37.** $\frac{7}{c + 7}$ **39.** $\frac{w^2 - 36}{8w}$

Exercises 7.3

1. $\dfrac{2x}{x+2}$ **3.** $\dfrac{x-2}{2(x+2)}$ **5.** $\dfrac{3x+8}{x+2}$ **7.** 1 **9.** $\dfrac{y+2}{4}$ **11.** $\dfrac{13x+10}{2x(x+2)}$ **13.** $\dfrac{10}{x(x+2)}$ **15.** $\dfrac{5x+12}{(x+2)(x+3)}$

17. $\dfrac{2a}{(a+7)(a+5)}$ **19.** $\dfrac{-2x+12}{3x^2(x+3)}$ **21.** $\dfrac{2a^2+9a-27}{12a^2(a-3)}$ **23.** $\dfrac{2x-24}{x(x+4)(x-4)}$ **25.** $\dfrac{x-2}{x-1}$ **27.** $\dfrac{2x}{x-1}$ **29.** $\dfrac{3}{2x(x-4)}$

31. $\dfrac{8}{x(x+2)}$ **33.** $\dfrac{x^2+2x+3}{(x+1)(x+3)^2}$ **35.** $\dfrac{3(a+1)}{3a+2}$ **37.** $\dfrac{15x^2-31x+9}{3x(x-2)}$ **39.** $\dfrac{4}{x(x-2)}$ **41.** $\dfrac{x+2}{2}$ **43.** $\dfrac{1-x}{2}$ **45.** a

47. $\dfrac{a^2b^2}{4(2a+b)}$

Exercises 7.4

1. $x=5$ **3.** $a=-7$ **5.** $x=7$ **7.** $r=4$ **9.** $x=\dfrac{3}{2}$ **11.** $t=\dfrac{10}{13}$ **13.** $x=5$ **15.** No solution   **17.** $m=0$

19. $x=4$ **21.** No solution   **23.** $a=1$ **25.** $x=8$ **27.** $x=\dfrac{14}{3}$ **29.** $\dfrac{13x}{12}$ **31.** $\dfrac{x^3}{24}$ **33.** $x=6$ **35.** $x=\dfrac{1}{3}$

Exercises 7.5

1. $x=\dfrac{3y+12}{4}$ **3.** $y=\dfrac{6-x}{2}$ **5.** $a=-2b-5$ **7.** $m=\dfrac{-2p}{7}$ **9.** $z=\dfrac{4x-y+4}{3}$ **11.** $x=\dfrac{-y-2}{2}$ **13.** $x=\dfrac{d-b}{a-2}$

15. $x=\dfrac{c-ay-ab}{y+b}$ **17.** $x=\dfrac{y-b}{m}$ **19.** $h=\dfrac{2A}{b_1+b_2}$ **21.** $P=\dfrac{A}{1+rt}$ **23.** $F=\dfrac{9}{5}C+32$ **25.** $v_0=\dfrac{2S-2S_0-gt^2}{2t}$

27. $x<2s+\mu$ **29.** $f=\dfrac{f_1 f_2}{f_1+f_2}$

Exercises 7.6

We include the equation used to solve each exercise.

1. Let $3x=$ width, $7x=$ length; $2(3x)+2(7x)=100$; ⟨width = 15 meters, length = 35 meters⟩

3. Let $n=$ # of people who preferred brand X, $200-n=$ # of people who did not prefer brand X; $\dfrac{n}{200-n}=\dfrac{13}{12}$; ⟨104⟩

5. Let $x=$ # of additional hits; $\dfrac{120+x}{450}=.400$; $x=60$. It cannot be done.

7. Let $x=$ # of hours to mow the lawn together. $\dfrac{x}{3}+\dfrac{x}{2}=1$; ⟨$1\frac{1}{5}$ hr.⟩

9. Let $x=$ # of hours electrician works, $2x=$ # of hours apprentice works. $\dfrac{6}{x}+\dfrac{6}{2x}=1$; ⟨electrician works 9 hr., apprentice works 18 hr.⟩

11. Let $r=$ rate of train, $r+100=$ rate of plane. $\dfrac{500}{r+100}=\dfrac{300}{r}$; ⟨rate of train = 150 kph, rate of plane = 250 kph⟩

13. Let $r=$ rate of current. $\dfrac{8}{4-r}=\dfrac{24}{4+r}$; ⟨$r=2$ mph⟩

15. $d=$ distance to the friend's house; $\dfrac{d}{6}+\dfrac{d}{14}=3$; ⟨12.6 km.⟩

17. Let the numbers be x and $3x$; $\dfrac{1}{x}+\dfrac{1}{3x}=\dfrac{5}{3}$; ⟨numbers are $\dfrac{4}{5}$ and $\dfrac{12}{5}$.⟩

19. Let $x=$ the number. $\dfrac{3+x}{5+x}=\dfrac{5}{6}$; ⟨$x=7$⟩   **21.** $\dfrac{x}{x+2}=\dfrac{x+2}{x}$; ⟨fraction is $\dfrac{-1}{1}$⟩   **23.** $S=252\pi$

Chapter 7 Review Exercises

1. Cannot be reduced   **2.** $\dfrac{x+6}{3x}$ **3.** $\dfrac{x-1}{x-4}$ **4.** $\dfrac{x+5}{2x+3}$ **5.** $\dfrac{a+4}{a+2}$ **6.** $\dfrac{xy}{x+2y}$ **7.** $\dfrac{z-2}{z+1}$ **8.** Cannot be reduced

9. $\dfrac{x^2+4x}{2(x+2)}$ **10.** $\dfrac{a-4}{a}$ **11.** $\dfrac{3x+12}{2x(x+2)}$ **12.** $\dfrac{x}{4}$ **13.** $\dfrac{4x^2}{x+1}$ **14.** $\dfrac{14y+12}{y(y+2)(y+6)}$ **15.** $\dfrac{2}{z(z-2)}$ **16.** $\dfrac{1}{(x+y)(x+2y)}$

17. $\dfrac{2x^2 + 6x - 2}{x(x + 2)}$ **18.** $\dfrac{x + 3}{6}$ **19.** $x = 2$ **20.** $a = -1$ **21.** $y = -\dfrac{1}{2}$ **22.** $x = 0$ **23.** No solution   **24.** $a = 10$

25. $x = \dfrac{3y + 4}{5}$ **26.** $a = \dfrac{2cx}{c - 3x}$ **27.** Let $b = $ # of black marbles, $b + 80 = $ # of red marbles. $\dfrac{b + 80}{b} = \dfrac{7}{5}$; $\boxed{480 \text{ marbles all together}}$

28. Let $r = $ rate of current, $\dfrac{8}{4 - r} = 2 \cdot \dfrac{8}{4 + r}$; $\boxed{r = 1\dfrac{1}{3} \text{ mph}}$ **29.** Let $x = $ # of hours for John to do the job alone. $\dfrac{4}{6} + \dfrac{4}{x} = 1$; $\boxed{12 \text{ hours}}$

30. Let $x = $ the number. $\dfrac{8 + x}{13 + x} = \dfrac{4}{5}$; $\boxed{x = 12}$

Chapter 7 Practice Test

1. $\dfrac{4x}{x - 4}$ **2.** $\dfrac{x + 3y}{3x}$ **3.** $\dfrac{5x + 12}{(x + 3)(x + 2)}$ **4.** $\dfrac{x - 2}{4}$ **5.** $\dfrac{5}{2(x + 4)}$ **6.** $\dfrac{2(x + 4)}{x}$ **7.** $\dfrac{5}{x}$ **8.** $x = 6$ **9.** $t = \dfrac{14 - 2b}{2a - 3}$

10. No solution   **11.** Let $d = $ distance from town A to town B; $\dfrac{d}{45} + \dfrac{d}{60} = 14$; $\boxed{360 \text{ km.}}$ **12.** $b_2 = 60$ cm.

Exercises 8.1

1.–19. 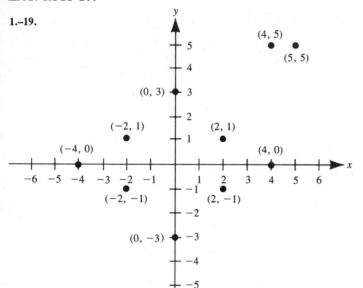   **21.** I   **23.** III   **25.** y-axis   **27.** II   **29.** Origin   **31.** IV

Exercises 8.2

1. $(7, 0)$ and $(0, -7)$ **3.** $(4, 0)$ and $(0, -4)$ **5.** $(6, 0)$ and $(0, 4)$ **7.** $(0, 0)$ **9.** $(0, 0)$ **11.** $(5, 0)$, no y-intercept

13. $(4, 0)$ and $\left(0, -\dfrac{4}{3}\right)$

15. 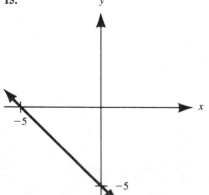   **17.** 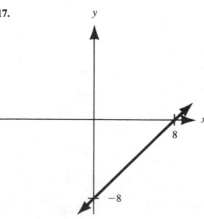   **19.**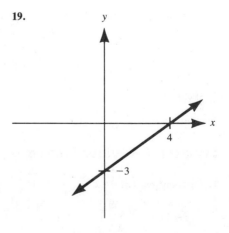

21.

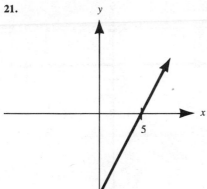

23.

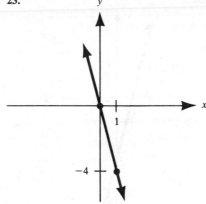

25.

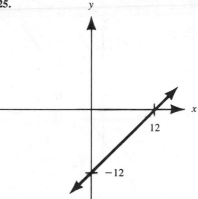

27.

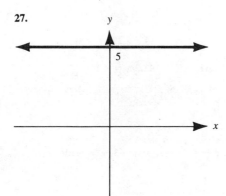

29.

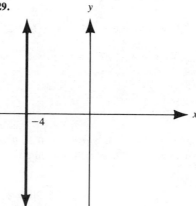

31.

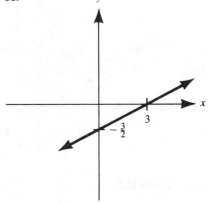

33.

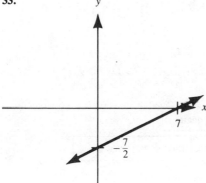

35.

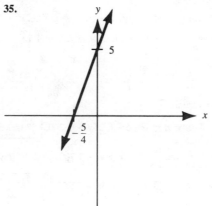

Exercises 8.3

1. $\frac{4}{3}$ **3.** $-\frac{3}{2}$ **5.** 1   **7.** $-\frac{2}{3}$ **9.** 0   **11.** No slope (slope is undefined)

13.

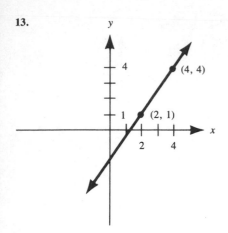

15.

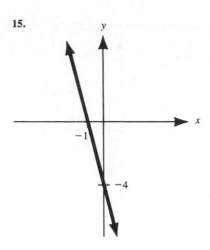

17.

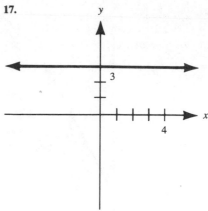

19. $y - 5 = 3(x - 1)$ **21.** $y - 4 = \frac{1}{2}(x + 1)$ **23.** $y = 5x + 6$ **25.** $y = -\frac{3}{4}(x + 2)$ **27.** $y + 2 = x - 4$ **29.** $y = 6$

31. $y - 3 = 2(x - 2)$ **33.** $y - 4 = -2(x + 1)$ **35.** $x = 4$ **37.** $y = -\frac{2}{5}x + 2$ **39.** 5 **41.** $-\frac{3}{2}$ **43.** $m = 4.14$

45. $m = -1.02$

Exercises 8.4

1. $x = 3, y = 1$ **3.** $x = 1, y = 0$ **5.** $\{(x, y) | 3x + y = 6\}$ **7.** $x = 0, y = 0$ **9.** $x = 2, y = 0$ **11.** $x = -3, y = -10$
13. $x = -5, y = -4$ **15.** No solution; the lines are parallel   **17.** $x = 14, y = 36$

Exercises 8.5

1. $x = 2, y = -1$ **3.** $x = 3, y = -1$ **5.** $x = 1, y = 4$ **7.** $x = 6, y = 3$ **9.** $x = -5, y = 2$ **11.** $r = -8, t = 9$
13. $x = \frac{1}{2}, y = 3$ **15.** No solution   **17.** $a = 2, b = -4$ **19.** $x = 0, y = \frac{3}{4}$ **21.** $x = 1, y = -2$ **23.** $x = 8, y = -9$
25. No solution   **27.** $x = 10, y = 20$ **29.** $x = 5, y = -4$ **31.** $x = 5.7, y = 1.1$

Exercises 8.6

We include the system of equations used to solve each exercise.

1. $x + y = 130, x - y = 28;$ $\boxed{x = 79, y = 51}$ **3.** $n + q = 80, 5n + 25q = 1,360;$ $\boxed{\text{32 nickels, 48 quarters}}$

5. $r_2 = r_1 + 15, 5r_1 + 5r_2 = 275;$ $\boxed{r_1 = 20 \text{ kph}, r_2 = 35 \text{ kph}}$ **7.** $x + y = 1,700, .07x + .06y = 110;$ $\boxed{\$800 \text{ at } 7\%, \$900 \text{ at } 6\%}$

9. Let $c =$ price of a cassette, $r =$ price of an LP; $4c + 6r = 48.80, 5c + 3r = 32.65;$ $\boxed{c = \$2.75, r = \$6.30}$

11. $L = 2W, 2W + 2L = 28;$ $\boxed{L = \frac{28}{3}, W = \frac{14}{3}}$

13. Let $r =$ cost of a regular selection, $s =$ cost of a special selection; $2r + 3s = 23, 3r + 4s = 32.50;$ $\boxed{r = \$5.50, s = \$4}$

15. $\frac{x}{y} = \frac{6}{5}, x - y = 8;$ $\boxed{x = 48, y = 40}$

17. $x =$ rate of slower plane, $y =$ rate of faster plane; $y = 2x, 4x + 4y = 1,800;$ $\boxed{x = 150 \text{ mph}, y = 300 \text{ mph}}$

19. $r =$ price of receiver, $t =$ price of turntable; $8r + 4t = 2,060, 5r + 6t = 1,690;$ $\boxed{r = \$200, t = \$115}$

21. $p =$ cost of plain donut, $f =$ cost of filled donut; $10p + 5f = 370, 5p + 10f = 410;$ $\boxed{p = 22¢}$

23. $x + y = 35, 7x + 9y = 271;$ $\boxed{\text{twenty-two } \$7 \text{ books, thirteen } \$9 \text{ books}}$

25. x = speed of slower car, y = speed of faster car; $y = x + 40$, $4x + 4y = 480$; $\boxed{x = 40 \text{ kph}, y = 80 \text{ kph}}$

27. p = speed of plane, w = speed of wind; $p + w = 150$, $p - w = 90$; $\boxed{p = 120 \text{ mph}, w = 30 \text{ mph}}$

29. x = number of pounds of \$3.35 candy, y = number of pounds of \$2.75 candy; $x + y = 60$, $3.35x + 2.75y = 60(3.10)$; $\boxed{x = 35, y = 25}$

Chapter 8 Review Exercises

1.–6.

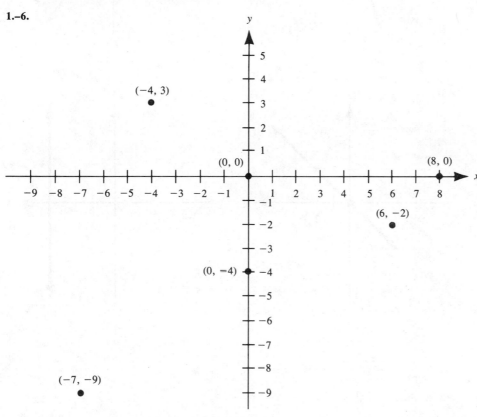

7. $y = 2$ **8.** $y = \dfrac{-8}{3}$ **9.** $y = 0$ **10.** $x = 5$

11.

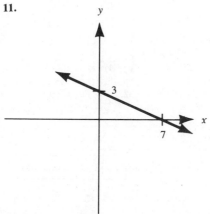

12.

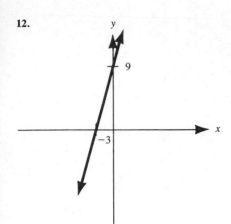

13.

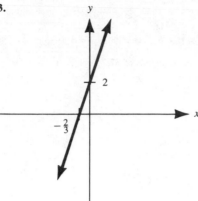

14.

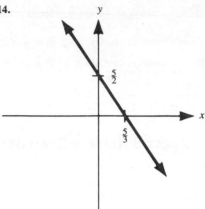

15.

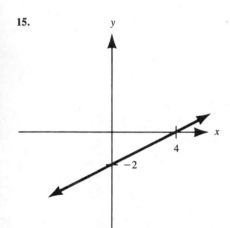

16.

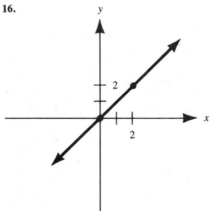

17.

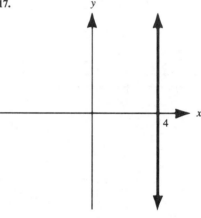

18.

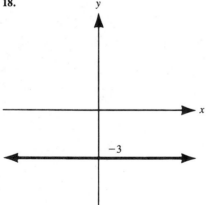

19. -2 **20.** $-\dfrac{7}{3}$ **21.** 0 **22.** No slope   **23.** $y - 2 = 4(x - 3)$ **24.** $y + 1 = \dfrac{3}{4}(x + 5)$ **25.** $y = -6(x - 1)$

26. $y = \dfrac{1}{7}x + 3$ **27.** $y = 8$ **28.** $x = 3$ **29.** $x = 2, y = 2$ **30.** $x = 2, y = 0$ **31.** $x = 1, y = 0$ **32.** $x = -2, y = 2$

33. $x = 1, y = -2$ **34.** $\{(x, y) \mid 2x - 4y = 3\}$ **35.** No solutions   **36.** $x = 4, y = 8$

37. Let x = amount of pure water to be added, y = amount of 30% solution to be used; $x + y = 30$, $.30y = .25(30)$; $\boxed{5 \text{ gal. of water}}$

38. Let a = number of adults' tickets sold, c = number of children's tickets sold; $a + c = 4,300$, $4.50a + 4c = 18,800$; $\boxed{a = 3,200, c = 1,100}$

39. Let w = rate walking, j = rate jogging; $w + j = 17$, $2w + \dfrac{1}{2}j = 16$; $\boxed{w = 5 \text{ kph}, j = 12 \text{ kph}}$

Chapter 8 Practice Test

1. No   **2. (a)** $y = 3$ **(b)** $x = 4$ **(c)** $y = 3$

3. (a)   **(b)**   **(c)**

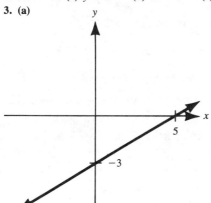
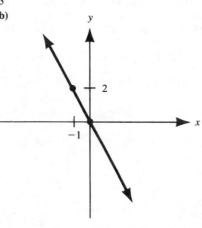
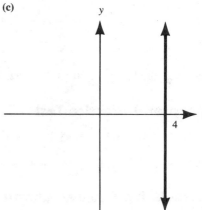

4. $\dfrac{2}{3}$ **5.** $y + 1 = \dfrac{4}{3}(x - 4)$ **6.** $y = -\dfrac{5}{2}(x - 4)$ **7.** Horizontal: $y = 5$, vertical: $x = 3$ **8.** $x = -2$, $y = 4$

9. (a) $x = 2$, $y = -1$ **(b)** $x = 2$, $y = 0$ **(c)** No solution

10. Let $x = $ price of orchestra seat, $y = $ price of balcony seat; $5x + 3y = 60$, $6x + 2y = 64$; $\boxed{x = \$9, \ y = \$5}$

Exercises 9.1

1. 2   **3.** -2 **5.** Undefined   **7.** 5   **9.** -10 **11.** 8   **13.** 11   **15.** 13   **17.** 15   **19.** 17   **21.** 19   **23.** 3   **25.** 29
27. 11   **29.** 33   **31.** x **33.** 49   **35.** $49\sqrt{7}$ **37.** 4   **39.** 2   **41.** 4   **43.** 16   **45.** 20.62   **47.** 25.24   **49.** 8.58
51. 1.45   **53.** .19

Exercises 9.2

1. 8   **3.** $3\sqrt{2}$ **5.** $4\sqrt{2}$ **7.** $5\sqrt{2}$ **9.** 20   **11.** x^3 **13.** $x^3\sqrt{x}$ **15.** $4x^8$ **17.** $3x^4\sqrt{x}$ **19.** $2x^4\sqrt{10}$ **21.** $5x^3\sqrt{x}$
23. $2x^2\sqrt{3x}$ **25.** ab^2 **27.** Cannot be simplified   **29.** $7a^4b^6$ **31.** $2x^4y^3\sqrt{7x}$ **33.** $5m^3n^5\sqrt{2mn}$ **35.** $4x^3y^4z^4\sqrt{3z}$ **37.** $\dfrac{2}{3}$
39. $\dfrac{\sqrt{7}}{5}$ **41.** $\dfrac{\sqrt{30}}{6}$ **43.** $\dfrac{\sqrt{2}}{2}$ **45.** $\dfrac{9\sqrt{10}}{5}$ **47.** $\dfrac{3\sqrt{x}}{x}$ **49.** $\dfrac{2\sqrt{6}}{5}$ **51.** $4\sqrt{2x}$ **53.** $\dfrac{x\sqrt{xy}}{y}$ **55.** $\dfrac{2\sqrt{3}}{3}$

Exercises 9.3

1. $6x$ **3.** $6\sqrt{5}$ **5.** $3x$ **7.** $3\sqrt{6}$ **9.** Cannot be simplified   **11.** Cannot be simplified   **13.** x^2 **15.** 5   **17.** $2x$ **19.** $2\sqrt{5}$
21. $-3\sqrt{5} - 2\sqrt{7}$ **23.** $7\sqrt{3} - \sqrt{2}$ **25.** $-x + y$ **27.** $-\sqrt{5} + \sqrt{3}$ **29.** $23\sqrt{6} + 6\sqrt{7}$ **31.** $3\sqrt{m} - 12\sqrt{n}$ **33.** $5\sqrt{2}$
35. $5 + 2\sqrt{6}$ **37.** $\sqrt{5}$ **39.** $3\sqrt{3}$ **41.** $2\sqrt{5} + 2\sqrt{10} + 2\sqrt{15}$ **43.** $-2\sqrt{2}$ **45.** $30 + 4\sqrt{30}$ **47.** $11\sqrt{x}$ **49.** $5\sqrt{3w}$
51. $\sqrt{5x}$ **53.** $-y\sqrt{5y}$ **55.** $5xy\sqrt{7xy}$ **57.** 0   **59.** $\dfrac{3\sqrt{2}}{2}$ **61.** $\dfrac{13\sqrt{3}}{3}$ **63.** $\dfrac{53\sqrt{2}}{14}$ **65.** $\dfrac{9\sqrt{14}}{14}$

Exercises 9.4

1. $\sqrt{33}$ **3.** $\sqrt{195}$ **5.** $3\sqrt{6}$ **7.** 12   **9.** $3\sqrt{10}$ **11.** $\sqrt{15} + 3\sqrt{2}$ **13.** 24   **15.** $6 - 3\sqrt{6}$ **17.** $5x - 10\sqrt{5x}$
19. $4 - 7\sqrt{2}$ **21.** $x - 2\sqrt{2x}$ **23.** $-7 - 3\sqrt{11}$ **25.** $x + 2\sqrt{3x} + 3$ **27.** $x - 3$ **29.** $38 - 12\sqrt{10}$ **31.** $9x - 7$
33. $6x - \sqrt{x} - 12$ **35.** $26 - 4\sqrt{42}$ **37.** $2t + 18\sqrt{t} + 90$ **39.** $4\sqrt{x} + 2$ **41.** $\dfrac{10\sqrt{11}}{11}$ **43.** $3\sqrt{2}$ **45.** $\dfrac{y}{x}$ **47.** $\dfrac{\sqrt{ab}}{b^2}$
49. $2(4 + \sqrt{11})$ **51.** $\dfrac{6(\sqrt{x} - \sqrt{3})}{x - 3}$ **53.** $2\sqrt{3} - 3$ **55.** $4 + \sqrt{15}$ **57.** $4\sqrt{5}$ **59.** 9   **61.** $20 + 8\sqrt{5}$ **63.** $\dfrac{2 + \sqrt{2}}{3}$
65. $\dfrac{6 - \sqrt{5}}{5}$

Chapter 9 Review Exercises

1. 7 **2.** -10 **3.** Undefined **4.** $3\sqrt{10}$ **5.** $4\sqrt{6}$ **6.** $4x^8$ **7.** $3x^4\sqrt{x}$ **8.** $2x^3y^5\sqrt{5x}$ **9.** $\frac{2}{3}$ **10.** $\frac{\sqrt{3}}{2}$ **11.** $\frac{\sqrt{15}}{5}$

12. $\frac{2\sqrt{5}}{5}$ **13.** $2\sqrt{7}$ **14.** $-4\sqrt{5} - \sqrt{3}$ **15.** $\sqrt{5}$ **16.** $15\sqrt{2}$ **17.** $7\sqrt{3x}$ **18.** $x^2\sqrt{6x}$ **19.** $5\sqrt{3x}$ **20.** $\frac{12\sqrt{35}}{35}$

21. $15 + \sqrt{10}$ **22.** $4x - 9\sqrt{3x} + 6$ **23.** $12 + 11\sqrt{21}$ **24.** 4 **25.** $x - 6\sqrt{x} + 9$ **26.** $a - 10$ **27.** $\frac{7\sqrt{3}}{3}$ **28.** $\frac{5\sqrt{6}}{3}$

29. $x\sqrt{x}$ **30.** $\frac{7\sqrt{6}}{18}$ **31.** $3\sqrt{3}$ **32.** $\frac{2m^2}{3}$ **33.** $2(3 + \sqrt{2})$ **34.** $\frac{7\sqrt{2}}{3}$ **35.** $\frac{7 + 4\sqrt{5}}{31}$ **36.** $3\sqrt{10}$ **37.** $-2\sqrt{7x}$

Chapter 9 Practice Test

1. $5x^8y^3$ **2.** $10\sqrt{3}$ **3.** $x\sqrt{2x}$ **4.** $x^2y^5\sqrt{6}$ **5.** $5x^4y^4\sqrt{5y}$ **6.** $\frac{x\sqrt{6x}}{2}$ **7.** $2x + 5\sqrt{5x} - 15$ **8.** 16 **9.** $-8\sqrt{x} + 20$

10. $\frac{5(\sqrt{7} + \sqrt{3})}{2}$ **11.** It does satisfy the equation.

Cumulative Review: Chapters 7–9

1. $6x^8y^6$ (§9.2) **2.** $\frac{x}{x + 4}$ (§7.1) **3.** $\frac{t - 2}{t - 3}$ (§7.1) **4.** $3x^3y^2\sqrt{x}$ (§9.2) **5.** $\frac{7\sqrt{6}}{6}$ (§9.2) **6.** $\frac{9(\sqrt{6} + 2)}{2}$ (§9.4)

7. $5(3 + \sqrt{5})$ (§9.4) **8.** $3\sqrt{5}$ (§9.2) **9.** $2\sqrt{30}$ (§9.2) **10.** $3xy^4\sqrt{2x}$ (§9.2) **11.** $\frac{23x + 12}{4x(x + 4)}$ (§7.3) **12.** $\sqrt{5t}$ (§9.3)

13. $\frac{10x - 21y^3}{12x^2y^3}$ (§7.3) **14.** $8\sqrt{6} - 16\sqrt{3}$ (§9.4) **15.** $\frac{8\sqrt{21}}{7}$ (§9.3) **16.** $\frac{x^2}{10(x + 5)}$ (§7.2) **17.** $2x + 7\sqrt{x} - 15$ (§9.4)

18. $\frac{6r(r + t)}{t(r - t)}$ (§7.2) **19.** $\frac{2x - 20}{x(x + 2)(x - 2)}$ (§7.3) **20.** $x^3y^2\sqrt{3y}$ (§9.3) **21.** $\frac{3\sqrt{5}}{2}$ (§9.3) **22.** $\frac{31u - 24}{6u(u + 3)}$ (§7.3)

23. $64 - 22\sqrt{10}$ (§9.4) **24.** $\frac{2(a + 4)}{a(a - 4)}$ (§7.3) **25.** $\frac{-2t - 30}{t + 2}$ (§7.2) **26.** $2x + 6\sqrt{x} + 12$ (§9.4) **27.** $3\sqrt{7} - 2\sqrt{2}$ (§9.4)

28. $\frac{1}{2z + 6}$ (§7.3) **29.** $x = 4$ (§7.4) **30.** $a = -\frac{5}{3}$ (§7.4) **31.** $t = -\frac{4x + 12}{3}$ (§7.5) **32.** $y = \frac{30 - 20a}{17}$ (§7.5)

33. No solutions (§7.4) **34.** $x = 0$ (§7.4) **35.** $x = 2$ (§7.4) **36.** $c = \frac{1}{6}$ (§7.4)

37. (§8.2)

38. (§8.2)

39. (§8.2)

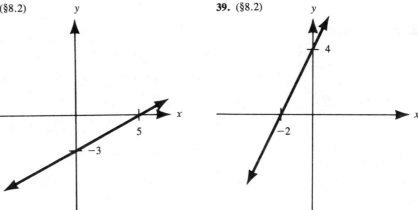

40. (§8.2)

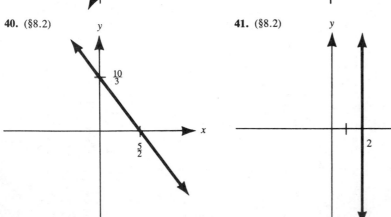

41. (§8.2)

42. (§8.2)

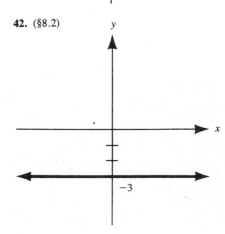

43. (§8.2)

44. (§8.2)

45. -1 (§7.3)   **46.** $\dfrac{5}{2}$ (§7.3)   **47.** 0 (§7.3)   **48.** No slope (§7.3)   **49.** $y - 3 = 4(x - 2)$ (§7.3)   **50.** $y + 1 = \dfrac{1}{2}(x + 4)$ (§7.3)

51. $y = -\dfrac{3}{4}x + 3$ (§7.3)   **52.** $x = 0$ (§7.3)   **53.** $y - 5 = -\dfrac{7}{5}(x + 3)$ (§7.3)   **54.** $y = -x + 4$ (§7.3)   **55.** $x = 4, y = 1$ (§7.4)

56. $x = 3, y = -1$ (§7.4)   **57.** $x = 4, y = 1$ (§7.5)   **58.** $x = 3, y = -1$ (§7.5)   **59.** $x = \dfrac{1}{2}, y = \dfrac{2}{3}$ (§7.5)   **60.** $x = 4, y = 1$ (§7.5)

61. $x = 0, y = -4$ (§7.5)   **62.** $x = 2, y = \dfrac{3}{4}$ (§7.5)   **63.** $\{(x, y) \mid 2x + y = 10\}$ (§7.5)   **64.** No solutions (§7.5)

65. Let x = number of days to plow the field together; $\dfrac{x}{6} + \dfrac{x}{4} = 1$; $\boxed{2\dfrac{2}{5}\ \text{days}}$ (§7.6)

66. Let t = number of hours for Pat to overhaul the engine alone; $\dfrac{5}{8} + \dfrac{5}{t} = 1$; $\boxed{13\dfrac{1}{3}\ \text{hours}}$ (§7.6)

67. Let the original fraction be $\dfrac{x - 4}{x}$; $\dfrac{x - 1}{x + 1} = \dfrac{1}{2}$; $\boxed{-\dfrac{1}{3}}$ (§7.6)

68. Let s = price of a shirt, t = price of a tie; $6s + 2t = 88.68$, $4s + 3t = 68.27$; $\boxed{\text{Shirt costs \$12.95, tie costs \$5.49}}$ (§8.6)

Cumulative Practice Test: Chapters 7–9

1. $8x^8$ (§9.2)   **2.** $\dfrac{3x}{x + 3}$ (§7.1)   **3.** $2x^3y^5\sqrt{10x}$ (§9.2)   **4.** $\dfrac{10\sqrt{6}}{3}$ (§9.2)   **5.** Cannot be reduced (§7.1)   **6.** $\dfrac{4(4 + \sqrt{7})}{3}$ (§9.4)

7. $\dfrac{x}{(x + 5)(x + 4)}$ (§7.3)   **8.** 0 (§9.4)   **9.** $x + 2\sqrt{xy} + y$ (§9.4)   **10.** $\dfrac{w}{4(w - 5)}$ (§7.2)   **11.** $\dfrac{3a - 27}{a(a + 3)(a - 3)}$ (§7.3)   **12.** $5x^3\sqrt{3x}$ (§9.3)

13. $6z - 7\sqrt{z} - 21$ (§9.4)   **14.** $\dfrac{1}{u^2(u + 9)}$ (§7.2)   **15.** $\dfrac{x}{2x - 1}$ (§7.3)   **16.** 10 (§9.4)   **17.** $t = 5$ (§7.4)

18. $u = \dfrac{15x + 35}{2 - 5a}$ (§7.5)   **19.** No solutions (§7.4)

20. (§8.2)

21. (§8.2)

22. (§8.2)

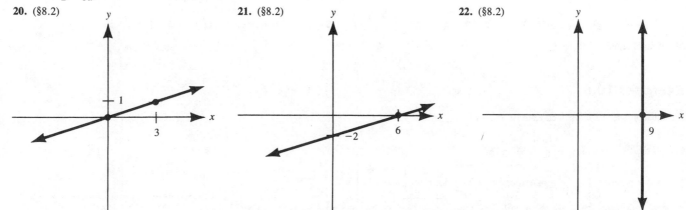

23. $\dfrac{4}{5}$ (§8.3)   **24.** $y + 4 = -\dfrac{7}{3}(x - 2)$ (§8.3)   **25.** $x = -1, y = -5$ (§8.5)   **26.** No solutions (§8.5)

27. Let x = number of hours for Susan to finish painting the house, $\dfrac{5}{12} + \dfrac{x}{9} = 1$; $\boxed{5\dfrac{1}{4} \text{ days}}$ (§7.6)

28. Let r = regular price, s = special price, $4r + 3s = 55.30$, $2r + 5s = 50.40$; $\boxed{r = \$8.95,\ s = \$6.50}$ (§8.6)

Exercises 10.1

1. $2, -3$ **3.** $-4, 3$ **5.** $0, 4$ **7.** $6, -2$ **9.** $3, -2$ **11.** $5, -2$ **13.** $4, -2$ **15.** $8, 1$ **17.** 0 **19.** $\dfrac{3}{2}, 4$ **21.** $0, -3$

23. $-5, 2$ **25.** Not factorable   **27.** $3, -3$ **29.** Not factorable   **31.** $\dfrac{1}{2}$ **33.** $3, -7$ **35.** 5 **37.** 1 **39.** 3 **41.** 5

Exercises 10.2

1. ± 5 **3.** ± 4 **5.** $\pm\dfrac{4}{3}$ **7.** No real solutions   **9.** $\pm\dfrac{2}{5}$ **11.** $\dfrac{\pm\sqrt{15}}{6}$ **13.** $\dfrac{\pm\sqrt{33}}{3}$ **15.** ± 2 **17.** $\dfrac{\pm 2\sqrt{5}}{3}$ **19.** $\dfrac{\pm 4\sqrt{6}}{3}$

21. $\pm\sqrt{5}$ **23.** $\pm\sqrt{3}$ **25.** No real solutions   **27.** $\pm 2\sqrt{6}$ **29.** $5, -1$ **31.** $-5 \pm \sqrt{7}$ **33.** $6 \pm 2\sqrt{3}$ **35.** $\dfrac{-2 \pm \sqrt{3}}{5}$

37. -1 **39.** ± 4 **41.** $\pm\sqrt{22}$ **43.** $\dfrac{1}{2}, -\dfrac{5}{2}$ **45.** $x = \pm 2.65$ **47.** $k = \pm 2.58$ **49.** $x = \pm 2.22$ **51.** $a = 1.52, -1.22$

Exercises 10.3

1. $-4 \pm \sqrt{10}$ **3.** $2 \pm \sqrt{7}$ **5.** $5 \pm 2\sqrt{10}$ **7.** $10, -2$ **9.** $-6 \pm \sqrt{30}$ **11.** $3 \pm \sqrt{7}$ **13.** $-2 \pm \sqrt{6}$ **15.** $\dfrac{-5}{2} \pm \dfrac{\sqrt{33}}{2}$

17. No real solutions   **19.** $\pm\sqrt{13}$ **21.** $\dfrac{3 \pm \sqrt{3}}{2}$ **23.** $-3 \pm \sqrt{6}$ **25.** No real solutions   **27.** $-9, 1$ **29.** $2, \dfrac{2}{3}$

Exercises 10.4

1. $a = 1, b = 3, c = -5$ **3.** $a = 1, b = -7, c = -6$ **5.** $a = 2, b = -8, c = 0$ **7.** $a = 3, b = 0, c = -11$ **9.** $\dfrac{-3 \pm \sqrt{29}}{2}$

11. $-2 \pm \sqrt{10}$ **13.** No real solutions   **15.** $\dfrac{7 \pm \sqrt{73}}{2}$ **17.** $\dfrac{3 \pm \sqrt{17}}{4}$ **19.** $\dfrac{1 \pm \sqrt{41}}{10}$ **21.** $\dfrac{-7 \pm \sqrt{77}}{2}$ **23.** $\dfrac{3 \pm 2\sqrt{6}}{5}$

25. Contradiction   **27.** $2 \pm \sqrt{2}$ **29.** $\dfrac{3 \pm \sqrt{15}}{2}$ **31.** $1, \dfrac{1}{2}$ **33.** $0, \dfrac{4}{9}$ **35.** $\pm\dfrac{2}{3}$ **37.** $-1 \pm \sqrt{7}$ **39.** $2, -\dfrac{1}{2}$

41. $t = 8.22, -1.22$ **43.** $w = -1.72, -4.06$ **45.** $x = 3.06, -1.17$

Exercises 10.5

1. $-5, -1$ **3.** $-3 \pm \sqrt{14}$ **5.** $\dfrac{1}{2}, 1$ **7.** $5, -1$ **9.** $-\dfrac{5}{2}$ **11.** Contradiction   **13.** No real solutions   **15.** $1 \pm \sqrt{5}$

17. $\dfrac{7 \pm 3\sqrt{5}}{2}$ **19.** ± 4 **21.** $0, 9$ **23.** $9, -5$ **25.** $\dfrac{-3 \pm \sqrt{13}}{2}$ **27.** 3 **29.** $6, -\dfrac{2}{3}$ **31.** 1 **33.** $2, \dfrac{1}{2}$ **35.** ± 2

37. Contradiction   **39.** $-3, 2$

Exercises 10.6

The equation used to solve each exercise is given.

1. n = the number, $n + \dfrac{1}{n} = \dfrac{13}{6}$; $\boxed{n = \dfrac{2}{3} \text{ or } n = \dfrac{3}{2}}$

3. x = one number, $20 - x$ = other number; $x(20 - x) = 96$; $\boxed{8 \text{ and } 12}$

5. W = width, $2W + 3$ = length; $W(2W + 3) = 90$; $\boxed{6 \text{ by } 15}$

7. x = length of the leg; $x^2 + 7^2 = 15^2$; $\boxed{x = 4\sqrt{11} \text{ cm.}}$

9. d = length of diagonal; $8^2 + 8^2 = d^2$; $\boxed{d = 8\sqrt{2} \text{ in.}}$

11. h = height; $h^2 + 8^2 = 30^2$; $\boxed{h = 2\sqrt{209} \text{ ft.} \approx 28.9 \text{ ft.}}$

13. Let the integers be x, $x + 1$, and $x + 2$; $x^2 + (x + 1)^2 = (x + 2)^2$; $\boxed{3, 4, 5}$

15. Let the fraction be $\dfrac{x}{x + 1}$; $\dfrac{x + 3}{x + 1} = \dfrac{x}{x + 1} + 1$; $\boxed{\dfrac{2}{3}}$

17. x = # of seats per row, $x - 8$ = # of rows; $x(x - 8) = 768$; $\boxed{32 \text{ seats per row}}$

19. The rates are r and $r - 20$; $\dfrac{120}{r} + \dfrac{30}{r - 20} = 2$; $\boxed{80 \text{ kph}}$

21. \$15,000; \$9,000   **23. (a)** Approximately 4.3 seconds   **(b)** Approximately 7.9 seconds
25. (a) Approximately 27 radios   **(b)** Approximately 26 radios

Chapter 10 Review Exercises

1. $\dfrac{7 \pm \sqrt{73}}{2}$ **2.** $12, -2$ **3.** No real solutions   **4.** $-5, 1$ **5.** $6 \pm \sqrt{13}$ **6.** No real solutions   **7.** $\dfrac{3}{2}, -5$ **8.** $\dfrac{3 \pm \sqrt{17}}{4}$

9. $\dfrac{1}{3}, 1$ **10.** $\dfrac{1 \pm \sqrt{21}}{5}$ **11.** $\dfrac{51}{10}$ **12.** $5 \pm 2\sqrt{5}$ **13.** $\dfrac{2}{3}, \dfrac{3}{2}$ **14.** $6, -3$ **15.** $\dfrac{\pm 3\sqrt{2}}{2}$ **16.** $\dfrac{2 \pm \sqrt{10}}{2}$ **17.** 0 **18.** $2, \dfrac{9}{7}$

19. $\dfrac{7}{2} \pm \dfrac{\sqrt{37}}{2}$ **20.** $2 \pm \sqrt{6}$

Chapter 10 Practice Test

1. $-7, 4$ **2.** $\dfrac{3 \pm \sqrt{33}}{4}$ **3.** $7, -2$ **4.** $-5 \pm \sqrt{10}$ **5.** No real solutions   **6.** $\dfrac{2}{3}, -5$ **7.** $3 \pm \sqrt{6}$ **8.** 4 **9.** $2 \pm \dfrac{\sqrt{57}}{3}$

10. W = width, $2W + 7$ = length, $W(2W + 7) = 30$; $\boxed{\dfrac{5}{2} \text{ cm. by 12 cm.}}$

Sample Final Exam, Form A

1. -6 **2.** -24 **3.** 13 **4.** $4a^3 - 17a^2b$ **5.** $3m^6n^2$ **6.** $9x - 24$ **7.** $6x^2 + 1$ **8.** $2x^2 - 6x + 1$ **9.** 2 **10.** 12.5

11. $\dfrac{9}{a^2}$ **12.** $6mn(3m - 1 + 2n)$ **13.** $(x - 9)(x + 4)$ **14.** $4(z - 4)(z - 4)$ **15.** $a(a - 4)(a + 4)$ **16.** $(x + 4)(x + y)$ **17.** $\dfrac{(x - 9)}{(x - 1)}$

18. 199 **19.** $3x^2\sqrt{6x}$ **20.** $\dfrac{-x}{12}$ **21.** $\dfrac{5x^2}{8}$ **22.** $\dfrac{8x^2y}{25z^4}$ **23.** $\dfrac{2x}{(x + 4)(x + 2)}$ **24.** $\dfrac{5x + 27}{3x(x + 3)}$ **25.** 0 **26.** 6 **27.** $x = -3$

28. $x = 4$ **29.** $x = \pm 2$ **30.** $x = \dfrac{3 \pm \sqrt{21}}{6}$ **31.** $x = \dfrac{7y + 5}{3}$

32. $x \le -2$ **33.** $x = 5, y = 2$

34.

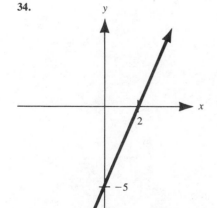

35. $y - 4 = -\frac{7}{3}(x + 1)$ **36.** 8 cm. by 19 cm.   **37.** 10.6 ounces   **38.** $2\frac{2}{3}$ hours   **39.** $2,400 at 8% and $4,800 at 10%

40. 9 books at $12 each and 6 books at $9 each

Sample Final Exam, Form B

1. -10 **2.** -24 **3.** -35 **4.** $5a^3 - 8a^3b$ **5.** $6m^6n$ **6.** $5a - 24$ **7.** $12x^3 + 7x^2 - 28x + 12$ **8.** $3x^2 + x + 2$ **9.** 2

10. 45,000   **11.** $\frac{64}{a^6}$ **12.** $5uv(3v - 2uv^2 + 1)$ **13.** $(x + 12)(x - 4)$ **14.** $6(a - 2)(a - 2)$ **15.** $2c(c + 5)(c - 5)$

16. $(x + 6)(x - y)$ **17.** $\frac{2x}{x + 3}$ **18.** 57   **19.** $2x^4\sqrt{6x}$ **20.** $\frac{-5x}{24}$ **21.** $\frac{7x^2}{32}$ **22.** $\frac{7xy^2}{27z^6}$ **23.** $\frac{2x}{(x - 5)(x - 3)}$ **24.** $\frac{6}{x(x - 2)}$

25. 0   **26.** 6   **27.** $x = -6$ **28.** $x = 2$ **29.** $x = 8, -1$ **30.** $x = \frac{1 \pm \sqrt{57}}{4}$ **31.** $x = \frac{3y - 7}{5}$

32. $-7 \le x < -3$ **33.** $x = -2, y = 1$

34.

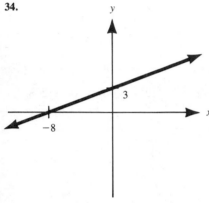

35. $y = \frac{1}{2}x - 2$ **35.** 5   **37.** 5 cm. by 13 cm.   **38.** 200 women   **39.** 25 minutes   **40.** 120 liters

Index